Silicon-Germanium Heterojunction Bipolar Transistors

John D. Cressler
Guofu Niu

Artech House
Boston • London
www.artechhouse.com

Library of Congress Cataloging-in-Publication Data
Cressler, John D.
　Silicon-germanium heterojunction bipolar transistors / John D. Cressler, Guofu Niu.
　　p. cm.
　Includes bibliographical references and index.
　ISBN 1-58053-361-2 (alk. paper)
　1. Bipolar transistors. 2. Germanium. I. Niu, Guofu. II. Title.

TK7871.96.B55 C74 2002
621.3815'28—dc21   2002038276

British Library Cataloguing in Publication Data
Cressler, John D.
　Silicon-germanium heterojunction bipolar transistors
　1. Bipolar transistors 2. Silicon 3. Germanium
　I. Title II. Niu, Guofu
　621.3'81528

　ISBN 1-58053-361-2

Cover design by Christina Stone

© 2003 ARTECH HOUSE, INC.
685 Canton Street
Norwood, MA 02062

All rights reserved. Printed and bound in the United States of America. No part of this book may be reproduced or utilized in any form or by any means, electronic or mechanical, including photocopying, recording, or by any information storage and retrieval system, without permission in writing from the publisher.
　All terms mentioned in this book that are known to be trademarks or service marks have been appropriately capitalized. Artech House cannot attest to the accuracy of this information. Use of a term in this book should not be regarded as affecting the validity of any trademark or service mark.

International Standard Book Number: 1-58053-361-2
Library of Congress Catalog Card Number: 2002038276

10 9 8 7 6 5 4 3 2 1

Silicon-Germanium Heterojunction Bipolar Transistors

For a listing of related titles, turn to the back of this book.

For Maria:
My beautiful wife, best friend, and soul mate for 20 years.
For Matthew John, Christina Elizabeth, and Joanna Marie:
God's awesome creations, and our precious gifts.
May your journey of discovery never end.

J.D.C.

For my wife:
Ying Li.
For my parents:
Pinzhang Niu and Xuehua Feng.

G.N.

Honesty of Thought
And Speech and Written Word
Is a Jewel,
And They Who Curb Prejudice
And Seek Honorably
To Know and Speak the Truth
Are the Only Builders of a Better Life.

John Galsworthy

知之为知之，不知为不知，是知也。
——孔子

Contents

Preface   xv

1 Introduction   1
    1.1 The Magic of Silicon 1
    1.2 IC Needs for the Twenty-First Century 6
    1.3 Application-Induced Design Constraints 7
    1.4 The Dream: Bandgap Engineering in Silicon 10
    1.5 The SiGe HBT 12
    1.6 A Brief History of SiGe Technology 14
    1.7 SiGe HBT Performance Trends 17
    1.8 The IC Technology Battleground: Si Versus SiGe Versus III-V . . 22
    References 26

2 SiGe Strained-Layer Epitaxy   35
    2.1 SiGe Alloys 35
       2.1.1 Pseudomorphic Growth and Film Relaxation 36
       2.1.2 Putting Strained SiGe into SiGe HBTs 40
       2.1.3 The Challenge of SiGe Epitaxy 42
    2.2 SiGe Growth 44
       2.2.1 Surface Preparation 45
       2.2.2 Growth Techniques 46
    2.3 Stability Constraints 48
       2.3.1 Theory 50
       2.3.2 Experimental Results 54
       2.3.3 Stability Calculations 55
    2.4 Band Structure 56
       2.4.1 Density-of-States 58
       2.4.2 Band Offsets 59
    2.5 Transport Parameters 61

		2.5.1	Hole Mobility	64
		2.5.2	Electron Mobility	65
		2.5.3	Choice of SiGe Parameter Models	66
	2.6		Open Issues	67
			References	68
3	**SiGe HBT BiCMOS Technology**			**73**
	3.1		The Technology Playing Field	73
	3.2		Integration of SiGe HBTs with CMOS	79
	3.3		Carbon Doping	82
	3.4		Passives	85
	3.5		Reliability and Yield Issues	89
			References	92
4	**Static Characteristics**			**95**
	4.1		Intuitive Picture	95
	4.2		Collector Current Density and Current Gain	98
		4.2.1	J_C in SiGe HBTs	98
		4.2.2	Relevant Approximations	103
		4.2.3	Nonconstant Base Doping	104
		4.2.4	Other SiGe Profile Shapes	105
		4.2.5	Implications and Optimization Issues for β	108
	4.3		Output Conductance	109
		4.3.1	V_A Trade-offs in Si BJTs	109
		4.3.2	V_A in SiGe HBTs	111
		4.3.3	Relevant Approximations	113
		4.3.4	Current Gain – Early Voltage Product	115
		4.3.5	Other SiGe Profile Shapes	116
		4.3.6	Implications and Optimization Issues for V_A and βV_A	117
	4.4		Equivalent Circuit Models	118
		4.4.1	Basic Ebers-Moll Model	118
		4.4.2	Transport Version	119
		4.4.3	Small-Signal Equivalent Circuit Model	120
	4.5		Avalanche Multiplication	121
		4.5.1	Carrier Transport and Terminal Currents	121
		4.5.2	Forced-V_{BE} Measurement of $M-1$	122
		4.5.3	Forced-I_E Measurement of $M-1$	124
		4.5.4	Effects of Self-Heating	127
		4.5.5	Impact of Current Density	128
		4.5.6	Si Versus SiGe	128

4.6	Breakdown Voltages	131
	4.6.1 BV_{CBO}	132
	4.6.2 BV_{CEO}	132
	4.6.3 Circuit Implications	135
	References	135

5 Dynamic Characteristics — 139

5.1	Intuitive Picture	139
5.2	Charge Modulation Effects	141
5.3	Basic RF Performance Factors	143
	5.3.1 Current Gain and Cutoff Frequency	143
	5.3.2 Current Density Versus Speed	147
	5.3.3 Base Resistance	149
	5.3.4 Power Gain and Maximum Oscillation Frequency	150
5.4	Linear Two-Port Parameters	154
	5.4.1 Z-Parameters	154
	5.4.2 Y-Parameters	155
	5.4.3 H-Parameters	155
	5.4.4 S-Parameters	155
5.5	Stability, MAG, MSG, and Mason's U	158
	5.5.1 Stability	158
	5.5.2 Power Gain Definitions	160
	5.5.3 MAG and MSG	160
	5.5.4 Mason's Unilateral Gain	162
	5.5.5 Which Gain Is Better?	162
	5.5.6 f_T Versus f_{max} Versus Digital Switching Speed	164
5.6	Base and Emitter Transit Times	165
	5.6.1 τ_b in SiGe HBTs	165
	5.6.2 Relevant Approximations	168
	5.6.3 τ_e in SiGe HBTs	170
	5.6.4 Other SiGe Profile Shapes	170
	5.6.5 Implications and Optimization Issues for f_T	173
5.7	ECL Gate Delay	174
	5.7.1 ECL Design Equations	175
	5.7.2 ECL Power-Delay Characteristics	177
	5.7.3 Impact of SiGe on ECL Power Delay	179
	References	182

6 Second-Order Phenomena — 185

- 6.1 Ge Grading Effect — 186
  - 6.1.1 Bandgap Reference Circuits — 189
  - 6.1.2 Theory — 194
  - 6.1.3 Measured Data and SPICE Modeling Results — 198
  - 6.1.4 The Bottom Line — 201
- 6.2 Neutral Base Recombination — 204
  - 6.2.1 Theory — 205
  - 6.2.2 Experimental Results — 212
  - 6.2.3 Impact of NBR on Early Voltage — 216
  - 6.2.4 Identifying the Physical Location of the NBR Traps — 222
  - 6.2.5 Circuit-Level Modeling Issues — 225
  - 6.2.6 Device Design Implications — 229
  - 6.2.7 The Bottom Line — 231
- 6.3 Heterojunction Barrier Effects — 232
  - 6.3.1 High-Injection in SiGe HBTs — 234
  - 6.3.2 Experimental Results and Simulations — 238
  - 6.3.3 Profile Optimization Issues — 241
  - 6.3.4 Compact Modeling — 244
  - 6.3.5 The Bottom Line — 254
- References — 256

7 Noise — 261

- 7.1 Fundamental Noise Characteristics — 262
  - 7.1.1 Thermal Noise — 262
  - 7.1.2 Shot Noise in a *pn* Junction — 263
  - 7.1.3 Shot Noise in Bipolar Transistors — 263
- 7.2 Linear Noisy Two-Port Network Theory — 264
  - 7.2.1 Two-Port Network — 264
  - 7.2.2 Input Noise Voltage and Current in BJTs — 267
  - 7.2.3 Noise Figure of a Linear Two-Port — 270
  - 7.2.4 Associated Gain — 272
  - 7.2.5 Y-Parameter Based Modeling — 273
- 7.3 Analytical Modeling — 274
  - 7.3.1 Noise Resistance — 277
  - 7.3.2 Optimum Source Admittance — 278
  - 7.3.3 Minimum Noise Figure — 278
  - 7.3.4 Associated Gain — 279
- 7.4 Optimal Sizing and Biasing for LNA Design — 280
  - 7.4.1 Emitter Width Scaling at Fixed J_C — 280

Contents

		7.4.2	Emitter Length Scaling at Fixed J_C	281
		7.4.3	Simultaneous Impedance and Noise Matching	284
		7.4.4	Current Density Selection	287
		7.4.5	A Design Example	288
		7.4.6	Frequency Scalable Design	288
	7.5	SiGe Profile Design Trade-offs		291
		7.5.1	Input Noise Current Limitations	292
		7.5.2	Input Noise Voltage Limitations	293
		7.5.3	Approaches to Noise Improvement	295
		7.5.4	SiGe Profile Optimization	296
		7.5.5	Experimental Results	296
	7.6	Low-Frequency Noise		301
		7.6.1	Upconversion to Phase Noise	301
		7.6.2	Measurement Methods	303
		7.6.3	Bias Current Dependence	306
		7.6.4	Geometry Dependence	310
		7.6.5	$1/f$ Noise Figures of Merit	312
	7.7	Substrate and Cross-Talk Noise		314
		7.7.1	Noise Grounding Using Substrate Contacts	315
		7.7.2	Noise Grounding Using n^+ Buried Layers	315
		References		316
8	**Linearity**			**321**
	8.1	Nonlinearity Concepts		322
		8.1.1	Harmonics	323
		8.1.2	Gain Compression and Expansion	324
		8.1.3	Intermodulation	324
	8.2	Physical Nonlinearities		329
		8.2.1	The I_{CE} Nonlinearity	329
		8.2.2	The I_{BE} Nonlinearity	331
		8.2.3	The I_{CB} Nonlinearity	332
		8.2.4	The C_{BE} and C_{BC} Nonlinearities	335
	8.3	Volterra Series		338
		8.3.1	Fundamental Concepts	338
		8.3.2	First-Order Transfer Functions	340
		8.3.3	Second-Order Transfer Functions	343
		8.3.4	Third-Order Transfer Functions	345
	8.4	Single SiGe HBT Amplifier Linearity		348
		8.4.1	Circuit Analysis	349
		8.4.2	Distinguishing Individual Nonlinearities	351

		8.4.3	Collector Current Dependence	352
		8.4.4	Collector Voltage Dependence	353
		8.4.5	Load Dependence	354
		8.4.6	Dominant Nonlinearity Versus Bias	358
		8.4.7	Nonlinearity Cancellation	358
	8.5	Cascode LNA Linearity		359
		8.5.1	Optimization Approach	360
		8.5.2	Design Equations	365
	References			369

9 Temperature Effects — 371

	9.1	The Impact of Temperature on Bipolar Transistors		372
		9.1.1	Current-Voltage Characteristics	372
		9.1.2	Transconductance	375
		9.1.3	Resistances and Capacitances	376
		9.1.4	Current Gain	379
		9.1.5	Frequency Response	380
		9.1.6	Circuit Performance	381
	9.2	Cryogenic Operation of SiGe HBTs		383
		9.2.1	Evolutionary Trends	384
		9.2.2	SiGe HBT Performance Down to 77 K	385
		9.2.3	Design Constraints at Cryogenic Temperatures	388
	9.3	Optimization of SiGe HBTs for 77 K		391
		9.3.1	Profile Design and Fabrication Issues	392
		9.3.2	Measured Results	394
	9.4	Helium Temperature Operation		396
		9.4.1	dc Characteristics at LHeT	397
		9.4.2	Novel Collector Current Phenomenon at LHeT	399
	9.5	Nonequilibrium Base Transport		406
		9.5.1	Theoretical Expectations	409
		9.5.2	Experimental Observations	411
		9.5.3	Interpretation of Results	414
	9.6	High-Temperature Operation		416
	References			419

10 Other Device Design Issues — 423

	10.1	The Design of SiGe *pnp* HBTs		423
		10.1.1	Simulation of *pnp* SiGe HBTs	425
		10.1.2	Profile Optimization Issues	426
		10.1.3	Stability Constraints in *pnp* SiGe HBTs	429

10.2	Arbitrary Band Alignments	431
	10.2.1 Low-Injection Theory	432
	10.2.2 Impact of High Injection	434
	10.2.3 Profile Optimization Issues	438
10.3	Ge-Induced Collector-Base Field Effects	441
	10.3.1 Simulation Approach	442
	10.3.2 Influence on Impact Ionization	443
	10.3.3 Influence on the Base Current Bias Dependence	445
	10.3.4 Experimental Confirmation	446
	References	448

11 Radiation Tolerance 453

11.1	Radiation Concepts and Damage Mechanisms	455
11.2	The Effects of Radiation on SiGe HBTs	460
	11.2.1 Transistor *dc* Response	460
	11.2.2 Spatial Location of the Damage	463
	11.2.3 Transistor *ac* Response	464
	11.2.4 Si versus SiGe and Structural Aspects	465
	11.2.5 Proton Energy Effects	465
	11.2.6 Low-Dose-Rate Gamma Sensitivity	467
	11.2.7 Broadband Noise	470
	11.2.8 Low-Frequency Noise	471
11.3	Technology Scaling Issues	476
	11.3.1 SiGe HBT Scaling	476
	11.3.2 Si CMOS Scaling	480
11.4	Circuit-Level Tolerance	483
	11.4.1 The Importance of Transistor Bias	483
	11.4.2 Bandgap Reference Circuits	485
	11.4.3 Voltage Controlled Oscillators	486
	11.4.4 Passive Elements	487
11.5	Single Event Upset	488
	11.5.1 Transistor Equivalent Circuit Under SEU	490
	11.5.2 SEU Simulation Methodology	492
	11.5.3 Charge Collection Characteristics	493
	11.5.4 Circuit Architecture Dependence	496
	References	502

12 Device Simulation — 509
- 12.1 Semiconductor Equations — 510
  - 12.1.1 Carrier Statistics — 510
  - 12.1.2 Drift-Diffusion Equations — 513
  - 12.1.3 Energy Balance Equations — 515
  - 12.1.4 Boundary Conditions — 517
  - 12.1.5 Physical Models — 520
  - 12.1.6 Numerical Methods — 523
- 12.2 Application Issues — 527
  - 12.2.1 Device Structure Specification — 527
  - 12.2.2 Mesh Specification and Verification — 529
  - 12.2.3 Model Selection and Coefficient Tuning — 532
  - 12.2.4 dc Simulation — 532
  - 12.2.5 High-Frequency Simulation — 534
  - 12.2.6 Qualitative Versus Quantitative Simulations — 537
- 12.3 Probing Internal Device Operation — 538
  - 12.3.1 Current Transport Versus Operating Frequency — 538
  - 12.3.2 Quasi-Static Approximation — 538
  - 12.3.3 Regional Analysis of Transit Time — 541
  - 12.3.4 Case Study: High-Injection Barrier Effect — 544
  - References — 548

13 Future Directions — 549
- 13.1 Technology Trends — 549
- 13.2 Performance Limits — 551
- References — 555

Appendix A: Properties of Silicon and Germanium — 557
- References — 558

About the Authors — 561

Index — 565

Preface

While the idea of combining the semiconductor silicon and the semiconductor germanium for use in transistor engineering is an old one, only in the past decade has this concept been reduced to practical reality. The fruit of that effort is the silicon-germanium heterojunction bipolar transistor (SiGe HBT). The implications of the SiGe success story contained in this book are far-ranging and likely to be quite lasting and influential in determining the future course of the electronics infrastructure fueling the miraculous communications explosion of the twenty-first century.

This book is intended for a number of different audiences and venues. It should prove to be a useful resource as: 1) a hands-on reference for practicing engineers and scientists working on various aspects of SiGe technology, including: characterization, device design, fabrication, modeling, and circuit design; 2) a textbook for graduate or advanced undergraduate students in electrical and computer engineering (ECE), physics, or materials science who are interested in cutting-edge integrated circuit (IC) device and circuit technologies; or 3) a reference for technical managers and even technical support / technical sales personnel in the semiconductor industry. It is assumed that the reader has some modest background in semiconductors and bipolar devices (say, at the advanced undergraduate ECE level), but we have been careful to build "from-the-ground-up" in our treatment.

The spirit and vision for this book from day one was that it be "SiGe HBT from A to Z." That is, the book is intentionally very broad as well as very deep, and proceeds from a basic motivation and history of the subject, to materials, to technology and fabrication issues, to a detailed discourse on a wide range of fundamental aspects of SiGe HBT operation and design, spanning *dc* and *ac* characteristics, including noise and linearity. These fundamental topics are then supplemented by an even closer look at some of the "fine points" which might be confronted by experts in the field, including second-order phenomena, temperature effects, radiation tolerance, and numerical simulation. We conclude with a brief glimpse at likely future directions for SiGe technology. While we recognize that not all readers have need for exposure to all of these subjects, we like the notion of having a complete reference on the subject contained under one cover.

We have written this book in a manner consistent with our own preferences. Hence, it contains detailed, careful expositions of theory, a discussion of key device design trade-offs and constraints, "bottom-line" arguments on how important this or that phenomenon may be in the overall scheme of things, supporting data to bear out the various claims and theoretical arguments presented, and sufficient breadth and depth to be useful to both the novice and the expert. We also prefer a fairly informal writing style to ensure reader friendliness, and believe it is important to grasp the historical background, trends, and evolution of *any* subject. We have gone to considerable length to carefully reexamine and explicitly state assumptions in our theoretical treatments, and we have also intentionally not skipped the intermediate steps in some of the more important derivations – they are not often seen and deserve to be appreciated. We have highlighted what we feel to be the "open issues" associated with SiGe research that are in need of increased attention by the academic and industrial communities. This book contains a fairly substantial body of previously unpublished data and theory, as well as many careful and critical reinterpretations of the various nuances of the theory of SiGe HBTs. We have found again and again that while some particular theoretical discourse may previously reside in the literature (and even be widely cited), the existing presentation is often either confusing, is not correctly applied, does not fit the facts, or in some way is in need of a closer look. We have done that here. As with any in-depth work of this sort, there will be some among you who may disagree with our theory or interpretations. That's what science is all about! Feel free to contact us with any questions (or gripes!).

As any honest professor will readily concede, our graduate students play an absolutely essential role in our research. We professors may supply ideas, give encouragement, and guide interpretations (okay, and chip in some dollars as well!), but in the end, the really hard work belongs to our students. No exception here. Perhaps the greatest pleasure for us as professors is to behold the blooming of our students and the career successes they enjoy once they "leave the nest." We would like to take this opportunity to thank our graduate students, past and present, including *(J.D.C.)* — David Richey, Alvin Joseph, Bill Ansley, Juan Roldán, Stacey Salmon, Lakshmi Vempati, Jeff Babcock, Suraj Mathew, Mike Hamilton, Kartik Jayanaraynan, Greg Bradford, Usha Gogineni, Gaurab Banerjee, Shiming Zhang, Krish Shivaram, Dave Sheridan, Gang Zhang, Ying Li, Zhenrong Jin, QingQing Liang, Ramkumar Krithivasan, Zhiyun Luo, Tianbing Chen, Yuan Lu, and Chendong Zhu; and *(G.N.)* — Jin Tang, Jun Pan, Yan Cui, Yun Shi, Muthu Varadharajaperumal, and Seema Hegde. A special debt is owed to some of our students, since we have borrowed (unpublished) passages from several of their dissertations and theses (thanks especially to Alvin, David, and Stacey).

Much of the work presented in this book would not have occurred if our funding sponsors had not embraced our vision of what is important in SiGe research. Special thanks are due to *(industry sponsors)* — IBM (Alvin Joseph, Dave Harame, Jim Dunn, Seshu Subbanna, Dave Ahlgren, Greg Freeman, Dean Herman, and Bernie Meyerson), Texas Instruments (John Erdeljac, Lou Hutter, Badih El-Kareh, and Dennis Buss), On Semiconductor (Joe Neel and Julio Costa), Maxim Semiconductor (Stewart Taylor), Analog Devices (John Yasaitis and Brad Scharf), Northrop Grumman (Harvey Nathanson, Bill Hall, and Rowan Messham) Hughes Electronics, now Boeing (Kay Jobe and Dave Sunderland), and the Semiconductor Research Corporation (Justin Harlow, Dale Edwards, and Jim Hutchby); and *(government sponsors)* — NASA Goddard Space Flight Center (Robert Reed, Cheryl Marshall, Paul Marshall, Hak Kim, Ken LaBel), the Defense and Threat Reduction Agency (Lew Cohn), DARPA, the U.S. Army Space and Missile Defense Command (Charlie Harper and Aaron Corder), NAVSEA Crane (Steve Clark and Dave Emily), Mission Research Corporation (Dave Alexander and Mary Dyson), the Office of Naval Research (Al Goodman), the NASA Center for Space Power and Advanced Electronics (Henry Brandhorst), the Naval Research Laboratory (Fritz Kub), the National Science Foundation, EPSCOR, and the Jet Propulsion Laboratory (Jagdish Patel).

It is also true that much of the work presented in this book would have been impossible without the generous support of the SiGe team at IBM, which provided our research group with ready access to their state-of-the-art hardware. In fact, the lion's share of the data presented in this book was measured on IBM devices, and we are especially grateful to Bernie Meyerson, Dean Herman, David Harame, Alvin Joseph, and Seshu Subbana, in particular, for making that happen. Many current members of the IBM SiGe team contributed directly to various aspects of our research, including Greg Freeman, Dave Ahlgren, Rob Groves, Jim Dunn, Chuck Webster, Fernando Guarin, Lou Lanzerotti, Kim Newton, Herschel Ainspan, and Mehmet Soyuer. John Cressler would like to reach back and thank his colleagues at IBM Yorktown who played a pivotal role in his early education in bipolar devices and SiGe HBTs, including Denny Tang, Tak Ning, Hans Stork, Jim Comfort, Emmanuel Crabbé, Achim Burghartz, Jack Sun, Dave Harame, Keith Jenkins, Kent Chuang, Jim Warnock, and Scott Stiffler.

We would like to thank Mike Palmer and Charles Ellis of the AMSTC for assistance in wire-bonding and fabrication, as well as Dick Jaeger and Dave Irwin for their constant support during the arduous writing process. Several individuals scattered across the globe contributed to various ideas or data presented in this book, including: Stewart Taylor, Larry Larson, Joerg Osten, Dieter Knoll, Robert Plana, and Frank Herzel. Our apologies if we have left anyone out. The body of knowledge contained in this work truly represents the efforts of a great many dedicated engineers and scientists.

Whew! This book has been a year-long labor of love, and although "fun" might be too harsh a word, given the countless hours required, SiGe is a subject we care deeply about, love to talk about, and remain fascinated by. There are many interesting puzzles left in SiGe! It has been immensely satisfying to see both the dream of SiGe and this book become a reality. We hope our efforts please you. Enjoy!

John D. Cressler
School of Electrical and Computer Engineering
Georgia Institute of Technology
December 2002

Guofu Niu
Department of Electrical and Computer Engineering
Auburn University
December 2002

Chapter 1

Introduction

Simply stated, silicon-germanium is "an idea whose time has come."[1] While the concept of combining silicon (Si) and germanium (Ge) into an alloy for use in transistor engineering is an old one, only in the past decade has this concept been reduced to practical reality. The implications of this success story are far ranging and likely to be quite lasting and influential in determining the future course of the communications explosion during the twenty-first century. This introductory chapter sets the stage for the detailed look at the silicon-germanium heterojunction bipolar transistor (SiGe HBT) presented in this book. We first examine the compelling features of the semiconductor Si, look at integrated circuit (IC) needs to support the emerging Information Age, and then examine application-induced design constraints. Armed with this background, the notion of using bandgap-engineering in Si to create the SiGe HBT is introduced, and we address why SiGe HBT BiCMOS technology has emerged as an important enabler for twenty-first century communications systems. We conclude with an historical perspective of this fascinating field, some performance trends, and a view of the looming technology battleground between Si, SiGe, and III-V technologies.

1.1 The Magic of Silicon

We live in a silicon world. This statement is literally as well as figuratively true. Silicates, the large family of silicon-oxygen bearing compounds such as feldspar ($NaAlSi_3O_6$), beryl ($BeAl_2(Si_6O_{18})$), and olivine (($MgFe)_2SiO_4$), make up 74% of the earth's crust. Si is the third most abundant element on planet Earth (15% by weight), after iron (35%) and oxygen (30%). One need go no further than the

---

[1] "There is one thing stronger than all the armies of the world, and that is an idea whose time has come." Victor Hugo

beach and scoop up some sand to hold Si in your hand. More important, however, Si, with its many compelling characteristics, has almost single-handedly fueled the emergence of the Information Age. Global semiconductor sales, of which Si captured well over 90%, totaled $204,400,000,000 in 2000 [1]. We humans owe a significant debt to this unique element. Indeed, it is the very existence of Si microelectronics that has enabled emergence of the Information Age, which is so profoundly reshaping the way we live and work and play and communicate. Why Si? This profound market dominance of Si rests on a number of surprisingly practical advantages Si has over other competing semiconductors, including the following.

- Si is wonderfully abundant (there are a lot of beaches in the world), and can be easily purified to profoundly low background impurity concentrations (below 10^{10} impurities / cm^3). Given that the atomic density of Si is 5×10^{22} atoms / cm^3, this means that in a production-grade Si wafer, the impurities are smaller than 1 part in 10^{12} (0.000001 ppm), making them some of the purest materials on Earth.

- Si crystals can be grown in amazingly large, virtually defect-free single crystals (200 mm diameter wafers are in production today worldwide, and are rapidly evolving to 300 mm). The resultant large Si wafer size translates directly into more ICs per wafer, effectively lowering the cost per IC. Given that a 200 mm Si boule is roughly 6 feet long, Si crystals are literally the largest and most perfect on the face of planet Earth.

- Si has excellent thermal properties, allowing for the efficient removal of dissipated heat. The thermal conductivity of Si at 77 K is actually larger than that of copper.

- Si can be controllably doped with both n-type and p-type impurities to extremely high dynamic range (less than 10^{14} to greater than 10^{21} cm^{-3}), at moderate incorporation temperatures (e.g., < 1000 °C). In addition, the ionization energies of the three principal dopants for Si (boron, phosphorus, and arsenic) are all at shallow levels in the bandgap (< 50 meV), making them essentially 100% ionized (electrically active) at room temperature.

- Si can be very easily grown or deposited in three different material forms: crystalline Si, polycrystalline Si ("poly" Si), or amorphous Si, each of which finds different uses in IC technologies.

- Si can be etched relatively easily, using either "wet" chemistries (e.g., KOH) or with "dry" chemistries (e.g., with reactive ion etching using CF_6).

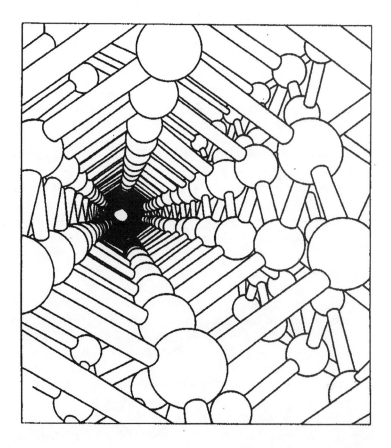

Figure 1.1 End-on view of the Si lattice along the <110> axis (after [2]).

- Like most of the technologically important semiconductors, Si crystallizes in the diamond lattice structure (Figure 1.1 and Figure 2.1). The crystal structure of Si is a direct consequence of its electron orbital configuration ($1s^2 2s^2 2p^6 3s^2 3p^2$), and is thus the underlying reason why Si has so many desirable mechanical and thermal properties. (It's a shame that crystallized Si, despite the fact that it shares an identical lattice structure with that of crystallized C (our beautiful diamonds), ends up with an opaque, fairly uninspiring, greyish-silver appearance.)

- The energy bandgap of Si is of moderate magnitude (1.12 eV at 300 K). If the bandgap were too small (< 0.5 eV), the intrinsic carrier density would be too large at 300 K, making parasitic off-state leakage currents too large. If, instead, the bandgap were too large (> 2.0 eV), then typically it becomes difficult to etch and diffuse dopant impurities (bandgap is a reflection of atomic bonding strength).

- Si is nontoxic and highly stable, making it in many ways the ultimate green material (although its common dopant sources of di-borane, phosphine, and arsine fall decidedly into the "nasty" category).

- Si has excellent mechanical properties, facilitating ease of mechanical handling during the fabrication process. For a 200-mm diameter crystal, for instance, this allows the Si wafers to be cut to roughly 600-μm thickness, maximizing the number of wafers per Si boule. This mechanical stability also minimizes wafer warpage with fabrication, and in addition allows processing to occur under very large thermal gradients without serious consequences (e.g., under rapid-thermal annealing conditions, ramping from 25°C to 1,000°C in 10 seconds).

- It is remarkably easy to form very low resistance ohmic contacts to Si, using a wide variety of metals and doping conditions. Specific contact resistances below 10–20 $\Omega\mu m^2$ can be achieved, for instance, with a heavily doped polysilicon emitter contact, minimizing parasitic device resistances.

- The damage and resultant interface states associated with cleaving or truncating a Si crystal to produce a crystalline surface are not excessively numerous and, importantly, can be easily passivated to manageable levels (e.g., with hydrogen). In device terms, this results in a low surface recombination velocity for Si, and a reduction in parasitic leakage currents and noise associated with surface leakage phenomena.

- The diffusion coefficients of the common Si dopants are "reasonable," meaning that these dopants can be ion-implanted, and then effectively moved to active substitutional sites using comparatively small thermal cycles (temperature and time). This modest annealing cycle also very efficiently restores the crystalline integrity of the Si lattice. This fact is crucial for allowing the formation of shallow junctions, and the maintenance of the thin doping profiles needed for making high-speed transistors.

- Perhaps most importantly, an extremely high-quality dielectric can be trivially grown on Si, simply by flowing oxygen across the wafer surface at an elevated temperature (or even sitting it on the shelf for a few short minutes). This dielectric, silicon-dioxide (SiO_2, "quartz" to geologists) is one of nature's most perfect insulators (it possesses a breakdown strength greater than

10 MV/cm) and can be used for electrical isolation, surface passivation, a planarization layer, an etch stop, or as an active layer (e.g., gate oxide) in the device. SiO_2 also acts as a wonderful diffusion and ion-implantation barrier to dopants, and thus functions as an ideal masking material for layer-by-layer stenciling of the features of our integrated circuits.

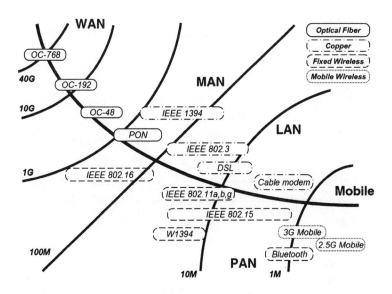

Figure 1.2 The global communications landscape in 2002, broken down by the various communications standards, and spanning the range of: wireless to wireline; fixed to mobile; copper to fiber; low data rate to broadband; and local area to wide area networks. WAN is *wide area network*, MAN is *metropolitan area network*, the so-called "last mile" access network, LAN is *local area network*, and PAN is *personal area network*, the emerging in-home network. (Used with the permission of Kyutae Lim, Georgia Tech.)

Simply put, it is a remarkable fact that nature blessed us with a single material embodying the features one might naively wish for when building low-cost transistors and ICs. From a semiconductor manufacturing standpoint, Si is literally a dream come true. Why is Si the driver of the Information Age? There is literally no other semiconductor that so nicely "fits the bill" as a material from which to construct the roughly 2×10^{20} transistors that currently reside today on planet Earth. Interestingly, the wonderful selling points of Si as a fundamental enabler of the

Information Age have little to do with the device or circuit designer's desires, or needs, and in fact are largely driven by manufacturing, yield, and ultimately cost issues. That is, mundane, but nonetheless compelling, economic issues command the driver's seat. They still do, and clearly will far into the future.

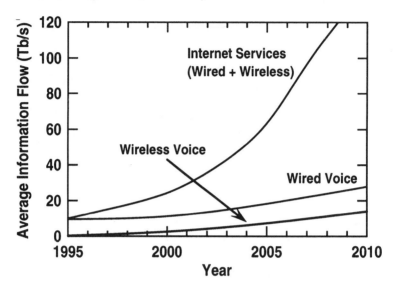

Figure 1.3 Projected global growth in information flow for wired voice, wireless voice, and Internet services (after [3]).

1.2 IC Needs for the Twenty-First Century

Despite the relative infancy of the Information Age, the requirement for integrated circuits and systems is undergoing explosive growth in the global marketplace, a growth that is unlikely to abate in the foreseeable future (Figure 1.2). Indeed, there are few people today who would project any kind of saturation, at least until the physical limits of our conventional semiconductor devices are reached in the 2010–2015 time frame. Even as those horizons inexorably come into view, the frantic search for faster and more complex circuits will only shift directions; it will not cease. Clear evidence for these trends can be found in the growth in average global information flow for wired voice, wireless voice, and Internet applications (Figure 1.3). As can be seen in the evolutionary path of Internet-based services, the Information Age is rapidly evolving into what might be appropriately termed the *Internet Age*, since the Internet appears to be the predominant enabling

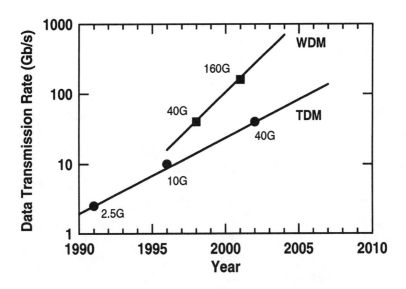

Figure 1.4 Trends in data transmission rates for optical fiber backbone networks. TDM is *time division multiplexing*, and WDM is *wavelength division multiplexing* (after [3]).

medium. The wireline data transmission rates along the global fiber backbone network required to support this projected growth in Internet services are increasing exponentially, fueling what can be termed the *Communications Revolution* (Figure 1.4).

Because of this relentless pace in global information generation, manipulation, storage, and transmission, an insatiable appetite for exponentially greater system-level computational complexity and performance has resulted, translating at the IC level into a demand for increasingly faster logic, increasingly higher memory density, and increasingly higher carrier frequencies for communications channels, as embodied in the well-known Moore's Law growth patterns in the various IC metrics. All at a lower price! Faster, denser, cheaper, the motto of the IC marketer in the twenty-first century. Often a disturbing oxymoron to us IC designers.

1.3 Application-Induced Design Constraints

Where does this evolutionary juggernaut of faster, denser, cheaper ICs leave us poor device and circuit designers? Ask anyone working in the IC trenches, and they will tell you that as IC operational throughput rises, life as a device and circuit designer gets exponentially more difficult! To appreciate why this is so, one simply needs to consider the design constraints imposed by the various types of IC venues that are required to support emerging Information Age applications. By way of illustration, consider simultaneously a classical digital IC (e.g., a microprocessor), a classical analog IC (e.g., a data converter), and a classical RF or microwave IC (e.g., a low noise amplifier). If we deconvolve the various constraints a device and circuit designer necessarily confronts when designing, modeling, laying out, fabricating, testing, packaging, and selling such ICs, some fundamental observations can be made. (Cost is clearly a primary constraint for all application sectors.) These application-induced IC design constraints include:

- Digital circuits (e.g., a microprocessor):
    - switching speed;
    - power consumption;
- Analog circuits (e.g., a data converter):
    - frequency response;
    - output conductance;
    - current gain;
    - 1/f noise;
    - power consumption;
    - temperature coefficient;
    - device-to-device matching;
    - resistor tolerance;
- RF and microwave circuits (e.g., a low-noise amplifier):
    - broadband noise;
    - 1/f noise;
    - linearity;
    - power gain;
    - power consumption;
    - Q of inductors and capacitors;
    - impedance matching;
    - transmission lines;
    - modulation scheme (e.g., GSM versus CDMA, etc.).

Introduction

A cursory glance at these three disparate application arenas paints a very clear picture. The performance requirements at the device and circuit level vary radically depending on the intended application. For instance, the key driving force in low-noise amplifier (LNA) design might be transistor noise figure, but a logic designer on a microprocessor design team most likely could care less about noise figure. This design constraint disparity translates to the system level as well. If we consider a generic radio frequency (RF) transceiver, which might, for instance, make up a cell phone, we see that multiple device technologies are required, ranging from: an RF power amplifier capable of large voltage swings, an RF LNA with very low noise capability, RF mixers and oscillators, memory, passives for matching and filtering, data converters, and digital complementary-metal-oxide-semiconductor (CMOS) for baseband processing (Figure 1.5). In today's cell phones, these individual functional blocks are typically packaged as separate ICs using distinct IC technologies in order to achieve acceptable system performance at the lowest possible cost (e.g., GaAs metal-semiconductor field effect transistor (MESFET) or HBT technology for the low-noise amplifier (LNA) and power amplifier (PA), Si BJT technology for the mixer and oscillator and converters, and Si CMOS technology for baseband processing and digital signal processing (DSP)).

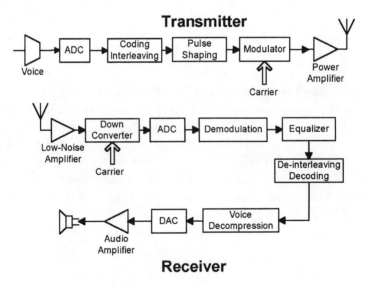

Figure 1.5 A generic RF transceiver architecture.

Given the over-arching theme of cost constraints at the IC level, however, we are led to a logical conclusion. It would be nice if a *single* IC device technology was capable of simultaneously supporting all of the types of self-conflicting circuit design needs: digital, analog, and RF. That is, a "one-technology-fits-all" approach would seem to offer compelling advantages from a cost standpoint, potentially enabling "system-on-a-chip" (SoC) integration (Figure 1.6). While the extent to which SoC will dominate the global communications market over the long haul remains a contentious issue, clearly the trend in most foreseeable communications applications favors an increased level of functional integration in order to achieve reduced form factor, lower chip count, longer battery life, reduced packaging complexity, and ultimately lower total system cost.

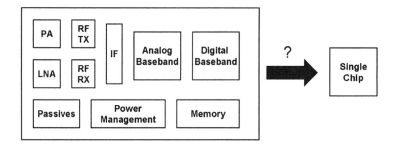

Figure 1.6 Block diagram of a generic cell phone, suggesting a path to single chip integration.

The system-level SoC dream can quickly translate, however, into a device designer's nightmare. Any practicing device engineer will tell you that a single transistor technology simultaneously capable of delivering low-power, high-linearity, low-noise, and high-speed operation for RF, analog, memory, and digital circuits all at a low cost just doesn't exist. Or does it? If we scan the entire field of available IC technologies, we are led inexorably to a logical conclusion. As SoC IC designers we would ideally like to combine the superior RF and analog performance properties of III-V technologies with Si CMOS for digital and memory functions, all married together with the economy of scale and low cost associated with Si IC manufacturing. A Si-compatible, III-V device technology? You bet!

1.4 The Dream: Bandgap Engineering in Silicon

As wonderful as Si is from a fabrication viewpoint, from a device designer's perspective, Si is hardly the ideal semiconductor. The carrier mobility for both electrons and holes in Si is comparatively small, and the maximum velocity that these

Introduction 11

carriers can attain under high electric fields is limited to about 1×10^7 cm/sec under normal conditions. Since the speed of a device ultimately depends on how fast the carriers can be transported through the device under sustainable operating voltages, Si can thus be regarded as a somewhat "slow" semiconductor. In addition, because Si is an indirect gap semiconductor, light emission is painfully inefficient, making active optical devices such as diode lasers impractical. Many of the III-V compound semiconductors (e.g., GaAs or InP), on the other hand, enjoy far higher mobilities and saturation velocities, and because of their direct gap nature, generally make efficient optical devices. In addition, III-V devices, by virtue of the way they are grown, can be compositionally altered for a specific need or application (e.g., to tune the light output of a diode laser to a specific wavelength). This atomic-level custom tailoring of a semiconductor is called *bandgap engineering*, and yields a large performance advantage for III-V technologies over Si [4]. Unfortunately, these benefits commonly associated with III-V semiconductors pale in comparison to the practical deficiencies associated with making highly integrated, low-cost ICs from these materials. There is no robust thermally grown oxide for GaAs or InP, for instance, and wafers are smaller with much higher defect densities, more prone to breakage, poorer heat conductors, etc. These deficiencies translate into generally lower levels of integration, more difficult fabrication, lower yield, and ultimately higher cost. In truth, of course, III-V materials such as GaAs and InP fill important niche markets today (e.g., GaAs MESFETs for cell phones, AlGaAs or InP-based lasers), but III-V semiconductor technologies will never become mainstream if Si-based technologies can do the job.

While Si ICs are well suited to high-transistor-count, high-volume microprocessors and memory applications, RF and microwave circuit applications, which by definition operate at significantly higher frequencies, generally place much more restrictive performance demands on the transistor building blocks. In this regime, the poorer intrinsic speed of Si devices becomes problematic. That is, even if Si ICs are cheap, they must deliver the required device and circuit performance to produce a competitive system at a given frequency. If not, the higher-priced but faster III-V technologies will dominate (as they indeed have until very recently in the RF and microwave markets).

The fundamental question then becomes simple and eminently practical: is it possible to improve the performance of Si transistors enough to be competitive with III-V devices for RF and microwave applications, while preserving the enormous yield, cost, and manufacturing advantages associated with conventional Si fabrication? The answer is clearly yes, and this book addresses the many nuances associated with using strained SiGe alloys to practice bandgap engineering in the

Si material system, a process culminating in the SiGe HBT. [2]

While the basic idea of using SiGe alloys to bandgap-engineer Si devices dates to the 1950s (Shockley considered it early in the transistor game), the synthesis of defect-free SiGe films proved surprisingly difficult, and device-quality SiGe films were not successfully produced until the mid-1980s. This difficulty has a very obvious physical underpinning. While Si and Ge can be combined to produce a chemically stable alloy, their lattice constants differ by roughly 4.2% and thus SiGe alloys grown on Si substrates are compressively strained. This process is referred to as *pseudomorphic* growth of strained SiGe on Si, with the SiGe film adopting the underlying Si lattice constant. These SiGe strained layers are subject to a fundamental stability criterion limiting their thickness for a given Ge concentration [5, 6]. Deposited SiGe films that lie below the stability curve are thermodynamically stable, and can be processed using conventional furnace or rapid-thermal annealing, or ion-implantation without generating defects. Deposited SiGe films that lie above the stability curve, however, are "metastable" and will relax to their natural lattice constant (> Si) if exposed to temperatures above the original growth temperature, generating device-killing defects in the process. For a manufacturable SiGe technology, it is obviously key that the SiGe films remain stable after processing. Stability of SiGe strained layers will be discussed at length in Chapter 2.

1.5 The SiGe HBT

Introducing Ge into Si has a number of consequences. First and most important, because Ge has a larger lattice constant than Si, the energy bandgap of Ge is smaller than that of Si (0.66 eV vs 1.12 eV), and thus SiGe will have a bandgap smaller than that of Si, making it a suitable candidate for bandgap engineering in Si. The compressive strain associated with SiGe alloys produces an additional bandgap shrinkage, and the net result is a bandgap reduction of approximately 75 meV for each 10% of Ge introduced. This Ge-induced "band offset" occurs predominantly in the valence band, making it conducive for use in tailoring *npn* bipolar transistors. In addition, the compressive strain lifts the conduction and valence band degeneracies at the band extremes, effectively reducing the density-of-states and improving

---

[2]It is technically correct to refer to silicon-germanium alloys according to their chemical composition, $Si_{1-x}Ge_x$, where x is the Ge mole fraction. Following standard usage, such alloys are usually referred to as "SiGe" alloys. Note, however, that it is common in the material science community to also refer to such materials as "Ge:Si" alloys. In this book we will follow standard usage and denote these materials as SiGe alloys. Believe it or not, this field also has its own set of slang pronunciations. The colloquial usage of \'sig-ee\ to refer to "SiGe" (begun at IBM in the late 1990s) has come into vogue recently, although we remain purists in this regard, sticking with the more traditional "silicon-germanium."

the carrier mobilities with respect to pure Si (the latter due to a reduction in carrier scattering). Because a practical SiGe film must be very thin if it is to remain stable and hence defect free, it is a natural candidate for use in the base region of a bipolar transistor (which by definition must be thin to achieve high-frequency operation). The resultant device contains an n-Si / p-SiGe emitter-base heterojunction and a p-SiGe / n-Si base-collector heterojunction, and thus this device is properly called an "SiGe double-heterojunction bipolar transistor," although for clarity we will continue the standard usage of "SiGe heterojunction bipolar transistor" (SiGe HBT).[3] The SiGe HBT represents the first practical bandgap-engineered transistor in the Si material system.

Perhaps most importantly, SiGe HBTs can be quite easily teamed with best-of-breed Si CMOS to form a monolithic SiGe HBT BiCMOS technology. While this might seem at first glance to be a mundane advantage, it is in fact a fundamental enabler for SiGe's long-term success, provided SiGe HBTs can be realized without an excessive cost penalty compared to standard Si ICs. The integration of SiGe HBTs with Si CMOS is also the fundamental departure point between SiGe technology and III-V technologies. If SiGe technology is to be successful in the long haul, it must bring to the table the RF and analog performance advantages of the SiGe HBT, and the low-power logic, integration level, and memory density of Si CMOS, into a single cost-effective IC that enables SoC integration (i.e., SiGe HBT BiCMOS). This merger appears to be the path favored by most companies today. Typically, SiGe HBTs (often with multiple breakdown voltages) exist as an "adder" to a basic CMOS IC building-block core, to be swapped in or out as the application demands, without excessive cost burden. Typical state-of-the-art SiGe HBT BiCMOS technologies generally have a roughly 20% adder in mask count compared to "vanilla" digital CMOS, and are viewed by many as an acceptable compromise between performance benefit and cost, depending on the application. In truth, SiGe HBT BiCMOS technologies are the future of the SiGe HBT, since it enables system-on-a-chip solutions across a very broad market base for both wired and wireless applications, all at an acceptable cost. This is clearly the evolutionary path being traveled today by almost all companies with commercially viable SiGe

---

[3] A common misconception persists in the literature that the SiGe HBT is not a "true" HBT, but rather some sort of "mutant" bipolar junction transistor (BJT). While it is true that the fundamental doping profile design of most SiGe HBTs in production today does not follow the lines of their III-V HBT brethren, the SiGe HBT is still an HBT. Traditional III-V HBTs exploit a wide bandgap emitter to reduce the back-injected base current (i.e., improve the emitter injection efficiency), thereby allowing an acceptable current gain while using a lightly doped emitter and a very heavily doped base. SiGe HBTs, on the other hand, typically employ a graded-Ge-base design with a heavily doped emitter and moderately doped base, similar to what might be found in a conventional Si BJT. Nevertheless, SiGe HBTs do in fact still contain dual SiGe/Si heterojunctions and thus should be properly referred to as HBTs.

technologies.

1.6 A Brief History of SiGe Technology

The concept of the HBT is an old one, dating to the fundamental BJT patent issued to Shockley in 1951 [7]. Given that the first bipolar transistor was built from Ge, it seems quite likely that Shockley even envisioned the combination of Si and Ge to form a SiGe HBT (he was a bright guy!). The basic formulation and operational theory of the HBT was pioneered by Kroemer, and was in place by 1957 [8, 9].[4] Reducing the SiGe HBT to practical reality, however, took 30 years due to material growth limitations. Once device-quality SiGe films were achieved in the mid-1980s, progress was quite rapid from that point forward. An interesting historical discussion of early SiGe HBT development is contained in [10]. Table 1.1 summarizes the key steps in the evolution of SiGe HBT technology.

The first functional SiGe HBT was demonstrated in December of 1987 [16],[5] but worldwide attention became squarely focused on SiGe technology in June of 1990 with the demonstration of a non-self-aligned SiGe HBT grown by ultra-high vacuum/chemical vapor deposition (UHV/CVD), with a peak cutoff frequency of 75 GHz [18, 19]. At the time, this SiGe result was roughly twice the performance of state-of-the-art Si BJTs (Figure 1.7), and clearly demonstrated the future performance potential of the technology. Eyebrows were lifted, and work to develop SiGe as a practical circuit technology began in earnest in a large number of laboratories around the world.

In December of 1990, the first emitter-coupled-logic (ECL) ring oscillators using self-aligned, fully integrated SiGe HBTs were produced [20]. The first SiGe BiCMOS technology was reported in December of 1992 [22], and the first LSI SiGe HBT circuit (a 1.2 GSample/s 12-bit digital-to-analog converter) was demonstrated in December of 1993 [23]. The first SiGe HBTs with frequency response greater than 100 GHz were described in December of 1993 [24, 25], and the first

---

[4]Kroemer was awarded the Nobel Prize in 2000 for his work in bandgap engineering.

[5]It is an interesting and often overlooked historical point that at least three independent groups were simultaneously racing to demonstrate the first functional SiGe HBT, all using the molecular beam epitaxy (MBE) growth technique: an IBM team [33], a Bell Labs team [34], and a Linköping University team [35]. The IBM team is fairly credited with the victory, since it presented (and published) its results in early December 1987 at the IEDM (it would have been submitted to the conference for review in the summer of 1987) [16]. Even for the published journal articles, the IBM team was the first to submit their paper for review (on November 17, 1987), followed by the Bell Labs team (on November 23, 1987), and the Linköping University team (on February 22, 1988). All three papers appeared in print in the spring of 1988. The first SiGe HBT demonstrated using (the more manufacturable) CVD growth technique followed shortly thereafter [17].

Introduction

Table 1.1 Key Steps in the Evolution of SiGe HBT Technology

Historical Event	Year	Reference
Fundamental HBT patent	1951	[7]
Drift-base HBT concept	1954	[8]
Basic HBT theory	1957	[9, 12, 13]
First growth of SiGe strained layers	1975	[11]
First growth of SiGe epitaxy by MBE	1985	[14]
First growth of SiGe epitaxy by UHV/CVD	1986	[15]
First SiGe HBT	1987	[16]
First ideal SiGe HBT grown by CVD	1989	[17]
First high-performance SiGe HBT	1990	[18, 19]
First self-aligned SiGe HBT	1990	[20]
First SiGe HBT ECL ring oscillator	1990	[20]
First *pnp* SiGe HBT	1990	[21]
First SiGe HBT BiCMOS technology	1992	[22]
First LSI SiGe HBT Integrated Circuit	1993	[23]
First SiGe HBT with peak f_T above 100 GHz	1993	[24, 25]
First SiGe HBT technology in 200-mm manufacturing	1994	[26]
First SiGe HBT technology optimized for 77 K	1994	[27]
First SiGeC HBT	1996	[28]
First high power SiGe HBTs	1996	[29, 30]
First sub-10 psec SiGe HBT ECL circuits	1997	[31]
First SiGe HBT with peak f_T above 200 GHz	2001	[32]

SiGe HBT technology entered commercial production on 200-mm wafers in December of 1994 [26]. The 200-GHz peak f_T barrier was broken in November of 2001 for a non-self-aligned device [32], and for a self-aligned device in February of 2002 [36]. SiGe HBT technologies with f_T above 300 GHz are clearly a realistic goal at this point, making SiGe HBTs quite competitive in performance with competing III-V HBT technologies.

To date, the IC with the highest SiGe HBT device count on a single chip is a 69 × 69 cross-point switch containing greater than 100,000 0.5-μm SiGe HBTs [37]. The highest demonstrated level of SiGe HBT BiCMOS integration to date is a 10.8 × 10.8 mm^2 mixed-signal, single-chip OC-192 10 Gb/s SONET/SDH mapper with integrated serializer/deserializer, clock and data recovery circuits, and synthesis unit, containing 6,000 0.5-μm SiGe HBTs and 1,200,000 CMOS transistors

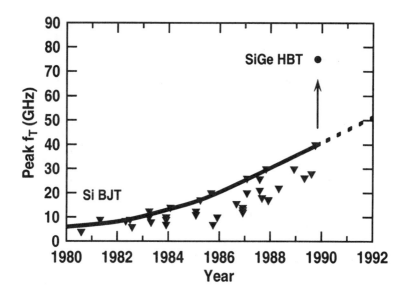

Figure 1.7 Historical trends in published peak cutoff frequency values for various Si BJT technologies compared with the first high-performance SiGe HBT result.

[36, 38].

Not surprisingly, research and development activity in SiGe devices, circuits, and technologies in both industry and at universities worldwide has grown rapidly since the first demonstration of a functional SiGe HBT in 1987. This global interest is nicely reflected in the number of SiGe HBT technical publications in IEEE journals and conferences from 1987 until present, as shown in Figure 1.8.

During the evolutionary path of SiGe HBTs, a large number of SiGe HBT device technologies have been demonstrated at laboratories throughout the world, using a variety of different SiGe epitaxial growth techniques. Commercial SiGe HBT technologies now exist in companies around the world, including: IBM [36],[6] Hitachi [39], Conexant (Jazz) [40], Infineon [41], NEC [42], IHP [43], IMEC [44], TI [45], Philips [46], Lucent [47], ST Microelectronics [48], TEMIC [49], and CNET [50].[7] In recent years, these various SiGe HBT technologies have been leveraged to demonstrate a large number of impressive digital, analog, RF, and microwave circuit results for wireless and wireline communications applications [51]–[100].

---

[6]For fascinating historical insight into the development of SiGe technology at IBM, see [10].

[7]Only the most recently published version of the SiGe technology from each respective company is given. For the interested reader, each paper contains relevant references to earlier versions of that respective company's SiGe technologies.

Introduction

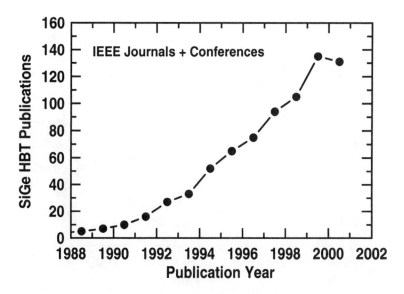

Figure 1.8 Historical trends in the yearly number of SiGe HBT papers published in IEEE journals and conferences (source: IEEE Xplore).

A large number of commercial products using SiGe HBTs are currently on the market, and a foundry service through MOSIS for SiGe HBT BiCMOS technology is available [101], all healthy signs for a new device technology. A variety of review papers on SiGe materials, devices, circuits, and technologies can be found in the literature [102]–[119], and four books (excluding the one you are reading) dealing in one way or another with SiGe materials and devices have been published [120]–[123].

1.7 SiGe HBT Performance Trends

While performance trend charts should always be taken with a grain of salt as to their predictive power, it is nonetheless instructive to examine how SiGe HBT performance has progressed from 1987 until present. For reasons that may or may not meet with approval from all quarters, we have chosen to limit these SiGe HBT trend data in the following manner:

- Consider only results published in the peer-reviewed technical literature.
- Consider either SiGe HBTs or SiGeC HBTs.[8]

---
[8] A SiGe HBT that has carbon-doping (e.g., less than 0.20% C) in the base to suppress boron

18   Silicon-Germanium Heterojunction Bipolar Transistors

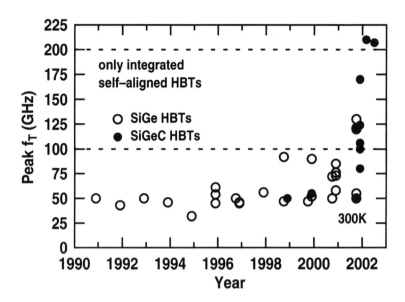

Figure 1.9 Historical trends in peak cutoff frequency for integrated, self-aligned SiGe HBT and SiGeC HBT technologies.

- Consider only self-aligned, fully integrated, Si-processing-compatible SiGe HBT technologies. This eliminates, for instance, non-self-aligned device results that were primarily intended as profile demonstrations. It also eliminates III-V-like mesa-isolated technologies, which cannot be easily integrated with high-transistor-count IC processes, although such device technologies clearly have merit for certain microwave and millimeter wave applications.
- Consider either SiGe HBT or SiGe HBT BiCMOS technologies.
- Consider only room-temperature (300 K) results.

This definition captures greater than 95% of published SiGe HBT results, and limits the trend data to SiGe device technologies that are at least potentially manufacturable and hence in principle commercially viable.

We note that while peak cutoff frequency (f_T) is reasonably straightforward to measure using standard S-parameter techniques (assuming proper calibration and

---

out-diffusion is properly referred to as a SiGe:C HBT, or simply SiGeC HBT (pronounced "silicon germanium carbon," *not* "silicon germanium carbide"). This class of devices should be viewed as optimized SiGe HBTs, and is distinct from HBTs fabricated using SiGeC alloys with a much higher C content (e.g., 2–3% C) needed to lattice-match SiGeC alloys to Si.

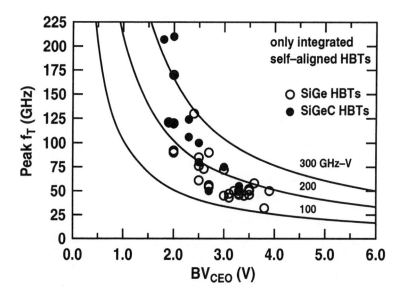

Figure 1.10 Historical trends in peak cutoff frequency as a function of collector-to-emitter breakdown voltage for integrated, self-aligned SiGe HBT and SiGeC HBT technologies.

parasitic de-embedding is performed), the same cannot be said for f_{max}. It has become common practice in the literature to cite f_{max} numbers using unilateral gain (U) extrapolations, which often gives more optimistic numbers than those determined from extrapolations of the maximum available gain (MAG). The f_{max} data presented does not distinguish between the two techniques, and thus adds a level of uncertainty to the f_{max} data presented.

Figure 1.9 shows the historical trends in peak f_T from the first self-aligned device demonstration in December of 1990 [20] until present. It is interesting to note that until about 1998, peak f_T remained in the 50–75 GHz range, suggesting that most research groups were on a profile design and fabrication learning curve, or else attempting to migrate their technologies from research-level demonstrations into commercial IC technologies, and thus worrying less about transistor performance than manufacturability, qualification, and yield. It is interesting to note that both the 100-GHz and 200-GHz f_T barriers were broken within 6 months of each other, 100 GHz being reached in September 2001 by four separate groups [48, 124, 125, 126], and 200 GHz being reached in February of 2002 [36]! [9] It is

---

[9] Note that non-self-aligned SiGe HBTs with $f_T > 100$ GHz were demonstrated as early as 1993 [24, 25], and a 210-GHz non-self-aligned SiGe HBT was demonstrated in November of 2001 [32].

also clear from Figure 1.9 that after 2001, many groups began migrating towards C-doping of their SiGe HBTs to reduce boron out-diffusion in the base profile, and thereby improve f_T.

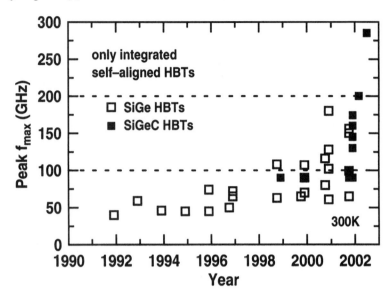

Figure 1.11 Historical trends in peak maximum oscillation frequency for integrated, self-aligned SiGe HBT and SiGeC HBT technologies.

It has been appreciated since 1965 that a fundamental reciprocal relationship exists between transistor peak f_T and BV_{CEO} [127], and the SiGe HBT data qualitatively bear out this trend (Figure 1.10). Most published SiGe HBT results are centered upon an $f_T \times BV_{CEO}$ product of about 200 GHz-V, slightly higher than original "Johnson limit" for Si of 170 GHz-V. More recent results suggest that higher values of the $f_T \times BV_{CEO}$ product are attainable as SiGe device technologies evolve, and have been clustered in the 250–300-GHz-V range over the 1999–2001 time frame. The present record for the $f_T \times BV_{CEO}$ product for a SiGe HBT is 420 GHz-V [36], substantially higher than one might naively expect given past trend data.[10]

Whether this is indirect evidence of an alternative transport mechanism (i.e., ballistic transport) at this level of vertical scaling remains to be seen. It is also worth noting that the current density at which peak f_T is reached has also been steadily rising over time, from about 1.5 mA/μm^2 in 1990 (50 GHz at $BV_{CEO} =$

---

[10]Note added in press: the demonstration of a SiGe HBT with 350-GHz peak f_T ($BV_{CEO} = 1.4$ V), to be presented at the IEDM in December of 2002, increases this number to 490 GHz-V [128]!

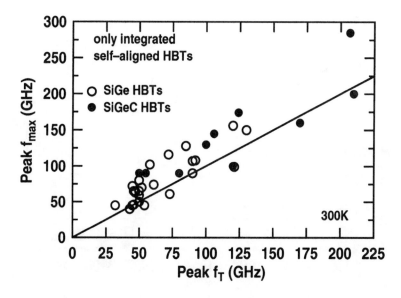

Figure 1.12 Historical trends in peak maximum oscillation frequency as a function of peak cutoff frequency for integrated, self-aligned SiGe HBT and SiGeC HBT technologies.

3.2 V) to about 10.0 mA/μm^2 in 2002 (210 GHz at $BV_{CEO} = 2.0$ V). This J_C rise over time clearly presents a host of challenges in terms of device reliability and technology metalization needs, not to mention system-level, voltage-compression design constraints induced by the ever-shrinking BV_{CEO}.

While peak f_T is very useful as a technology figure-of-merit, f_{max} is a more relevant circuit-level performance metric. Figure 1.11 shows that the SiGe HBT peak f_{max} data has risen over time is a similar manner to that of peak f_T, as might be naively expected. In 2002, best-of-breed SiGe HBTs have attained peak f_{max} in the 200-GHz range (the present record being 285 GHz [129]), quite impressive even by III-V HBT standards. Figure 1.12 shows that for most SiGe technologies, peak f_{max} is generally comparable to or even exceeds peak f_T.[11] Achieving comparable f_T and f_{max} is highly desirable for many types of high-speed circuit applications, and this trend for SiGe is different from that seen in most traditional III-V HBT designs, where f_{max} greatly exceeds f_T. This ability of SiGe HBTs to simultaneously maintain both high f_T and high f_{max} is a direct result of the inherently low-parasitic nature of highly-scaled, self-aligned Si device structures, and is a decided advantage from an application standpoint.

---

[11] Please see the above cautionary note concerning the interpretation of f_{max} data extrapolations.

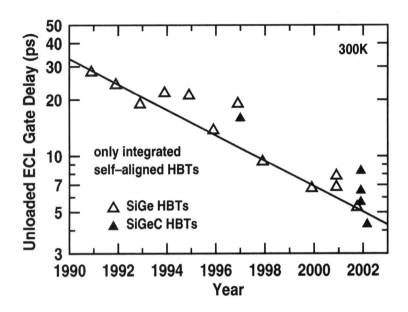

Figure 1.13 Historical trends in unloaded ECL gate delay for integrated, self-aligned SiGe HBT and SiGeC HBT technologies.

Unloaded ECL ring oscillator gate delay has historically been used as a "one-step-better" technology performance metric, since it is easy to implement and is the simplest "real" circuit demonstration vehicle, with a delay that depends strongly on both f_T as well as the resistive and capacitive device parasitics. Figure 1.13 shows the SiGe HBT ECL gate delay trend data. The fact that this data is roughly following a linear decrease on log-linear scales indicates that this performance data is following a classical Moore's Law exponential growth pattern. The long-standing 10 psec ECL delay barrier was broken in December of 1997 [31], and has since marched steadily downward to the present record of 4.3 psec [36].

1.8 The IC Technology Battleground: Si Versus SiGe Versus III-V

And the winner is? From the very beginning it has been, and remains to this day, a highly contentious issue as to whether SiGe technology will be able to successfully position itself to dominate existing and future IC market sectors across a broad array of application fronts. Even broad-brushed comparisons of the relative merits of the competing IC device technologies can be perilous, given that there is no such

Introduction

Figure 1.14 Headline News: "SiGe 'Pac-Man' gobbles up GaAs competition!" (Used with the permission of Michael W. Davidson, Florida State University.) This SiGe Pac-Man was found on a SiGe RFIC designed by TEMIC Semiconductors. Pac-Man was originally designed by Toru Iwantani and programmed by Hideyuki Mokajima and his associates. The name Pac-Man is derived from the Japanese slang "Paku-paku," which means "to eat." Originally, the Japanese named the game "Puckman," but it was changed to "Pac-Man" upon launching in the United States. Pac-Man is the best-selling video game in history.

thing as a true "apples-to-apples" comparison, and one will inevitably be accused of a personal bias of this or that sort, or be charged with comparing one inferior example of a given technology with a superior example of a competing technology, thus artificially skewing the result. In addition, the potential circuit applications of any given technology are so diverse, some favoring one performance metric, others favoring another performance metric, that sweeping generalizations are simply impossible, and the reader should be wary when they are attempted. Given this disclaimer, however, it is nonetheless instructive in this context to make some general comparisons of the performance metrics that might be encountered, for instance, while designing an radio-frequency integrated circuit (RFIC) using each of the various device topologies (Table 1.2).

From an RF viewpoint, state of the art SiGe HBTs offer frequency response, noise figure, and linearity comparable to current-generation III-V devices, and better than both Si BJTs and Si CMOS (even highly scaled CMOS). SiGe HBTs offer better low-frequency (1/f) and phase noise than all of the competition, with the

Table 1.2 Relative Performance Comparisons of Various Device Technologies for RFICs (Excellent: ++; Very good: +; Good: 0; Fair: –; Poor: – –)

Performance Metric	SiGe HBT	Si BJT	Si CMOS	III-V MESFET	III-V HBT	III-V HEMT
Frequency response	+	0	0	+	++	++
1/f and phase noise	++	+	–	– –	0	– –
Broadband Noise	+	0	0	+	+	++
Linearity	+	+	+	++	+	++
Output conductance	++	+	–	–	++	–
Tranconductance/area	++	++	– –	–	++	–
Power dissipation	++	+	–	–	+	0
CMOS integration	++	++	N/A	– –	– –	– –
IC cost	0	0	+	–	–	– –

possible exception of Si BJTs. Being a bipolar transistor, the transconductance per unit area of a SiGe HBT is much higher than for either Si or III-V FETs, and for profile designs with a graded Ge base, the output conductance of a SiGe HBT is also superior to either Si or III-V FETs. SiGe HBTs also have the beneficial feature that their broadband noise is minimized at very low current densities (typically 10× lower than peak f_T), in direct contrast to FETs (Si or III-V), making them very attractive from a power dissipation point of view for portable applications.[12] III-V devices, especially high-electron-mobility transistors (HEMTs), will continue to provide the very best noise performance, albeit at a higher cost, and given their larger bandgaps and hence breakdown voltages, III-V devices will make the best power devices. The long-term advantage of SiGe HBTs over the competition is a strong function of system-level integration and cost. That is, the ability of SiGe HBTs to integrate easily with conventional CMOS distinguishes them fundamentally from all III-V technologies. SiGe technology is essentially equivalent to Si technology in that sense, and enjoys all of the advantages associated with the economy of scale of Si IC manufacturing, including yield and die cost.

Which device technology is likely to walk away from the IC technology battleground? The SiGe advocates obviously embrace the notion depicted in Figure 1.14, in which SiGe swallows the GaAs competition. Wishful thinking? The III-V ad-

---

[12]That the SiGe HBT exhibits far lower power dissipation than CMOS at fixed RF noise figure is wonderfully ironic, given that the large power dissipation associated with ECL is ultimately what doomed Si BJT technology to CMOS domination in the digital world. Sweet revenge!

vocates clearly see no need for a SiGe upstart! As a interested spectator on the device technology battleground, however, it has been mildly amusing to witness the III-V camp at technical conferences slowly but surely change from a "SiGe is no threat at all, please go away" mentality in the early to mid-1990s, to a grudging but gradual acceptance of SiGe as a serious contender in the late 1990s, to a recent near-paranoia of being supplanted and marginalized by SiGe technology. SiGe technology is indeed evolving rapidly, and given its happy marriage to conventional high-volume, low-cost Si fabrication (and CMOS), it does embody the best of both the III-V and Si worlds, a decided advantage. The CMOS advocates, of course, confidently and zealously maintain that it is simply inevitable that scaled CMOS will "conquer the world," leading to little or no need for either SiGe or III-V devices. Perhaps. While the CMOS tsunami did in fact effectively gobble up Si BJT-based ECL in the high-end server market in the early 1990s, and now dominates the digital microprocessor world, the wireless and high-data-rate wireline domains are another matter entirely, and place far more stringent demands on the devices than simple digital logic.

To this question of long-term market dominance, there is simply no easy answer: only time will tell. In the end, the outcome is likely to be the obvious one: SiGe, III-V, and CMOS will all be around 10 years from now, and each will continue to hold important market share for the foreseeable future, in sectors that value their respective strengths. Clearly CMOS will continue to dominate the digital world, and will grow in importance in the low-end wireless sector. SiGe will make steady inroads into a broad array of both wireless and wireline markets, particularly as frequency bands and data rates continue to rise. III-V technologies will continue to dominate the small but important microwave market and the RF power amplifier market. Given the fact that the global electronics market is already enormous ($1,128,000,000,000 in 2000 [1]), and growing rapidly with no real end in sight, holding even a small niche market is likely to be sufficient to provide long-term sustenance for a variety of device technologies.

The worldwide interest in SiGe as a commercial IC technology is growing rapidly. Very rapidly. For those with lingering doubts as to the beautiful efficacy of the SiGe solution for a wide variety of twenty-first IC needs, it should be noted that there are virtually no companies in the world with a vested interest in communications ICs that do not at present have SiGe technology either in production or under development, or at least in use via foundry services. That message in itself is instructive, and should be considered carefully by SiGe pundits. With this requisite background, we now dive into the deep, rich, and fascinating subject of SiGe HBTs. Enjoy!

References

[1] Source: Cahners Instat Group (www.instat.com), 2001.
[2] L. Pauling and R. Hayward, *The Architecture of Molecules*, W.H. Freeman and Company, New York, 1964.
[3] M. Nakamura, "Challenges in semiconductor technology for multi-megabit network services," *Tech. Dig. IEEE Int. Solid-State Circ. Conf.*, pp. 16-18, 1998.
[4] F. Capasso, "Band-gap engineering: from physics and materials to new semiconductor devices," *Science*, vol. 235, pp. 172-176, 1987.
[5] J.W. Matthews and A.E. Blakeslee, "Defects in epitaxial multilayers– I: misfit dislocations in layers," *J. Cryst. Growth*, vol. 27, pp. 118-125, 1974.
[6] J.W. Matthews and A.E. Blakeslee, "Defects in epitaxial multilayers– II: dislocation pile-ups, threading dislocations, slip lines and cracks," *J. Cryst. Growth*, vol. 32, pp. 265-273, 1975.
[7] W. Shockley, U.S. Patent 2,569,347, issued 1951.
[8] H. Kroemer, "Zur theorie des diffusions und des drifttransistors, part III," *Arch. Elektr. Ubertragung*, vol. 8, pp. 499-504, 1954.
[9] H. Kroemer, "Theory of a wide-gap emitter for transistors," *Proc. IRE*, vol. 45, pp. 1535-1537, 1957.
[10] D.L. Harame and B.S. Meyerson, "The early history of IBM's SiGe mixed signal technology," *IEEE Trans. Elect. Dev.*, vol. 48, pp. 2555-2567, 2001.
[11] E. Kasper, H.J. Herzog, and H. Kibbel, "A one-dimensional SiGe superlattice frown by UHV epitaxy," *J. Appl. Phys.*, vol. 8, pp. 1541-1548, 1975.
[12] H. Kroemer, "Heterostructure bipolar transistors and integrated circuits," *Proc. IEEE*, vol. 70, pp. 13-25, 1982.
[13] H. Kroemer, "Heterostructure bipolar transistors: what should we build?," *J. Vacuum Sci. Tech.: B1*, vol. 2, pp. 112-130, 1983.
[14] R. People, "Indirect bandgap of coherently strained $Si_{1-x}Ge_x$ bulk alloys on <001> silicon substrates," *Physical Review B*, vol. 32, pp. 1405-1408, 1985.
[15] B.S. Meyerson, "Low-temperature silicon epitaxy by ultrahigh vacuum / chemical vapor deposition," *Appl. Phys. Lett.*, vol. 48, pp. 797-799, 1986.
[16] S.S. Iyer et al., "Silicon-germanium base heterojunction bipolar transistors by molecular beam epitaxy," *Tech. Dig. IEEE Int. Elect. Dev. Meeting*, pp. 874-876, 1987.
[17] C.A. King et al., "$Si/Si_{1-x}/Ge_x$ heterojunction bipolar transistors produced by limited reaction processing," *IEEE Elect. Dev. Lett.*, vol. 10, pp. 52-54, 1989.

[18] G.L. Patton et al., "63-75 GHz f_T SiGe-base heterojunction-bipolar technology," *Tech. Dig. IEEE Symp. VLSI Tech.*, pp. 49-50, 1990.

[19] G.L. Patton et al., "75 GHz f_T SiGe base heterojunction bipolar transistors," *IEEE Elect. Dev. Lett.*, vol. 11, pp. 171-173, 1990.

[20] J.H. Comfort et al., "Profile leverage in a self-aligned epitaxial Si or SiGe-base bipolar technology," *Tech. Dig. IEEE Int. Elect. Dev. Meeting*, pp. 21-24, 1990.

[21] D.L. Harame et al., "SiGe-base PNP transistors fabrication with n-type UHV/CVD LTE in a "NO DT" process," *Tech. Dig. IEEE Symp. VLSI Tech.*, pp. 47-48, 1990.

[22] D.L. Harame et al., "A high-performance epitaxial SiGe-base ECL BiCMOS technology," *Tech. Dig IEEE Int. Elect. Dev. Meeting*, pp. 19-22, 1992.

[23] D.L. Harame et al., "Optimization of SiGe HBT technology for high speed analog and mixed-signal applications," *Tech. Dig. IEEE Int. Elect. Dev. Meeting*, pp. 71-74, 1993.

[24] E. Kasper, A. Gruhle, and H. Kibbel, "High speed SiGe-HBT with very low base sheet resistivity," *Tech. Dig. IEEE Int. Elect. Dev. Meeting*, pp. 79-81, 1993.

[25] E.F. Crabbé et al., "Vertical profile optimization of very high frequency epitaxial Si- and SiGe-base bipolar transistors," *Tech. Dig. IEEE Int. Elect. Dev. Meeting*, pp. 83-86, 1993.

[26] D.L. Harame et al., "A 200 mm SiGe-HBT technology for wireless and mixed-signal applications," *Tech. Dig. IEEE Int. Elect. Dev. Meeting*, pp. 437-440, 1994.

[27] J.D. Cressler et al., "An epitaxial emitter cap SiGe-base bipolar technology for liquid nitrogen temperature operation," *IEEE Elect. Dev. Lett.*, vol. 15, pp. 472-474, 1994.

[28] L. Lanzerotti et al., "Si/Si$_{1-x-y}$Ge$_x$C$_y$/Si heterojunction bipolar transistors," *IEEE Elect. Dev. Lett.*, vol. 17, pp. 334-337, 1996.

[29] A. Schüppen et al., "1-W SiGe power HBTs for mobile communications," *IEEE Micro. Guided Wave Lett.*, vol. 6, pp. 341-343, 1996.

[30] P.A. Potyraj et al., "A 230-Watt S-band SiGe heterojunction junction bipolar transistor," *IEEE Trans. Micro. Theory Tech.*, vol. 44, pp. 2392-2397, 1996.

[31] K. Washio et al., "A selective-epitaxial SiGe HBT with SMI electrodes featuring 9.3-ps ECL-Gate Delay," *Tech. Dig. IEEE Int. Elect. Dev. Meeting*, pp. 795-798, 1997.

[32] S.J. Jeng et al., "A 210-GHz f_T SiGe HBT with non-self-aligned structure," *IEEE Elect. Dev. Lett.*, vol. 22, pp. 542-544, 2001.

[33] G.L. Patton et al., "Silicon-germanium-base heterojunction bipolar transistors by molecular beam epitaxy," *IEEE Elect. Dev. Lett.*, vol. 9, pp. 165-167, 1988.

[34] H. Temkin et al., "Ge_xSi_{1-x} strained-layer heterostructure bipolar transistors," *Appl. Phys. Lett.*, vol. 52, pp. 1089-1091, 1988.

[35] D.-X. Xu et al., "$n - Si/p - Si_{1-x}Ge_x/n - Si$ double-heterojunction bipolar transistors," *Appl. Phys. Lett.*, vol. 52, pp. 2239-2241, 1988.

[36] A.J. Joseph et al., "0.13 μm 210 GHz f_T SiGe HBTs - expanding the horizons of SiGe BiCMOS," *Tech. Dig. IEEE Int. Solid-State Circ. Conf.*, pp. 180-182, 2002.

[37] AMCC S2090 preliminary data (www.amcc.com), 2002.

[38] AMCC S4803 preliminary data (www.amcc.com), 2002.

[39] K. Oda et al., "Self-aligned selective-epitaxial-growth $Si_{1-x-y}Ge_xC_y$ HBT technology featuring 170-GHz f_{max}," *Tech. Dig. IEEE Int. Elect. Dev. Meeting*, pp. 332-335, 2001.

[40] M. Racanelli et al., "Ultra high speed SiGe npn for advanced BiCMOS Technology," *Tech. Dig. IEEE Int. Elect. Dev. Meeting*, pp. 336-339, 2001.

[41] J. Böck et al., "High-speed SiGe:C bipolar technology," *Tech. Dig. IEEE Int. Elect. Dev. Meeting*, pp. 344-347, 2001.

[42] T. Hashimoto et al., "A 73 GHz f_T 0.18μm RF-SiGe BiCMOS technology considering thermal budget trade-off and with reduced boron-spike effect on HBT characteristics," *Tech. Dig. IEEE Int. Elect. Dev. Meeting*, pp. 149-152, 2000.

[43] B. Heinemann et al., "Cost-effective high-performance high-voltage SiGe:C HBTs with 100 GHz f_T and $BV_{CEO} \times f_T$ products exceeding 220 VGHz," *Tech. Dig. IEEE Int. Elect. Dev. Meeting*, pp. 348-352, 2001.

[44] S. Decoutere et al., "A 0.35 μm SiGe BiCMOS process featuring a 80 GHz f_{max} HBT and integrated high-Q RF passive components," *Proc. IEEE Bipolar/BiCMOS Circ. Tech. Meeting*, pp. 106-109, 2000.

[45] F.S. Johnson et al., "A highly manufacturable 0.25 μm RF technology utilizing a unique SiGe integration," *Proc. IEEE Bipolar/BiCMOS Circ. Tech. Meeting*, pp. 56-59, 2001.

[46] P. Deixler et al., "Explorations for high performance SiGe-heterojunction bipolar transistor integration," *Proc. IEEE Bipolar/BiCMOS Circ. Tech. Meeting*, pp. 30-33, 2001.

[47] M. Carroll et al., "COM2 SiGe Modular BiCMOS technology for digital, mixed-signal, and RF applications, *Tech. Dig. IEEE Int. Elect. Dev. Meeting*, pp. 145-148, 2000.

[48] H. Baudry et al., "High performance 0.25 μm SiGe and SiGe:C HBTs using non-selective epitaxy," *Proc. IEEE Bipolar/BiCMOS Circ. Tech. Meeting*, pp. 52-55, 2001.

[49] A. Schüppen et al., "SiGe-technology and components for mobile communication systems," *Proc. IEEE Bipolar/BiCMOS Circ. Tech. Meeting*, pp. 130-133, 1996.

[50] A. Chantre et al., "A high performance low complexity SiGe HBT for BiCMOS integration," *Proc. IEEE Bipolar/BiCMOS Circ. Tech. Meeting*, pp. 93-96, 1998.

[51] M. Soda et al., "Si-analog ICs for 20 Gb/s optical receiver," *Tech. Dig. IEEE Int. Solid-State Circ. Conf.*, pp. 170-171, 1994.

[52] T. Hashimoto et al., "SiGe bipolar ICs for a 20Gb/s optical transmitter," *Proc. IEEE Bipolar/BiCMOS Circ. Tech. Meeting*, pp. 167-170, 1994.

[53] F. Sato et al., "Sub-20ps ECL circuits with high-performance super self-aligned selectively grown SiGe base (SSSB) bipolar transistors," *IEEE Trans. Elect. Dev.*, vol. 42, pp. 483-488, 1995.

[54] M. Case et al., "A 23 GHz static 1/128 frequency divider implemented in a manufacturable Si/SiGe HBT process," *Proc. IEEE Bipolar/BiCMOS Circ. Tech. Meeting*, pp. 121-123, 1995.

[55] W. Gao et al., "A 5-GHz SiGe HBT return-to-zero comparator," *Proc. IEEE Bipolar/BiCMOS Circ. Tech. Meeting*, pp. 166-169, 1995.

[56] J. Glenn et al., "12-GHz Gilbert mixers using a manufacturable Si/SiGe epitaxial-base bipolar technology," *Proc. IEEE Bipolar/BiCMOS Circ. Tech. Meeting*, pp. 186-189, 1995.

[57] H. Schumacher et al., "A 3 V supply voltage, DC-18 GHz SiGe HBT wideband amplifier," *Proc. IEEE Bipolar/BiCMOS Circ. Tech. Meeting*, pp. 190-193, 1995.

[58] A. Gruhle et al., "Monolithic 26 GHz and 40 GHZ VCO's with SiGe heterojunction bipolar transistors," *Tech. Dig. IEEE Int. Elect. Dev. Meeting*, pp. 725-728, 1995.

[59] F. Sato et al., "The optical terminal IC: a 2.4 Gb/s receiver and a 1:16 demultiplexer in one chip," *Proc. IEEE Bipolar/BiCMOS Circ. Tech. Meeting*, pp. 162-165, 1995.

[60] L. Larson et al., "Si/SiGe HBT technology for low-cost monolithic microwave integrated circuits," *Tech. Dig. IEEE Int. Solid-State Circ. Conf.*, pp. 80-81, 1996.

[61] F. Sato et al., "A 2.4Gb/s receiver and a 1:16 demultiplexer in one chip using a super self-aligned selectively grown SiGe base (SSSB) bipolar transistor," *IEEE J. Solid-State Circ.*, vol. 31, 1451-1457, 1996.

[62] M. Soyuer et al., "An 11 GHz 3V SiGe voltage-controlled oscillator with integrated resonator," *Proc. IEEE Bipolar/BiCMOS Circ. Tech. Meeting*, pp. 169-172, 1996.

[63] R. Götzfried et al., "Zero power consumption Si/SiGe HBT SPDT T/R antenna switch," *Tech. Dig. IEEE MTT-S Int. Micro. Symp.*, pp. 651-653, 1996.

[64] J. Long et al., "RF analog and digital circuits in SiGe technology," *Tech. Dig. IEEE Int. Solid-State Circ. Conf.*, pp. 82-83, 1996.

[65] W. Gao, W. Snelgrove, and S. Kovacic, "A 5-GHz SiGe HBT return-to-zero comparator for RF A/D conversion," *IEEE J. Solid-State Circ.*, vol. 31, pp. 1502-1506, 1996.

[66] M. Soda et al., "A 1Gb/s 8-channel array OEIC with SiGe photodetector," *Tech. Dig. IEEE Int. Solid-State Circ. Conf.*, pp. 120-121, 1997.

[67] M. Wurzer et al., "42 GHz static frequency divider in Si/SiGe bipolar technology," *Tech. Dig. IEEE Int. Solid-State Circ. Conf.*, pp. 122-123, 1997.

[68] P. Xiao et al., "A 4b 8GSample/s A/D converter in SiGe bipolar technology," *Tech. Dig. IEEE Int. Solid-State Circ. Conf.*, pp. 124-125, 1997.

[69] M. Soyuer et al., "A 5.8 GHz 1 V low-noise amplifier in SiGe bipolar technology," *Proc. IEEE MTT RFIC Symp.*, pp. 19-22, 1997.

[70] H. Ainspan et al., "A 6.25 GHz low DC power low-noise amplifier," *Proc. IEEE Cust. Int. Circ. Conf.*, pp. 177-180, 1997.

[71] S. Shiori et al., "A 10Gb/s SiGe bipolar framer/demultiplexer for SDH system," *Tech. Dig. IEEE Int. Solid-State Circ. Conf.*, pp. 202-203, 1998.

[72] T. Masuda et al., "40Gb/s analog IC chipset for optical receiver using SiGe HBTs, *Tech. Dig. IEEE Int. Solid-State Circ. Conf.*, pp. 314-315, 1998.

[73] J. Sevenhans et al., "Silicon germanium and silicon bipolar RF circuits for 2.7 V single chip radio transceiver integration," *Proc. IEEE Cust. Int. Circ. Conf.*, pp. 409-412, 1998.

[74] J.-O. Plouchart et al., "A 5.2 GHz 3.3 V I/Q SiGe RF transceiver," *Proc. IEEE Cust. Int. Circ. Conf.*, pp. 217-220, 1999.

[75] Y.M. Greshishchev and P. Schvan, "A 60 dB gain 55 dB dynamic range 10 Gb/s broadband SiGe HBT limiting amplifier," *Tech. Dig. IEEE Int. Solid-State Circ. Conf.*, pp. 382-383, 1999.

[76] M. Bopp et al., "A DECT transceiver chip set using SiGe technology," *Tech. Dig. IEEE Int. Solid-State Circ. Conf.*, pp. 68-69, 1999.

[77] T. Morikawa et al., "A SiGe single chip 3.3 V receiver IC for 10Gb/s optical communication systems," *Tech. Dig. IEEE Int. Solid-State Circ. Conf.*, pp. 380-381, 1999.

[78] D. Goren et al., "2.5 GHz pin electronics SiGe driver for IC test equipment," *Proc. IEEE Bipolar/BiCMOS Circ. Tech. Meeting*, pp. 31-33, 1999.

[79] P.D. Tseng et al., "A monolithic SiGe power amplifier for dual-mode (CDMA/AMPS) cellular handset applications," *Proc. IEEE Bipolar/BiCMOS Circ. Tech. Meeting*, pp. 153-156, 1999.

[80] G. Grau et al., "A current-folded up-conversion mixer and VCO with center-tapped inductor in a SiGe-HBT technology for 5 GHz wireless LAN applications," *Proc. IEEE Bipolar/BiCMOS Circ. Tech. Meeting*, pp. 161-164, 1999.

[81] H. Ainspan and M. Soyuer, "A fully-integrated 5-GHz frequency synthesizer in SiGe BiCMOS," *Proc. IEEE Bipolar/BiCMOS Circ. Tech. Meeting*, pp. 165-168, 1999.

[82] Y.M. Greshishchev and P. Schvan, "SiGe clock and data recovery IC with linear type PLL for 10 Gb/s SONET application," *Proc. IEEE Bipolar/BiCMOS Circ. Tech. Meeting*, pp. 169-172, 1999.

[83] Y.M. Greshishchev et al., "A fully integrated SiGe receiver IC for 10Gb/s data rate," *Tech. Dig. IEEE Int. Solid-State Circ. Conf.*, pp. 52-53, 2000.

[84] M. Meghelli et al., "SiGe BiCMOS 3.3 V clock and data recovery circuits for 10 Gb/s serial transmission systems," *Tech. Dig. IEEE Int. Solid-State Circ. Conf.*, pp. 56-57, 2000.

[85] T. Masuda et al., "45 GHz transimpedance 32 dB limiting amplifier and 40 Gb/s 1:4 high-sensitivity demultiplexer with decision circuit using SiGe HBTs for 40 Gb/s optical receiver," *Tech. Dig. IEEE Int. Solid-State Circ. Conf.*, pp. 60-61, 2000.

[86] H. Pretl et al., "A SiGe-bipolar down-conversion mixer for a UTMS zero-IF receiver," *Proc. IEEE Bipolar/BiCMOS Circ. Tech. Meeting*, pp. 40-43, 2000.

[87] M. Roßberg et al., "11 GHz SiGe circuits for ultra wideband radar," *Proc. IEEE Bipolar/BiCMOS Circ. Tech. Meeting*, pp. 70-73, 2000.

[88] K. Ettinger et al., "An integrated 20 GHz SiGe bipolar differential oscillator," *Proc. IEEE Bipolar/BiCMOS Circ. Tech. Meeting*, pp. 161-164, 2000.

[89] H.-I. Cong et al., "A 10 Gb/s 16:1 multiplexer and 10 GHz clock synthesizer in 0.25 μm SiGe BiCMOS," *Tech. Dig. IEEE Int. Solid-State Circ. Conf.*, pp. 80-81, 2001.

[90] S. Ueno et al., "A single-chip 10 Gb/s transceiver LSI using SiGe SOI/BiCMOS," *Tech. Dig. IEEE Int. Solid-State Circ. Conf.*, pp. 82-83, 2001.

[91] M. Reinhold et al., "A fully-integrated 40Gb/s clock and data recovery / 1:4 DEMUX IC in SiGe technology," *Tech. Dig. IEEE Int. Solid-State Circ. Conf.*, pp. 84-86, 2001.

[92] D. Belot et al., "A DCS1800/GSM900 RF to digital fully integrated receiver in SiGe 0.35 *mu*m BiCMOS," *Proc. IEEE Bipolar/BiCMOS Circ. Tech. Meeting*, pp. 86-89, 2001.

[93] G. Babcock, "SiGe-HBT active receiver mixers for base-station applications," *Proc. IEEE Bipolar/BiCMOS Circ. Tech. Meeting*, pp. 90-93, 2001.

[94] D. Johnson and S. Raman, "A packaged SiGe x2 sub-harmonic mixer for U-NII band applications," *Proc. IEEE Bipolar/BiCMOS Circ. Tech. Meeting*, pp. 159-162, 2001.

[95] S. Hackl et al., "A 45 GHz SiGe active frequency multiplier," *Tech. Dig. IEEE Int. Solid-State Circ. Conf.*, pp. 82-83, 2002.

[96] T. Masuda et al., "Single-chip 5.8 GHz ETC transceiver IC with PLL and demodulation circuits using SiGe HBT/CMOS," *Tech. Dig. IEEE Int. Solid-State Circ. Conf.*, pp. 96-97, 2002.

[97] T. Masuda et al., "Single-chip 5.8 GHz ETC transceiver IC with PLL and demodulation circuits using SiGe HBT/CMOS," *Tech. Dig. IEEE Int. Solid-State Circ. Conf.*, pp. 96-97, 2002.

[98] V. Aparin et al., "A highly-integrated tri-band/quad-mode SiGe BiCMOS TF-to-baseband receiver for wireless CDMA/WCDMA/AMPS applications with GPS capability," *Tech. Dig. IEEE Int. Solid-State Circ. Conf.*, pp. 234-235, 2002.

[99] A. Bellaouar et al., "A highly-integrated SiGe BiCMOS WCDMA transmitter IC," *Tech. Dig. IEEE Int. Solid-State Circ. Conf.*, pp. 238-239, 2002.

[100] M. Meghelli, A. Rylyakov, and L. Shan, "50 Gb/s SiGe BiCMOS 4:1 multiplexer and 1:4 demultiplexer for serial communication systems," *Tech. Dig. IEEE Int. Solid-State Circ. Conf.*, pp. 260-261, 2002.

[101] MOSIS IC Foundry Service (www.mosis.org).

[102] R. People, "Physics and applications of Ge_xSi_{1-x}/Si strained layer heterostructures," *IEEE J. Quan. Elect.*, vol. 22, p. 1696-1710, 1986.

[103] S.S. Iyer et al., "Heterojunction bipolar transistors using Si-Ge alloys," *IEEE Trans. Elect. Dev.*, vol. 36, pp. 2043-2064, 1989.

[104] G.L. Patton et al., "SiGe-base heterojunction bipolar transistors: physics and design issues," *Tech. Dig. IEEE Int. Elect. Dev. Meeting*, pp. 13-16, 1990.

[105] B. Meyerson, "UHV/CVD growth of Si and SiGe alloys: chemistry, physics, and device applications," *Proc. IEEE*, vol. 80, p. 1592-1608, 1992.

[106] J.D. Cressler et al., "Silicon-germanium heterojunction bipolar technology: the next leap in silicon?," *Tech. Dig. IEEE Int. Solid-State Circ. Conf.*, pp. 24-27, 1994.

[107] C. Kermarrec et al., "SiGe HBTs reach the microwave and millimeter-wave frontier," *Proc. IEEE Bipolar/BiCMOS Circ. Tech. Meeting*, pp. 155-162, 1994.

[108] J.D. Cressler, "Re-engineering silicon: Si-Ge heterojunction bipolar technology," *IEEE Spectrum*, pp. 49-55, 1995.

[109] D.L. Harame et al., "Si/SiGe epitaxial-base transistors: part I - materials, physics, and circuits," *IEEE Trans. Elect. Dev.*, vol. 40, pp. 455-468, 1995.

[110] D.L. Harame et al., "Si/SiGe epitaxial-base transistors: part II - process integration and analog applications," *IEEE Trans. Elect. Dev.*, vol. 40, pp. 469-482, 1995.

[111] D.L. Harame, "High-performance BiCMOS process integration: trends, issues, and future directions," *Proc. IEEE Bipolar/BiCMOS Circ. Tech. Meet.*, pp. 36-43, 1997.

[112] J.D. Cressler, "SiGe HBT technology: a new contender for Si-based RF and microwave circuit applications," *IEEE Trans. Micro. Theory Tech.*, vol. 46, pp. 572-589, 1998.

[113] L.E. Larson, "High-speed Si/SiGe technology for next-generation wireless system applications," *J. Vacuum Sci. Tech. B*, vol. 16, pp. 1541-1548, 1998.

[114] L.E. Larson, "Integrated circuit technology options for RFIC's - present status and future directions," *IEEE J. Solid-State Circ.*, vol. 33, pp. 387-399, 1998.

[115] H.J. Osten et al., "Carbon-doped SiGe heterojunction bipolar transistors for high-frequency applications," *Proc. IEEE Bipolar/BiCMOS Circ. Tech. Meeting*, pp. 109-116, 1999.

[116] B.S. Meyerson, "Silicon:germanium-based mixed-signal technology for optimization of wired and wireless telecommunications," *IBM J. Res. Dev.*, vol. 44, pp. 391-407, 2000.

[117] A. Gruhle, "Prospects for 200 GHz on silicon with SiGe heterojunction bipolar transistors," *Proc. IEEE Bipolar/BiCMOS Circ. Tech. Meeting*, pp. 19-25, 2001.

[118] D.L. Harame et al., "Current status and future trends of SiGe BiCMOS technology," *IEEE Trans. Elect. Dev.*, vol. 48, pp. 2575-2594, 2001.

[119] D.L. Harame and J.D Cressler, "The SiGe HBT: A Primer," *IEEE Trans. Elect. Dev.*, in press, 2002.

[120] E. Kaspar, editor, *Properties of Strained and Relaxed Silicon Germanium*, EMIS Datareviews Series No. 12, INSPEC, London, 1995.

[121] J.S. Yuan, *SiGe, GaAs, and InP Heterojunction Bipolar Transistors*, John Wiley and Sons, Inc., New York, 1999.

[122] C.K. Maiti and G.A. Armstrong, *Applications of Silicon-Germanium Heterostructure Devices*, Institute of Physics Publishing, London, 2001.

[123] C.K. Maiti, N.B. Chakrabarti, and S.K. Ray, *Strained Silicon Heterostructures: Materials and Devices*, The Institution of Electrical Engineers, London, 2001.

[124] E. Ohue et al., "5.3-ps ECL and 71-GHz static frequency divider in self-aligned SEG SiGe HBT," *Proc. IEEE Bipolar/BiCMOS Circ. Tech. Meeting*, pp. 26-29, 2001.

[125] A. Joseph et al., "A 0.18 μm BiCMOS technology featuring 120/100 GHz (f_T/f_{max}) HBT and ASIC-compatible CMOS using copper interconnect," *Proc. IEEE Bipolar/BiCMOS Circ. Tech. Meeting*, pp. 143-146, 2001.

[126] K. Schuegraf et al., "0.18 μm SiGe BiCMOS technology for wireless and 40 Gb/s communication products," *Proc. IEEE Bipolar/BiCMOS Circ. Tech. Meeting*, pp. 147-150, 2001.

[127] E.O. Johnson, "Physical limitations on frequency and power parameters of transistors," *RCA Review*, pp. 163-177, 1965.

[128] J.-S. Rieh et al., "SiGe HBTs with Cut-off Frequency of 350 GHz," to appear in the *Tech. Dig. IEEE Int. Elect. Dev. Meeting*, December 2002.

[129] B. Jagannathan et al., "Self-aligned SiGe NPN transistors with 285 GHz f_{max} and 207 GHz f_T in a manufacturable technology," *IEEE Elect. Dev. Lett.*, vol. 23, pp. 258-260, 2002.

Chapter 2

SiGe Strained-Layer Epitaxy

It is ironic, in the present context, that the first functional transistors were in fact fabricated from Ge. Little time elapsed until the recognition that Si would prove to be a much better commercial platform for the emerging transistor field than Ge, and except for a few niche applications, Ge dropped out of vogue and was soon forgotten as a viable device material. Interestingly, however, it was appreciated very early in the game that the appropriate combination of Si and Ge, being chemically compatible semiconductors with differing bandgaps, would present interesting device engineering opportunities. Unfortunately, it took nearly 30 years to reduce that idea to the practical reality of device-quality SiGe strained-layer epitaxy. In this chapter we examine the creation of strained-layer epitaxy from Si and Ge, and explore the stability constraints that the SiGe world is governed by. We then address the resultant band structure and transport parameters of SiGe alloys, followed by a brief discourse on remaining open issues that merit further attention.

2.1 SiGe Alloys

Si and Ge are both Group IV elemental semiconductors, and crystallize in the diamond lattice structure, as depicted in Figure 2.1. For a comprehensive table of the bulk structural, mechanical, and electrical properties of both Si and Ge, refer to the Appendix. Si and Ge are completely miscible over their entire compositional range, giving rise to chemically stable SiGe alloys that preserve their parent diamond crystal structure and that have a linearly interpolated lattice constant given to first order by Vegard's rule,

$$a(Si_{1-x}Ge_x) = a_{Si} + x(a_{Ge} - a_{Si}), \qquad (2.1)$$

where a is the lattice constant, and x is the Ge fraction. Diffraction measurements

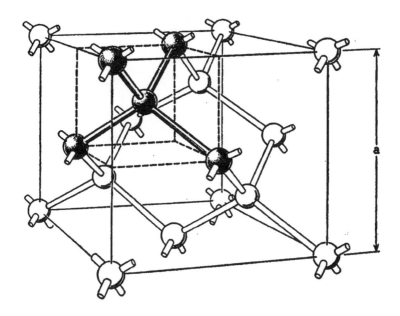

Figure 2.1 Unit cell of the diamond lattice (after [1]).

of actual SiGe films show minor departures of this linear dependence and can be fit by a parabolic relationship of the form

$$a(Si_{1-x}Ge_x) = 0.002733\, x^2 + 0.01992\, x + 0.5431\ (nm), \qquad (2.2)$$

as depicted in Figure 2.2 [2].

2.1.1 Pseudomorphic Growth and Film Relaxation

The lattice mismatch between pure Si ($a = 5.431$ Å) and pure Ge ($a = 5.658$ Å) is 4.17% at 300 K, and increases only slightly with increasing temperature. When SiGe epitaxy [1] is grown (actually it is more properly said to be deposited) onto a thick Si substrate host, this inherent lattice mismatch between the SiGe film and the underlying Si substrate can be accommodated in only one of two ways.

First, the lattice of the deposited SiGe alloy distorts in such a way that it adopts the underlying Si lattice constant, resulting in perfect crystallinity across the growth interface. In essence, the SiGe film is forced to adopt its host's smaller

---
[1] The word "epitaxy" is derived from the Greek word *epi*, meaning "upon" or "over."

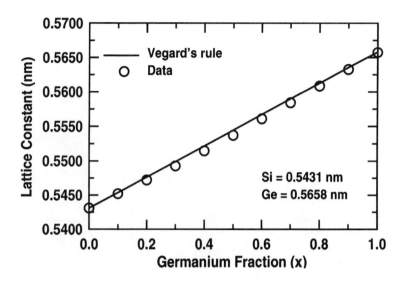

Figure 2.2 Theoretical and experimental lattice constant of a $Si_{1-x}Ge_x$ alloy as a function of Ge fraction.

lattice constant. This scenario is known as "pseudomorphic" [2] growth, and is the desired result for most device applications. Under processing conditions that favor pseudomorphic growth, the SiGe film is forced into biaxial (in-plane) compression. In this case, the SiGe lattice constant in the growth plane is determined by the Si substrate, and the result is a tetragonal distortion (extension) of the normally cubic SiGe crystal in the orthogonal direction, in accordance with the Poisson ratio. The SiGe alloy is now under strain, and "SiGe strained-layer epitaxy" results. Because of the additional strain energy contained in the SiGe film during pseudomorphic growth, it embodies a higher energy state than for an unstrained film, and hence nature does not favor this growth condition except under a very narrow range of conditions, as discussed below.

Second and alternatively, the SiGe film can "relax" during growth to the natural lattice constant determined by its Si and Ge fraction, as given by (2.2). The SiGe film relaxes via misfit dislocation formation, resulting in a break in crystallinity across the growth interface, and a defected film unsuitable for high-yielding device applications. Relaxation during SiGe growth occurs when the pent-up strain energy is sufficiently large that misfit dislocations nucleate and then glide (move).

---

[2]The word "pseudo" is derived from the Greek word *pseudēs*, meaning "false," and the word "morphic" is derived from the Greek word *morphē*, meaning "form." Hence, pseudomorphic literally means *false-form*.

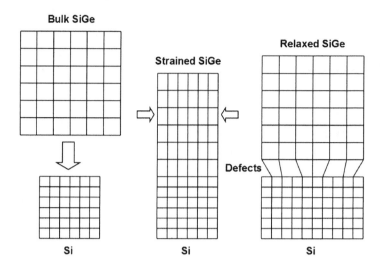

Figure 2.3 Schematic 2-D representation of both strained and relaxed SiGe on a Si substrate.

In essence, when the strain energy in the film exceeds the activation energy required for misfit formation and movement, the film will relax, releasing the stored strain energy. Not surprisingly, this relaxation mechanism is complex, and varying degrees of residual (post-growth) strain can reside in relaxed SiGe films. The misfit dislocations formed during the relaxation process may be either confined to the original growth interface plane, or "thread" their way up through the overlaying SiGe epitaxy, or both, and in either case represent a bad situation from a device design perspective, since such defects can act as generation/recombination (G/R) trapping centers, and high-diffusivity pipes for dopants, which are well-known yield "killers" in bipolar technologies. [3] These two growth scenarios are depicted schematically in Figure 2.3 and Figure 2.4.

Practically speaking, if we imagine an unrestricted SiGe growth process for some arbitrary Ge fraction, it would proceed as follows. Since the Si substrate is very thick (about 600 μm for a 200-mm wafer), and very stiff, it remains essentially unchanged during the epitaxial growth process. Assuming a pristine initial growth interface, the growth of the SiGe film will begin pseudomorphically, adopting the

---

[3]It is worth pointing out that SiGe-based FET device technologies are inherently less sensitive to such strain relaxation induced defects, simply because the FET is a majority carrier device. Minority carrier devices such as the pn junction and bipolar transistor will always be less tolerant of growth induced defects.

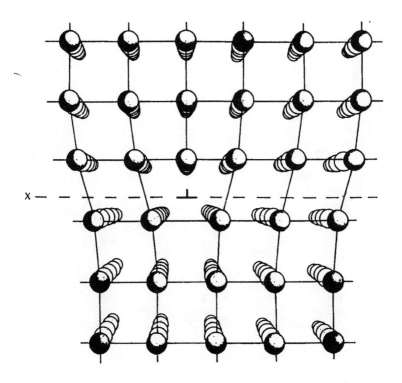

Figure 2.4 Schematic representation of misfit dislocation formed at the Si/SiGe growth interface.

underlying Si lattice constant, but when a given "critical thickness" is reached, the strain energy becomes too large to maintain local equilibrium and the SiGe film will relax to its natural lattice constant, with the excess strain energy being released via misfit formation. In practice, it is also common for the film to remain pseudomorphic until the end of the growth cycle, even though it may have exceeded the critical thickness. Such a SiGe layer is said to be "metastable." Metastable films will relax during subsequent thermal processing steps that add energy to the system, and thus are not suitable for use in Si-fabrication-compatible SiGe technologies. During the relaxation process, whenever it occurs, chaos results, as can be clearly seen in the plan-view TEM micrograph of a relaxed and heavily defected SiGe film shown in Figure 2.5.

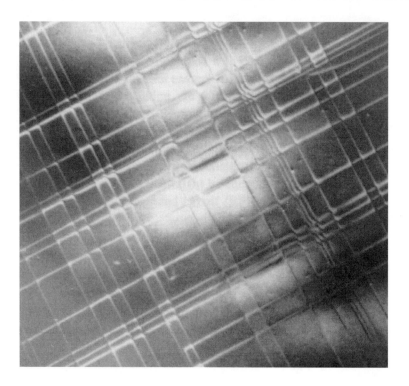

Figure 2.5 Plan-view TEM (top down image) of an unstable SiGe film that has been annealed and undergone relaxation. The visible linear structures are misfit dislocations.

2.1.2 Putting Strained SiGe into SiGe HBTs

Regardless of the growth technique used, or the structure and self-alignment schemes employed in the transistor, strained SiGe films found in today's commercially viable SiGe HBTs all have a similar form. As depicted in Figure 2.6, the deposited SiGe film actually consists of a three-layer composite structure:

- A thin, undoped Si buffer layer;
- The actual boron-doped SiGe active layer;
- A thin, undoped Si cap layer.

The Si buffer layer is used to start the growth process off on the right foot, and serves two purposes. First, the Si buffer layer helps ensure that a pristine SiGe epitaxial growth interface is preserved between the original Si substrate, which was

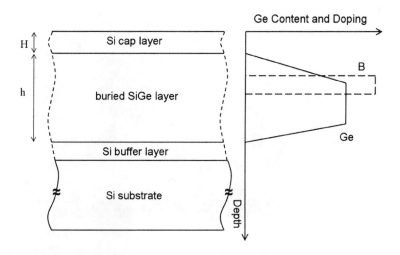

Figure 2.6 Schematic epitaxial SiGe film for use in a SiGe HBT. The film consists of a thin Si buffer layer, the compositionally graded SiGe layer of thickness (h), and a Si cap layer of thickness (H). The boron base doping is contained within the SiGe layer.

grown by a high-temperature Si epitaxy process, and the coming SiGe strained layer that will be grown by a more difficult low-temperature epitaxy process. Maintaining a contaminant-free growth interface with perfect crystallinity is essential for obtaining device-quality SiGe films. Second, this Si buffer layer also frequently plays a role in device design, since it allows the incorporation of intrinsic layers (i-layers) to be easily embedded in the collector-base junction, and can be used to decrease the junction field and aid in breakdown voltage tailoring [3].

The active SiGe layer, of thickness h, has a position-varying Ge composition, and an embedded boron doping spike, typically deposited as a boron box profile of say 10 nm by $2-4 \times 10^{19}$ cm^{-3}, for an integrated base charge of roughly $2-4 \times 10^{13}$ cm^{-2}. The SiGe layer forms the active region of the bandgap-engineered device, and the specific shape, thickness, and placement of the Ge profile with respect to the boron base profile will in large measure determine the resultant performance of the transistor. The *dc* and *ac* trade-offs and implications of Ge and doping profile design in SiGe HBTs are discussed at length in subsequent chapters.

Finally, the Si cap layer, of thickness H, serves four purposes. First, it provides a Si termination to the SiGe composite. This is particularly important since most SiGe HBT fabrication approaches involve some form of oxidation step to form the emitter-base spacer used in self-alignment, and SiGe does not oxidize well.

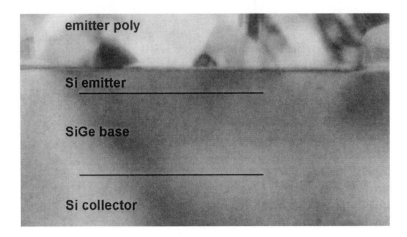

Figure 2.7 Cross sectional TEM showing the active device region of a fabricated SiGe HBT. The (stable) strained SiGe base layer has a peak Ge content of 10% and is defect free, and cannot be delineated from the Si matrix.

Second, the Si cap provides additional space to allow the modest out-diffusion of the boron base profile during processing, while at the same time providing room for the emitter out-diffusion. Third, as with the Si buffer layer, a Si cap layer can be used to introduce an active i-layer into the emitter-base junction to lower the junction electric field and thereby reduce the parasitic EB tunneling current, which typically limits the base current ideality at low-injection, and hence degrades device reliability. Finally, an unintentional but nonetheless important consequence of having this Si cap layer is that it helps improve the overall stability of the film (discussed at length below), increasing the thickness and Ge fraction of the layer to levels higher than might otherwise be expected.

Proper surface cleaning and careful growth of such a three-layer SiGe film can result in beautiful device-quality films, as can be seen in the cross-sectional TEM of a commercial SiGe HBT technology shown in Figure 2.7. In this case, perfect crystalline structure is obtained and the original growth interface cannot even be determined by casual viewing. For all intents and purposes the SiGe film has become part of the Si host crystal, resulting in a perfect crystal.

2.1.3 The Challenge of SiGe Epitaxy

In the earliest use of Si epitaxy, films were grown to provide well-controlled regions of uniformly doped material upon the less-uniform bulk-grown silicon wafer, which served only as a growth template. All subsequent definition and fabrication of active device regions was based upon lithographic patterning and ion implantation. The use of epitaxy to form *active* layers in silicon-based devices (e.g., the base region of a SiGe HBT) is a relatively recent development.

Employing Si epitaxy in device fabrication enables one to overcome the fundamental limitations faced by ion implantation, which include: the implantation energy-dependent Gaussian distribution of dopants as a function of depth, ion channeling of the implanted dopant species, and the need for high temperature annealing to remove implant damage and activate the dopants. Although progress has been made in reducing ion-implanted base-widths in Si BJTs, base widths in the sub-100-nm regime are difficult to control in practice, limiting Si BJT performance to below about 50-GHz peak f_T. In principle, however, using Si epitaxy, the dopant profiles can be grown into the epitaxial layer at a precise location, tailored in shape within the vertical section of the device, be made atomically abrupt, be electrically active as grown, and may be combined within an alloy such as SiGe. In theory. Reality dictates that the thermal budget of conventional Si epitaxy, combined with that of subsequent routine device processing (e.g., additional implant activation steps or oxidation), makes attaining these theoretical ideals difficult.

The high thermal budget associated with conventional Si epitaxy may be understood as follows. The successful realization of device-quality Si epitaxy requires that we first prepare an atomically "clean" Si surface that will serve as the epitaxial growth template. The historical approach taken in growing Si epitaxy has been to bake the Si wafer in a hydrogen atmosphere at temperatures high enough (generally in excess of 1,000°C) to evaporate surface oxides, as well as remove (or dissolve) surface carbon and dopant contamination. Film growth is then commenced at high temperatures, assuring that residual growth system contaminants such as oxygen or carbon do not incorporate into the growing epitaxial layers. Because both the cleaning and growth temperatures for conventional Si epitaxy are in the range of 1,000°C, they are incompatible with the requirements of epitaxy for the purposes of advanced device applications. At such temperatures, any advantage obtained from precise device layer formation by epitaxy is lost in the subsequent diffusion of dopants away from their intended locations. In the more extreme case of working with a strained material set such as SiGe alloys, a further risk is the potential for high-temperature relaxation of such layers by defect formation. The key to the successful use of Si (or SiGe) epitaxy to make advanced devices is thus to affect high-quality film growth at very low temperatures (< 600°C). It is the inherent difficulty of this feat that delayed progress in SiGe HBT development until the mid-1980s.

2.2 SiGe Growth

A number of growth techniques have been developed over the past 25 years with a demonstrated capability to produce device-quality SiGe films. In this context, by "device-quality" we mean visibly defect-free SiGe films that have been used to produce ideal or near-ideal individual SiGe HBTs. We do not imply that all such tools are commercially viable when considered for entering high-volume, large-scale production of a SiGe HBT technology.

While we will intentionally avoid an exhaustive discussion of the contentious issues surrounding which SiGe growth technique is best for producing SiGe HBTs, it is nonetheless instructive to consider the various metrics that delineate the various growth techniques, and which should be considered when selecting a growth tool for commercial production of SiGe HBTs. These factors include:

- Cross-wafer profile and doping control;
- Wafer-to-wafer, and run-to-run profile and doping control;
- Wafer throughput;
- Background contaminant levels;
- Ease of changing Ge profiles;
- Ease of growth surface preparation and cleaning;
- Ease of wafer size scale-up;
- Required vacuum level;
- Film growth temperature and time;
- Patterned wafer versus blanket wafer deposition;
- Batch versus single wafer processing;
- Tool operating and maintenance costs;
- Selective versus nonselective epi deposition;
- Loading effects associated with isolation pattern fill;
- Compatibility with CMOS toolsets and fabrication facilities;
- Ability to incorporate C doping;

- Capability of both n-type and p-type doping polarities.

Each available SiGe growth tool has pluses and minuses when measured against such a list. Perhaps the ultimate measure, however, of which tools are best suited for building SiGe HBTs is to see which tools are actually being used in the field to produce commercial SiGe HBT products, and have thus stood the test of time. Of these, the ultra-high vacuum/chemical vapor deposition (UHV/CVD) technique [4] and the atmospheric pressure chemical vapor deposition (APCVD) technique [5]–[7] are clearly at the forefront today. [4] Both tools are commercially available as "turn-key" SiGe growth systems, and are in use around the world. Due to space constraints, only UHV/CVD and APCVD will be described in detail. The interested reader is referred to [10] for more detail on other growth techniques.

2.2.1 Surface Preparation

In any form of Si-based epitaxy, we can consider the film growth process as two distinct phases, the first being preparation of the initial growth interface, the second being film growth itself. Both are equally important in obtaining device-quality SiGe films. Successful growth of SiGe films clearly begins with a obtaining a pristine growth interface [11]. When considering the means of growth surface preparation, one must first identify the nature of the surface to be prepared. In classical high temperature Si epitaxy, the surface being prepared was that of an unpatterned, bulk-grown Si wafer. This is the most straightforward growth interface to prepare, since the absence of any subsurface patterning allows exposure to a high thermal budget without detrimental effects. If patterned and implanted regions were present during the thermal cycles employed in classical Si epitaxy, where temperatures in excess of 1,000°C for 10 minutes are typical, dopant redistribution and reincorporation during subsequent film growth would both occur at high levels. Limiting oneself to nonpatterned substrate materials forces the preparation of a class of devices known as "mesa" transistors, where blanket Si layers are deposited, and subsequently etched back to isolate the varied regions required for device fabrication. Such mesa-isolated device structures are generally inconsistent with highly integrated, CMOS-compatible IC fabrication. This fact drove the development of growth techniques and cleaning procedures that are compatible with the use of prepatterned substrates, as dictated by the needs of subsequent device integration.

---

[4] We note, however, that the earliest device-quality SiGe films [8], and in fact the first SiGe HBT demonstration [9], were actually grown by MBE, a technique pioneered (and still in heavy use today) for producing bandgap engineered III-V materials of various types.

2.2.2 Growth Techniques

The UHV/CVD technique eliminates the high thermal budget of conventional epitaxy by chemical means. It was discovered [4] that one could passivate a Si surface with hydrogen by employing a simple wet chemical procedure: a 10–15 second etch in a dilute 10:1 H_2O/HF solution. The hydrogen adlayer created during this wet etch reduces the reactivity of the growth interface approximately 13 orders of magnitude from that of a bare Si surface with respect to its oxidation rate in ambient air. This passive behavior extends to the adsorption of dopant species as well. It was further found that such hydrogen passivated Si surfaces de-wet completely when extracted from the *HF* bath, so that the handling of blanket film growth wafer preparation is particularly straightforward. Wafers are dry as pulled from the etch, and may thus be loaded directly into the growth chamber. Patterned wafers having a variety of exposed surface materials, both semiconductor and dielectric, may also be prepared in this manner. A liability of working with patterned wafers is that they may not de-wet completely when materials other than Si are present, requiring a blow-dry of these surfaces prior to their insertion into the film growth chamber. However, in most SiGe HBT processes, simple structures have been employed in which the top-most exposed surface is everywhere silicon, rendering the wafer fully hydrophobic, and thus allowing patterned wafers to de-wet completely. The resultant hydrogen terminated Si surface, as employed in UHV/CVD, is robust from a number of viewpoints.

A long-standing difficulty in the deposition of active device regions, such as in epitaxial base technology, stems from the presence of electrically active impurities at the initial growth interface. It is common to find a boron dose in excess of 10^{12} cm^{-2} at the initial growth interface, even in the UHV conditions employed in MBE. A variety of methods are employed to reduce the magnitude and impact of this contamination, a common method being the deposition of a buffer layer of material to bury the contamination well below the active device region. However, if one is depositing layers on patterned substrates, this is not a viable approach. In particular, when the epitaxial base is being deposited upon a wafer containing patterned collector regions, one is in effect growing the base-collector junction, and little unintended dopant is tolerable. In the instance of UHV/CVD, the residual boron dose at the growth interface is in the range of 10^9–10^{10} cm^{-2}, and is thus of no consequence in the resultant device. It is important to note that the boron contamination under consideration is sourced from the ambient. In the instance where boron doped regions are existent upon the wafer itself prior to film growth, care must be taken in preceding processing to avoid the accidental cross contamination of wafer regions. Having addressed the means employed to prepare the initial Si surface for device-quality epitaxy, one must consider the epitaxial growth itself.

The utilization of high temperatures for conventional Si epitaxy has frequently been ascribed to the need to provide for adatom mobility, such that a high-quality epitaxial layer would result. Furthermore, high temperature growth was known to suppress the inclusion of undesirable dopant species in the films being deposited. A subtlety of chemical vapor deposition is that one in fact may grow epitaxial Si at room temperature if one considers the trivial case, where a monolayer of the Si source gas adsorbs on a Si lattice site. This is in fact the case for growth by UHV/CVD, where the gaseous species silane (SiH_4) is employed as the Si source. For an additional layer to deposit, thermal energy is required to drive off residual surface hydrogen, enabling the adsorption of an additional monolayer of film. As such, adatom mobility is not a key issue in setting the lower bound on the temperature at which Si may be deposited epitaxially by UHV/CVD. In the limit of CVD techniques employing film growth at higher temperatures and pressures, it is possible to grow films at rates exceeding the rate of adatom ordering, such that a transition from epitaxial to either polycrystalline or amorphous film growth takes place. Regardless, for films ranging from pure Ge to pure Si, temperatures in the range 400–500°C, respectively, have been shown adequate for the deposition of high-quality epitaxial layers by UHV/CVD. Therefore, the thermal budget of the UHV/CVD process does not contribute measurably to that of the overall transistor fabrication process, a key advantage.

To achieve adequate film purity during low temperature epitaxy, several divergent approaches have been employed. Best known are the UHV techniques associated with MBE, where excellent ultimate vacuum in the range 10^{-11} torr is commonly achieved. This vacuum level degrades significantly during film growth, but films of high purity and perfection are achieved after many hours of film growth. To reduce the complexity and expense of such an apparatus, UHV/CVD utilizes a chemically selective form of the UHV technique. Recognizing the need to eliminate only those species that are chemically active with Si, a simplified UHV methodology employing O-ring seals and quartz reaction tubes is employed. Although relatively "soft" levels of UHV are achieved, typically in the range of 10^{-9} torr, the preponderance of the residual gas is hydrogen, which is unimportant in such trace quantities. Oxygen and water levels are reduced to the range of 10^{-11} torr partial pressure. Carbon-bearing species are not detectable owing to the use of turbo-molecular pumping. This selective chemical approach to purity has subsequently become the rule, where systems with base pressures from that found in UHV/CVD to those operating in the absence of vacuum pumps (e.g., APCVD) commonly utilize load-locks, gas scrubbers, and other methods to selectively eliminate potentially problematic impurities from the growth environment.

Films are deposited by UHV/CVD at temperatures in the range of 400–500°C, those temperatures corresponding to the growth of pure Ge, and pure Si, respectively. Wafers are HF-passivated as described above, and then loaded into the load-lock of the UHV/CVD apparatus. Note that the UHV/CVD system is a batch tool, and SiGe films can be deposited on multiple device wafers at the same time, greatly enhancing throughput. After pump-down below 10^{-6} torr, wafers are transferred under flowing hydrogen into the UHV section of the apparatus, and growth is commenced immediately. The gaseous sources employed are silane (SiH_4), germane (GeH_4), diborane (B_2H_6), and phosphine (PH_3). Film growth rates may be varied from 0.1–100 Å/minute as a function of temperature and film Ge content, with typical rates of 4–40 Å/minute. These growth rate limits are used to ensure a precise a cross-wafer dimensional control on the order of 1–2 atomic layers in this instance. This level of precision is required if one is to compete effectively with the control of ion implantation, the benchmark for doping control in Si processing. The deposition of compositionally graded Ge profiles with peak Ge content of 10–30% over dimensions of 50–150 nm can be routinely practiced with UHV/CVD with excellent cross-wafer, wafer-to-wafer, and run-to-run control.

The APCVD SiGe deposition tools [5]–[7] have recently emerged as an important commercially viable technique for the growth of SiGe films. The APCVD technique deposits Si and SiGe at atmospheric pressure using SiH_2Cl_2 and GeH_4 gas sources. The SiGe film deposition is typically carried out in a conventional induction-heated, air-cooled Si epi reactor. Unlike for UHV/CVD, an *in-situ* RCA preclean is used on the starting wafers, followed by a short prebake (e.g., 1,070°C for 10 minutes), followed by a gaseous HCl etch for a short additional time. Gas purifiers and a loadlock system are used in the place of UHV to control oxygen and carbon contamination. APCVD systems are single wafer tools, with the wafers placed horizontally on a quartz holder. While little wafer-to-wafer and run-to-run tracking data are available in the literature, the fact that such APCVD reactors are being used to produce commercial SiGe HBT technologies (e.g., [12]) bodes well for its capabilities.

2.3 Stability Constraints

The thickness of the SiGe-bearing layer is clearly a key variable in SiGe HBT device design. The maximum thickness for obtaining pseudomorphic growth post-fabrication (i.e., after any thermal anneals or ion-implantation steps which might relax an overstable film) is known as the "critical thickness" (h_{crit}). The original concept of critical thickness in strained-layer epitaxy was introduced in [13] based on equilibrium theory, and defined to be the film thickness below which it is energetically favorable to contain ("freeze in") a given misfit by storing the elastic energy in the distorted (strained) crystal, and above which it is favorable to release (part of) that elastic energy by generating misfit dislocations at the heteroepitaxial growth interface. This theoretical approach to stability calculations in SiGe strained layers can be termed "energy minimization" [13, 14]. Alternatively, the much-cited work of Matthews and Blakeslee [15, 16] defined critical thickness in terms of the mechanical equilibrium of a preexisting threading dislocation. In this case, the force of the dislocation segment residing at the hetero-interface is balanced with the component of the force per unit length acting on the threading component of the dislocation in growth plane. The thickness at which these two forces are equal is defined to be the critical thickness. This theoretical approach to stability calculations in SiGe strained layers can be termed "force balance" [15]–[18]. The interested reader is referred to a review article on stability calculations in SiGe strained layers [19].

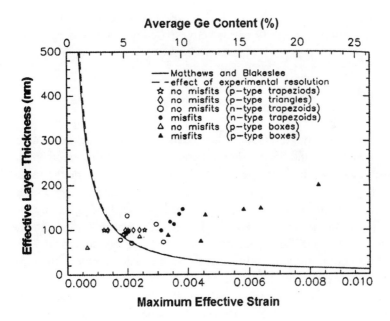

Figure 2.8 SiGe strained-layer thermodynamic stability diagram comparing UHV/CVD experimental data to Matthews and Blakeslee's theoretical result (after [21]).

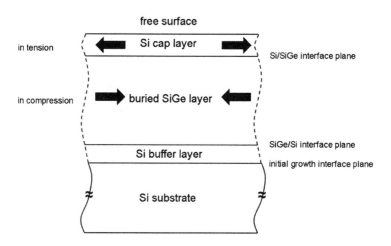

Figure 2.9 Schematic epitaxial SiGe film for use in a SiGe HBT. The film consists of a thin Si buffer layer, the active SiGe layer, and a Si cap layer.

2.3.1 Theory

Despite the large body of literature that exists on critical thickness calculations and measurements in the SiGe system, there is significant disagreement between those basic theories and experimental results on actual device-quality SiGe films (Figure 2.8 [20, 21]). The origin of this discrepancy apparently lies in the use of a Si cap layer on top of the SiGe strained layer in practical SiGe films used in SiGe HBTs, as shown schematically in Figure 2.9. The difficulties encountered in modeling such a Si/SiGe/Si multilayer arise because the theoretical approaches [13]–[18] do not properly account for the stored elastic energy in the top Si cap layer (which helps improve overall film stability). Recent work [22], however, has recently expanded existing force-balance theory to properly account for the effects of the Si cap layer, and shows excellent agreement between theory and experiment, for both CVD and MBE grown films. That theoretical approach to stability calculations in device-compatible SiGe films will be outlined below, following [22]. For simplicity, only the calculation of the equilibrium critical thickness is attempted, since partially relaxed films (i.e., films with residual strain) are significantly more complex to model. The primary physical assumptions include: a force balance solution for the in-plane stress, isotropic behavior, and equal elastic moduli. Under these conditions, far from any lateral free surface, the state of stress is uniformly biaxial of magnitude σ, and is related to the film strain (ϵ) by

$$\sigma = \frac{2G(1+v)\epsilon}{(1-v)} \tag{2.3}$$

along the in-plane axes, with G the shear modulus of the SiGe layer, and v its Poisson ratio. Referenced to the $\{111\}$ slip plane surfaces and the $<110>$ slip directions, the shear stress component (τ) of σ is given by

$$\tau = \cos\lambda \, \cos\phi \, \sigma, \tag{2.4}$$

where λ is the angle between the Burger's vector (the magnitude and direction of the slip from the motion of a single dislocation) and the direction of the interface, normal to the dislocation line, and ϕ is the angle between the slip plane and the normal vectors to the strained interface.

The in-plane strain ϵ arising from the lattice mismatch between Si and Ge is $0.042x$, where x is the Ge fraction. In a Si/SiGe/Si heterostructure, the strain that acts to generate a misfit dislocation is

$$\epsilon = 0.042\,x - \frac{b\cos\lambda}{2(h+H)}, \tag{2.5}$$

where b is the magnitude of the Burger's vector, and h and H are the thicknesses of the buried SiGe layer and the Si cap layer, respectively (refer to Figure 2.6).

Dislocation self-stress terms associated with the strain energy required to create straight misfit dislocation segments in both the interface planes, and with the dislocation-dislocation interaction energy between the two parallel segments, act to counteract (balance) this shear stress component. The self-stress of a strain-relaxing 60° misfit dislocation on the slip plane of the lower Si buffer/SiGe interface is given by

$$\tau_{SiGe} = \frac{Gb\left(1-\frac{v}{4}\right)}{4\pi(1-v)h} \ln\frac{h+H}{b}. \tag{2.6}$$

Similarly, the self-stress associated with the upper SiGe/Si cap interface can be written as

$$\tau_{Si} = \frac{Gb\left(1-\frac{v}{4}\right)}{4\pi(1-v)h} \ln\frac{H}{b}, \tag{2.7}$$

and the self-stress term connecting the bottom and top interfaces (i.e., Si buffer/SiGe and SiGe/Si cap) is

$$\tau_{Si-SiGe} = -\frac{Gb\left(1-\frac{v}{4}\right)}{4\pi(1-v)h} \ln\frac{h+H}{h}. \tag{2.8}$$

Note that in the latter term, the negative interaction energy represents a decrease in free-energy of the film.

The ratio of the stored misfit strain in the Si cap layer to that in the buried SiGe layer of the Si/SiGe/Si multilayer enables a determination of the respective amount of stored elastic strain energy in both the SiGe and Si cap layers. The total misfit strain (ϵ_t) in the SiGe/Si cap composite is

$$\epsilon_t = 0.042\, x \left\{ \frac{h}{h+H} \right\}, \qquad (2.9)$$

where $x\, h/(h+H)$ is the average fractional Ge content. Note that by letting $H = 0$, we are considering only the misfit strain of the SiGe layer (ϵ_b), and obtain the well-known expression

$$\epsilon_b = 0.042\, x. \qquad (2.10)$$

The misfit strain in the Si cap layer (ϵ_c) is then given by the difference between the ϵ_t and ϵ_b, as

$$\epsilon_c = -0.042\, x \left\{ \frac{H}{h+H} \right\}. \qquad (2.11)$$

Intuitively, (2.11) indicates that if the SiGe buried layer incorporates the nucleation energy of a single misfit dislocation, the Si cap layer holds energy only for $H/(h+H)$ misfit dislocations. This energy density factor (δ) can vary from zero for a SiGe layer with no Si cap, to unity, for an infinitely thick Si cap. Hence, the presence of a Si cap layer, and its associated stored energy, is expected to improve film stability.

Finally, the excess resolved shear stress (τ_{exc}) driving the bending of threading dislocations to form single and/or double misfit dislocation segments will then be given by the difference between the external stress (τ) and the internal stress components (τ_{SiGe}, τ_{Si}, and τ_{Si-Ge}) according to

$$\tau_{exc} = \tau - \tau_{SiGe} - \delta\tau_{Si} - \delta\tau_{Si-SiGe}, \qquad (2.12)$$

where δ is the energy density factor. Substituting (2.3)–(2.11) into (2.12) under equilibrium conditions (i.e., $\tau_{exc} = 0$), this force balance solution yields an implicit expression of the critical strained layer thickness (h_{crit}) as a function of Ge fraction (x) of

$$x = \frac{b \cos\lambda}{0.084 h_{crit}} \left[\gamma + \frac{1 - \frac{\nu}{4}}{4\pi \cos^2\lambda \cos\phi (1+\nu)} \left[\ln\frac{h_{crit}+H}{b} + \delta\ln\frac{H}{b} - \delta\ln\frac{1}{\gamma} \right] \right] \qquad (2.13)$$

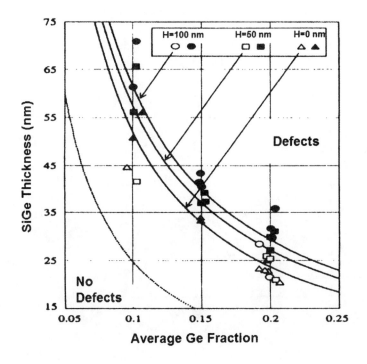

Figure 2.10 SiGe strained-layer thermodynamic stability diagram comparing Fischer's theory and experimental data (after [22]). The dashed line represents the Matthews and Blakeslee theoretical result. Open symbols represent stable (defect-free) films, and closed symbols represent relaxed films.

where

$$\delta = \frac{H}{h_{crit} + H} \qquad (2.14)$$

and

$$\gamma = \frac{h_{crit}}{h_{crit} + H}. \qquad (2.15)$$

This theoretical result gives a generalized statement of the balance between external and internal forces acting in a realistic Si/SiGe/Si multilayer of arbitrary geometry, and is a measure of the transition between a thermodynamically stable state and a metastable state that will relax upon heat treatment or any other common fabrication step that adds energy to the system.

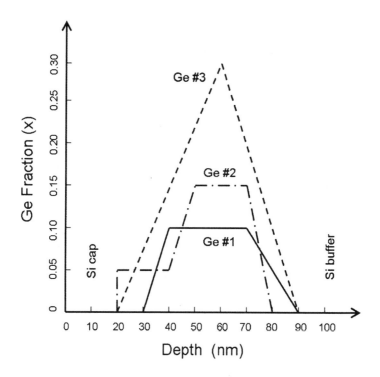

Figure 2.11 Hypothetical SiGe profiles for stability analysis.

2.3.2 Experimental Results

Inserting appropriate values for the various material parameters into (2.13), one can now calculate stability curves for various Si cap layer thicknesses. Generally accepted values for growth on the (001) Si surface are: $cos\,\lambda = 0.50$, $cos\,\phi = 0.82$, $b = 3.84$ Å, and $v = 0.36$. Figure 2.10 shows a comparison between (2.13) and a variety of CVD and MBE grown SiGe samples. The results are also qualitatively consistent with the earlier UHV/CVD results shown in Figure 2.8, which include a wide variety of Ge profile shapes and doping profiles. Agreement between theory and experiment is excellent and serves to confirm the approach. These results are particularly nice in that they confirm what has been long suspected by practitioners of SiGe technology: namely, that the original Matthews and Blakeslee stability result is quite conservative for practical (Si-capped) SiGe films. That is, thicker SiGe layers with more Ge content than might otherwise be expected can be used without violating film stability constraints.

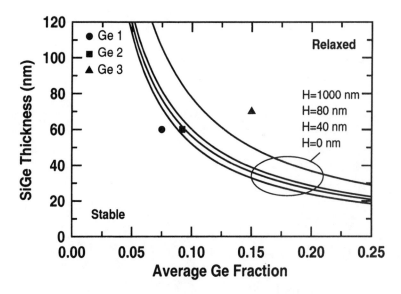

Figure 2.12 SiGe strained-layer thermodynamic stability diagram comparing theoretical stability calculations with example SiGe profiles.

2.3.3 Stability Calculations

As a practical issue, one often needs to determine if a given SiGe profile design is expected to be thermodynamically stable or not. That is, one needs to place a given SiGe profile data point on a stability diagram, and compare it to the calculated stability curve. This experimental stability point can be placed on the stability diagram according to the following prescription. Let h be the thickness of the Ge layer (not including the Si buffer or Si cap layer), as measured by secondary ion mass spectroscopy (SIMS) or other techniques. For a position-dependent Ge profile, the stability point must be calculated as an average value of Ge fraction across h according to [5]

$$\bar{x} = \frac{1}{h} \int_0^h x(z)\, dz. \qquad (2.16)$$

The ordered-pair $(\bar{x}, h)$ can then be placed on the stability diagram and com-

---

[5]There are clearly untested assumptions associated with taking an average value of the Ge fraction, since the theoretical analysis above only explicitly holds for constant Ge content over h (i.e., a Ge box profile).

pared to theoretical calculations of h_{crit} versus $\bar{x}$, for variable values of Si cap layer thicknesses H.

As an example, consider the three hypothetical Ge profiles shown in Figure 2.11. Following the determination of $(\bar{x}, h)$ for each Ge profile, each point can be placed on the stability diagram (Figure 2.12). For each profile, one then compares the calculated stability point with the critical thickness curve calculated using (2.13) with the appropriate Si cap thickness (H) for the profile in question. In this case, we see that profile 1 is clearly thermodynamically stable, profile 2 is close to the stability boundary, and profile 3 is metastable, and can thus be expected to relax during fabrication.

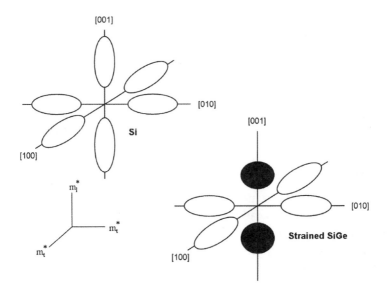

Figure 2.13 Schematic deformation of the SiGe conduction band constant energy ellipsoids with compressive strain.

2.4 Band Structure

The resultant energy band structure obtained in a strained SiGe alloy with respect to its original Si constituent is clearly key to its usefulness in transistor engineering. For the purposes of designing a SiGe HBT, we desire a SiGe alloy which:

- Has a smaller bandgap than that of Si;

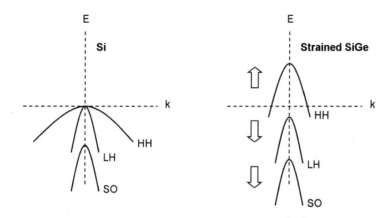

Figure 2.14 Schematic deformation of the SiGe valence bands with compressive strain.

- Has a band offset that is predominantly in the valence band;
- Either improves or at least does not substantially degrade the carrier transport parameters (mobilities or lifetime) with respect to Si.

As will be seen below, strained SiGe fulfills all of these requisite conditions.

Both Si and Ge are indirect energy gap semiconductors (see the Appendix for a list of material parameters, energy band structure, and effective mass parameters for bulk Si and Ge). Si has a principal bandgap of 1.12 eV at 300 K, located in the Γ–X (<100>) equivalent k-space directions, and thus there are six equivalent principal conduction bands in Si (one each for [100], [010], [001], [$\bar{1}$00], [0$\bar{1}$0], and [00$\bar{1}$]). Ge, on the other hand, has a principal bandgap of 0.66 eV at 300 K, located in the Γ – L (<111>) equivalent k-space directions, and thus there are eight equivalent principal conduction bands in Ge (one each for [111], [$\bar{1}$11], [1$\bar{1}$1], [11$\bar{1}$], [1$\bar{1}\bar{1}$], [$\bar{1}$1$\bar{1}$], [$\bar{1}\bar{1}$1], and $\bar{1}\bar{1}\bar{1}$]). There are three principal valence bands in Si and Ge: the heavy-hole (*hh*), light-hole (*lh*), and split-off (*so*) bands. The *hh* and *lh* bands are degenerate at the Γ point ($\vec{k} = 0$) in both Si and Ge.

Since Ge has a significantly smaller bandgap than Si (primarily due to its larger lattice constant), it is not surprising that the bandgap of SiGe will be smaller than that of Si. The strain in a pseudomorphic SiGe alloy, however, also plays an important role in shaping the final band structure and carrier mobilities.

As predicted by early theoretical calculations [23]–[25], in a Si-rich (low Ge fraction) pseudomorphic (strained) SiGe alloy, there are a number of consequences to the band structure of Si that result from the addition of a small amount of Ge, including:

- The 6-fold conduction band degeneracy of Si is lifted, resulting in a 2-fold (out-of-plane) and 4-fold (in-plane) band splitting (Figure 2.13).
- The 4-fold degenerate conduction bands move downward in energy, resulting in a small net conduction band offset ($\Delta E_C > 0$).
- The 2-fold degenerate conduction bands move upward in energy.
- The heavy-hole and light-hole valence band degeneracy of Si is lifted (Figure 2.14).
- The heavy-hole valence band moves upward in energy resulting in a net valence band offset ($\Delta E_V > 0$). This heavy-hole band movement is the most substantial in the strained SiGe system, resulting in a valence band offset large enough to be useful in transistor engineering.
- The light-hole and split-off valence bands move downward in energy.
- The film strain produces band-edge curvature distortion that perturbs the carrier effective masses and hence the conduction and valence band density-of-states, as well as the carrier mobilities and lifetime parameters.

2.4.1 Density-of-States

It is generally agreed upon that the effective conduction and valence band density-of-states product ($N_C N_V$) is reduced strongly due to strain-induced distortion of both the valence and conduction band extrema, a consequence of which is the reduction in the electron and hole effective masses [26]. A comprehensive set of effective mass parameters as a function of Ge content and doping is very difficult to experimentally determine since it involves tedious cyclotron resonance studies, and is not available in the literature. It is far easier to infer the change in the density-of-states product (which is proportional to the effective masses) with Ge content by using transistor collector current measurements [27], although this technique cannot discriminate between the strain-induced changes to the hole and electron masses. Even here, the uncertainty in the carrier mobilities and the inherent position-dependence of the doping profiles makes it difficult to minimize the experimental error bars involved. Figure 2.15 shows representative results for $N_C N_V$ as a function of Ge content [28]. The substantial reduction in $N_C N_V$ with increasing Ge content can be considered undesirable since it translates directly to

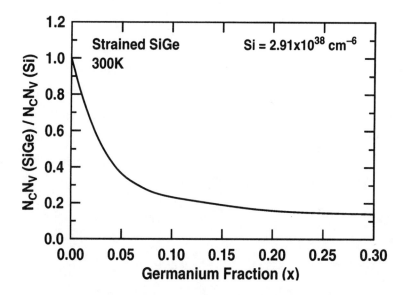

Figure 2.15 Reduction in the valence band and conduction band effective density-of-states product due to SiGe.

a reduction in collector current in the SiGe HBT (refer to Chapter 4), and hence reduces the available current gain. Fortunately, however, the same reduction in effective masses that produces the decrease in $N_C N_V$ also increases the carrier mobilities, which partially offset the impact on the collector current.

2.4.2 Band Offsets

The final band-edge alignments in strained SiGe are shown in Figure 2.16, yielding a Type-I band alignment scheme: both ΔE_C and ΔE_V are positive, and thus the SiGe conduction and valence band edges are contained within the original Si band edges. In the case of practical SiGe films, the valence band offset is by far the largest and most significant, as desired.

From a device design perspective, accurate experimental knowledge of the band offsets in SiGe is crucial. While the body of literature is reasonably large in this context (see, for instance, [29]–[34]), the precise determination of the band offsets in SiGe are in general very difficult to measure due to several experimental complexities. For instance, ΔE_g determined from transistor measurements [33] (typically from temperature measurements of I_C) require assumptions on the magnitude and temperature dependence of various parameters compared to that in Si (e.g., mobility). In addition, the base region of SiGe HBTs is doped heavily

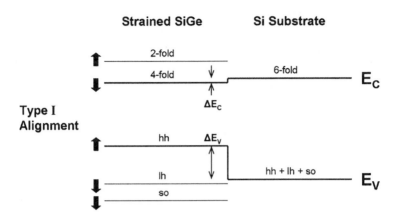

Figure 2.16 Schematic band alignments of strained SiGe grown on a Si substrate.

(typically above 10^{18} cm^{-3}), and has position-dependence near the EB and CB junctions. This practical complexity adds a (position-dependent) contribution of doping-induced bandgap narrowing to the problem, which unfortunately is indistinguishable electrically from Ge-induced bandgap narrowing, thereby forcing an unjustified assumption to be made (usually that the two effects are additive [34]).

Capacitance-voltage measurements on lightly-doped p-type SiGe MOS capacitors, though still nontrivial, and requiring a theoretical data fit, are generally considered more straightforward to interpret than those made on HBTs [29]–[32]. Figure 2.17 shows one of the most current examples of C-V determined valence band offset data in strained SiGe on Si [29]. Even here, the sample scatter is nonnegligible, but nevertheless clearly demonstrates the widely cited trend of a near-linear valence band offset as a function of Ge fraction for low-Ge content SiGe alloys. The standard rule of thumb for SiGe device design of 74 meV / 10% Ge content appears to represent an acceptable approximation across the practical range of 0–30% Ge content used in transistor design. Even so, it is worth noting that such band offset measurements are performed on constant Ge content films (or superlattices of such films), and thus their extrapolation to the position-dependent Ge profiles used in actual SiGe HBTs requires that additional (often ignored) assumptions be made.

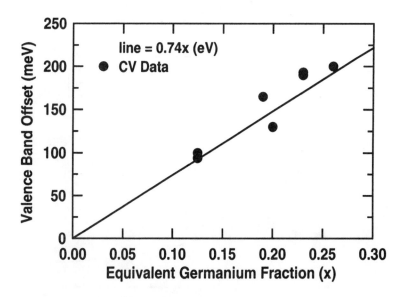

Figure 2.17 Measured valence band offset as a function of Ge fraction (data after [29]).

2.5 Transport Parameters

Given the known Ge- and strain-induced changes to the energy, degeneracy, and local curvature of both the conduction and valence bands in Si, it is to be expected that both the carrier effective masses will be significantly altered in strained SiGe compared to their original Si values. Because carrier transport parameters (the carrier mobilities μ_n and μ_p, and the carrier lifetimes τ_n and τ_p) depend intimately on the band structure and the resultant carrier effective masses (m_n^* and m_p^*), all of the carrier transport parameters can be expected to change with the addition of Ge to Si.

These changes to the device transport parameters in SiGe are important because: 1) the collector current in a SiGe HBT is proportional to the minority electron mobility in the base (μ_{nb}); 2) the base current is proportional to the minority hole mobility in the emitter (μ_{pe}); 3) the base transit time is reciprocally proportional to μ_{nb}; 4) the base resistance, which is important in both dynamic switching and noise performance, is proportional to the hole mobility in the base (μ_{pb}); and 5) the recombination statistics and parasitic leakage currents that influence both current gain and output conductance depend reciprocally on τ_n and τ_p. Accurate knowledge of how these transport parameters depend on doping, temperature, and Ge fraction is particularly important in achieving predictive 2-D simulation of SiGe

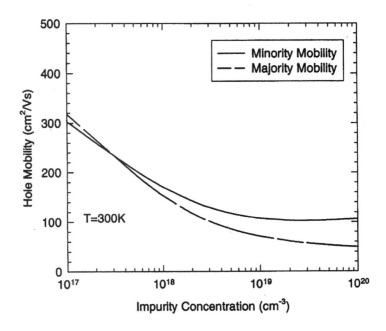

Figure 2.18 Majority and minority hole mobility in Si using Klaassen's mobility model [35, 36].

HBTs, as will be discussed at length in Chapter 12.

Given the obvious importance of determining the precise influence of Ge on the various transport parameters, arriving at quantitative values for them is a notoriously difficult problem. For example, SiGe films suitable for transport parameter measurements are generally incompatible with those found in practical SiGe HBTs, which require very thin and heavily doped regions, often with compositionally graded Ge content. In addition, accurate measurements of *minority* carrier transport parameters in heavily doped SiGe is a particularly challenging problem even in Si, requiring accurate knowledge of bandgap narrowing, as well as independent lifetime data. Not surprisingly, available transport parameter data in the literature is sparse and often inconsistent. We offer here some guidelines regarding generally agreed-upon information.

It is reasonably well-established via calibration of data with 2-D simulations using various interchangeable mobility models that the so-called "Philips unified mobility model" (the "Klaassen model" [35, 36]) for both electrons and holes works reasonably well in modeling Si BJTs. This temperature-dependent model

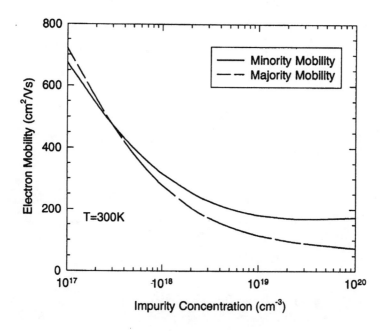

Figure 2.19 Majority and minority electron mobility in Si using Klaassen's mobility model [35, 36].

has a companion set of consistent bandgap narrowing parameters (refer to the discussion in Chapter 12) and distinguishes between majority and minority carrier mobilities. Figure 2.18 and Figure 2.19 show the respective minority and majority hole and electron mobilities at 300 K over a doping range of practical interest in SiGe HBTs. Observe that in the range of base doping of interest to SiGe HBTs ($10^{18} - 10^{19}$ cm^{-3}), the minority carrier mobility is higher than the majority carrier mobility, a distinct advantage from a dynamic performance point of view.

In strained SiGe films, the carrier mobilities will be altered from their Si values due to local distortion to the band extrema due to strain effects, as well as the additional influence of alloy scattering on the carriers. Because of the nonisotropic nature of the strain in pseudomorphic SiGe on Si, it can also be expected that the changes to the mobilities will depend on the transport direction (either parallel to the original SiGe growth interface or orthogonal to the original SiGe growth interface).

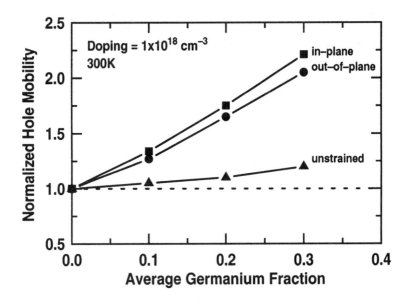

Figure 2.20 Normalized hole mobility as a function of doping and Ge content (after [38]). The squares represent the mobility component parallel to the growth interface (in-plane), the circles represent the component transverse to the growth interface (out-of-plane), and the triangles represent the unstrained mobility.

2.5.1 Hole Mobility

The largest body of literature exists for the majority hole mobility in p-type SiGe (see, for instance, [37]–[39]). There is general agreement that the addition of Ge to Si increases the majority hole mobility, although the exact extent of the mobility enhancement as a function of both doping and Ge content is still debated. Representative results for both in-plane (parallel to the growth interface) and out-of-plane (orthogonal to the growth interface) are shown in Figure 2.20 [38]. In the context of SiGe HBTs, the in-plane hole mobility is by far the most important since it largely determines the base resistance, and thereby directly impacts the frequency response (via f_{max}), the dynamic switching speed, and the broadband noise performance. In a typical SiGe HBT having a polysilicon emitter contact and a partially realigned poly-to-Si interface (high poly-to-Si surface recombination velocity S_{pe}), the emitter is very heavily doped ($> 10^{20}$ cm^{-3}) and the hole lifetime is thus far more important in determining the base current than the out-of-plane hole mobility. Observe that for the in-plane hole mobility there is about a 35% improvement in μ_p at 10% Ge content at 10^{18} cm^{-3}, and the dependence on

Ge fraction is near-linear. Ge-induced strain also clearly plays a large role in the observed hole mobility improvement.

2.5.2 Electron Mobility

There is less available experimental data and less agreement on the effects of strain on the electron mobility in SiGe compared to comparably-doped Si. Experimental results are more difficult to obtain in this context, since it is the out-of-plane *minority* electron mobility (in the p-type base) that is of the greatest importance. Figure 2.21 shows representative electron mobility results for both the in-plane (parallel to the growth interface) and the out-of-plane (orthogonal to the growth interface) transport directions [40]–[42]. Observe that for the out-of-plane electron mobility there is about a 15% degradation in μ_n at 10% Ge content at 10^{18} cm^{-3}, and the dependence on Ge fraction is reasonably linear at low Ge fraction.

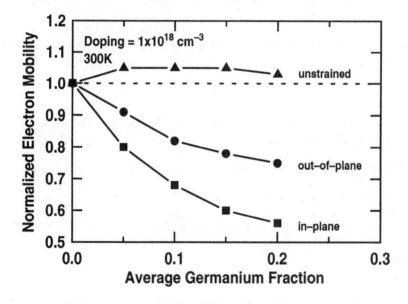

Figure 2.21 Normalized hole mobility as a function of doping and Ge content (after [41]). The squares represent the mobility component parallel to the growth interface (in-plane), the circles represent the component transverse to the growth interface (out-of-plane), and the triangles represent the unstrained mobility.

2.5.3 Choice of SiGe Parameter Models

The over-arching principle in SiGe parameter model selection is to bear in mind that practical SiGe HBTs employ low-Ge content films (typically less than 30% peak Ge and less than 15% average Ge), and thus are for the most part Si-like. Hence, proven Si parameter models should always be used as a starting point in SiGe HBT simulations. Our recommendations for parameter model selection for SiGe HBT simulation include the following:

- Use Klaassen's unified mobility model (and associated bandgap narrowing model) for Si as a starting point for SiGe HBT simulations.

- If needed, add a linear scale factor (K_p) to account for the enhancement of the hole mobility with increasing Ge fraction (x) according to [28]:

$$\mu_p(SiGe)(x) = (1 + K_p\, x)\, \mu_p(Si). \qquad (2.17)$$

For instance, if $K_p = 5$, then we have a 50% enhancement of $\mu_p(SiGe)$ over $\mu_p(Si)$ at 10% Ge content.

- If needed, add a linear scale factor (K_n) to account for the degradation of the electron mobility with increasing Ge fraction according to [28]:

$$\mu_n(SiGe)(x) = (1 + K_n\, x)\, \mu_n(Si). \qquad (2.18)$$

For instance, if $K_n = -5$, then we have a 50% degradation of $\mu_n(SiGe)$ over $\mu_n(Si)$ at 10% Ge content.

- Use the default Si lifetime models [43].

- Use the default Si velocity saturation model.

- Use the default Si bandgap model, and account for Ge-induced band offsets by using the simple expression:

$$\Delta E_g \cong \Delta E_V = 0.74\, x. \qquad (2.19)$$

Make sure the band offset is placed in the valence band (the default for most simulators).

- Assume that the doping-induced bandgap narrowing and the Ge-induced band offsets are additive. Experience shows that this assumption, while reasonable in the absence of data to the contrary, nevertheless is problematic from a simulation-to-model calibration standpoint, and should be revisited as needed.

2.6 Open Issues

In this chapter we have highlighted the practical aspects of introducing Ge into Si to produce device-quality SiGe films for use in bandgap engineering, the stability constraints one is forced to live under when working in a lattice-mismatched strained-layer system like SiGe, and the influence of Ge content and strain on the various parameters that influence SiGe HBT operation and optimization. One might naively conclude that this is the end of the SiGe epi story: that all is understood, and the remaining battleground lies only in producing commercially viable SiGe HBT technologies. Not so. Like many new technologies, SiGe has entered commercial production with what could be argued is a number of important unknowns that have been effectively swept under the rug. While clearly not showstoppers for present-generation SiGe HBTs, these issues could clearly resurface at a future date in some to-be-determined context, producing serious reliability (or other) problems, and thus should not be forgotten.

We thus conclude this chapter with a brief (and obviously personal) list of what can be considered open issues with respect to epitaxial SiGe strained layers. We would argue that these issues demand additional focused attention, careful analysis, and particularly extensive experimentation, in order to shed light on them for future generations of SiGe technologists. These open issues include (in no particular order):

- The interaction of doping-induced bandgap narrowing versus Ge-induced band offsets must be quantified. These effects are difficult to distinguish experimentally, but there is mounting evidence that the two distinct band-edge phenomena are not additive, as is often naively assumed.

- Both the majority and minority carrier electron and hole mobilities, as well as the carrier lifetimes, in both p-type and n-type SiGe, must be studied comprehensively as a function of both doping, Ge fraction, and temperature. Valid parameter models must be established.

- The impact of compositional Ge grading (with its induced quasi-electric fields) on mobility and lifetime must be quantified.

- The impact of compositional Ge grading on stability analysis, and the appropriateness of using averaged Ge content in stability calculations must be quantified.

- The impact of Si cap thickness on stability analysis must be quantified in SiGe profiles with compositional Ge grading.

- Comprehensive stability data for a variety of SiGe growth techniques and conditions (e.g., UHV/CVD versus APCVD), as well as for varying Ge profile shape, should be undertaken (and published).

- The interaction between pattern shape and pattern density on SiGe film stability must be quantified. This should include an analysis of the (potential) differences in stability results on blanket wafers versus patterned wafers.

- If defects generated from over-stable (relaxed) SiGe films can be safely confined and buried in the collector region, do they cause problems from a device reliability point of view?

- The interaction between isolation-induced stress (e.g., from the deep and/or shallow trench oxides) and Ge-induced strain must be quantified, and its impact on device performance established.

- What happens to the magnitude of the band offsets in SiGe profiles that are grown over-stable and only partially relax during later processing?

- The width of the stability curves must be quantified. That is, how close can one come to a given stability curve without generating a detrimental (yield-killing) density of defects? Is the slope around a given stability curve gentle, or does one fall "off a cliff?"

- Does high-field transport differ between Si and SiGe for very thin base, high Ge content devices (e.g., these new >200-GHz SiGe HBTs)? As a related question, when will drift-diffusion or hydrodynamic simulations of SiGe HBTs fail to capture the requisite physics needed for meaningful transistor design and optimization?

References

[1] W. Shockley, *Electrons and Holes in Semiconductors*, New York, D. van Nostrand Company, Inc., 1950.

[2] E. Kaspar, editor, *Properties of Strained and Relaxed Silicon Germanium*, EMIS Datareviews Series No. 12, INSPEC, London, 1995.

[3] D.D. Tang and P.-F. Lu, "A reduced-field design concept for high-performance bipolar transistors," *IEEE Elect. Dev. Lett.*, vol. 10, pp. 67-69, 1989.

[4] B.S. Meyerson, "Low-temperature silicon epitaxy by ultrahigh vacuum/chemical vapor deposition," *Appl. Phys. Lett.*, vol. 48, pp. 797-799, 1986.

[5] T.O. Sedgwick, M. Berkenbilt, and T.S. Kuan, "Low-temperature selective epitaxial growth of silicon at atmospheric pressure," *Appl. Phys. Lett.*, vol. 54, pp. 2689-2691, 1989.

[6] W. De Boer and D.J. Meyer, "Low-temperature chemical vapor deposition of epitaxial Si and SiGe at atmospheric pressure," *Appl. Phys. Lett.*, vol. 58, pp. 1286-1288, 1991.

[7] T.I. Kamins and D.J. Meyer, "Kinetics of silicon-germanium deposition at atmospheric pressure chemical vapor deposition," *Appl. Phys. Lett.*, vol. 59, pp. 178-180, 1991.

[8] J.C. Bean et al., "Pseudomorphic growth of Ge_xSi_{1-x} on silicon by molecular bean epitaxy," *Appl. Phys. Lett.*, vol. 44, pp. 102-104, 1984.

[9] S.S. Iyer et al., "Silicon-germanium base heterojunction bipolar transistors by molecular beam epitaxy," *Tech. Dig. IEEE Int. Elect. Dev. Meeting*, pp. 874-876, 1987.

[10] C.K. Maiti et al., "Growth and characterization of group IV binary alloy films," *Def. Sci. J.*, vol. 50, pp. 299-315, 2000.

[11] D.L. Harame et al., "Si/SiGe epitaxial-base transistors: part I – materials, physics, and circuits," *IEEE Trans. Elect. Dev.*, vol. 40, pp. 455-468, 1995.

[12] F.S. Johnson et al., "A highly manufacturable 0.25 μm RF technology utilizing a unique SiGe integration," *Proc. IEEE Bipolar/BiCMOS Circ. Tech. Meeting*, pp. 56-59, 2001.

[13] J.H. van der Merwe, "Crystal interfaces. part I: semi-infinite crystals," *J. Appl. Phys.*, vol. 34, pp. 117-125, 1963.

[14] R. People and J.C. Bean, "Calculation of critical layer thickness versus lattice mismatch in Ge_xSi_{1-x}/Si strained layer heterostructures," *Appl. Phys. Lett.*, vol. 47, pp. 322-324, 1985.

[15] J.M. Mathews and A.E. Blakeslee, "Defects in epitaxial multilayers: I. misfit dislocations in layers," *J. Cryst. Growth*, vol. 27, pp. 118-125, 1974.

[16] J.M. Mathews and A.E. Blakeslee, "Defects in epitaxial multilayers: II. dislocation pileups, threading dislocations, slip lines, and cracks," *J. Cryst. Growth*, vol. 32, pp. 265-273, 1975.

[17] J.M. Mathews, "Defects associated with the accommodation of misfits between crystals," *J. Vacuum Sci. Tech.*, vol. 12, pp. 126-133, 1975.

[18] B.W. Dodson and J.Y. Tsao, "Relaxation of strained layer semiconductor structures via plastic flow," *Appl. Phys. Lett.*, vol. 51, pp. 1325-1327, 1987.

[19] S.C. Jain and W. Hayes, "Structure, properties, and applications of Ge_xSi_{1-x} strained layers and superlattices," *Semi. Sci. Tech.*, vol. 6, pp. 547-576, 1991.

[20] S.R. Stiffler et al., "The thermal stability of SiGe films deposited by ultrahigh-vacuum chemical vapor deposition," *J. Appl. Phys.*, vol. 70, pp. 1416-1420, 1991.

[21] S.R. Stiffler et al., "Erratum: The thermal stability of SiGe films deposited by ultrahigh-vacuum chemical vapor deposition," *J. Appl. Phys.*, vol. 70, p. 7194, 1991.

[22] A. Fischer, H.-J. Osten, and H. Richter, "An equilibrium model for buried SiGe strained layers," *Solid-State Elect.*, vol. 44, pp. 869-873, 2000.

[23] R. People, "Indirect bandgap of coherently strained Ge_xSi_{1-x}/Si alloys on <001> silicon substrates," *Phys. Rev. B*, vol. 32, pp. 1405-1408, 1985.

[24] C.G. Van de Walle and R.M. Martin, "Theoretical calculations of heterojunction discontinuities in the Si/Ge system," *Phys. Rev. B*, vol. 34, pp. 5621-5634, 1986.

[25] R. People and J.C. Bean, "Band alignments of coherently strained Ge_xSi_{1-x}/Si heterostructures on <001> Ge_ySi_{1-y} substrates," *Appl. Phys. Lett.*, vol. 48, pp. 538-540, 1986.

[26] T. Fromherz and G. Bauer, "Energy gaps and band structure of SiGe and their temperature dependence," pp. 87-93, *Properties of Strained and Relaxed Silicon Germanium*, E. Kaspar, editor, EMIS Datareviews Series No. 12, INSPEC, London, 1995.

[27] E.J. Prinz et al., "The effect of base-emitter spacers and strain-dependent densities of states in Si/SiGe/Si heterojunction bipolar transistors," *Tech. Dig. IEEE Int. Elect. Dev. Meeting*, pp. 639-642, 1989.

[28] D.J. Richey, J.D. Cressler, and A.J. Joseph, "Scaling issues and profile optimization in advanced UHV/CVD SiGe HBTs," *IEEE Trans. Elect. Dev.*, vol. 44, pp. 431-440, 1997.

[29] J.L. Hoyt et al., *Thin Solid Films*, vol. 321, pp. 41-46, 1998.

[30] C.H. Gan et al., "$Si/Si_{1-x}Ge_x$ valence band discontinuity measurements using a semiconductor-insulator-semiconductor (SIS) heterostructure," *IEEE Trans. Elect. Dev.*, vol. 41, p. 2430-2439, 1994.

[31] K. Nauka et al., "Admittance spectroscopy measurements of band offsets in $Si/Si_{1-x}Ge_x/Si$ heterostructures," *Appl. Phys. Lett.*, vol. 60, pp. 195-197, 1992.

[32] J.C. Brighten et al., "The determination of valence band discontinuities in $Si/Si_{1-x}Ge_x/Si$ heterojunctions by capacitance-voltage techniques," *J. Appl. Phys.*, vol. 74, pp. 1894-1899, 1993.

[33] Z. Matutinovic-Krstelj et al., "Base resistance and effective bandgap reduction in n-p-n $Si/Si_{1-x}Ge_x/Si$ HBTs with heavy base doping," *IEEE Trans. Elect. Dev.*, vol. 43, pp. 457-466, 1996.

[34] J.C. Sturm, "Si/SiGe/Si heterojunction bipolar transistors," pp. 193-204, *Properties of Strained and Relaxed Silicon Germanium*, E. Kaspar, editor, EMIS Datareviews Series No. 12, INSPEC, London, 1995.

[35] D.M.B. Klaassen, "A unified mobility model for device simulation – I. Model equations and concentration dependence," *Solid-State Elect.*, vol. 35, pp. 953-959, 1992.

[36] D.M.B. Klaassen, "A unified mobility model for device simulation – II. Temperature dependence of carrier mobility and lifetime," *Solid-State Elect.*, vol. 35, pp. 961-967, 1992.

[37] J.M. Hinckley and J. Singh, "Hole transport theory in pseudomorphic $Si_{1-x}Ge_x$ alloys grown on Si(100) substrates," *Phys. Rev. B*, vol. 41, pp. 2912-2926, 1990.

[38] T. Manku et al., "Drift hole mobility in strained and unstrained doped $Si_{1-x}Ge_x$ alloys," *IEEE Trans. Elect. Dev.*, vol. 40, pp. 1990-1996, 1993.

[39] T.K. Carns et al., "Hole mobility measurements in heavily doped $Si_{1-x}Ge_x$ strained layers," *IEEE Trans. Elect. Dev.*, vol. 41, pp. 1273-1281, 1994.

[40] L.E. Kay and T.-W. Tang, "Monte Carlo calculation of strained and unstrained $Si_{1-x}Ge_x$ using an improved ionized-impurity model," *J. Appl. Phys.*, vol. 70, pp. 1483-1488, 1991.

[41] T. Manku and A. Nathan, "Electron drift mobility model for devices based on unstrained and coherently strained Si_{1-x}/Ge_x grown on <001> silicon substrate," *IEEE Trans. Elect. Dev.*, vol. 39, pp. 2082-2089, 1992.

[42] F.M. Bufler et al., "Low- and high-field electron transport parameters for unstrained and strained $Si_{1-x}Ge_x$," *IEEE Elect. Dev. Lett.*, vol. 18, p. 264, 1997.

[43] J. Dziewior and W. Schmid, "Auger coefficients for highly-doped and highly excited silicon," *Appl. Phys. Lett.*, vol. 31, pp. 346-348, 1977.

Chapter 3

SiGe HBT BiCMOS Technology

The device structure, requisite fabrication steps and thermal cycles, and ultimately the transistor performance of any advanced integrated circuit technology are all rapidly moving targets. It can be frustratingly difficult to meaningfully compare two different companies' IC technologies, even though their performance may be quite similar. SiGe is no exception to this rule. Given the inherent futility of such technology comparisons, we instead seek to accomplish only two things in this chapter: 1) allow the reader to get a feel for the numbers, for multiple SiGe technology generations; and 2) examine broad technology issues that are common to all practitioners of the art, regardless of the specifics of their approach to the problem of building a manufacturable SiGe technology.

We first examine the current SiGe technology landscape (the playing field), in order to develop a firm sense of specifics of the various SiGe technology generations currently in existence. We next address the issues associated with integrating SiGe HBTs with Si CMOS to form SiGe HBT BiCMOS technology, followed by a discussion of carbon doping as a key technology enabler. We then examine the integration issues associated with obtaining high-Q passive elements, and conclude with a brief look at device reliability and yield issues.

3.1 The Technology Playing Field

The goal of this section is to help readers develop a feel for the numbers of the various SiGe technology generations currently in existence globally. Of interest in this context is what a generic SiGe HBT of a given technology generation looks like in cross section, what its doping profiles are likely to be, and what level of transistor performance can be expected. We are after rules of thumb from which one can more easily compare and contrast the various SiGe technologies that either currently exist or will in the future. For reasons outlined in Chapter 1, we limit our discussion here to only self-aligned, fully integrated, Si-processing-compatible SiGe HBT technologies that have been reported in the literature. An interesting discussion of the evolution of both non-self-aligned and self-aligned SiGe HBT device structures, as well as their relative merits can be found in [1].

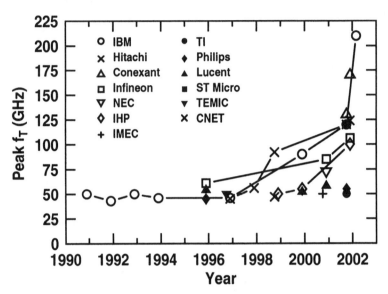

Figure 3.1 Reported cutoff frequency versus year for a variety of different industrial SiGe HBT technologies.

Figure 3.1 shows the historical trend in peak f_T from the first self-aligned device demonstration in 1990 until present. Unlike in Chapter 1, this historical data is broken down by individual company.[1] Due to space constraints, it is most efficient to focus on a single company's SiGe technology, and we have chosen IBM's SiGe technology suite to facilitate the present discussion [2]–[6]. While this choice may seem overly preferential to some, IBM's selection as a technology paradigm for SiGe is both logical and pragmatic since: 1) they were the first to commercialize SiGe technology; 2) they have published substantially more on the subject than anyone else, and hence their data is widely available; and 3) they have three distinct versions of SiGe technology currently in production, with a fourth on the way, and hence can be used to examine technology scaling issues. While in no

---

[1]This is clearly a dynamic field, and hence the specifically cited list of companies with SiGe technologies should not be considered exhaustive. The most recently published version of the SiGe technology from each respective company can be found in [7]–[18].

way do we mean to imply that IBM's SiGe technologies are superior to others in the world, we think it can be safely agreed upon by all that their technologies fairly represent the state of the art in SiGe.

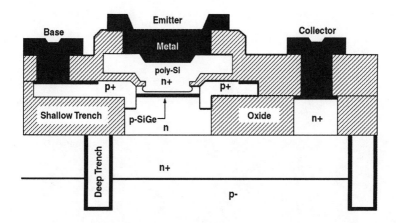

Figure 3.2 Schematic cross section of a representative first generation SiGe HBT, drawn through first metal. Drawing is not to scale.

It is meaningful here to distinguish between different SiGe technology generations, as defined by the *ac* performance of the SiGe HBT (e.g., peak f_T, which is a very strong function of the vertical profile and hence nicely reflects the degree of sophistication in structural design, thermal cycle, epi growth, etc.). We thus label a SiGe HBT technology having a SiGe HBT with a peak f_T of 45–55 GHz as "first generation," that with a peak f_T of 100–120 GHz as "second generation," and that with a peak f_T of 200+ GHz as "third generation." [2]

Regardless of the integration approach and processing steps employed, there are numerous common fabrication elements and modules which exist among the various SiGe HBT technologies, and include for a typical first generation SiGe HBT:

---

[2]Note that in IBM's terminology, they actually have four distinct SiGe HBT BiCMOS technology generations:
- 5HP — 0.5 μm, 50-GHz peak f_T
- 6HP — 0.25 μm, 50-GHz peak f_T (a laterally scaled version of 5HP)
- 7HP — 0.18 μm, 120-GHz peak f_T (a laterally and vertically scaled version of 6HP)
- 8HP — 0.12 μm, 210-GHz peak f_T (a laterally and vertically scaled version of 7HP)

In this book we distinguish SiGe technology generations only by the performance (vertical profile) of the SiGe HBT, and hence would label 5HP and 6HP as "first generation" SiGe, 7HP as "second generation" SiGe, and 8HP as "third generation" SiGe.

- A starting n^+ subcollector (e.g., 5–10 $\Omega/\square$) on a p^- substrate (e.g., 10–15 Ω-cm), probably utilizing a patterned subcollector to allow CMOS integration;

- A high-temperature, lightly doped n-type collector epi (e.g., 0.4–0.6 μm thick at 5×10^{15} cm^{-3});

- Polysilicon-filled deep trenches for isolation of adjacent device subcollectors (e.g., 0.8–1.2 μm wide and 7–10 μm deep);

- Oxide-filled shallow trenches (or perhaps LOCOS) for local device isolation (e.g., 0.4–0.6 μm thick and planarized using chemical-mechanical-polishing (CMP));

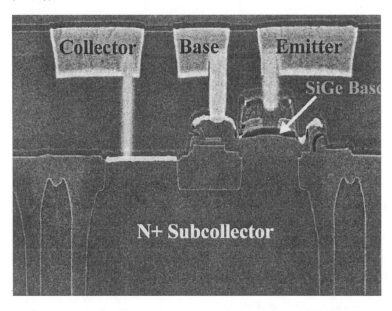

Figure 3.3 Cross sectional SEM of a representative second generation SiGe HBT (after [4]).

- An implanted collector "sinker" or "reach-through" to the subcollector (e.g., 10–20 $\Omega\mu$m^2);

- A composite SiGe epi layer consisting of a Si buffer, boron-doped SiGe (with or without C doping) active layer, and a Si cap. For example, the Si buffer layer might be 10–20 nm thick, followed by a boron-containing ($1 - 3 \times 10^{13}$ cm^{-2} integral boron charge) SiGe (or SiGeC) layer 70–100 nm thick, and a Si cap layer 10–30 nm thick;

SiGe HBT BiCMOS Technology 77

- A variety of emitter-base self-alignment schemes "borrowed" from Si BJT technology to be used depending on the device structure and SiGe deposition approach (single-poly, double-poly, etc.). All self-alignment schemes employ some type of emitter-base "spacer" (e.g., 0.1–0.3 μm wide);

- A local collector implantation used to improve high-J_C performance and enable breakdown voltage tuning (e.g., 0.5 – 1×10^{17} cm^{-3} at the metallurgical CB junction, and graded upward toward the subcollector). This is the self-aligned, selectively implanted collector (SIC) long used in Si BJT technology;

- Polysilicon extrinsic base contacts (usually the SiGe epi layer deposited over the shallow trench) with additional self-aligned extrinsic base implants to lower the total sheet resistance;

- A silicided extrinsic base (e.g., 5–10 Ω/□);

- A heavily-doped (e.g., > 5×10^{20} cm^{-3}) polysilicon emitter, either implanted or *in-situ* doped (e.g., 150–200 nm thick);

- A variety of multilevel back-end-of-the-line (BEOL) metalization schemes (either Al-based or Cu). These are typically "borrowed" from existing CMOS processes, and might include 3 to 6 levels. They usually consist of small tungsten (W) studs between metal layers, using CMP-planarized oxide interlayers;

These technology elements can be located in the schematic cross section of a first generation SiGe HBT shown in Figure 3.2, as well as the cross sectional SEM of a fabricated second generation SiGe HBT (Figure 3.3).

A representative first generation SIMS doping and Ge profile is shown in Figure 3.4. The metallurgical base width is about 90 nm (about 65-nm neutral base width under forward-active bias), the metallurgical emitter junction depth is about 35 nm (from the Si surface), and the peak Ge content is about 8% (it is thermodynamically stable). The emitter polysilicon layer is doped to solid-solubility limits, multiple self-aligned phosphorus implants are used to locally tailor the collector doping profile, and the peak base doping is about 4×10^{18} cm^{-3} ($R_{bi} \cong 6$ kΩ/□). The Ge profile is trapezoidal in shape, with substantial grading across the neutral base. This vertical profile design can be considered quite conservative by today's standards, but it nonetheless achieves a peak f_T of 50 GHz (70-GHz peak f_{max}) at a BV_{CEO} of 3.3 V, solidly in the range of a first generation technology. Cross-company typical profile numbers for first generation SiGe technologies are: $W_{b0} = 60 - 90$ nm, $W_e = 20 - 40$ nm, peak Ge = 8–15%.

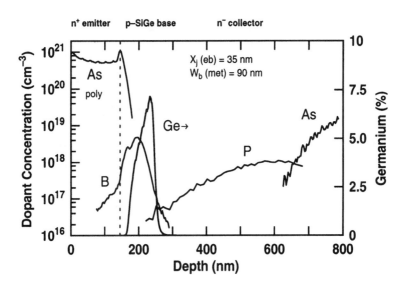

Figure 3.4 Measured SIMS profile of a representative first generation SiGe HBT.

Those acquainted with Si BJT technologies will recognize the striking similarity in doping profiles between this SiGe HBT and advanced ion-implanted Si BJTs (just removing the Ge makes it look like a high-speed Si BJT). The key difference between this SiGe HBT and a conventional ion-implanted double-poly Si BJT lies in the base profile, which can be much more heavily doped at a given base width using epitaxial growth (leading to much lower base resistance and better dynamic response). The observed broadening of the final boron profile in this SiGe HBT (Figure 3.4) is a direct measure of the total process thermal cycle the post-deposited epi-layer sees (the boron is deposited as an atomically-abrupt box about 10 nm wide), which is usually gated by the requisite oxidation steps, and the emitter/extrinsic base anneal (typically shared and done with RTA). It is also key to appreciate that this epi-base scheme employed in SiGe HBTs is extendable to much more aggressive dimensions as the technology scales for higher performance, whereas an implanted base Si BJT would be nearly at its practical scaling limit at 90-nm base width.

Table 3.1 compares the resultant SiGe HBT performance of IBM's three SiGe technology generations (5HP, 7HP, and 8HP). Within some reasonable error bar, all existing SiGe technologies, no matter the company, are reasonably similar in performance to the values shown. This fact, which might seem initially surprising at first glance, actually makes sense, given that the target application markets (and hence the required transistor-level performance) are basically the same, indepen-

Table 3.1 Representative SiGe HBT Parameters for Three Distinct SiGe HBT BiCMOS Technology Generations (after [2], [4], and [6])

Parameter	First	Second	Third
$W_{E,eff}$ (μm)	0.42	0.18	0.12
peak β	100	200	400
V_A (V)	65	120	> 150
BV_{CEO} (V)	3.3	2.5	1.7
BV_{CBO} (V)	10.5	7.5	5.5
peak f_T (GHz)	47	120	207
peak f_{max} (GHz)	65	100	285
min. NF_{min} (dB)	0.8	0.4	< 0.3

dent of company. As a general rule of thumb, first generation SiGe technologies are being currently used to support circuit needs for the global 900-MHz and 2.4-GHz RF cellular markets (both GSM and CDMA), for both handsets and base stations, 1–2.5-Gbit/sec Ethernet applications, Bluetooth, 4–6-GHz WLAN, GPS, and 10-Gbit/sec (OC-192) synchronous optical networks (SONET) transmit-and-receive (T/R) modules (to name a few) [20]. Second generation SiGe technologies are being targeted for 40-Gbit/sec networks and X-band (10 GHz) microwave systems, while emerging third generation SiGe technologies are being positioned for 80-Gbit/sec networks and ISM-band (60 GHz) communications systems.

3.2 Integration of SiGe HBTs with CMOS

The natural ability of SiGe HBTs to integrate seamlessly with conventional Si CMOS is perhaps the single most important advantage SiGe HBT technology has over competing III-V HBT technologies. While it remains contentious in some circles as to the long-term role BiCMOS technology will play compared to CMOS-only or HBT-only integration schemes for mixed-signal applications, it is clear that from a tool and process compatibility, yield, and ultimately a cost-savings viewpoint, being able to realize SiGe technology within a conventional CMOS fabrication facility represents an enormous advantage. One should never compete head-to-head with Si CMOS. Rather, coexistence, coupled with a shrewd "borrowing" of existing CMOS process schemes (e.g., deep and shallow trench isolation, W stud contacts, back-end-of-the-line (BEOL) metalization, etc.) can translate into a large cost savings. This is exactly the path taken by the most successful practitioners of

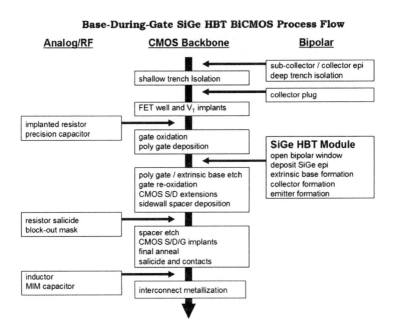

Figure 3.5 Schematic process flow for a BDG first generation SiGe HBT BiCMOS technology (after [20]).

SiGe technology.

The stated compatibility of SiGe HBTs and Si CMOS does obviously require careful up-front structural and process flow design to produce a robust SiGe HBT BiCMOS technology. Key in this context is to integrate the SiGe HBT into a best-of-breed Si CMOS core technology (as a "plug-in" module) without: 1) degrading the SiGe HBT performance; or 2) perturbing the CMOS device characteristics. The latter is especially important given that there is substantial incentive to preserve the preexisting CMOS design libraries and modeling tools. Historically, two different integration schemes have been used to produce SiGe HBT BiCMOS, each with relative pluses and minuses: "base-during-gate" (BDG) integration and "base-after-gate" (BAG) integration (for a detailed discussion of the various BiCMOS integration schemes, refer to [19, 20]). The BDG scheme (or close derivatives) has been widely used to produce first generation SiGe HBT BiCMOS technologies, while the BAG scheme is more easily extendable to second and third generation technologies.

In the BDG scheme (also known in the literature as "base = gate"), the SiGe HBT shares the CMOS layers and thermal cycles in order to reduce structural

SiGe HBT BiCMOS Technology

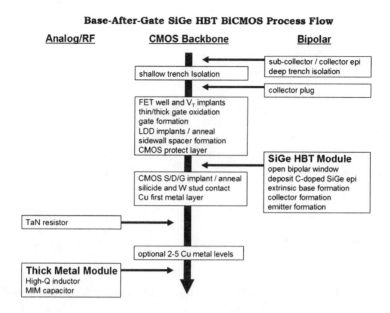

Figure 3.6 Schematic process flow for a BAG second generation SiGe HBT BiCMOS technology (after [4]).

complexity (Figure 3.5). In this case, the SiGe epitaxial base (and importantly, the boron doping it contains) sees all of the CMOS thermal cycle, which can be substantial, leading to a broadening of base profile (this can be easily seen in Figure 3.4).

For further technology scaling (to second and third generations), the BDG approach (Figure 3.6) becomes increasingly problematic due to the inherently large thermal cycles associated with the CMOS process. In the BAG integration scheme, the CMOS devices are completed *before* the SiGe epitaxial base is deposited, effectively decoupling the fabrication of the two device types. This facilitates a step-by-step copying of the underlying CMOS fabrication steps from the pre-existing CMOS-only technology. The BAG approach also more easily allows the incorporation of CMOS device derivatives (e.g., higher voltage CMOS for I/O's and/or analog/RF circuits). The only thermal cycle shared between the HBT and CMOS is the final emitter anneal, making it ideal for SiGe HBT profile optimization. These attractive features of the BAG scheme do not come for free of course. The BAG scheme is inherently more complex, largely because the bipolar layers are deposited on top of the CMOS topography, and have to be removed.

Even with the substantially reduced thermal cycles afforded by a second gen-

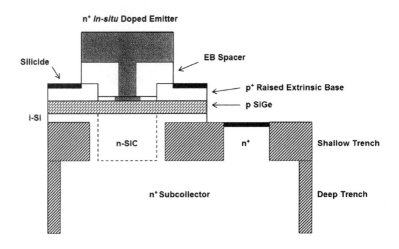

Figure 3.7 Schematic cross section of a representative third generation SiGe HBT. Drawing is not to scale (after [6]).

eration BAG integration scheme, evolution from second generation to third generation SiGe HBT performance levels (i.e., > 200 GHz peak f_T) necessitates device structural changes that eliminate any extrinsic base implantation steps into the deposited SiGe-bearing epi layer during the HBT module. Such implantations are known to introduce interstitials into the active device region, yielding enhanced boron diffusion (even with C-doping), making it very difficult to decouple the achievable f_T (i.e., boron base profile) from the extrinsic base design. The so-called "raised-extrinsic base" structure (Figure 3.7) appears to offer several advantages in this context, and has been used to demonstrate impressive SiGe HBT performance levels [6]. The CMOS integration still follows a modular HBT-after-CMOS (BAG) scheme.

Table 3.2 compares the resultant Si CMOS performance of IBM's three SiGe HBT BiCMOS technology generations. Within some reasonable error bar, all existing SiGe technologies, no matter the company, are reasonably similar in performance to the values shown.

3.3 Carbon Doping

Carbon doping (C-doping) of epitaxial SiGe layers as a means to effectively suppress boron out-diffusion during fabrication is rapidly becoming the preferred ap-

Table 3.2 Representative CMOS Parameters for Three Distinct SiGe HBT BiCMOS Technology Generations (after [2], [4], and [5])

Parameter	First nFET	First pFET	Second nFET	Second pFET	Third nFET	Third pFET
L_{eff} (μm)	0.36	0.36	0.14	0.15	0.092	0.092
V_{DD} (V)	3.3	3.3	1.8	1.8	1.5	1.5
t_{ox} (nm)	7.8	7.8	4.2	4.2	2.2	2.2
$V_{T,lin}$ (mV)	580	-550	326	-415	250	-210
$I_{D,sat}$ (μA/μm)	468	231	600	243	500	210

proach for commercial SiGe technologies, particularly in second and third generation processes. C-doping of SiGe HBTs (yielding a SiGe:C or just SiGeC HBT) has its own interesting history, dating back to the serendipitous discovery [21] that incorporating small amounts of C into a SiGe epi layer strongly retards (by an order of magnitude!) the diffusion of the boron (B) base layer during subsequent thermal cycles. Given that maintaining a thin base profile during fabrication is perhaps *the* most challenging aspect of building a manufacturable SiGe technology, it is somewhat surprising that it took so long for the general adoption of C-doping as a technology element. It is fair to say that most SiGe practitioners at the time viewed C-doping with more than a small amount of skepticism, given that C can act as a deep trap in Si, and C contamination is generally avoided at all costs in Si processes, particularly for minority carrier devices such as the HBT. At the time of the discovery of C-doping of SiGe in the mid 1990s, most companies were focused on simply bringing up a SiGe process and qualifying it, relegating the potential use of C to the back burner. In fairness, most felt that C-doping was not necessary to achieve first generation HBT performance. The lone visionary group to solidly embrace C-doping of SiGe HBTs at the onset was the IHP team in Germany [22, 23], whose pioneering work eventually paid off and began to convince the skeptics of the merits of C-doping. The minimum required C concentration for effective out-diffusion suppression of B was empirically established to be in the vicinity of 0.2% C (1×10^{20} cm^{-3}). Early on, much debate ensued on the physical mechanism of *how* C impedes the B diffusion process, but general agreement for the most part now exists, and is briefly reviewed here (following [22]).

In short, C-doping produces an undersaturation of Si self-interstitials during the out-diffusion of C in thermal processing, which inhibits the conventional interstitial diffusion mechanism of B [24, 25]. Substitutional C in a B-containing SiGe layer can be used to both decrease the diffusion coefficient of B in Si by more than 10×, as well as suppress transient-enhanced diffusion (TED) of B in Si. Depending on the exact film growth conditions, C can be at least partially substitutionally dissolved in Si. Diffusion of C in Si occurs via a substitutional-interstitial diffusion mechanism. During thermal processing, mobile interstitial C atoms (C_I) are created through the reaction of Si self-interstitials (I) with immobile substitutional C (C_S)

$$C_S + I \rightleftarrows C_I \tag{3.1}$$

as well as in the dissociative reaction

$$C_S \rightleftarrows C_I + V, \tag{3.2}$$

where V is the vacancy concentration. In order to conserve the total number of atoms, the flux of interstitial C atoms out of the C-rich region must be balanced by either a flux of Si self-interstitials into this region, or a flux of vacancies out of this region. The individual atomic fluxes are determined by the products of the diffusion coefficient (D) and the concentration.

For C concentrations (C_C) greater than about 10^{18} cm^{-3}, the transport coefficient of the C may exceed the transport coefficients of the Si self-interstitials and Si vacancies, such that

$$D_C C_C > D_I C_I^{equil} \tag{3.3}$$

and

$$D_C C_C > D_V C_V^{equil}, \tag{3.4}$$

where C_I^{equil} and C_V^{equil} are the equilibrium concentrations of Si self-interstitials and vacancies, respectively. As a consequence, out-diffusion of supersaturated C from C-rich regions becomes limited by the compensating flux of Si point defects, leading to an undersaturation of self-interstitials in the C-rich region. Given that B diffusion in Si occurs via an interstitial mechanism, the effective diffusion coefficient of B in Si is proportional to the normalized concentration of self-interstitials, and hence interstitial undersaturation suppresses the B diffusion process.

An instructive experimental confirmation of this C-doping out-diffusion suppression mechanism can be seen in Figure 3.8, which shows the impact of putting supersaturated C into a B superlattice which is exposed to subsequent ion implantation and annealing, similar to what might be encountered in SiGe HBT fabrication. To date, no deleterious effects on device *dc* or *ac* or noise characteristics have been reported for the incorporation of moderate levels of C-doping in SiGe HBTs, but this is an issue that should be revisited as vertical scaling continues.

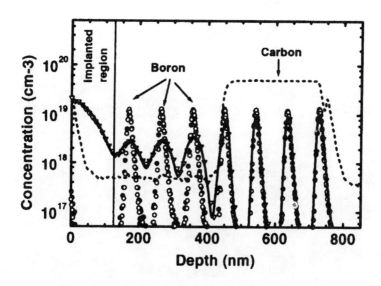

Figure 3.8 SIMS profiles of a B superlattice implanted with 1×10^{14} cm^{-2} 45 keV BF_2 ions and annealed in N_2 at 930°C for 30 sec. The open circles are as-grown, filled circles are after anneal, and the solid line calculated (after [22]).

3.4 Passives

The lack of high quality factor (Q) passive components and low-loss transmission lines in Si technologies has long been touted as a major reason to favor III-V technologies over Si for RF and microwave components, since they can easily bring to bear semi-insulating substrates and thick, low-resistivity gold (Au) metal layers. Si technologists have remained predictably obstinate in the face of these odds, however, and the Qs of the most problematic of the RF passives, the integrated inductor, have steadily risen over time to fairly respectable levels of 15–20, through the use of thick dielectrics (e.g., 3 μm), thick last-metal layers (e.g., 4 μm), optimized layout, effective shielding, and substrate resistivity tuning. In a typical SiGe HBT BiCMOS technology, the passive elements dwarf the underlying SiGe HBT, and reside far above the Si surface in the upper levels of the multilayer interconnect schemes, as shown in Figure 3.9 for a second generation SiGe technology. Table 3.3 shows that a full suite of high-quality resistors, capacitors, and inductors

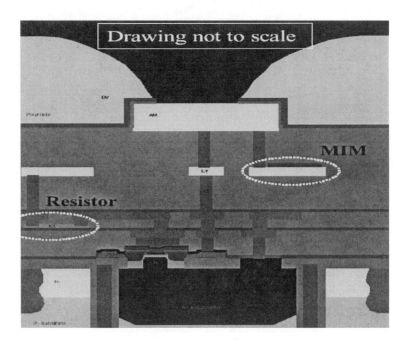

Figure 3.9 Schematic cross section of a representative second generation SiGe HBT showing the passives and metalization (after [4]).

can be supported in SiGe HBT BiCMOS technologies.

To better understand the inherent challenges associated with building high-Q passives in SiGe technology, one need only note that conventional technology scaling in Si naturally works against increasing Q since: 1) common Al-based BEOL metals are fairly resistive (10–100 mΩ/$\square$); 2) the metal layer and their associated interlayer dielectrics usually decrease in thickness with scaling in order to improve wiring density; and 3) bipolar-based substrate resistivities are naturally low (10–20 Ω-cm), and magnetic fields in the inductor, for instance, can thus more easily induce lossy parasitic eddy currents in the substrate.

As an example of optimization of passives in SiGe technology, we consider the inductor [20, 26]. Using a lumped-element model as a guide for the optimization of the inductors, the primary resistive lossy elements are the series resistance of the metal spiral and the substrate resistance, while the primary capacitive lossy element is the metal spiral-to-substrate capacitance. In view of this, four approaches can be taken to improve the Q of monolithic inductors in SiGe technology: 1) decrease the metal series resistance by using a thicker final metal layer [27]; 2) increase the effective substrate resistivity; 3) move the spiral further above the sub-

Table 3.3 Representative Passive Elements for a Second Generation SiGe HBT BiCMOS Technology (after [4])

Resistor	Sheet resistance ($\Omega/\square$)	TCR (ppm/°C)
subcollector	8.1	1,430
n^+ diffusion	72	1,910
p^+ diffusion	105	1,430
p^+ polysilicon	270	50
p polysilicon	1,600	-1,178
TaN	142	-750
Capacitor	Capacitance (fF/μm^2)	VCC (+5/-5 ppm/V)
MIM capacitor	1.0	< 45
MOS capacitor	2.6	-7,500 / -1,500
Varactor	Tuning range	Q at 500 MHz
CB junction	1.64:1	90
MOS accumulation	3.1:1	300
Inductor	Inductance (nH)	Q at 5 GHz
spiral (Al)	> 0.7	18

strate by thickening the dielectric layer before the final metal layer, to decrease the capacitive coupling to the substrate; and 4) using patterned ground planes (Faraday shielding [28]). There are pros and cons to each approach, of course, and one typically finds that techniques 1), 3), and 4) can be used in concert for improvement in Q (technique 2 often presents problems with maintaining CMOS compatibility). Employing such optimization schemes can produce usable inductors (L > 0.5 nH) with acceptable Qs (> 15 at 5 GHz) for many demanding RF applications (e.g., for VCOs). The increasingly common usage of Cu metalization schemes will further improve inductor performance, since the resistivity is as much as 2× lower than conventional Al-based schemes. In practice, however, the typically thinner Cu layer thicknesses found in CMOS Cu BEOL processes do not yield a full 2× improvement in series resistance, and further optimization using a thick top Cu layer is possible.

Monolithic capacitors in SiGe technology are important for both matching in RF circuits as well as in a variety of analog circuits, and typically take the form of either metal-insulator-metal (MIM) capacitors [29] or polysilicon-gate to substrate (i.e., MOS) capacitors formed over a collector plug. The MOS capacitor has a higher specific capacitance than the MIM capacitor, but lower Qs (a typical Q might be in the range of 20 at 2 GHz for an MOS capacitor, while it can be 70–80 for an MIM capacitor (Table 3.3)). Because of this, MIM capacitors are typically used in critical RF and analog circuits, while the MOS capacitor might be used for power supply bypassing and decoupling.

Achieving acceptable transmission line performance in SiGe generally follows similar guidelines to those needed for making high Q passives; move the lines as far away from the substrate as possible, and use thick low sheet resistance metal layers. While transmission line performance in SiGe is not at present competitive with III-V values, acceptable losses for 10–20-GHz components have been demonstrated, with attenuation numbers in the 1.5–2.0-dB/cm range at 10 GHz. In this context, SiGe need not be the best, but rather need only achieve acceptable levels of performance for most microwave applications, since it brings many other compelling virtues to the table. Moving to thick (e.g., 10 μm) spun-on organic dielectrics (e.g., benzo-cyclo-butene (BCB)), followed by Au metalization in order to achieve very low-loss lines is a demonstrated option in SiGe [30], although this approach by definition occurs postfabrication, and thus has some significant cost disadvantages.

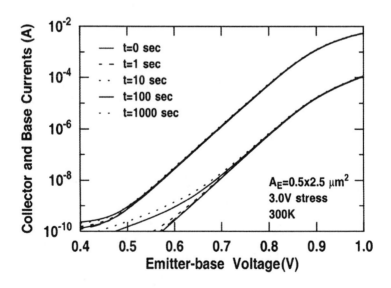

Figure 3.10 Gummel characteristics of a SiGe HBT showing the effects of reverse-bias emitter-base stress.

3.5 Reliability and Yield Issues

It is vital to the long-term viability of SiGe technology that it have a clearly demonstrated reliability and yield that are comparable to or better than existing Si technology. That is, any reliability or yield loss due to the incorporation of strained SiGe films are potential showstoppers. Although published data on commercial SiGe technologies is sparse, there is no evidence to date that the use of thermodynamically stable SiGe films imposes any such reliability risk.

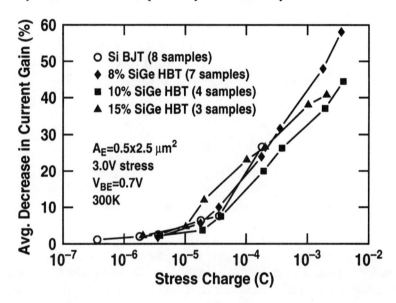

Figure 3.11 Current gain degradation as a function of injected stress charge for a variety of SiGe profiles and a comparably constructed epi-base Si BJT.

Reliability stress and burn-in of bipolar transistors historically proceeds along two different paths [31]: 1) reverse emitter-base (EB) stress, which is used to inject hot electrons (or holes [32]) into the EB spacer oxide, thereby introducing G/R center traps which lead to excess nonideal base current (Figure 3.10) and hence current gain degradation as well as increased low-frequency noise [33]; and 2) high forward-current stress, which also results in current gain degradation, but is generally attributed to electromigration-induced pressure on the emitter contact, resulting in a decrease in collector current with increasing stress time. Accelerated lifetime testing of SiGe HBTs using reverse-bias EB stress is generally conducted under high reverse EB bias (e.g., 3.0 V) at reduced temperatures (e.g., -40°C), where carrier velocities are higher due to reduced scattering, whereas high forward-current stress is conducted under a large J_C near peak f_T (e.g., 1.0–2.0 mA/μm^2) at elevated temperatures (e.g., 140°C), where electromigration is more severe.

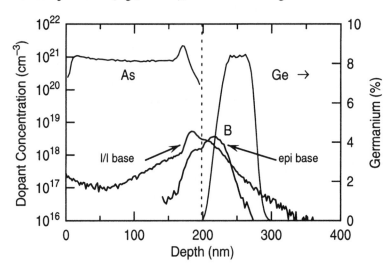

Figure 3.12 Comparison of representative base profiles for a state of the art implanted base Si BJT and a first generation SiGe HBT.

Typical reverse-bias EB burn-in data from first generation SiGe HBTs show less than 6% change in the current gain after a 500-hour, -40°C reverse-bias EB stress at 2.7 V [20]. Comparison of reverse-bias EB stress data of SiGe HBTs having various Ge profile shapes with an epi-base Si BJT control (Figure 3.11) suggests that there is no enhanced reliability risk associated with the SiGe layer [34].

Interestingly, the reverse-bias EB stress response of SiGe HBTs is actually substantially better than that for aggressively scaled ion-implanted Si BJTs. As can be seen in Figure 3.12, very shallow, low energy base implants needed to realize high-performance implanted Si BJTs inevitably place the peak of the base doping at the metallurgical EB junction, and thus increase the EB electric field. In contrast, for an epitaxial base device (Si or SiGe), the B can be placed inside the base region as a B box, and while the finite thermal cycle spreads the B during processing, a B retrograde is naturally produced at the EB junction, thereby lowering the EB electric field. [3] Since hot electron injection under reverse-bias EB stress conditions

---

[3]Observe that this B retrograde at the EB junction itself produces a doping-gradient-induced electric field that retards electron transport through the base under forward-bias, degrading f_T. In a SiGe HBT this doping-induced retarding field is more than compensated by the Ge-induced accelerating field, but in an epi-base Si BJT, a performance penalty is inevitable.

depends exponentially on the EB electric field [31], a transistor with an epitaxial base will have a fundamental and decided advantage over an implanted base device in terms of reliability.

Typical high forward-current burn-in data from first generation SiGe HBTs show less than 5% change in the current gain after a 500-hour, 140°C forward-current stress at 1.3 mA/μm^2 [20]. Using empirically determined acceleration factors, this result is theoretically equivalent to a more-than-acceptable 10% current gain degradation after 100,000 power-on-hours (POH) under "normal use" conditions (1.25 mA/μm^2 at 100°C). Given that technology scaling naturally leads to higher current density operation in bipolar devices, it will nonetheless be important to quantify these changes with each successive technology generation, as well as assess the capability of the BEOL infrastructure to support these higher current densities (i.e., is Cu required?).

High yield is key to the cost advantage Si enjoys over its III-V competition, and as in reliability, SiGe must not unfavorably impact device and circuit yield. It does not. CMOS yield in a SiGe HBT BiCMOS technology is typically evaluated using an static random-access-memory (SRAM) yield monitor (e.g., a 154k SRAM for IBM's first and second generation technologies). If any of the HBT films or residuals are not properly removed, then this will be reflected in the SRAM yield. Yield values can also be easily compared with CMOS-only processes to gauge the robustness of the CMOS section of the BiCMOS process. Typical yield numbers for the 154k SRAM in first and second generation SiGe technology are above 75% [20].

SiGe HBT yield is typically quantified using large chains of small transistors wired in parallel. A chain yield "failure" is defined as the intersection of emitter-to-collector shorts (pipes), high EB leakage, or high CB leakage (i.e., any of the three occurrences is defined as a "bad" or "dead" device chain). For instance, 4000 0.42 × 2.3 μm^2 SiGe HBTs is used as a yield monitor in IBM's first generation technology, and typically has greater than 85% yield. Both BDG and BAG integration methodologies show similar results. Interestingly, the primary failure mechanism in both the CMOS and SiGe HBTs is the same, and can be traced to the shallow trench isolation [20]. By assuming an ideal Poisson distribution relating defect density and emitter area, one can infer the net defect density associated with a given SiGe HBT BiCMOS technology, in this case yielding numbers in the range of 100–500 defects/cm^2. For orientation, a defect density of 426 defects/cm^2 would ideally produce a 60% yield on a IC containing 100,000 0.5 × 2.5 μm^2 SiGe HBTs, ample transistor count (and yield) to satisfy almost any imaginable application [1]. This is clearly excellent news.

References

[1] D.L. Harame et al., "Si/SiGe epitaxial-base transistors: part II – process integration and analog applications," *IEEE Trans. Elect. Dev.*, vol. 40, pp. 469-482, 1995.

[2] D.C. Ahlgren et al., "A SiGe HBT BiCMOS technology for mixed-signal RF applications," *Proc. IEEE Bipolar/BiCMOS Circ. Tech. Meeting*, pp. 195-198, 1997.

[3] S.A. St. Onge et al., "A 0.24 μm SiGe BiCMOS mixed-signal RF production technology featuring a 47 GHz f_T HBT and a 0.18 μm L_{eff} CMOS," *Proc. IEEE Bipolar/BiCMOS Circ. Tech. Meeting*, pp. 117-120, 1999.

[4] A. Joseph et al., "A 0.18 μm BiCMOS technology featuring 120/100 GHz (f_T/f_{max}) HBT and ASIC compatible CMOS using copper interconnect," *Proc. IEEE Bipolar/BiCMOS Circ. Tech. Meeting*, pp. 143-146, 2001.

[5] A. Joseph et al., "0.13 μm 210 GHz f_T SiGe HBTs – expanding the horizons of SiGe BiCMOS," *Tech. Dig. IEEE Int. Solid-State Circ. Conf.*, pp. 180-1821, 2002.

[6] B. Jagannathan et al., "Self-aligned SiGe NPN transistors with 285 GHz f_{max} and 207 GHz f_T in a manufacturable technology," *IEEE Elect. Dev. Lett.*, vol. 23, pp. 258-260, 2002.

[7] K. Oda et al., "Self-aligned selective-epitaxial-growth $Si_{1-x-y}Ge_xC_y$ HBT technology featuring 170-GHz f_{max}," *Tech. Dig. IEEE Int. Elect. Dev. Meeting*, pp. 332-335, 2001.

[8] M. Racanelli et al., "Ultra high speed SiGe npn for advanced BiCMOS Technology," *Tech. Dig. IEEE Int. Elect. Dev. Meeting*, pp. 336-339, 2001.

[9] J. Böck et al., "High-speed SiGe:C bipolar technology," *Tech. Dig. IEEE Int. Elect. Dev. Meeting*, pp. 344-347, 2001.

[10] T. Hashimoto et al., "A 73 GHz f_T 0.18 μm RF-SiGe BiCMOS technology considering thermal budget trade-off and with reduced boron-spike effect on HBT characteristics," *Tech. Dig. IEEE Int. Elect. Dev. Meeting*, pp. 149-152, 2000.

[11] B. Heinemann et al., "Cost-effective high-performance high-voltage SiGe:C HBTs with 100 GHz f_T and $BV_{CEO} \times f_T$ products exceeding 220 VGHz," *Tech. Dig. IEEE Int. Elect. Dev. Meeting*, pp. 348-352, 2001.

[12] S. Decoutere et al., "A 0.35 μm SiGe BiCMOS process featuring a 80 GHz f_{max} HBT and integrated high-Q RF passive components," *Proc. IEEE Bipolar/BiCMOS Circ. Tech. Meeting*, pp. 106-109, 2000.

[13] F.S. Johnson et al., "A highly manufacturable 0.25 µm RF technology utilizing a unique SiGe integration," *Proc. IEEE Bipolar/BiCMOS Circ. Tech. Meeting*, pp. 56-59, 2001.

[14] P. Deixler et al., "Explorations for high performance SiGe-heterojunction bipolar transistor integration," *Proc. IEEE Bipolar/BiCMOS Circ. Tech. Meeting*, pp. 30-33, 2001.

[15] M. Carroll et al., "COM2 SiGe Modular BiCMOS technology for digital, mixed-signal, and RF applications," *Tech. Dig. IEEE Int. Elect. Dev. Meeting*, pp. 145-148, 2000.

[16] H. Baudry et al., "High performance 0.25 µm SiGe and SiGe:C HBTs using non-selective epitaxy," *Proc. IEEE Bipolar/BiCMOS Circ. Tech. Meeting*, pp. 52-55, 2001.

[17] A. Schüppen et al., "SiGe-technology and components for mobile communication systems," *Proc. IEEE Bipolar/BiCMOS Circ. Tech. Meeting*, pp. 130-133, 1996.

[18] A. Chantre et al., "A high performance low complexity SiGe HBT for BiCMOS integration," *Proc. IEEE Bipolar/BiCMOS Circ. Tech. Meeting*, pp. 93-96, 1998.

[19] D.L. Harame, "High performance BiCMOS process integration: trends, issues, and future directions," *Proc. IEEE Bipolar/BiCMOS Circ. Tech. Meeting*, pp. 36-43, 1997.

[20] D.L. Harame et al., "Current status and future trends of SiGe BiCMOS technology," *IEEE Trans. Elect. Dev.*, vol. 48, pp. 2575-2594, 2001.

[21] L. Lanzerotti et al., "Si/Si$_{1-x-y}$Ge$_x$C$_y$/Si heterojunction bipolar transistors," *IEEE Elect. Dev. Lett.*, vol. 17, pp. 334-337, 1996.

[22] H.J. Osten et al., "Carbon doped SiGe heterojunction bipolar transistors for high frequency applications," *Proc. IEEE Bipolar/BiCMOS Circ. Tech. Meeting*, pp. 109-116, 1999.

[23] H.J. Osten et al., "Increasing process margin in SiGe heterojunction bipolar technology by adding carbon," *IEEE Trans. Elect. Dev.*, vol. 46, pp. 1910-1912, 1999.

[24] R. Scholz, *Appl. Phys. Lett.*, vol. 72, pp. 200-202, 1998.

[25] H. Rucker et al., "Dopant diffusion in C-doped Si and SiGe: physical model and experimental verification," *Tech. Dig. IEEE Int. Elect. Dev. Meeting*, pp. 345-348, 1999.

[26] J.N. Burghartz et al., "Integrated RF components in a SiGe bipolar technology," *IEEE J. Solid-State Circ.*, vol. 32, pp. 1440-1445, 1997.

[27] M. Soyuer et al., "Multilevel monolithic inductors in silicon technology," *Elect. Lett.*, vol. 31, pp. 359-360, 1995.

[28] C.P. Yu and S.S. Wong, "On-chip spiral inductors with patterned ground shields for Si-based RF ICs," *IEEE J. Solid-State Circ.*, vol. 33, pp. 743-752, 1998.

[29] K. Stein et al., "High reliability metal insulator metal capacitors for silicon germanium analog applications," *Proc. IEEE Bipolar/BiCMOS Circ. Tech. Meeting*, pp. 191-194, 1997.

[30] L. Larson et al., "Si/SiGe HBT technology for low-cost monolithic microwave integrated circuits," *Tech. Dig. IEEE Int. Solid-State Circ. Conf.*, pp. 80-81, 1996.

[31] D.D. Tang and E. Hackbarth, "Junction degradation in bipolar transistors and the reliability imposed constraints to scaling and design," *IEEE Trans. Elect. Dev.*, vol. 35, pp. 2101-2107, 1988.

[32] A. Neugroschel and C.T. Sah, "Comparison of time-to-failure of GeSi and Si bipolar transistors," *IEEE Elect. Dev. Lett.*, vol. 17, pp. 211-213, 1996.

[33] J.A. Babcock et al., "Correlation of low-frequency noise and emitter-base reverse-bias stress in epitaxial Si- and SiGe-base bipolar transistors," *Tech. Dig. IEEE Int. Elect. Dev. Meeting*, pp. 357-360, 1995.

[34] U. Gogineni et al., "Hot electron and hot hole degradation of SiGe heterojunction bipolar transistors," *IEEE Trans. Elect. Dev.*, vol. 47, pp. 1440-1448, 2000.

Chapter 4

Static Characteristics

Due to the presence of Si-SiGe heterojunctions in both the emitter-base and collector-base junctions, the device physics of the SiGe HBT fundamentally differs from that of the conventional Si BJT. In this chapter we examine these differences from a *dc* point of view, by first reviewing an intuitive picture of how the SiGe HBT operates, and importantly how its operation differs from that of a comparably constructed Si BJT. Using this insight, we then formally derive the collector current density, current gain, output conductance, and current gain – Early voltage product of an ideal SiGe HBT under low-injection conditions. Explicit and implicit assumptions as well as physically relevant approximations are highlighted throughout the theoretical development. Armed with this knowledge we then examine the *dc* Ge profile shape and optimization issues, develop a low-frequency equivalent circuit model, and address impact ionization and breakdown issues.

4.1 Intuitive Picture

The essential operational differences between the SiGe HBT and the Si BJT are best illustrated by considering a schematic energy band diagram. For simplicity, we consider an ideal, graded-base SiGe HBT with constant doping in the emitter, base, and collector regions. In such a device construction, the Ge content is linearly graded from 0% near the metallurgical emitter-base (EB) junction to some maximum value of Ge content near the metallurgical collector-base (CB) junction, and then rapidly ramped back down to 0% Ge. The resultant overlaid energy band diagrams for both the SiGe HBT and the Si BJT, biased identically in forward-active mode, are shown in Figure 4.1. Observe in Figure 4.1 that a Ge-induced reduction in base bandgap occurs at the EB edge of the quasi-neutral base ($\Delta E_{g,Ge}(x = 0)$), and at the CB edge of the quasi-neutral base ($\Delta E_{g,Ge}(x = W_b)$).

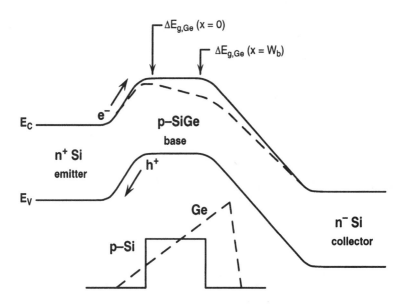

Figure 4.1 Energy band diagram for a Si BJT and graded-base SiGe HBT, both biased in forward active mode at low-injection.

This grading of the Ge across the neutral base induces a built-in quasi-drift field $((\Delta E_{g,Ge}(x = W_b) - \Delta E_{g,Ge}(x = 0))/W_b)$ in the neutral base that will impact minority carrier transport (Chapter 5).

A logical first question to ask is why the valence band offset due to the introduction of Ge into Si ends up in the *conduction band* of the *npn* SiGe HBT? To understand this, it is instructive to consider the introduction of the graded Ge layer into the p-type base as a two-step process, as depicted in Figure 4.2. For constant p-doping in the base of the Si BJT, we know that both the Fermi level and the energy difference between the Fermi level and the valence band edge is fixed. As Ge is introduced and graded across the neutral base, a valence band offset is induced, as depicted in Step 1 of Figure 4.2. We know, however, that the Fermi level must realign itself such that it is fixed in energy to its previous (Si) value, and further, that it must be constant (flat) if the system is in equilibrium. Thus, compared to the Si case, the Fermi level must decrease in energy and flatten via charge transport. Given that the total bandgap is fixed for a given Ge content at each position x, the consequence, as depicted in Step 2 of Figure 4.2, is that the conduction band edge in the neutral base region is forced downward in energy. Thus, the inherent valence band offset associated with the (position-dependent) Ge profile is effectively translated into the conduction band of the device. This valence band to conduction band translation is fortuitous, given that the induced drift field associated with the now position-dependent conduction band edge will positively influence the minority electron transport through the base, as desired. Note that as we move out of the neutral base and into the space charge region in the CB junction, we again return to the expected valence band offset. In a well-designed SiGe HBT, the SiGe-Si heterojunction on the CB side of the neutral base is intentionally buried in the strong band bending of the CB junction, and thus is not visible in the band diagram at low-injection. As will be seen in Chapter 6, however, under high-level injection, this is not the case, and SiGe-Si heterojunction will have important consequences on device performance at high J_C. To intuitively understand how

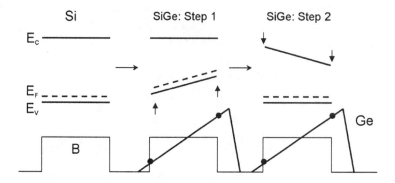

Figure 4.2 Illustration of the bandgap changes induced by the introduction of Ge into the base region of an *n-p-n* SiGe HBT.

these band edge changes affect the *dc* operation of the SiGe HBT, first consider the operation of the Si BJT. When V_{BE} is applied to forward bias the EB junction, electrons are injected from the electron-rich emitter into the base across the EB potential barrier (refer to Figure 4.1). The injected electrons diffuse across the base, and are swept into the electric field of the CB junction, yielding a useful collector current. At the same time, the applied forward bias on the EB junction produces a back-injection of holes from the base into the emitter. If the emitter region is doped heavily with respect to the base, however, the density of back-injected holes will be small compared to the forward-injected electron density, and hence a finite current gain $\beta \propto n/p$ results.

As can be seen in Figure 4.1, the introduction of Ge into the base region has two tangible *dc* consequences: 1) the potential barrier to injection of electrons from emitter into the base is decreased. Intuitively, this will yield exponentially more electron injection for the same applied V_{BE}, translating into higher collector current and hence higher current gain, provided the base current remains unchanged. Given that band edge effects generally couple strongly to transistor properties, we naively expect a strong dependence of J_C on Ge content. Of practical consequence, the introduction of Ge effectively decouples the base doping from the current gain, thereby providing device designers with much greater flexibility than in Si BJT design. If, for instance, the intended circuit application does not require high current gain (as a rule of thumb, $\beta = 100$ is usually sufficient for most circuits), we can effectively trade the higher gain induced by the Ge band offset for a higher base doping level, leading to lower net base resistance, and hence better dynamic switching and noise characteristics. 2) The presence of a finite Ge content in the CB junction will positively influence the output conductance of the transistor, yielding higher Early voltage. While it is more difficult to physically visualize why this is the case, in essence, the smaller base bandgap near the CB junction effectively weights the base profile (through the integral of intrinsic carrier density across the base), such that the backside depletion of the neutral base with increasing applied V_{CB} (Early effect) is suppressed compared to a comparably doped Si BJT. This translates into a higher Early voltage compared to an Si BJT.

4.2 Collector Current Density and Current Gain

To understand the inner workings of the SiGe HBT, we must first formally relate the changes in the collector current density and hence current gain to the physical variables of this problem. It is also instructive to carefully compare the *differences* between a comparably constructed SiGe HBT and a Si BJT. In the present analysis, the SiGe HBT and Si BJT are taken to be of identical geometry, and it is assumed that the emitter, base, and collector doping profiles of the two devices are identical, apart from the Ge in the base of the SiGe HBT. For simplicity, a Ge profile which is linearly graded from the EB to the CB junction is assumed, as depicted in Figure 4.3. The resultant expressions can be applied to a wide variety of practical SiGe profile designs, ranging from constant (box) Ge profiles, to triangular (linearly graded) Ge profiles, and including the intermediate case of the Ge trapezoid (a combination of box and linearly graded profiles) [1]. Unless otherwise stated, this analysis assumes standard low-injection conditions, negligible bulk and surface recombination, Boltzmann statistics, and holds for *npn* SiGe HBTs.

4.2.1 J_C in SiGe HBTs

The theoretical consequences of the Ge-induced bandgap changes to J_C can be derived in closed-form for a constant base doping profile ($p_b(x) = N_{ab}^-(x) = N_{ab}^-$

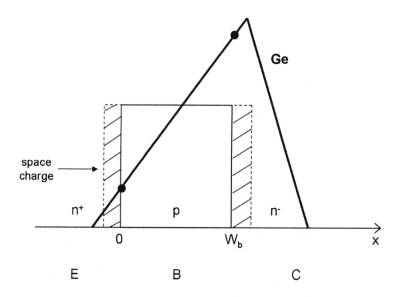

Figure 4.3 Schematic doping and Ge profiles used in the derivations.

= constant) by considering the generalized Moll-Ross collector current density relation, which holds for low-injection in the presence of both nonuniform base doping and nonuniform base bandgap at fixed V_{BE} and temperature (T) [2]

$$J_C = \frac{q(e^{qV_{BE}/kT} - 1)}{\int_0^{W_b} \frac{p_b(x)dx}{D_{nb}(x)n_{ib}^2(x)}}, \quad (4.1)$$

where $x = 0$ and $x = W_b$ are the neutral base boundary values on the EB and CB sides of the base, respectively. In this case, the base doping is constant, but both both n_{ib} and D_{nb} are position-dependent; the former through the Ge-induced band offset, and the latter due to the influence of the (position-dependent) Ge profile on the electron mobility ($D_{nb} = kT/q\mu_{nb} = $ f(Ge)). Note that J_C depends only on the Ge-induced changes in the base bandgap. In general, the intrinsic carrier density in the SiGe HBT can be written as

$$n_{ib}^2(x) = (N_C N_V)_{SiGe}(x) e^{-E_{gb}(x)/kT}, \quad (4.2)$$

where $(N_C N_V)_{SiGe}$ accounts for the (position-dependent) Ge-induced changes associated with both the conduction and valence band effective density-of-states (refer to Chapter 2). In (4.2), the SiGe base bandgap can be broken into its various contributions, as depicted in Figure 4.4.

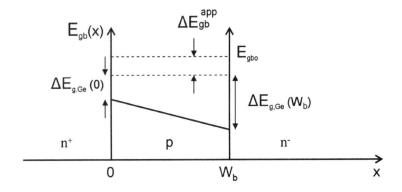

Figure 4.4 Schematic base bandgap in a linearly graded SiGe HBT.

In Figure 4.4, E_{gbo} is the Si bandgap under low-doping (1.12 eV at 300 K), ΔE_{gb}^{app} is the heavy-doping-induced apparent bandgap narrowing in the base region, $\Delta E_{g,Ge}(0)$ is the Ge-induced band offset at $x = 0$, and $\Delta E_{g,Ge}(W_b)$ is the Ge-induced band offset at $x = W_b$. We can thus write $E_{gb}(x)$ as

$$E_{gb}(x) = E_{gbo} - \Delta E_{gb}^{app} + [\Delta E_{g,Ge}(0) - \Delta E_{g,Ge}(W_b)]\frac{x}{W_b} - \Delta E_{g,Ge}(0). \quad (4.3)$$

Substitution of (4.3) into (4.2) gives

$$n_{ib}^2(x) = \gamma n_{io}^2 e^{\Delta E_{gb}^{app}/kT} e^{[\Delta E_{g,Ge}(W_b)-\Delta E_{g,Ge}(0)]x/(W_b kT)} e^{\Delta E_{g,Ge}(0)/kT}, \quad (4.4)$$

where we have made use of the fact that for Si, we can define a low-doping intrinsic carrier density for Si as

$$n_{io}^2 = N_C N_V e^{-E_{go}/kT}, \quad (4.5)$$

and we have defined an "effective density-of-states ratio" between SiGe and Si according to [3]

$$\gamma = \frac{(N_C N_V)_{SiGe}}{(N_C N_V)_{Si}} < 1. \quad (4.6)$$

Equation (4.4) can be inserted into the generalized Moll-Ross relation (4.1) to obtain

$$J_C = \frac{q\widetilde{D_nb}}{N_{ab}^-} \frac{\left(e^{qV_{BE}/kT} - 1\right) \widetilde{\gamma} n_{io}^2 e^{\Delta E_{gb}^{app}/kT} e^{\Delta E_{g,Ge}(0)/kT}}{\int_0^{W_b} e^{-[\Delta E_{g,Ge}(W_b)-\Delta E_{g,Ge}(0)](x/W_b kT)} dx}, \quad (4.7)$$

where we have defined $\widetilde{D}_{nb}$ and $\widetilde{\gamma}$ to be position-averaged quantities across the base profile, according to

$$\widetilde{D}_{nb} = \frac{\int_0^{W_b} \frac{dx}{n_{ib}^2(x)}}{\int_0^{W_b} \frac{dx}{D_{nb}(x)\, n_{ib}^2(x)}}. \qquad (4.8)$$

Using standard integration techniques, and defining

$$\Delta E_{g,Ge}(grade) = \Delta E_{g,Ge}(W_b) - \Delta E_{g,Ge}(0), \qquad (4.9)$$

we get

$$J_{C,SiGe} = \frac{q\widetilde{D}_{nb}}{N_{ab}^-} \frac{\left(e^{qV_{BE}/kT} - 1\right) \widetilde{\gamma} n_{io}^2 e^{\Delta E_{gb}^{app}/kT} e^{\Delta E_{g,Ge}(0)/kT}}{\frac{W_b kT}{\Delta E_{g,Ge}(grade)} \left\{1 - e^{-\Delta E_{g,Ge}(grade)/kT}\right\}}. \qquad (4.10)$$

Finally, by defining a minority electron diffusivity ratio between SiGe and Si as

$$\widetilde{\eta} = \frac{\left(\widetilde{D}_{nb}\right)_{SiGe}}{(D_{nb})_{Si}}, \qquad (4.11)$$

we obtain the final expression for $J_{C,SiGe}$ [1, 4]

$$\begin{aligned}J_{C,SiGe} &= \frac{q D_{nb}}{N_{ab}^- W_b} \left(e^{qV_{BE}/kT} - 1\right) n_{io}^2 e^{\Delta E_{gb}^{app}/kT} \\ &\quad \cdot \left\{\frac{\widetilde{\gamma}\widetilde{\eta} \Delta E_{g,Ge}(grade)/kT \; e^{\Delta E_{g,Ge}(0)/kT}}{1 - e^{-\Delta E_{g,Ge}(grade)/kT}}\right\}.\end{aligned} \qquad (4.12)$$

Within the confines of our assumptions stated above, this can be considered an exact result. As expected from our intuitive discussion of the band diagram, observe that J_C in a SiGe HBT depends exponentially on the EB boundary value of the Ge-induced band offset, and is linearly proportional to the Ge-induced bandgap grading factor. Given the nature of an exponential dependence, it is obvious that strong enhancement in J_C for fixed V_{BE} can be obtained for small amounts of introduced Ge, and that the ability to engineer the device characteristics to obtain a desired current gain is easily accomplished. Note as well that the thermal energy (kT) resides in the denominator of the Ge-induced band offsets. This is again expected from a simple consideration of how band edge effects generally couple to the device transport equations. The inherent temperature dependence in SiGe HBTs will be revisited in detail in Chapter 9.

If we consider a comparably constructed SiGe HBT and Si BJT with identical emitter contact technology, and further assume that the Ge profile on the EB side of the neutral base does not extend into the emitter enough to change the base current density, our experimental expectations are that for a comparably constructed SiGe HBT and Si BJT, the J_B should be comparable between the two devices, while J_C at fixed V_{BE} should be enhanced for the SiGe HBT. Figure 4.5 confirms this expectation experimentally. In this case, we note that the ratio of the current gain

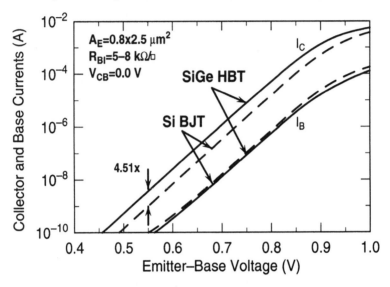

Figure 4.5 Comparison of current-voltage characteristics of a comparably constructed SiGe HBT and Si BJT.

between an identically constructed SiGe HBT and a Si BJT can be written as

$$\frac{\beta_{SiGe}}{\beta_{Si}} \cong \frac{J_{C,SiGe}}{J_{C,Si}}, \qquad (4.13)$$

and thus we can define a SiGe current gain enhancement factor as

$$\left.\frac{\beta_{SiGe}}{\beta_{Si}}\right|_{V_{BE}} \equiv \Xi = \left\{ \frac{\tilde{\gamma}\tilde{\eta} \Delta E_{g,Ge}(grade)/kT \, e^{\Delta E_{g,Ge}(0)/kT}}{1 - e^{-\Delta E_{g,Ge}(grade)/kT}} \right\}. \qquad (4.14)$$

Typical experimental results for Ξ are shown for a comparably constructed SiGe HBT and Si BJT in Figure 4.6. Theoretical calculations using (4.14) as a function of Ge profile shape are shown in Figure 4.7 at 300 K and 77 K (the integrated Ge content is held fixed, and the Ge profile varies from a 10% triangular (linearly graded) to a 5% box (constant) Ge profile) [5].

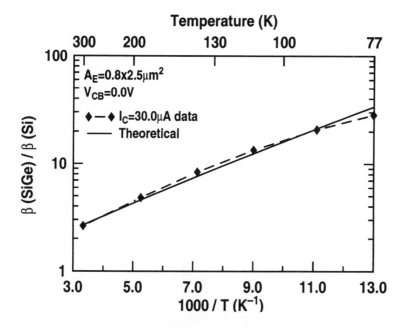

Figure 4.6 Measured and calculated current gain ratio as a function of reciprocal temperature for a comparably constructed SiGe HBT and Si BJT.

4.2.2 Relevant Approximations

Two physically relevant approximations can be made to obtain additional insight. First, we can assume that $\Delta E_{g,Ge}(grade) \gg kT$. This approximation can be termed the "strong Ge grading" scenario. In this case (4.12) reduces to

$$J_{C,SiGe} \simeq \frac{q D_{nb}}{N_{ab}^- W_b} \left(e^{qV_{BE}/kT} - 1 \right) n_{io}^2 e^{\Delta E_{gb}^{app}/kT}$$
$$\cdot \left\{ \tilde{\gamma}\tilde{\eta} \frac{\Delta E_{g,Ge}(grade)}{kT} e^{\Delta E_{g,Ge}(0)/kT} \right\}. \quad (4.15)$$

Note, however, that care should be exercised in applying this approximation. To check its validity for a realistic profile, assume that we have a 0% to 15% triangular Ge profile in a SiGe HBT operating at 300 K. Taking a band offset of roughly 75 meV per 10% Ge, we find that $\Delta E_{g,Ge}(grade)/kT = 4.3$ compared to unity, a reasonable but not overly compelling approximation. Clearly, however, as the temperature drops, the validity of this approximation improves rapidly as kT decreases. For example, in the case above, a 0% to 15% triangular Ge profile yields $\Delta E_{g,Ge}(grade)/kT = 17.0$ at 77 K.

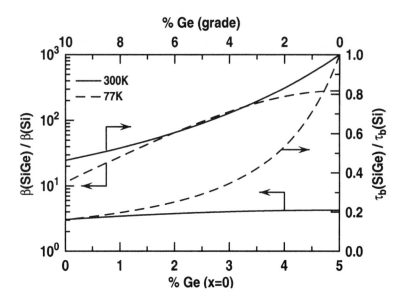

Figure 4.7 Theoretical calculations of the current gain ratio Ξ as a function of Ge profile shape.

In addition to the strongly graded profile, we also can define an approximation for a "weak Ge grading," that would be valid, for instance, in the case of a Ge box profile. In this case, $\Delta E_{g,Ge}(grade) \ll kT$. By expanding the exponential of the Ge grading factor in the denominator of (4.12) in a Taylor series and canceling terms we obtain

$$J_{C,SiGe} \simeq \frac{q D_{nb}}{N_{ab}^- W_b} \left(e^{qV_{BE}/kT} - 1 \right) n_{io}^2 e^{\Delta E_{gb}^{app}/kT} \left\{ \widetilde{\gamma} \widetilde{\eta} e^{\Delta E_{g,Ge}(0)/kT} \right\}. \quad (4.16)$$

4.2.3 Nonconstant Base Doping

In general there is no closed-form solution for $J_{C,SiGe}$ if both the base doping and the Ge profile are position-dependent. In the case of nonconstant base doping, we can, however, define an "effective Ge-induced bandgap reduction" ($\Delta E_{g,Ge}(\text{eff})$) according to

$$n_{ib}^2(x) = \widetilde{\gamma} n_{io}^2 e^{\widetilde{\Delta E_{gb}^{app}}/kT} e^{\widetilde{\Delta E_{g,Ge}}(\text{eff})/kT}, \quad (4.17)$$

where the tilde again refers to a position-averaged quantity. Physically, $\Delta E_{g,Ge}(\text{eff})$ can be thought of as an average Ge band offset across the neutral base. We then can define the intrinsic base sheet resistance (or pinched base sheet resistance) as

$$R_{bi} = \left\{ \int_0^{W_b} q\, \mu_{pb}(x)\, N_{ab}^-(x)\, dx \right\}^{-1}. \qquad (4.18)$$

Observe that R_{bi} is the integral of the neutral base charge, and is an important bipolar parameter because it is directly measurable from an independent on-wafer test structure. Thus, for a SiGe HBT with an arbitrary base doping profile and an effective Ge profile, we find an effective current gain enhancement factor to be

$$\Xi_{\text{eff}} = \frac{J_{C,SiGe}(\text{eff})}{J_{C,Si}} = \frac{\widetilde{\mu}_{nb,SiGe}\, \widetilde{\mu}_{pb,SiGe}}{\widetilde{\mu}_{nb,Si}\, \widetilde{\mu}_{pb,Si}} \frac{R_{bi,SiGe}}{R_{bi,Si}}$$
$$\cdot \left\{ e^{(\widetilde{\Delta E}_{gb,SiGe}^{app} - \widetilde{\Delta E}_{gb,Si}^{app})/kT}\, e^{\widetilde{\Delta E}_{g,Ge}(\text{eff})/kT} \right\}, \qquad (4.19)$$

where we have allowed for the possibility of a difference in base doping profile between the SiGe HBT and the Si BJT. Equation (4.19) is useful because it allows one to compare electrical data from fabricated SiGe HBTs and Si BJTs to infer information about the Ge profile. As a simple illustration, if, for instance, $R_{bi,SiGe} \cong R_{bi,Si}$, then (4.19) becomes

$$\Xi_{\text{eff}}(T) \simeq e^{\widetilde{\Delta E}_{g,Ge}(\text{eff})/kT}. \qquad (4.20)$$

Hence, if we measure $J_{C,SiGe}$ and $J_{C,Si}$ at fixed V_{BE} as a function of temperature for two identical emitter geometries, a plot of $log\, \Xi_{\text{eff}}(T)$ versus $1000/T$ will be linear, and thus will allow an experimental determination of $\Delta E_{g,Ge}(\text{eff})$ [6]. By comparison of Ξ_{eff} to Ξ for a specific Ge profile shape, we can electrically infer information about the Ge profile shape from the collector current data. For instance, for a triangular Ge profile, we find

$$\widetilde{\Delta E}_{g,Ge}(\text{eff}) \simeq \widetilde{\Delta E}_{g,Ge}(0) + kT\, ln\left\{ \Delta E_{g,Ge}(grade)/kT \right\}. \qquad (4.21)$$

4.2.4 Other SiGe Profile Shapes

The analysis above holds for a range of Ge profiles between the triangular (linearly graded Ge) and box (constant Ge) profiles. There also exists, however, a class of technologically important Ge profiles that can be considered hybrid combinations of the triangular and box Ge profiles, which we will call Ge trapezoids, as depicted schematically in Figure 4.8.

In this case, one takes a linearly graded profile and truncates the grading at some intermediate position x_T in the neutral base, and the Ge content is held constant from x_T to W_b, and is then ramped down to zero as usual. At constant Ge

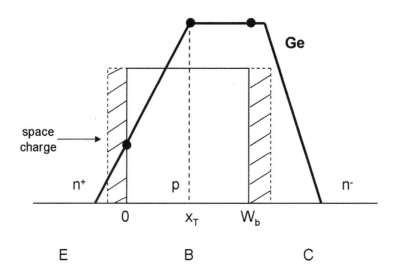

Figure 4.8 Schematic representation of the hybrid Ge trapezoidal profile.

stability and for fixed Ge profile width, this Ge trapezoidal profile design approach allows one to induce higher Ge grading across the more heavily doped EB end of the base, thereby maintaining good dynamic response, while using lower peak Ge content. The region of constant Ge in the neutral base, at least in principle, does not degrade *ac* performance since the CB side of the neutral base typically will have a doping-gradient-induced drift field in addition to the Ge-grading-induced drift field, which will aid electron transport. For the Ge trapezoid, one can also derive collector current density expressions in the presence of constant base doping. In this case, the Ge-induced band offsets can be written as [6]

$$\Delta E_{g,Ge}(x) = \begin{cases} \Delta E_{g,Ge}(0) + \Delta E_{g,Ge}(grade)\left(\frac{x}{x_T}\right) &, 0 \leq x \leq x_T \\ \Delta E_{g,Ge}(W_b) &, x_T \leq x \leq W_b \end{cases} \quad (4.22)$$

and the intrinsic carrier density is then

$$n_{ib}^2(x) = \begin{cases} \gamma n_{io}^2 \, e^{\Delta E_{gb}^{app}/kT} \, e^{\Delta E_{g,Ge}(0)/kT} \, e^{[\Delta E_{g,Ge}(grade)x/x_T]/kT} &, 0 \leq x \leq x_T \\ \gamma n_{io}^2 \, e^{\Delta E_{gb}^{app}/kT} \, e^{\Delta E_{g,Ge}(W_b)/kT} &, x_T \leq x \leq W_b \, . \end{cases} \quad (4.23)$$

If we split the Gummel integral in the Moll-Ross relation into two pieces, integrate, and compare the result with the Si BJT, we obtain [5]

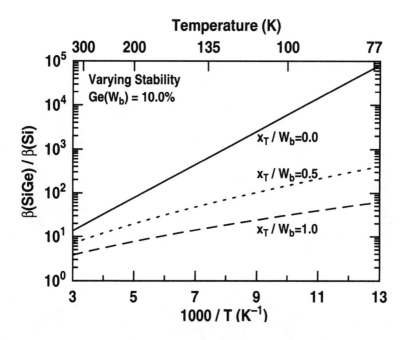

Figure 4.9 Current gain ratio as a function of reciprocal temperature for varying ξ values. Note that the integrated Ge content is not fixed in this case.

$$\frac{J_{C,SiGe}}{J_{C,Si}} = \frac{\widetilde{\gamma\eta}\, e^{\Delta E_{g,Ge}(0)/kT}}{\frac{\xi kT}{\Delta E_{g,Ge}(grade)} + \left\{1 - \xi\left(1 + \frac{kT}{\Delta E_{g,Ge}(grade)}\right)\right\} e^{-\Delta E_{g,Ge}(grade)/kT}}, \quad (4.24)$$

where we have defined $\xi = x_T/W_b < 1$ to be the normalized trapezoidal intermediate boundary point. In this case, $\xi = 0$ corresponds to the pure Ge box profile, and $\xi = 1$ corresponds to the pure triangular Ge profile. Again, assuming that the current gain ratio between a SiGe HBT and a Si BJT is simply equal to the ratio of J_C between the devices, we can plot current gain as a function of the reciprocal temperature for varying ξ values, as shown in Figure 4.9 (here we fix Ge content at $x = W_b$ to be 10%). As expected, the Ge trapezoid result lies between that of the pure Ge box and the pure Ge triangle profiles.

Note that by simply letting ξ vary, the integrated Ge content for the profiles also will change, and hence the stability of the Ge film will decrease as ξ decreases (we have assumed that the Ge content at $x = W_b$ is fixed). Alternatively, one can fix the Ge content at $x = 0$ and then allow the Ge content at $x = W_b$ to vary, such that the total integrated Ge content (i.e., stability) remains constant. The gain enhancement factor in this case (for a Ge content = 2% at $x = 0$) is shown in Figure 4.10.

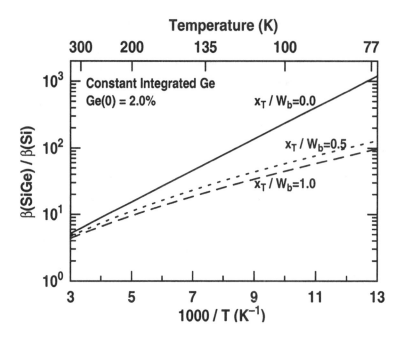

Figure 4.10 Current gain ratio as a function of reciprocal temperature for varying ξ values. Note that the integrated Ge content is held fixed in this case.

4.2.5 Implications and Optimization Issues for β

Based on the analysis above, we can make several observations regarding the effects of Ge on the collector current in a SiGe HBT:

- The presence of any Ge, in whatever shape, in the base of a bipolar transistor will enhance J_C at fixed V_{BE} over a comparably constructed Si BJT.

- The J_C enhancement depends exponentially on the EB boundary value of Ge-induced band offset, and linearly on the Ge grading across the base. This observed dependence will play a role in understanding the best approach to profile optimization.

- In light of that, for two Ge profiles of constant stability, a box Ge profile is better for current gain enhancement than a triangular Ge profile, everything else being equal.

- The Ge-induced J_C enhancement is thermally activated (exponentially dependent on reciprocal temperature), and thus cooling will produce a strong magnification of the enhancement.

4.3 Output Conductance

The dynamic output conductance ($\partial I_C/\partial V_{CE}$ at fixed V_{BE}) of a transistor is a critical design parameter for many analog circuits. Intuitively, from the transistor output characteristics, we would like the output current to be independent of the output voltage, and thus ideally have zero output conductance (infinite output resistance). In practice, of course, this is never the case. As we increase V_{CB}, we deplete the neutral base from the backside, thus moving the neutral base boundary value ($x = W_b$) inward. Since W_b determines the minority carrier density on the CB side of the neutral base, the slope of the minority electron profile, and hence the collector current, necessarily rises [7]. Thus, for finite base doping, I_C must increase as V_{CB} increases, giving a finite output conductance. This mechanism is known as the "Early effect," and for experimental convenience, we define the Early voltage (V_A) as

$$V_A = J_C(0) \left\{ \left.\frac{\partial J_C}{\partial V_{CB}}\right|_{V_{BE}} \right\}^{-1} - V_{BE} \simeq J_C(0) \left\{ \left.\frac{\partial J_c}{\partial W_b}\right|_{V_{BE}} \frac{\partial W_b}{\partial V_{CB}} \right\}^{-1}, \quad (4.25)$$

where $J_C(0) = J_C(V_{CB} = 0V)$. The Early voltage is a simple and convenient measure of the change in output conductance with changing V_{CB}. A schematic representation of the Early effect, and the definition of V_A, is shown in Figure 4.11 and Figure 4.12. As will be seen below, simultaneously maintaining high current gain, high frequency response, and high V_A is particularly challenging in a Si BJT.

4.3.1 V_A Trade-offs in Si BJTs

For a Si BJT, we can use (4.1) together with (4.25) to obtain

$$V_{A,Si} = \frac{\int_0^{W_b} p_b(x)\,dx}{p_b(W_b)\left\{\frac{\partial W_b}{\partial V_{CB}}\right\}} = \frac{Q_b(0)}{C_{cb}}, \quad (4.26)$$

where $Q_b(0)$ is the total base charge at $V_{CB} = 0V$, C_{cb} is the collector base depletion capacitance, and we have assumed V_{BE} is negligible compared to V_{CB}. Note that C_{cb} is dependent on both the ionized collector doping (N_{dc}^+) and the ionized base doping (N_{ab}^-). To estimate the sensitivity of V_A on N_{dc}^+ and N_{ab}^-, we can consider a Si BJT with constant base and collector doping profiles. In this case, we

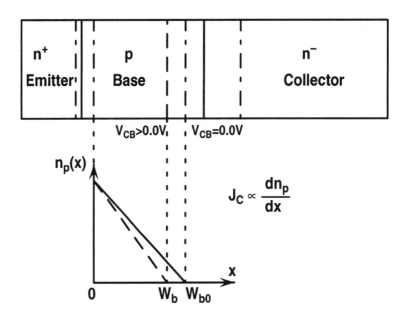

Figure 4.11 Schematic representation of the Early effect in a bipolar transistor.

can write

$$V_{A,Si} = -W_b(0) \left\{ \left. \frac{\partial W_b}{\partial V_{CB}} \right|_{V_{BE}} \right\}^{-1}, \quad (4.27)$$

where $W_b(0)$ is the neutral base width at $V_{CB} = 0$ V. The dependence of W_b on voltage and doping can be obtained from [8]

$$W_b \simeq W_m - \sqrt{\left(\frac{2\epsilon}{q}\right)(\phi_{bi} + V_{CB}) \left\{ \frac{N_{dc}^+}{N_{ab}^-(N_{ab}^- + N_{dc}^+)} \right\}}, \quad (4.28)$$

where W_m is the metallurgical base width, and ϕ_{bi} is the CB junction built-in voltage. Using (4.27) and (4.28) we can calculate V_A as a function of doping, as shown in Figure 4.13 ($W_m = 100$ nm, and $\Delta V_{CB} = 1.0$ V). As can be seen, if we fix N_{ab}^-, increasing N_{dc}^+ degrades V_A, physically because the amount of backside neutral base depletion per unit bias is enhanced for a higher collector doping. If we instead fix N_{dc}^+, increasing N_{ab}^- rapidly increases V_A, which makes intuitive sense given that the base is much more difficult to deplete as the base doping increases, everything else being equal. In real Si BJT designs, a given device generally has a specified collector-to-emitter breakdown voltage (BV_{CEO}) determined by the circuit requirements. To first order, this BV_{CEO} sets the collector doping level. While this may appear to favor achieving a high V_A, we must recall that the current gain is reciprocally related to the integrated base charge (refer to (4.1)). Hence, increas-

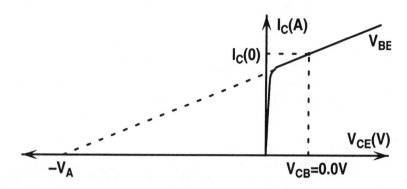

Figure 4.12 Schematic representation impact of the Early effect on the transistor output characteristics.

ing N_{ab}^- to improve V_A results in a strong decrease in β. In addition, for a Si BJT, for a fixed base width, increasing N_{ab}^- will degrade the cutoff frequency of the transistor (due to the reduction in the minority electron mobility). We might imagine we can then increase N_{dc}^+ to buy back the ac performance lost, but as can be seen in Figure 4.13, this in turn degrades V_A. This catch-22 represents a fundamental problem in Si BJT design: it is inherently difficult to simultaneously obtain high V_A, high β, and high f_T. In practice one must then find some compromise design for V_A, β, and f_T, and in the process the performance capabilities of a given analog circuit suffers. Intuitively, this Si BJT design constraint occurs because β and V_A are both coupled to the base doping profile. The introduction of Ge into the base region of a Si BJT can favorably alter this constraint by effectively decoupling β and V_A from the base doping profile.

4.3.2 V_A in SiGe HBTs

To formally obtain V_A in a SiGe HBT, we begin by combining (4.1) with (4.25) to obtain [9]

$$V_{A,SiGe} = \frac{-\int_0^{W_b} \frac{p_b(x)\,dx}{D_{nb}(x)\,n_{ib}^2(x)}}{\frac{\partial}{\partial V_{CB}} \left\{ \int_0^{W_b} \frac{p_b(x)\,dx}{D_{nb}(x)\,n_{ib}^2(x)} \right\}}, \qquad (4.29)$$

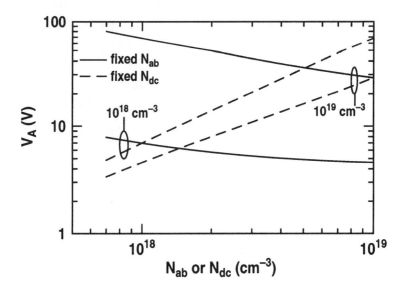

Figure 4.13 Dependence of V_A on both N_{dc}^+ and N_{ab}^- for a Si BJT.

from which we can write

$$V_{A,SiGe} = \left\{ \frac{-D_{nb}(W_b)\, n_{ib}^2(W_b)}{p_b(W_b)} \int_0^{W_b} \frac{p_b(x)\, dx}{D_{nb}(x)\, n_{ib}^2(x)} \right\} \left[\frac{\partial W_b}{\partial V_{CB}} \right]^{-1}. \quad (4.30)$$

Comparing (4.26) and (4.30) we can see that the fundamental difference between V_A in a SiGe HBT and a Si BJT arises from the variation of n_{ib}^2 as a function of position (the variation of W_b with V_{CB} is, to first order, similar between SiGe and Si devices). Observe that if n_{ib} is position-independent (i.e., for a box Ge profile), then (4.30) collapses (4.26) and there is no V_A enhancement due to Ge (albeit there will obviously still be a strong β enhancement). On the other hand, if n_{ib} is position-dependent (i.e., in a linearly graded Ge profile), V_A will depend exponentially on the difference in bandgap between $x = W_b$ and that region in the base where n_{ib} is smallest. That is, the base profile is effectively "weighted" by the increasing Ge content on the collector side of the neutral base, making it harder to deplete the neutral base for a given applied V_{CB}, all else being equal, effectively increasing the Early voltage of the transistor.

For a linearly graded Ge profile, we can use (4.4) and (4.30) to obtain the ratio of V_A between a comparably constructed SiGe HBT and Si BJT (Θ) to be [10]

$$\left.\frac{V_{A,SiGe}}{V_{A,Si}}\right|_{V_{BE}} \equiv \Theta \simeq e^{\Delta E_{g,Ge}(grade)/kT} \left[\frac{1 - e^{-\Delta E_{g,Ge}(grade)/kT}}{\Delta E_{g,Ge}(grade)/kT}\right]. \quad (4.31)$$

The important result is that the V_A ratio between a SiGe HBT and a Si BJT is an exponential function of Ge-induced bandgap grading across the neutral base. Typical experimental results for Θ are shown for a comparably constructed SiGe HBT and Si BJT in Figure 4.14. Theoretical calculations using (4.31) as a function

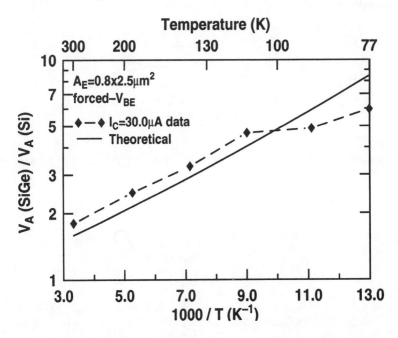

Figure 4.14 Measured and calculated Early voltage ratio as a function of reciprocal temperature for a comparably constructed SiGe HBT and Si BJT.

of Ge profile shape are shown in Figure 4.15 at 300 K and 77 K (the integrated Ge content is held fixed, and the Ge profile varies from a 10% triangular (linearly graded) to a 5% box (constant) Ge profile) [5].

4.3.3 Relevant Approximations

In a similar manner to that of J_C, two physically relevant approximations can be made to obtain additional insight. First, we can make the assumption that

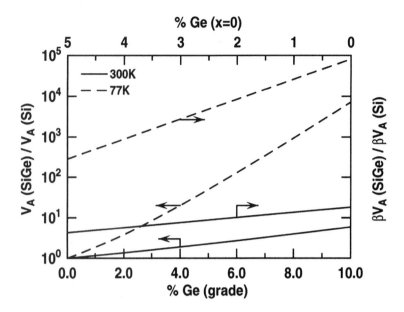

Figure 4.15 Theoretical calculations of the V_A ratio and βV_A ratio as a function of Ge profile shape.

$\Delta E_{g,Ge}(grade) \gg kT$. This approximation can be termed the "strong Ge grading" scenario. In this case (4.31) reduces to

$$\left.\frac{V_{A,SiGe}}{V_{A,Si}}\right|_{V_{BE}} \simeq \frac{e^{\Delta E_{g,Ge}(grade)/kT}}{\Delta E_{g,Ge}(grade)/kT}. \tag{4.32}$$

As discussed above, care should be exercised in applying this approximation. To check its validity for a realistic profile, assume that we have a 0% to 15% triangular Ge profile in a SiGe HBT operating at 300 K. Taking a band offset of roughly 75 meV per 10% Ge, we find that $\Delta E_{g,Ge}(grade)/kT = 4.3$ compared to unity, a reasonable but not overly compelling approximation. Clearly, however, as the temperature drops, the validity of this approximation improves rapidly as kT decreases. For example, in the case above, a 0% to 15% triangular Ge profile yields $\Delta E_{g,Ge}(grade)/kT = 17.0$ at 77 K.

In addition to the strongly graded profile, we also can define an approximation for a "weak Ge grading," which would be valid, for instance, in the case of a Ge box profile. In this case, $\Delta E_{g,Ge}(grade) \ll kT$. By expanding the exponentials of the Ge grading factor in (4.31) in a Taylor series and canceling terms we see that $\Theta = 1$, and thus we have *no* enhancement in V_A over a Si BJT, despite the presence of Ge in the base.

4.3.4 Current Gain – Early Voltage Product

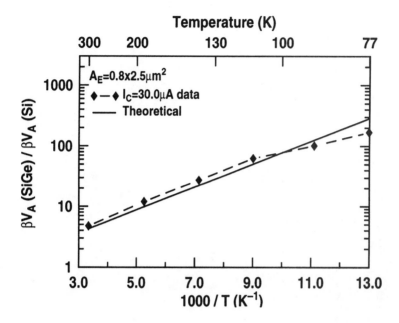

Figure 4.16 Measured and calculated ratio of the current gain – Early voltage product ratio as a function of reciprocal temperature for a comparably doped SiGe HBT and Si BJT.

In light of the discussion above regarding the inherent difficulties in obtaining high V_A simultaneously with high β, one conventionally defines a figure-of-merit for analog circuit design: the so-called "βV_A" product. In a conventional Si BJT, a comparison of (4.1) and (4.26) shows that βV_A is to first-order independent of the base profile, and is thus not favorably impacted by conventional technology scaling, as for instance, the transistor frequency response would be. For a SiGe HBT, however, both β and V_A are decoupled from the base profile, and can be independently tuned by changing the Ge profile shape. By combining (4.14) and (4.31) we find that the ratio of βV_A between a comparably constructed SiGe HBT and Si BJT can be written as [9]

$$\frac{\beta V_{A,SiGe}}{\beta V_{A,Si}} = \widetilde{\gamma \eta} \, e^{\Delta E_{g,Ge}(0)/kT} e^{\Delta E_{g,Ge}(grade)/kT}. \tag{4.33}$$

Typical experimental results for the βV_A ratio for a comparably constructed SiGe HBT and Si BJT are shown in Figure 4.16.

Observe that βV_A is a thermally activated function of *both* the Ge-induced band offset at the EB junction and the Ge-induced grading across the neutral base. As can be seen in Figure 4.16, βV_A in a SiGe HBT is significantly improved over a comparably designed Si BJT, regardless of the Ge profile shape chosen, although the triangular Ge profile remains the profile shape of choice for both V_A and βV_A optimization. Due to their thermally activated nature, both V_A and βV_A are strongly enhanced with cooling, yielding enormous values ($\beta V_A > 10^4$) at 77 K for a 10% Ge triangular profile [10].

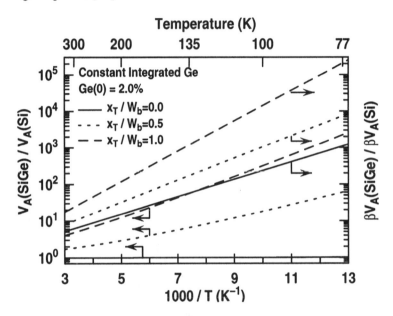

Figure 4.17 Early voltage and current gain – Early voltage product as a function of reciprocal temperature for varying ξ values. Note that the integrated Ge content is held fixed in this case.

4.3.5 Other SiGe Profile Shapes

For the Ge trapezoid discussed above (refer to Figure 4.8), one can also derive V_A expressions in the presence of constant base doping. We know that V_A depends on the ratio of the collector current density and the slope of the collector current density with respect to V_{CB} (i.e., (4.25)), which can be expressed generally as

$$\frac{\partial J_{C,SiGe}}{\partial V_{CB}} = \left\{ \Xi \left(\frac{\partial J_{C,SiGe}}{\partial W_b} \right) + J_{C,Si} \left(\frac{\partial \Xi}{\partial W_b} \right) \right\} \left(\frac{\partial W_b}{\partial V_{CB}} \right), \quad (4.34)$$

Static Characteristics

where $\xi = x_T/W_b < 1$, and Ξ is the current gain ratio (4.14). This can be then rewritten as

$$\frac{\partial J_{C,SiGe}}{\partial V_{CB}} = \Xi \left(\frac{\partial J_{C,Si}}{\partial V_{CB}}\right) \left[1 + J_{C,Si} \left(\frac{\partial J_{C,Si}}{\partial W_b}\right)^{-1} \left(\frac{1}{\Xi}\frac{\partial \Xi}{\partial W_b}\right)\right]. \quad (4.35)$$

From (4.24), we can express the variation in the current gain ratio on V_{CB} as

$$\frac{1}{\Xi}\frac{\partial \Xi}{\partial W_b} = \left[\frac{e^{\Delta E_{g,Ge}(grade)/kT} - 1 - \Delta E_{g,Ge}(grade)/kT}{\xi\left(e^{\Delta E_{g,Ge}(grade)/kT} - 1\right) + (1-\xi)\Delta E_{g,Ge}(grade)/kT}\right]$$
$$\cdot \left(\frac{\xi^2}{x_T}\right) \quad (4.36)$$

and finally obtain the expression of the V_A ratio for a general SiGe trapezoidal profile [5]

$$\left.\frac{V_{A,SiGe}}{V_{A,Si}}\right|_{V_{BE}} = 1 - \xi + \frac{\xi\left(e^{\Delta E_{g,Ge}(grade)/kT} - 1\right)}{\Delta E_{g,Ge}(grade)/kT}. \quad (4.37)$$

Again, applying the limits for the triangular and box profiles we can retrieve the standard expressions for V_A (4.31). Note that a similar expression can be trivially obtained for βV_A by combining (4.37) and (4.24). As before, one can fix the Ge content at $x = 0$ and then allow the Ge content at $x = W_b$ to vary, such that the total integrated Ge content (i.e., stability) remains constant. The V_A and βV_A enhancement factor in this case (for a Ge content = 2% at $x = 0$) is shown in Figure 4.17. As expected, the Ge trapezoid result lies between that of the pure Ge box and the pure Ge triangle profiles.

4.3.6 Implications and Optimization Issues for V_A and βV_A

Based on the analysis above, we can make several observations regarding the effects of Ge on both the Early voltage and current gain – Early voltage product in SiGe HBTs:

- Unlike for J_C, only the presence of a larger Ge content at the CB side of the neutral base than at the EB side of the neutral base (i.e., finite Ge grading) will enhance V_A at fixed V_{BE} over a comparably constructed Si BJT.

- This V_A enhancement depends exponentially on the Ge grading across the base. This observed dependence will play a role in understanding the best approach to profile optimization, generally favoring strongly graded (triangular) profiles.

- In light of this, for two Ge profiles of constant stability, a triangular Ge profile is better for Early voltage enhancement than a box Ge profile is, everything else being equal.

- The Ge-induced V_A enhancement is thermally activated (exponentially dependent on reciprocal temperature), and thus cooling will produce a strong magnification of the enhancement.

- Given that β and V_A have the exact opposite dependence on Ge grading and EB Ge offset, the βV_A product in a SiGe HBT enjoys an ideal win-win scenario. Putting any Ge into the base region of a device will exponentially enhance this key analog figure-of-merit, a highly favorable scenario given the discussion above of inherent difficulties of achieving high βV_A in a Si BJT.

- A reasonable compromise Ge profile design that balances the dc optimization needs of β, V_A, and βV_A would be a Ge trapezoid, with a small (e.g., 3–4%) Ge content at the EB junction, and a larger (e.g., 10–15% Ge content at the CB junction (i.e., finite Ge grading).

4.4 Equivalent Circuit Models

4.4.1 Basic Ebers-Moll Model

Historically speaking, the first and most basic equivalent circuit model for a bipolar transistor is the Ebers-Moll model shown in Figure 4.18. A fundamental assumption in this model is that the overall transistor operation can be viewed as a superposition of both the forward and the reverse (inverse) mode operation. Here, I_F represents the total emitter current for forward operation, and $\alpha_F I_F$ represents the electron current component of I_F, or the forward collector current. The parameter α_F is the forward common-base current gain. Similarly, I_R is the total "emitter current" for inverse operation, and $\alpha_R I_R$ represents the electron current component of I_R. The parameter α_R is the inverse common-base current gain. Both I_F and I_R have an exponential $I - V$ functional form

$$I_F = I_{F0} \left(e^{qV_{BE}/kT} - 1 \right)$$
$$I_R = I_{R0} \left(e^{qV_{BC}/kT} - 1 \right). \qquad (4.38)$$

Here I_{F0} and I_{R0} are the saturation currents of the forward and inverse emitter currents.

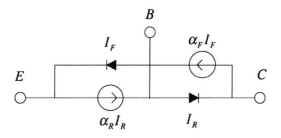

Figure 4.18 The basic Ebers-Moll model for a bipolar transistor.

4.4.2 Transport Version

Another equivalent circuit model for the bipolar transistor, which better describes carrier transport, is shown in Figure 4.19(a), which is also known as the "transport version" of the Ebers-Moll model. The collector current instead of the emitter current is chosen as the reference current. For forward-mode operation, the collector current is transported from emitter to collector, while the base current is injected into the emitter,

$$I_{CF} = I_S \left(e^{qV_{BE}/kT} - 1\right)$$

$$I_{BF} = \frac{I_{CC}}{\beta_F}, \qquad (4.39)$$

where β_F is the current gain (β) for forward operation. Similarly, for inverse operation, we have

$$I_{CR} = I_S \left(e^{qV_{BC}/kT} - 1\right)$$

$$I_{BR} = \frac{I_{CR}}{\beta_R}, \qquad (4.40)$$

where β_R is the inverse β. Note that the saturation current I_S is identical for the collector currents for both the forward and inverse mode operation. Figure 4.19(a) can be redrawn as Figure 4.19(b), which is better suited for common-emitter circuit analysis.

It can be easily shown that I_S is related to I_{F0} and I_{R0} by

$$I_S = \alpha_F I_{F0} = \alpha_R I_{R0}, \qquad (4.41)$$

and is also known as the "reciprocity" property of the bipolar transistor. Reciprocity can be easily understood to be the result of minority carrier transport. The minority

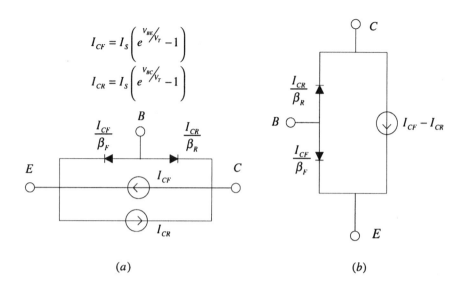

Figure 4.19 (a) Transport version of the Ebers-Moll model, and (b) common-emitter representation of the transport version Ebers-Moll model.

carrier (electron) current in the base is determined by the properties of the base region, which is shared by both the forward and inverse mode transistor operation. In general, reciprocity holds only approximately for SiGe HBTs.

4.4.3 Small-Signal Equivalent Circuit Model

The large-signal equivalent model shown in Figure 4.19(b) can be linearized for a given *dc* operating point. The resulting circuit is referred to as the "small-signal" equivalent circuit. Under forward-mode operation ($V_{BE} > 0$ and $V_{CB} > 0$), this model reduces to the well-known linear small-signal hybrid-π model shown in Figure 4.20(a). The transport current source I_{CF} becomes the transconductance current source $g_m v_{be}$, with $g_m = qI_C/kT$. The forward EB diode becomes the g_{be} conductance ($g_{be} = g_m/\beta$), and $1/g_{be}$ is popularly known as r_π. The reverse-biased CB diode becomes an open circuit, and at high currents, the voltage drops across the parasitic resistances can no longer be neglected. Their effects can be included by adding appropriate terminal resistances. The increase of I_C with increasing V_{CB} due to the Early effect can be accounted for by adding r_o in parallel with the $g_m v_{b'e'}$ current source. The final equivalent circuit is shown in Figure 4.20(b).

Static Characteristics

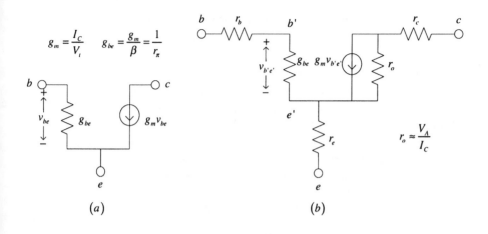

Figure 4.20 (a) Low-frequency small-signal equivalent circuit for an ideal transistor, and (b) low-frequency small-signal equivalent circuit including parasitic resistances and Early effect.

4.5 Avalanche Multiplication

To achieve the high f_T potential offered by smaller base and emitter transit times in SiGe HBTs, the collector current density J_C must be sufficiently high so that the charging time associated with depletion and parasitic capacitances is smaller than the sum of base and emitter transit time ($\tau_b + \tau_e$). Hence, the smaller the transit time, the higher J_C needs to be. To suppress Kirk effect at high J_C, SiGe HBTs are typically designed with heavily doped implanted collectors. High collector doping leads to high electric field in the (typically) reverse-biased CB junction, and thus a high rate of impact ionization. For practical SiGe HBT circuits operating at either high collector current density (J_C) or high collector-base bias (V_{CB}), avalanche multiplication is an important effect that must be accurately measured and modeled. In digital applications, the avalanche multiplication factor ($M - 1$) determines the breakdown voltage as well as the base current reversal voltage, which in turn determines the maximum useful V_{CE} for stable circuit operation.

4.5.1 Carrier Transport and Terminal Currents

Under reverse bias, the electric field in the space-charge region of the CB junction is large. Electrons injected from the emitter drift to the collector through the CB space-charge region. For a sufficiently high electric field, electrons can gain enough energy from the electric field to create an electron-hole pair upon impact with the lattice (simple analysis shows that the minimum threshold energy for impact ionization is $1.5 \times E_g$). This carrier generation process is known as "impact ionization." Electrons and holes generated by impact ionization can subsequently acquire energy from the strong electric field, and create additional electron-hole pairs by further impact ionization. This process of multiplicative impact ionization is known as "avalanche multiplication." The net effect is that the electron current leaving the CB space-charge region (the I_C observed at the collector) is larger than that entering the CB space-charge region (the I_C that would be observed without avalanche multiplication). The ratio of the two currents is known as the avalanche multiplication factor M

$$M = \frac{I_{n,out}}{I_{n,in}}, \qquad (4.42)$$

where $I_{n,in}$ and $I_{n,out}$ are the electron currents going into and out of the CB space-charge region. In practice, $M - 1$ instead of M is often used simply because $M - 1$ better describes the yield or efficiency of the resulting collector current increase. The net increase of electron current due to impact ionization is simply $(M-1)I_{n,in}$. Because electrons and holes are always generated in pairs, an equal amount of hole current is generated in this process, and flows into the p-type base. The net base current observed at the base terminal is thus reduced by $(M-1)I_{n,in}$

$$I_B = I_{p,e} - (M - 1)I_{n,in}, \qquad (4.43)$$

where $I_{p,e}$ is the base current component due to hole injection into the emitter. We have neglected the I_B component due to neutral base recombination, which is far less than that due to hole injection into emitter in modern SiGe HBTs [11]. The CB junction reverse leakage I_{CBO} is also neglected, since it is much smaller than $I_{p,e}$ under normal operation. Note that I_{CBO}, however, cannot be neglected in open-base breakdown voltage analysis (i.e., BV_{CEO}) when it is the only initiating current for avalanche multiplication. The relationships between the various electron and hole current components in the presence of avalanche multiplication in a SiGe HBT are illustrated in Figure 4.21 for normal operation.

4.5.2 Forced-V_{BE} Measurement of $M - 1$

Consider a transistor operated with a fixed V_{BE} and a variable V_{CB}. At $V_{CB} = 0$, I_B is dominated by hole injection into the emitter in modern SiGe HBTs ($I_{p,e}$). At higher V_{CB}, $I_{p,e}$ remains constant, while the M factor increases, causing a decrease of I_B. The net I_C increase due to avalanche multiplication $(M - 1)I_{n,in}$ can thus

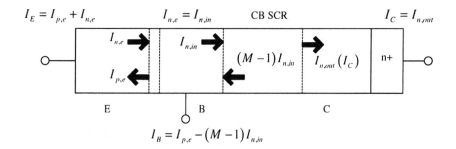

Figure 4.21 The avalanche multiplication process in a bipolar transistor under normal operation.

be obtained from the I_B difference

$$(M-1)I_{n,in} = \Delta I_B$$
$$\Delta I_B \equiv I_B(V_{CB} = 0) - I_B(V_{CB}), \qquad (4.44)$$

because the avalanche multiplication induces an equal number of electrons and holes. The electron current leaving the CB space-charge region is the measured I_C. The electron current entering the CB space-charge region is thus

$$I_{n,in} = I_C(V_{CB}) - \Delta I_B. \qquad (4.45)$$

Using (4.44), the $M-1$ factor can be expressed as:

$$M - 1 = \frac{\Delta I_B}{I_C(V_{CB}) - \Delta I_B}. \qquad (4.46)$$

Importantly, $M-1$ can be conveniently measured using (4.46). A common-base biasing configuration is naturally suited for sweeping V_{CB} at fixed V_{BE}, as shown in Figure 4.22. Note that $I_{n,in}$ is the electron current injected into the CB space-charge region, which increases with V_{CB} because of base width modulation (Early effect). In modern bipolar transistors, a single parameter, the Early voltage (V_A), is often insufficient to describe the I_C change due to Early effect [12]. In this case, we can then define an "Early effect factor" F_{Early} [13]

$$\begin{aligned}F_{Early} &= \frac{\text{Electron current injected into the CB SCR}}{\text{Electron current injected into the CB SCR at } V_{CB} = 0}\\ &= \frac{I_{n,in}(V_{CB})}{I_{n,in}(V_{CB} = 0)}\\ &= \frac{I_C(V_{CB}) - \Delta I_B}{I_C(V_{CB} = 0)},\end{aligned} \qquad (4.47)$$

where (4.45) is used for the electron current injected into the CB space-charge region.

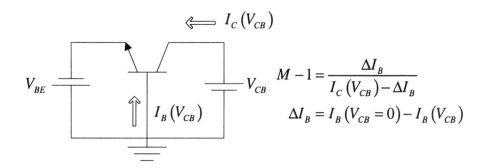

Figure 4.22 Forced V_{BE} setup for $M - 1$ measurement.

4.5.3 Forced-I_E Measurement of $M - 1$

At low J_C and low V_{CB}, the fixed-V_{BE} technique typically works well. Problems arise, however, at high J_C and high V_{CB}. Self-heating becomes significant, and often causes permanent damage to devices under test if (when) thermal runaway occurs. For fixed-V_{BE} biasing, increasing V_{CB} leads to higher $M-1$ and hence higher I_C. Both high V_{CB} and high I_C increase the local junction temperature, which in turn increases the electron current injected into the CB space-charge region ($I_{n,in}$). This positive feedback mechanism results in thermal runaway at sufficiently high junction temperatures.

A safer alternative is to use a fixed-I_E biasing configuration [13]. In this case, a current I_E is forced into the emitter, as shown in Figure 4.23. Here, V_{CB} is swept and the base-emitter voltage V_{BE} is measured. Recall that the I_B component due to hole injection is only dependent on V_{BE}, and is equal to the I_B measured at $V_{CB} = 0$ V. Therefore, the initial current for avalanche multiplication, $I_{n,in}$, can be expressed as

$$I_{n,in} = I_E - I_B(V_{BE})|_{V_{CB}=0}, \qquad (4.48)$$

and $M - 1$ is thus

$$M - 1 = \frac{I_C}{I_E - I_B(V_{BE})|_{V_{CB}=0}} - 1. \qquad (4.49)$$

Static Characteristics

In the measurement, V_{BE} is recorded during the V_{CB} sweep, and $I_B(V_{BE})|_{V_{CB}=0}$ is found from the $I_B - V_{BE}$ curve obtained with $V_{CB} = 0$ V. The Early effect factor can be determined to be

$$F_{Early} = \frac{I_{n,in}(V_{CB})}{I_{n,in}(V_{CB}) = 0}$$
$$= \frac{I_E - I_B(V_{BE})|_{V_{CB}=0}}{I_C(V_{BE})|_{V_{CB}=0}}. \quad (4.50)$$

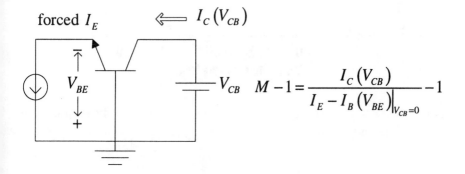

Figure 4.23 Forced I_E setup for $M - 1$ measurement.

The forced-I_E method makes the measurement of $M-1$ safer because the total amount of current injected into the CB space-charge region is always limited by I_E. The feedback mechanism for avalanche multiplication is thus effectively limited during the presence of self-heating. Figure 4.24 shows $M-1$ versus V_{CB} measured at low J_E using the fixed-I_E method for SiGe HBTs with different collector doping levels. Here $M-1$ increases significantly as the doping level increases, as expected.

The Early effect factor (F_{Early}) and the avalanche multiplication factor $M-1$, measured with forced I_E, can be used to reproduce $I_C - V_{CB}$ and $I_B - V_{CB}$ for fixed V_{BE},

$$I_C(V_{BE})|_{V_{CB}} = I_C(V_{BE})|_{V_{CB}=0} \times F_{Early} \times M, \quad (4.51)$$

and

$$I_B(V_{BE})|_{V_{CB}} = I_B(V_{BE})|_{V_{CB}=0} - I_C(V_{BE})|_{V_{CB}=0} \times F_{Early} \times (M-1). \quad (4.52)$$

Figure 4.25(a) and (b) show examples of $I_C - V_{CB}$ and $I_B - V_{CB}$ obtained in this manner, together with measured data. The individual contributions of Early

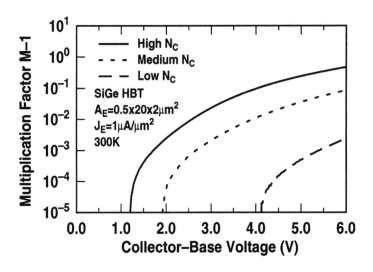

Figure 4.24 $M - 1$ versus V_{CB} measured at low injection for different collector doping profiles.

effect and avalanche multiplication to the increase of I_C with increasing V_{CB} can be easily distinguished using this technique.

One may wonder why we need to separate the Early effect from the avalanche multiplication effect since both act to cause an increase of I_C with increasing V_{CB}. The answer becomes clear if we examine their respective impact on I_E and I_B. Consider a transistor biased at fixed V_{BE}, and now imagine increasing V_{CB}. The Early effect results in an increase in both I_C and I_E or, more specifically, the electron current injected from the emitter into the CB space-charge region. It has no effect, however, on the base current of a transistor with negligible neutral base recombination. Avalanche multiplication, on the other hand, by its very nature, results in not only an additional increase in I_C, but also a decrease in I_B by the same amount. Note that I_E, however, is not affected by avalanche multiplication. The base current reduces to zero when

$$\beta|_{V_{CB}=0} \times F_{Early} \times (M - 1) = 1. \tag{4.53}$$

The V_{CB} at which (4.53) is satisfied is often referred to as the "base current reversal voltage."

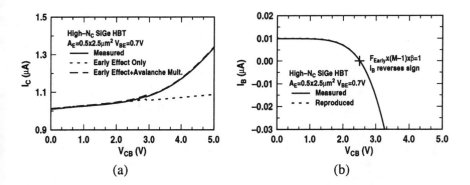

Figure 4.25 (a) $I_C - V_{CB}$, (b) and $I_B - V_{CB}$ at fixed $V_{BE} = 0.7$ V. Calculations using F_{Early} and $M - 1$ from a fixed I_E measurement reproduces $I_C - V_{CB}$ and $I_B - V_{CB}$ that were directly measured with constant V_{BE}.

4.5.4 Effects of Self-Heating

At high J_C and high V_{CB}, self-heating occurs, and substantial errors may result from using either the forced-voltage or forced-current techniques. The forced-I_E technique is generally much more resistant to thermal runaway, however, as discussed above. Caution must be exercised when interpreting the measured $M - 1$ result. The interpretational difficulty lies in the inaccurate estimation of the I_B component due to hole injection into the emitter, $I_{p,e}$. Without self-heating, $I_{p,e}$ is determined using the I_B measured at V_{BE}, the value of which is recorded during the fixed-I_E versus V_{CB} sweep. With significant self-heating, however, hole injection into the emitter depends not only on V_{BE}, but also on V_{CB} through the junction temperature. A simple fix exists if the $M - 1$ is higher than $1/\beta$, and this solution path can be identified by comparing the measured $I_C/I_E - 1$ with $1/\beta$. If this condition is valid, the initial current $I_{n,in}$ can then be approximated by I_E, and $I_C/I_E - 1$ can be approximated by $M - 1$. Insight can be gained by expressing M in a different manner

$$\begin{aligned} M &= \frac{I_C}{I_E - I_B(V_{BE})|_{V_{CB}=0,T(V_{CB})}} \\ &= \frac{I_C}{I_E \left(1 - 1/\beta(V_{BE})|_{V_{CB},T(V_{CB})}\right)} \\ &\approx \frac{I_C}{I_E}\left(1 + 1/\beta(V_{BE})|_{V_{CB},T(V_{CB})}\right). \end{aligned} \qquad (4.54)$$

One can safely neglect the $1/\beta\,(V_{BE})|_{V_{CB},T(V_{CB})}$ term when the following condition holds:

$$\frac{I_C}{I_E} \gg 1 + 1/\beta\,(V_{BE})|_{V_{CB},T(V_{CB})}. \tag{4.55}$$

Consider, for example, $\beta = 100$ and $I_C/I_E = 1.1$. The M value obtained by neglecting the $1/\beta\,(V_{BE})|_{V_{CB},T(V_{CB})}$ term is 1.1, which is a good approximation to its actual value $1.1 \times (1 + 1/100) = 1.111$. Another measurement concern is that $1/\beta\,(V_{BE})|_{V_{CB},T(V_{CB})}$ is in principle a function of J_C and V_{CB} due to both temperature rise and Early effect. However, as long as the measured $I_C/I_E - 1$ is far larger than $1/\beta\,(V_{BE})|_{V_{CB},T(V_{CB})}$, one can simply take I_E as the initial current, and take $I_C/I_E - 1$ as $M - 1$, with little error being introduced in the process. Self-heating may either increase or decrease the current gain in a SiGe HBT (see Chapter 9), but as long as the gain is high enough for the high V_{CB} region where $M - 1$ is large, a very good approximation of the actual M-1 is still obtained.

4.5.5 Impact of Current Density

In most cases, $M - 1$ is measured at low V_{BE} or low J_C. The resultant $M - 1$ is only a function of V_{CB}, as expected. The low J_C values of $M - 1$, however, are considerably higher than the $M - 1$ at the J_C where f_T is high (i.e., where circuits are normally designed). Figure 4.26 shows $M - 1$ versus the emitter current density for SiGe HBTs with different collector doping levels. The compensation of charges in the CB space-charge region by free carriers reduces the effective doping and electric field, thus decreasing $M - 1$ at higher J_E where many circuits are biased. For the high and medium N_C devices, the M-1 obtained in the useful J_E range of 0.1 to 1.0 mA/μm^2, at which these devices have optimum frequency response, is considerably smaller than its low J_E value. This difference demonstrates the importance of M-1 measurements at these practical operational current densities. For accurate linearity analysis, for instance, $M - 1$ must be modeled as a function of both current density and V_{CB}, as detailed in Chapter 8.

4.5.6 Si Versus SiGe

To suppress high-injection barrier effect (refer to Chapter 6), Ge is often retrograded into the collector. It is generally assumed that the unavoidable SiGe in the CB space-charge region does not inadvertently affect $M - 1$, because of the so-called "dead space" effect [14]. That is, the peak electron energy position is located deep in the Si region (outside of the SiGe profile), and thus impact ionization occurs mostly in the Si region, resulting in the same $M - 1$ for both SiGe and Si devices [1, 15]. The difference in impact ionization between Si and strained SiGe

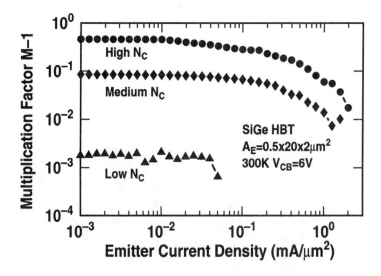

Figure 4.26 $M - 1$ versus J_E at $V_{CB} = 6$ V for three different collector doping profiles.

is thus not reflected in typical $M - 1$ measurements made on state-of-the-art SiGe HBTs.

To reveal how the avalanche multiplication effect is different between SiGe and Si, one can measure $M - 1$ at very high V_{CB}, such that secondary-hole avalanche multiplication is engaged. In this case, electrons are accelerated towards the collector side of the CB space-charge region, while holes are accelerated towards the base side of the CB space-charge region. Consequently, hot holes populate the base side of the CB space-charge region. During the carrier drift process, these secondary holes can gain energy, and create electron-hole pairs at the base end of the CB space-charge region. The above description is obviously a simplified view of the physical process, and its exact analysis requires self-consistent solution of the Boltzmann transport equation. The electrical signature for secondary hole impact ionization is an abrupt increase of the avalanche multiplication factor [16]. If the peak hole energy lies in the portion of the CB space-charge where the Ge content peaks, we can distinguish the impact ionization by hot holes between SiGe and Si. Figure 4.27 shows the electron and hole energy distribution in a SiGe HBT simulated using the energy transport advanced application module of MEDICI [17]. While the simulator does not have a specific model for energy transport in strained SiGe, the simulated hole temperature distribution is still useful in determining where the peak hole energy lies. The simulated results show that the peak hole energy position indeed lies in the region where Ge content peaks in these devices.

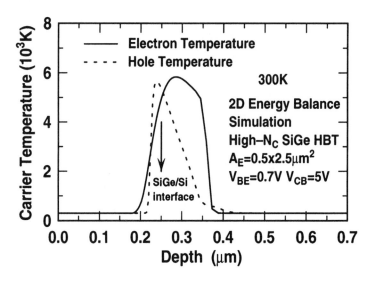

Figure 4.27 Simulated depth profile of carrier temperatures at $V_{BE} = 0.7$ V and $V_{CB} = 5$ V.

Figure 4.28 shows the measured $M - 1$ up to $V_{CB} = 12$ V for two SiGe HBTs with slightly different Ge gradings into the collector, together with an identically processed Si BJT control device. A safe extraction up to V_{CB} of 12 V and $M - 1$ of 10^3 is easily achieved using the fixed-I_E method. The $M - 1$ obtained for the SiGe HBTs and the Si BJT control are nearly identical for $V_{CB} < 9$ V due to the dead space effect. However, at higher V_{CB}, the secondary hole impact ionization becomes significant, the Si and SiGe devices show a clear difference in $M - 1$. The signature of this secondary hole impact ionization is the observed dramatic increase of $M - 1$. A higher onset voltage for the secondary hole impact ionization mechanism and a smaller value of $M - 1$ are observed in the SiGe HBT. Despite the smaller bandgap of SiGe, the impact ionization rate by holes in strained SiGe is smaller than in Si, which could be the result of a higher impact ionization threshold due to the in-plane strain in the SiGe.

It was shown in [18] that the threshold for impact ionization is dramatically increased if a layer is compressively strained without reducing its bandgap. An increase of the impact ionization threshold was later experimentally observed in a compressively strained layer with a wider bandgap [19]. The results shown in Figure 4.28 suggest that the threshold for impact ionization can be increased in a

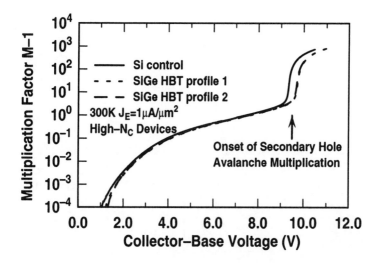

Figure 4.28 $M - 1$ versus V_{CB} comparison between SiGe HBTs with different Ge profiles and the Si BJT control.

compressively strained layer even if the bandgap is reduced. In addition to its obvious relevance for SiGe HBTs, this is clearly also good news for hetero-structure FETs utilizing strained SiGe channels.

4.6 Breakdown Voltages

Breakdown voltages are often characterized by applying a reverse bias across two of the three terminals, while leaving the third terminal open. For instance, BV_{CBO} typically refers to the collector-base breakdown voltage with an open emitter. Similarly, BV_{CEO} refers to the collector-emitter breakdown voltage with an open base. Both BV_{CBO} and BV_{CEO} are often quoted in IC technology specifications. Unfortunately, however, their significance to real circuit applications is often misunderstood. One persistent misconception is that SiGe HBTs cannot be used in implementing lightwave communication ICs because its BV_{CEO} is much less than the required power supply voltage. In practice, however, it is typical to use devices with $BV_{CEO} = 3.3$ V in SiGe HBT logic circuits operating with $V_{EE} = -5.2$ V.

During circuit operation, the maximum V_{CE} that can be supported by a SiGe HBT is much higher than BV_{CEO}, since the dc termination at the base is never truly electrically open. That is, the dc biasing network always presents a finite impedance between the base and ground. As a result, the collector-emitter breakdown voltage is considerably higher than BV_{CEO}. For RF signals, the impedance between base and ground is even smaller, making the effective breakdown voltage even larger. Another issue that must be considered is the current dependence of the avalanche multiplication factor $M - 1$. At high J_C where f_T is maximized, $M - 1$ is much smaller than at low J_C where BV_{CBO} and BV_{CEO} are typically measured.

4.6.1 BV_{CBO}

Defining where transistor breakdown begins to occur from an $I - V$ curve is somewhat arbitrary and there are no commonly accepted standards. An alternative and more meaningful way to define BV_{CBO} is to fit measured $M - 1$ versus V_{CB} data using the Miller equation

$$M = \frac{1}{1 - (V_{CB}/BV_{CBO})^m}, \qquad (4.56)$$

where BV_{CBO} and m are simply defined as fitting parameters. BV_{CBO} can then be simply viewed as a lumped representation of $M - 1$. In this case, a smaller BV_{CBO} simply corresponds to a higher $M - 1$ in the device. Figure 4.29 shows a sample $M - 1$ plot at $V_{BE} = 0.6$ V. Observe that $BV_{CBO} = 6.1265$ V can be unambiguously obtained from the measured $M - 1$ versus V_{CB} data ($m = 5.252$). Together with m, the extracted BV_{CBO} allows convenient evaluation of $M - 1$, because the definition of BV_{CBO} directly relates to $M - 1$. Equation (4.56), however, is of little use for RF linearity modeling, because avalanche multiplication strongly depends on operating current density J_C, which will be discussed in detail in Chapter 8.

4.6.2 BV_{CEO}

In general, BV_{CEO} is much smaller than BV_{CBO}. This can be physically understood by examining the avalanche multiplication process under open base conditions (Figure 4.30). The CB junction leakage current I_{CBO}, which was previously neglected in Figure 4.21, now must be considered because of the open base terminal. Now I_{CBO} appears as a hole current in the base, and can only flow into the emitter since the base is open. This current is amplified by β during this process, producing enhanced electron current flowing into the CB junction. This electron current, in turn, creates additional hole current via avalanche multiplication, which again flows into the EB junction as base current. A steady-state condition is reached, and the resulting "diode" current I_{CEO} is much higher than I_{CBO}. From the physical description above, I_{CEO} is expected to be closely related to the current amplification process, and hence β. Because of this positive feedback via β amplification, it takes only $M - 1 = 1/\beta$ for BV_{CEO} to occur, as intuitively expected and

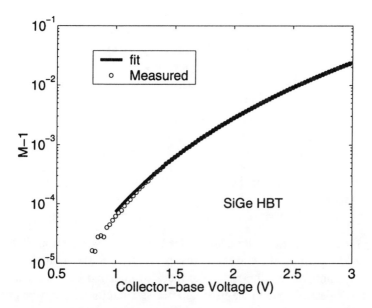

Figure 4.29 An example of $M - 1$ versus V_{CB} fitting using the Miller equation.

formally proved below. In comparison, an infinitely large $M - 1$ value is needed for BV_{CBO} to occur, since there is no feedback process.

The total collector current consists of I_{CBO} and $I_{n,out} = MI_{n,in}$. The total emitter current consists of $I_{p,e} = I_{n,in}/\beta$ and $I_{n,in}$, where $I_B = 0$ A because of the open base condition. Therefore,

$$I_E = I_C$$
$$I_{n,in}(1 + 1/\beta) = MI_{n,in} + I_{CBO}, \qquad (4.57)$$

and $I_{n,in}$ is thus given by

$$I_{n,in} = \frac{I_{CBO}}{1/\beta - (M - 1)}. \qquad (4.58)$$

At low V_{CE}, $M - 1 = 0$, and $I_{n,in} = \beta I_{CBO}$. Therefore, $I_{CEO} = (1 + \beta)I_{CBO}$. That is, the open base leakage current I_{CEO} is β times higher than the CB junction leakage I_{CBO} alone. With increasing V_{CE}, M increases, and hence breakdown occurs when $M - 1$ approaches $1/\beta$. We have neglected the Early effect in the above breakdown analysis, although it introduces negligible error. One thing that must be kept in mind, however, is that β has its own unique bias current dependence in SiGe HBTs (refer to Chapter 6). The β value at very low injection (e.g., for

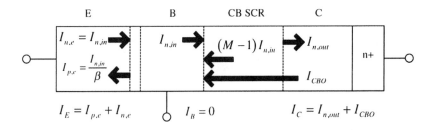

Figure 4.30 An illustration of the avalanche multiplication process with an open base.

$I_B = I_{CBO}$) can be much smaller than the β value under low-injection. In this case, when breakdown occurs, the current increases rapidly, and the β rapidly increases to its low-injection value. Similarly, β at medium injection can be much smaller than β at low-injection, due to Ge-grading effects. Therefore, β must be treated as bias dependent for accurate modeling of BV_{CEO} in SiGe HBTs.

One may have observed that $M - 1 = 1/\beta$ occurs at the base current reversal V_{CB}, according to (4.53). For BV_{CBO} to occur, $M - 1$ must approach infinity, while for BV_{CEO} to occur, $M - 1$ only needs to approach $1/\beta$, which clearly explains why BV_{CEO} is always much smaller than BV_{CBO}. An inherent trade-off between β and BV_{CEO} therefore exists at a fundamental level, and must be considered in SiGe device design if the circuit application requires high BV_{CEO}.

The increase of collector doping in SiGe HBTs is driven by the need to achieve high current density operation, a necessary condition to realize the full f_T potential offered by scaled SiGe HBTs featuring small transit times. Fundamentally, the time constant related to the charging of depletion capacitance is inversely proportional to J_C, and can only be made smaller than the transit time through the use of higher J_C. This point will be further developed in Chapter 5. Increasing the collector doping, however, increases the electric field in the CB junction and hence $M - 1$, thus reducing the transistor breakdown voltages. A high peak f_T transistor typically has higher collector doping and hence lower BV_{CBO} and BV_{CEO}. The $f_T \times BV_{CEO}$ product is often referred to as the *Johnson limit*. An often-quoted number for $f_T \times BV_{CEO}$ is 200 GHz-V. This value, however, should not be viewed as a "hard" or physical limit. Recently, SiGe HBTs with $f_T \times BV_{CEO} > 400$ GHz-V have been demonstrated (refer to Chapter 1). In addition, many commercial SiGe technologies offer SiGe HBTs with several different BV_{CEO} values by using selective collector implantation. Designers thus have additional freedom in circuit design and can choose the desired device with sufficient f_T and BV_{CEO}, and can thus take advantage of the lower CB capacitance and lower $M - 1$ than in devices with the highest f_T.

4.6.3 Circuit Implications

While BV_{CEO} is often viewed as a hard limit for the maximum collector voltage or maximum power supply voltage for circuit applications of SiGe HBTs, this is not the case. In practice, SiGe HBT circuits can operate with supply voltages significantly higher than BV_{CEO}. For instance, due to series gating, the V_{CE} of the SiGe HBTs used in high-speed logic (e.g., multiplexer (MUX) circuits) is only a portion of the total power supply voltage. In addition, the maximum operating V_{CE} can also be much higher than BV_{CEO} because the base terminal is not electrically open in practical circuit implementations. For instance, SiGe HBTs with $BV_{CEO} = 3.3$ V are used in AMCCs' commercial OC-768 (40 Gbit/s) components operating with $V_{CC} = 3.3$ V and $V_{EE} = -5.2$ V. The finite base termination of the dc biasing network, the low impedance RF base termination, and the decrease of $M - 1$ at high J_C all make the effective breakdown voltage much higher than BV_{CEO} for practical circuits and systems. An impedance of 100 kΩ, for example, between the base terminal and ground is sufficient to increase the breakdown voltage from BV_{CEO} to BV_{CBO}, which may be as high as 8–10 V for first generation SiGe technologies. Thus, careful optimization of the impedance seen by the base terminal during circuit design can modify the collector-emitter breakdown voltage to lie in the range of BV_{CBO} to BV_{CEO}. In fact, several proprietary circuit techniques along these lines have been recently developed to prevent adverse breakdown conditions from occurring without sacrificing the performance capability of SiGe HBTs. While the continuous decrease of BV_{CEO} with scaling for higher f_T clearly necessitates careful circuit design, it does not fundamentally exclude the application of SiGe HBT technologies for long-haul optical transmission systems, or other traditional high operating voltage applications.

References

[1] D.L. Harame et al., "Si/SiGe epitaxial-base transistors – part I: materials, physics, and circuits," *IEEE Trans. Elect. Dev.*, vol. 42, pp. 455-468, 1995.

[2] H. Kroemer, "Two integral relations pertaining to electron transport through a bipolar transistor with a nonuniform energy gap in the base region," *Solid-State Elect.*, vol. 28, pp. 1101-1103, 1985.

[3] E.J. Prinz et al., "The effect of emitter-base spacers and strain-dependent density-of-states in Si/SiGe/Si heterojunction bipolar transistors," *Tech. Dig. IEEE Int. Elect. Dev. Meeting*, pp. 639-642, 1989.

[4] S.S. Iyer et al., "Heterojunction bipolar transistors using Si-Ge alloys," *IEEE Trans. Elect. Dev.*, vol. 36, pp. 2043-2064, 1989.

[5] A.J. Joseph, "The physics, optimization, and modeling of cryogenically operated silicon-germanium heterojunction bipolar transistors," Ph.D. Dissertation, Auburn University, 1997.

[6] J.D. Cressler et al., "On the profile design and optimization of epitaxial Si- and SiGe-base bipolar technology for 77 K applications – part I: transistor dc design considerations," *IEEE Trans. Elect. Dev.*, vol. 40, pp. 525-541, 1993.

[7] J.M. Early, "Effects of space-charge layer widening in junction transistors," *Proc. IRE*, vol. 40, pp. 1401-1406, 1952.

[8] D.J. Roulston, *Bipolar Semiconductor Devices*, McGraw-Hill, New York, NY, 1990.

[9] E.J. Prinz and J.C. Sturm, "Current gain-Early voltage products in heterojunction bipolar transistors with nonuniform base bandgaps," *IEEE Elect. Dev. Lett.*, vol. 12, pp. 661-663, 1991.

[10] A.J. Joseph, J.D. Cressler, and D.M. Richey, "Optimization of Early voltage for cooled SiGe HBT precision current sources," *J. de Physique IV*, vol. 6, pp. 125-129, 1995.

[11] G. Niu, J.D. Cressler, and A.J. Joseph, "Quantifying neutral base recombination and the effects of collector-base junction traps in UHV/CVD SiGe HBT's," *IEEE Trans. Elect. Dev.*, vol. 45, pp. 2499-2504, 1998.

[12] C.C. McAndrew and L.W. Nagel, "Early effect modeling in SPICE," *IEEE J. Solid-State Circ.*, vol. 31, pp. 136-138, 1996.

[13] G. Niu et al., "Measurement of collector-base junction avalanche multiplication effect in advanced UHV/CVD SiGe HBTs," *IEEE Trans. Elect. Dev.*, vol. 46, pp. 1007-1015, 1999.

[14] G.E. Bulman, V.M. Robbins, and G.E. Stillman, "The determination of impact ionization coefficients in (100) gallium arsenide using avalanche noise and photocurrent measurements," *IEEE Trans. Elect. Dev.*, vol. 32, pp. 2454-2466, 1995.

[15] G. Niu et al., "Collector-base junction avalanche multiplication effects in advanced UHV/CVD SiGe HBTs," *IEEE Elect. Dev. Lett.*, vol. 19, pp. 288-290, 1998.

[16] E. Zanoni et al., "Extension of impact-ionization multiplication coefficient measurements to high electric fields in advanced Si BJT's," *IEEE Elect. Dev. Lett.*, vol. 14, pp. 69-71, 1993.

[17] MEDICI, 2-D Semiconductor Device Simulator, Avant!, Fremont, CA, 1997.

[18] J. Singh, "A theoretical study of electron impact ionization in pseudomorphic InGaAs on GaAs and InP substrates," *Int. Symp. GaAs Related Comp.*, 1991.

[19] K.W. Eisenbeiser et al., "Breakdown voltage improvement in strained InGaAs/GaAs FET's," *IEEE Elect. Dev. Lett.*, vol. 13, pp. 421-423, 1992.

Chapter 5

Dynamic Characteristics

Due to the presence of Si-SiGe heterojunctions in both the EB and CB junctions, the device physics of the SiGe HBT fundamentally differs from that of the conventional Si BJT. In this chapter we examine these differences from an *ac* point of view, by first reviewing an intuitive picture of how the SiGe HBT operates dynamically and, importantly, how its operation differs from that of a comparably constructed Si BJT. We then address *ac* charge storage phenomena, make some fundamental parameter definitions, as well as discuss high-frequency equivalent circuit models for SiGe HBTs. Using linear two-port theory, we then define the relevant *ac* figures-of-merit for assessing and comparing SiGe HBT dynamic performance. We next formally derive the base and emitter transit times of an ideal SiGe HBT under low-injection conditions. Explicit and implicit assumptions as well as physically relevant approximations are highlighted throughout the theoretical development. Armed with this knowledge we examine the *ac* Ge profile shape and optimization issues, and conclude with a discussion on ECL gate delay and the impact of SiGe on digital bipolar circuit performance.

5.1 Intuitive Picture

The essential operational differences between the SiGe HBT and the Si BJT are best illustrated by considering a schematic energy band diagram. For simplicity, we consider an ideal, graded-base SiGe HBT with constant doping in the emitter, base, and collector regions. In such a device construction, the Ge content is linearly graded from 0% near the metallurgical EB junction to some maximum value of Ge content near the metallurgical CB junction, and then rapidly ramped back down to 0% Ge. The resultant overlaid energy band diagrams for both the SiGe HBT and the Si BJT, biased identically in forward-active mode, are shown in Figure 5.1.

Observe in Figure 5.1 that the grading of the Ge across the neutral base induces a built-in quasi-drift field in the neutral base, which will (favorably) impact minority carrier transport.

To intuitively understand how these band edge changes affect the *ac* operation of the SiGe HBT, first consider the dynamic operation of the Si BJT. Electrons injected from the emitter into the base region must diffuse across the base (for constant doping), and are then swept into the electric field of the CB junction, yielding a useful (time-dependent) collector current. The time it takes for the electrons to traverse the base (base transit time) is significant, and typically is the limiting transit time that determines the overall transistor *ac* performance (e.g., peak f_T). At the same time, the applied forward bias on the EB junction dynamically produces a back-injection of holes from the base into the emitter. For fixed collector bias current, this dynamic storage of holes in the emitter (emitter charge storage delay time) is reciprocally related to the *ac* current gain of the transistor (β_{ac}).

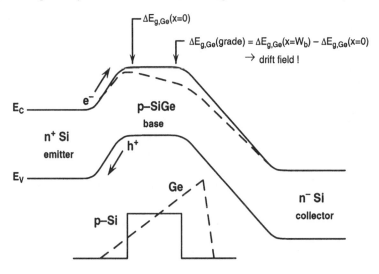

Figure 5.1 Energy band diagram for a Si BJT and graded-base SiGe HBT, both biased in forward active mode at low-injection.

As can be seen in Figure 5.1, the introduction of Ge into the base region has an important consequence, since the Ge-gradient-induced drift field across the neutral base is aligned in a direction (from collector to emitter) such that it will accelerate the injected minority electrons across the base. We are thus able to add a large drift field component to the electron transport, effectively speeding up the diffusive transport of the minority carriers and thereby decreasing the base transit time. Even though the band offsets in SiGe HBTs are typically small by III-V technology standards, the Ge grading over the short distance of the neutral base can translate into large electric fields. For instance, a linearly graded Ge profile with a modest peak Ge content of 10%, graded over a 50-nm neutral base width, yields 75 mV / 50 nm = 15 kV/cm electric field, sufficient to accelerate the electrons to near saturation velocity ($v_s \simeq 1 \times 10^7$ cm/sec). Because the base transit time typically limits the frequency response of a Si BJT, we would expect that the frequency response should be significantly improved by introducing this Ge-induced drift field. In addition, we know that the Ge-induced band offset at the EB junction will exponentially enhance the collector current density (and thus β) of a SiGe HBT compared to a comparably constructed Si BJT. Since the emitter charge storage delay time is reciprocally related to β, we would also expect the frequency response to a SiGe HBT to benefit from this added emitter charge storage delay time advantage.

For low-injection, the unity-gain cutoff frequency (f_T) in a bipolar transistor can be written generally as

$$f_T = \frac{1}{2\pi \tau_{ec}} = \frac{1}{2\pi} \left[\frac{kT}{qI_C}(C_{te} + C_{tc}) + \tau_b + \tau_e + \frac{W_{CB}}{2v_{sat}} + r_c C_{tc} \right]^{-1}, \quad (5.1)$$

where $g_m = kT/qI_C$ is the intrinsic transconductance at low-injection ($g_m = \partial I_C / \partial V_{BE}$), C_{te} and C_{tc} are the EB and CB depletion capacitances, τ_b is the base transit time, τ_e is the emitter charge storage delay time, W_{cb} is the CB space-charge region width, v_{sat} is the saturation velocity, and r_c is the dynamic collector resistance. In (5.1), τ_{ec} is the total emitter-to-collector delay time, and sets the ultimate limit of the switching speed of a bipolar transistor. Thus, we see that for fixed bias current, improvements in τ_b and τ_e due to the presence of SiGe will directly translate into an enhanced f_T and f_{max} of the transistor at fixed bias current.

5.2 Charge Modulation Effects

At a deep level, transistor action, be it for a bipolar or field-effect transistor, is physically realized by voltage modulation of the charges inside the transistor, that in turn leads to voltage modulation of the output current. The voltage modulation of the charges results in a capacitive current which increases with frequency. The bandwidth of the transistor is thus ultimately limited by various charge storage effects in both the intrinsic and extrinsic device structure. Exact analysis of charge storage effects requires the solution of semiconductor transport equations in the frequency domain. In practice, charge storage effects are often taken into account by assuming that the charge distributions instantly follow the changes of terminal voltages under dynamic operation (i.e., a "quasi-static" assumption).

The first charge modulation effect in a SiGe HBT is the modulation of space charges associated with the EB and CB junctions. Voltage changes across the EB and CB junctions lead to changes of the space-charge (depletion) layer thicknesses and hence the total space charge. The capacitive behavior is similar to that of a parallel plate capacitor, because the changes in charge occur at the opposing faces of the space charge layer (which is depleted of carriers under reverse bias) to neutral region transition boundaries. The resulting capacitances are referred to as EB and CB "depletion" capacitances. Under high-injection conditions, the modulation of charges inside the space charge layer becomes significant. The resulting capacitance is referred to as the "transition" capacitance, and is important for the EB junction since it is forward biased. Under low-injection conditions, the CB capacitance is similar to that of a reverse biased pn junction, and is a function of the CB biasing voltage. At high injection, however, even in forward-active mode, the CB capacitance is also a function of the collector current, because of charge compensation by mobile carriers as well as base push-out at very high injection levels.

The second charge modulation effect is due to injected minority carriers in the neutral base and emitter regions. To maintain charge neutrality, an equal amount of excess majority carriers are induced by the injected minority carriers. Both minority and majority carriers respond to EB voltage changes, effectively producing an EB capacitance. This capacitance is historically referred to as "diffusion" capacitance, because it is associated with minority carrier diffusion in an ideal bipolar transistor with uniform base doping.

What is essential in order to achieve transistor action is modulation of the output current by an input voltage. The modulation of charge is just a means of modulating the current, and must be minimized in order to maintain ideal transistor action at high frequencies. For instance, a large EB diffusion capacitance causes a large input current which increases with frequency, thus decreasing current gain at higher frequencies. At a fundamental level, for a given output current modulation, a decreased amount of charge modulation is desired in order to achieve higher operating frequency. A natural figure-of-merit for the efficiency of transistor action is the ratio of output current modulation to the total charge modulation

$$\tau_{ec} = \frac{\partial I_C}{\partial Q_n}, \qquad (5.2)$$

which has dimensions of time and is thus called "transit time." Here, Q_n refers to the integral electron charge across the whole device, and can be broken down into various components for regional analysis. The partial derivative in (5.2) indicates that there is modulation of both charge and current, and is thus necessary. A popular but *incorrect* definition of transit time leaves out the derivatives in (5.2), and instead uses the simple ratio of charge to collector current. The problem with this common formulation can be immediately deduced if we consider the resultant τ_{ec} of an npn bipolar transistor, where Q_n is dominated by the total number of emitter dopants. The use of $\tau_{ec} = Q_n/I_C$ thus leads to an incorrect transit time definition, since it produces a transit time that is independent of the base profile design, and is clearly nonphysical. Equation (5.2) can be rewritten using the input voltage as an intermediate variable

$$\tau_{ec} = \frac{\partial I_C/\partial V_{BE}}{\partial Q_n/\partial V_{BE}} = \frac{g_m}{C_i}, \qquad (5.3)$$

where C_i is the total input capacitance, and g_m is the transconductance. C_i can be divided into two components $C_{be} = \partial Q_n/\partial V_{BE}$ and $C_{bc} = \partial Q_n/\partial V_{BC}$. The transit times related to the neutral base and neutral emitter charge modulation are the base transit time and the emitter transit time, respectively. The base charge modulation required to produce a given amount of output current modulation can be decreased by introducing a drift field via Ge grading, thereby reducing the base transit time and extending transistor functionality to much higher frequencies. This Ge-grading-induced reduction in charge modulation is the fundamental reason why SiGe HBTs have better frequency response than Si BJTs. Ge grading is simply a convenient means by which we reduce the charge modulation.

5.3 Basic RF Performance Factors

Figure 5.2 shows a small-signal high-frequency equivalent circuit for a bipolar transistor, which we use here to discuss transistor RF performance. For simplicity, we have neglected the emitter resistance, the collector resistance, and the output resistance due to Early effect. Here, the EB capacitance C_{be} is the sum of the EB diffusion capacitance $g_m \tau_f$ and the EB depletion capacitance C_{te}, while $g_m = qI_C/kT$, and $g_{be} = g_m/\beta$, relationships which hold for an ideal transistor.

5.3.1 Current Gain and Cutoff Frequency

The high-frequency current amplification capability of a SiGe HBT is typically measured by the small-signal current gain for a shorted output termination (i.e., h_{21}). Imagine driving the base terminal with a small-signal current $i_b = i_0 e^{j\omega t}$, and now short-circuit the output (collector), as shown in Figure 5.3. The node voltage v_b then equals

$$v_b = \frac{1}{g_{be} + j\omega(C_{be} + C_{bc})} i_b. \qquad (5.4)$$

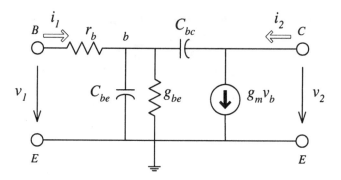

Figure 5.2 A simple high-frequency equivalent circuit model.

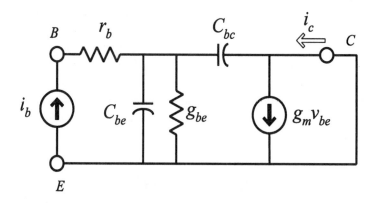

Figure 5.3 Equivalent circuit model used for the h_{21} derivation.

The effective capacitive load for the input due to Miller capacitance C_{bc} is still C_{bc} because of the "zero" voltage gain resulting from the short-circuited output. Because the reverse-biased CB junction capacitance is far smaller than the forward-biased EB junction capacitance, we can neglect its contribution to the output current i_c

$$i_c \approx g_m v_b = \frac{g_m}{g_{be} + j\omega(C_{be} + C_{bc})} i_b. \quad (5.5)$$

Therefore, we have

$$h_{21} = \left.\frac{i_c}{i_b}\right|_{v_c=0} = \frac{g_m}{g_{be} + j\omega(C_{be} + C_{bc})} = \frac{\beta}{1 + j\omega(C_{be} + C_{bc})/g_{be}}. \quad (5.6)$$

Note that h_{21} is constant at low frequencies, and then decreases at higher frequencies. Obviously, the imaginary part increases with ω, and dominates at high frequencies. Under these conditions the above equation becomes

$$h_{21} = \frac{g_m}{j\omega(C_{be} + C_{bc})}, \qquad (5.7)$$

which is equivalent to

$$h_{21} \times f = \frac{f_T}{j}, \qquad (5.8)$$

$$f_T = \frac{g_m}{2\pi(C_{be} + C_{bc})}. \qquad (5.9)$$

The $|h_{21} \times f|$ product is a constant over the frequency range where these assumptions hold. This constant is referred to as f_T, the transition frequency, or more commonly, the cutoff frequency. In practice, f_T is extracted by extrapolating the measured $|h_{21}|$ versus frequency data in a range where a slope of -20 dB/decade is observed. The frequency at which the extrapolated $|h_{21}|$ reduces to unity is defined to be f_T (i.e., the unity gain cutoff frequency). Practically speaking, the extrapolation is necessary here because we are usually not interested in operating transistors at the frequency of unity current gain, which can be different from the extrapolated f_T, depending on parasitics and other factors. Instead, we are interested in the gain available at much lower frequencies where the current gain is much higher than unity. In the frequency range where $|h_{21}|$ rolls off at -20 dB/decade, $|h_{21}|$ can be easily estimated as f_T/f.[1]

State-of-the-art SiGe HBTs exhibit f_T values above 200 GHz [1], which is much higher than the operating frequencies of the bulk of existing wireless systems, which are typically below 10 GHz. In this case, caution must be exercised in estimating $|h_{21}|$ from f_T, because the operating frequency f may be below the frequency range over which $|h_{21} \times f| = f_T$. In this case, we then need to resort to (5.6) which applies to all frequencies below f_T and can be rewritten as follows using (5.9)

$$h_{21} = \frac{\beta}{1 + jf/f_\beta}, \qquad (5.10)$$

$$f_\beta = \frac{f_T}{\beta}.$$

Here, $|h_{21}|$ is equal to β at low frequencies, reduces by 3 dB at $f = f_\beta = f_T/\beta$, and then drops off with increasing f at a theoretical slope of -20 dB/decade. Hence, for

---
[1]We note that for the very high f_T SiGe HBTs being realized today (200+ GHz), instrumentation limitations place a practical upper bound on directly measuring f_T in any case, since the highest reliable measurement frequencies are in the 110-GHz range for commercially available test systems.

a SiGe HBT with $f_T = 100$ GHz and $\beta = 100$, the 3 dB frequency is $f_\beta = 1$ GHz. For a design frequency of 2 GHz, which is close to f_β, (5.11) needs to be used for $|h_{21}|$ estimation instead of f_T/f. Figure 5.4 shows an example of measured h_{21} versus frequency from 2 to 110 GHz for a SiGe HBT. The extrapolated f_T is 117 GHz. A noticeable deviation from the 20-dB/decade straight line fit is observed below 7 GHz, necessitating the use of (5.11) for h_{21} estimation. Note as well that a deviation from the 20-dB/decade slope is observed above 40 GHz.

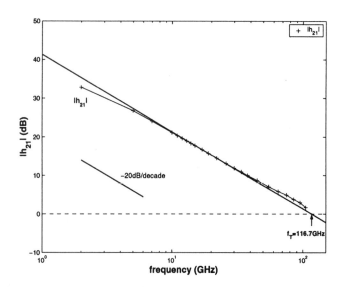

Figure 5.4 Measured $|h_{21}|$ versus frequency for a state-of-the-art SiGe HBT.

The simple model presented above captures the magnitude of h_{21} at high frequencies with reasonable accuracy, but does not describe the phase of h_{21} particularly well, because of non-quasi-static effects. The phase of h_{21} is important, for instance, for the design of feedback amplifiers and oscillators. A simple yet effective fix is to replace the transconductance g_m in Figure 5.2 with a complex transadmittance y_m

$$y_m = \frac{g_m}{1 + jf/f_N}, \qquad (5.11)$$

where $f_N \approx 3/2\pi\tau_f$, with τ_f being the forward transit time (the sum of base, emitter, and collector transit times: $\tau_b + \tau_e + \tau_c$). In the case of the data above, the overall phase correction to h_{21} is approximately 18 degrees at $f = f_T$.

5.3.2 Current Density Versus Speed

The fundamental nature of SiGe HBTs require the use of high operating current density in order to achieve high speed. The operating current density dependence of f_T is best illustrated by examining the inverse of f_T using (5.9)

$$\frac{1}{2\pi f_T} = \frac{C_{be} + C_{bc}}{g_m}. \quad (5.12)$$

Since $C_{be} = g_m \tau_f + C_{te}$, $C_{bc} = C_{tc}$, and $g_m = qI_C/kT$, (5.12) can be rewritten as

$$\frac{1}{2\pi f_T} = \tau_f + \frac{kT}{qI_C}C_t, \quad (5.13)$$

where $C_t = C_{te} + C_{tc}$. Since both C_{te} and C_{tc} are proportional to emitter area, (5.13) can be rewritten in terms of the biasing current density J_C as

$$\frac{1}{2\pi f_T} = \tau_f + \frac{kT}{qJ_C}C'_t, \quad (5.14)$$

where $C'_t = C_t/A_E$ is the total EB and CB depletion capacitances per unit emitter area, and $J_C = I_C/A_E$ is the collector operating current density. Thus, the cutoff frequency f_T is fundamentally determined by the biasing current density J_C, independent of the transistor emitter length. For very low J_C, the second term is very large, and f_T is very low regardless of the forward transit time τ_f. With increasing J_C, the second term decreases, and eventually becomes smaller than τ_f. At high J_C, however, base push-out (Kirk effect, refer to Chapter 6) occurs, and τ_f itself increases with J_C, leading to f_T roll-off. A typical f_T versus J_C characteristic is shown in Figure 5.5 for a first generation SiGe HBT.

The values of τ_f and C'_t can be easily extracted from a plot of $1/2\pi f_T$ versus $1/J_C$, as shown in Figure 5.6. Near the peak f_T, the $1/2\pi f_T$ versus $1/J_C$ curve is nearly linear, indicating that C'_t is close to constant for this biasing range at high f_T. Thus, C'_t can be obtained from the slope, while τ_f can be determined from the y-axis intercept at infinite current ($1/J_C = 0$).

To improve f_T in a SiGe HBT, the transit time τ_f must be decreased by using a combination of vertical profile scaling as well as Ge grading across the base. At the same time, the operating current density J_C must be increased in proportion in order to make the second term in (5.14) negligible compared to the first term (τ_f). That is, the high f_T potential of small τ_f transistors can only be realized by using sufficiently high operating current density. This is a fundamental criterion for high-speed SiGe HBT design. The higher the peak f_T, the higher the required operating J_C. For instance, the minimum required operating current density has increased from 1.0 mA/μm^2 for a first generation SiGe HBT with 50-GHz peak f_T to 8–10

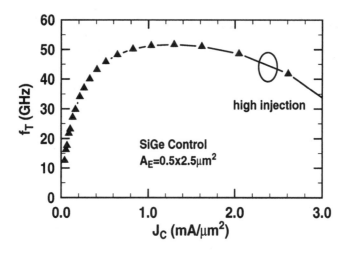

Figure 5.5 A typical $f_T - J_C$ behavior for a SiGe HBT.

mA/μm^2 for >200-GHz peak f_T third generation SiGe HBTs [1]. Higher current density operation naturally leads to more severe self-heating effects, which must be appropriately dealt with in compact modeling and circuit design [2]. Electromigration and other reliability constraints associated with very high J_C operation have also produced an increasing need for copper metalization schemes.

In order to maintain proper transistor action under high J_C conditions, the collector doping must be increased in order to delay the onset of high injection effects. This requisite doping increase obviously reduces the breakdown voltage. At a fundamental level, trade-offs between breakdown voltage and speed are thus inevitable for all bipolar transistors (Si, SiGe, or III-V). Since the collector doping in SiGe HBT is typically realized by self-aligned collector implantation (as opposed to during epi growth in III-V), devices with multiple breakdown voltages (and hence multiple f_T) can be trivially obtained in the same fabrication sequence, giving circuit designers added flexibility.

Another closely related manifestation of (5.14) is that the minimum required J_C to realize the full potential of a small τ_f transistor depends on C'_t. Both C'_{te} and C'_{tc} thus must be minimized in the device and are usually addressed via a combination of structural design, ground-rule shrink, and doping profile tailoring via selective collector implantation. This reduction of C'_{tc} is also important for increasing the power gain (i.e., maximum oscillation frequency – f_{max}).

Dynamic Characteristics 149

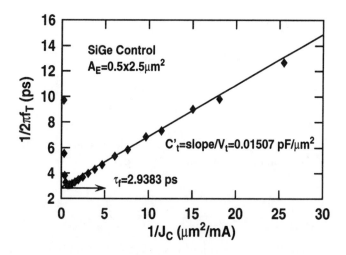

Figure 5.6 Illustration of C'_i and τ_f extraction in a SiGe HBT.

5.3.3 Base Resistance

Observe that the base resistance r_b does not directly enter the h_{21} expressions, simply because r_b is in series with the ideal transistor (without r_b). In practice, however, r_b limits transistor power gain and noise performance, because it consumes input power and produces thermal noise directly at the base terminal, the worst possible place for the location of a noise source! As a result, minimization of the various components of the base resistance is a major challenge in SiGe HBT structural design, fabrication, and process integration. The base resistance is a key parameter for both process control and circuit design, and deserves careful attention. Unlike many bipolar parameters, base resistance is particularly challenging (and time consuming) to extract in a robust manner.

A popular technique to extract r_b is to use the input impedance with a shorted output, which by definition is equal to h_{11}. An inspection of Figure 5.3 shows

$$h_{11} = Z_{in}|_{v_c=0} = r_b + \frac{1}{g_{be} + j\omega C_i},$$
$$C_i = C_{be} + C_{bc}. \tag{5.15}$$

The real and imaginary parts of h_{11} are

$$x = \Re(h_{11}) = r_b + \frac{g_{be}}{g_{be}^2 + (\omega C_i)^2}$$

$$y = \Im(h_{11}) = -\frac{\omega C_i}{g_{be}^2 + (\omega C_i)^2}. \tag{5.16}$$

Using (5.16), one can easily prove that the (x, y) ordered pairs at different frequencies form a semicircle on the complex impedance plane

$$(x - x_0)^2 + y^2 = r^2, \tag{5.17}$$

$$x_0 = r_b + 1/2g_{be} \quad r = 1/2g_{be}.$$

The (x, y) impedance point moves clockwise with increasing frequency. The base resistance is then determined to be the high frequency intercept between the fitted impedance semicircle and the real axis, which appears on the left. This is the so-called "circle impedance" base resistance extraction method. In the above analysis, the emitter resistance r_e is neglected for simplicity, but it can be shown that the extracted r_b is actually the sum of the transistor r_b and r_e. Figure 5.7 shows an example of such an r_b extraction for a typical first generation SiGe HBT with an effective emitter area of 0.5 × 40 μm^2. The h_{11} data was measured from 0.5 to 15 GHz in order to make a meaningful fit to a semicircle. Choosing a proper measurement frequency range is important in reliable r_b extraction, as can be seen from Figure 5.7. In this case, had we used a frequency range of 15–50 GHz, the data would have formed only a tiny portion of the semicircle, making fitting and r_b extraction much more difficult. Deviation from circular behavior is often observed at frequencies close to f_T, and those data should be discarded in the r_b extraction. Given the I_C dependence of f_T, the frequency range over which r_b extraction is made can be varied with I_C to order to obtain an accurate I_C dependence of r_b, which is needed in compact modeling.

5.3.4 Power Gain and Maximum Oscillation Frequency

The base resistance directly reduces the transistor power gain because the input current flows through r_b, resulting in a loss of input power. In the previous discussion on current gain, the output termination is a short circuit, which gives the highest *ac* current gain. The highest *ac* power gain, however, is achieved when the transistor output is terminated with an impedance that is the conjugate of the output impedance. Several power gains can be defined in this case, depending on the choice of power levels for the input and output. Let us consider the operating

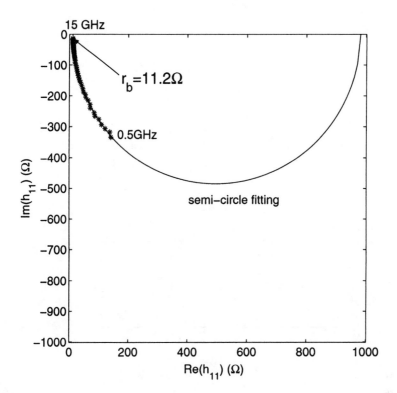

Figure 5.7 Extraction of r_b using the circle impedance method. The measured h_{11} data forms a semicircle. The frequency increases clockwise.

power gain first, which is defined as the ratio between the power delivered to the load and the power delivered to the transistor base.

We first determine the transistor output impedance by applying a test voltage v_{test} to the transistor output, as shown in Figure 5.8(a). At high frequencies, the conduction between node B' and E mainly occurs through C_{be}, and the $r_b + Z_s$ can be approximated by an open circuit. This is clearly an approximation that holds only at sufficiently high frequencies. Under these conditions, v_b is simply related to v_{test} through voltage division by C_{be} and C_{bc}

$$v_b = \frac{C_{bc}}{C_{bc} + C_{be}} \approx \frac{C_{bc}}{C_{be}} v_{test}, \qquad (5.18)$$

where the approximation is justified since $C_{be} \gg C_{bc}$. We further assume that the current through C_{bc} remains small and negligible compared to $g_m v_b$. Using (5.18), the output impedance is found to be

$$Z_{out} = \frac{v_{test}}{i_{test}} \approx \frac{v_{test}}{g_m v_b} = \frac{1}{g_m}\frac{C_{be}}{C_{bc}}. \tag{5.19}$$

The output impedance at sufficiently high frequencies is *resistive*, and inversely proportional to C_{bc}. Therefore, C_{bc} needs to be minimized in order to increase the output impedance and thus increase the maximum power gain.

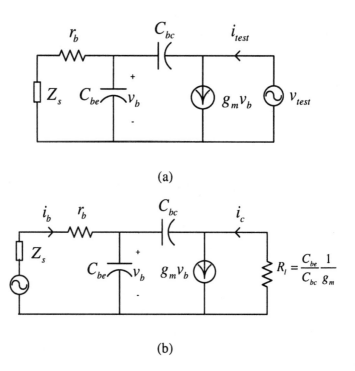

(a)

(b)

Figure 5.8 (a) High frequency output impedance calculation used in the f_{max} derivation, and (b) the equivalent circuit used for the f_{max} derivation. The output is conjugate-matched.

Next we terminate the output with the conjugate of Z_{out}, which remains a resistance, and drive the input with a source impedance of Z_s, as shown in Figure 5.8(b). At sufficiently high frequency, the input power looking into the base is mainly the result of power consumption by the base resistance. The operating

Dynamic Characteristics

power gain is thus

$$G_{p,max} = \frac{P_{out}}{P_{in}} = \left(\frac{i_c}{i_b}\right)^2 \frac{R_l}{r_b} = \left(\frac{i_c}{i_b}\right)^2 \frac{1}{g_m} \frac{C_{be}}{C_{bc} r_b}. \quad (5.20)$$

Once again, we assume the current through C_{bc} is negligible compared to $g_m v_b$, which leads to

$$i_c = g_m v_b, \quad (5.21)$$

$$v_c = -g_m v_b R_l = -g_m v_b \frac{1}{g_m} \frac{C_{be}}{C_{bc}} = -\frac{C_{be}}{C_{bc}} v_b. \quad (5.22)$$

The input current i_b is the sum of currents through C_{be} and C_{bc}

$$i_b = j\omega C_{be} v_b + j\omega C_{bc}(v_b - v_c) = j\omega 2 C_{be} v_b, \quad (5.23)$$

where (5.22) was used. Substitution of (5.23) and (5.21) into (5.20) gives

$$G_{p,max} = \frac{1}{4\omega^2} \frac{g_m}{C_{be}} \frac{1}{C_{bc} r_b} = \frac{1}{4\omega^2} \omega_T \frac{1}{C_{bc} r_b}, \quad (5.24)$$

where $f_T = g_m/2\pi C_{be}$ was used. Equation (5.24) can be rewritten in terms of f and f_T instead of ω and ω_T to give

$$G_{p,max} = \frac{f_{max}^2}{f^2}, \quad (5.25)$$

$$f_{max} = \sqrt{\frac{f_T}{8\pi C_{bc} r_b}}. \quad (5.26)$$

The maximum operating power gain $G_{p,max}$ is thus inversely proportional to f^2, and decreases to unity when $f = f_{max}$, the so-called "maximum oscillation frequency." A larger f_T, a smaller r_b, and a smaller C_{bc} are clearly desired to increase the maximum power gain, which is realized by conjugate matching of the output impedance. In practice, f_{max} is determined from the maximum available power gain $(G_{a,max})$, which can be proven to be identical to the maximum operating power gain $G_{p,max}$, as well as the maximum transducer power gain $G_{t,max}$, when there exists simultaneous conjugate matching at both the input and the output. More detail on the significance of these various power gains will be described in the following sections.

5.4 Linear Two-Port Parameters

The small-signal RF performance of a SiGe HBT is typically described by a set of two-port parameters, including Z-, Y-, H-, $ABCD$-, and S-parameters. These parameters can be easily converted from one to another through matrix manipulation. For equivalent circuit based analysis, the Y-parameters are usually the most convenient, while for RF and microwave measurements, the scattering parameters (S-parameters) are almost exclusively used for practical reasons. Note that all of the various two-port parameters represent the same electrical network, but simply use different dependent and independent variables, including voltages, currents, or traveling waves, as shown in Figure 5.9.

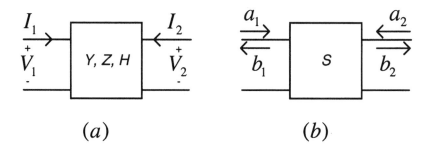

Figure 5.9 (a) Y-, Z- or H-parameters describe the relations among terminal currents and voltages of a linear network. (b) S-parameters describe the relations between the voltage waves, defined as independent linear combinations of terminal currents and voltages.

5.4.1 Z-Parameters

Using I_1, I_2 as independent variables and V_1, V_2 as dependent variables, the Z-parameters are defined by

$$\begin{pmatrix} V_1 \\ V_2 \end{pmatrix} = \begin{pmatrix} Z_{11} & Z_{12} \\ Z_{21} & Z_{22} \end{pmatrix} \begin{pmatrix} I_1 \\ I_2 \end{pmatrix}. \quad (5.27)$$

Note that I_1 or I_2 can be set to zero by terminating the input or output with an open circuit. The Z-parameters can then be determined. For instance, $Z_{11} = V_1/I_1|_{I_2=0}$.

Dynamic Characteristics 155

5.4.2 Y-Parameters

Using V_1, V_2 as independent variables, and I_1, I_2 as dependent variables, the Y-parameters are defined by

$$\begin{pmatrix} I_1 \\ I_2 \end{pmatrix} = \begin{pmatrix} Y_{11} & Y_{12} \\ Y_{21} & Y_{22} \end{pmatrix} \begin{pmatrix} V_1 \\ V_2 \end{pmatrix}. \qquad (5.28)$$

The Y-parameters can be determined using short-circuit terminations at the input or the output.

5.4.3 H-Parameters

Using I_1, V_2 as independent variables, and I_2, V_1 as dependent variables, the H-parameters are defined by

$$\begin{pmatrix} V_1 \\ I_2 \end{pmatrix} = \begin{pmatrix} H_{11} & H_{12} \\ H_{21} & H_{22} \end{pmatrix} \begin{pmatrix} I_1 \\ V_2 \end{pmatrix}. \qquad (5.29)$$

H_{11} is essentially the input impedance with the output short circuited ($V_2 = 0$), and H_{21} is the current gain I_2/I_1 with the output short circuited. H_{11} is used to extract the base resistance, and H_{21} is used to extract f_T. Measurement of the H-parameters involves setting I_1 and V_2 to zero.

5.4.4 S-Parameters

At high frequencies, accurate open and short circuits are extremely difficult to achieve because of the inherent parasitic inductances and capacitances. Consequently, the device under test (DUT) often oscillates with open or short terminations. The interconnection between the DUT and test equipment is also comparable to the wave length, requiring the consideration of distributive effects. Because of these practical difficulties, S-parameters were developed and are almost exclusively used to characterize transistor RF and microwave performance.

S-parameters contain no more and no less information than the Z-, Y-, or H-parameters introduced above. The only difference is that the independent and dependent variables are no longer simple voltages and currents. Instead, linear combinations of the simple variables are used to produce four "voltage waves," which contain the same information since they are chosen to be linearly independent. These combinations are chosen such that they can be physically measured at high frequencies using transmission line techniques. One can understand this formulation as a simple transform of the Y-, Z- or H-parameters into a new form, just like one can transform an impedance Z to a voltage reflection coefficient Γ

$$\Gamma = \frac{Z - Z_0}{Z + Z_0}, \tag{5.30}$$

where Z_0 is a characteristic impedance. Such a transform from Z to Γ is extremely useful in studying transmission lines, and the various definitions of two-port parameters provide a similar utility.

The newly defined voltage wave variables a_1, b_1, a_2, and b_2 are shown in Figure 5.9(b), where a indicates incident, and b indicates reflection or scattering. The waves are related to port voltages and currents by

$$a_1 = \frac{V_1 + Z_0 I_1}{2\sqrt{Z_0}}, \tag{5.31}$$

$$b_1 = \frac{V_1 - Z_0 I_1}{2\sqrt{Z_0}}, \tag{5.32}$$

$$a_2 = \frac{V_2 + Z_0 I_2}{2\sqrt{Z_0}}, \tag{5.33}$$

$$b_2 = \frac{V_2 - Z_0 I_2}{2\sqrt{Z_0}}. \tag{5.34}$$

The voltage waves are defined using voltages and currents for a characteristic impedance Z_0, similar to the definition of Γ in transmission lines. These voltage waves are not "voltages" per se, but voltages normalized to a $2\sqrt{Z_0}$ term such that when squared they have dimensions of power. The voltages a_1 and a_2 are called the incident waves, and b_1 and b_2 are called the scattered waves. The scattered waves are related to the incident waves by a set of linear equations, just as the port voltages are related to the port currents by the Z-parameters

$$\begin{pmatrix} b_1 \\ b_2 \end{pmatrix} = \begin{pmatrix} S_{11} & S_{12} \\ S_{21} & S_{22} \end{pmatrix} \begin{pmatrix} a_1 \\ a_2 \end{pmatrix}. \tag{5.35}$$

The coefficients of these relationships are the S-parameters. One can mathematically prove that the resulting S-parameters are unique for a given linear network, just as they are for the Z-, Y-, and H-parameters.

The measurement of S-parameters involves setting a_1 and a_2 to zero, which is easily accomplished by terminating the ports with Z_0. For instance, to set $a_2 = 0$, we terminate port 2 with Z_0. As a result, $v_2 = -I_2 Z_0$, and thus $a_2 = 0$ according to the definition of a_2. Using the definitions of a_1 and b_1, S_{11} is then obtained as

$$S_{11} = \frac{b_1}{a_1} = \frac{V_1 - I_1 Z_0}{V_1 + I_1 Z_0} = \frac{Z_{in,0} - Z_0}{Z_{in,0} + Z_0}, \tag{5.36}$$

Dynamic Characteristics 157

where $Z_{in,0} = V_1/I_1$ is the input impedance with $Z_l = Z_0$. We see then that S_{11} is therefore simply the reflection coefficient corresponding to the input impedance when the output is terminated with Z_0. The required condition for S-parameter measurements is hence termination with the proper characteristic impedance, just as for short-circuit termination for Y-parameters, or open-circuit termination for Z-parameters. Similarly,

$$S_{21} = \frac{b_2}{a_1} = \frac{V_2 - I_2 Z_0}{V_1 + I_1 Z_0} = 2\frac{V_2}{V_1 + I_1 Z_0} = 2\frac{V_2}{V_s}, \quad (5.37)$$

where $V_s = V_1 + I_1 Z_0$ is equal to the source voltage if a source impedance Z_s is chosen to be Z_0, as illustrated in Figure 5.10. We note that $Z_s = Z_0$ is indeed used in practical S-parameter measurements. We see that S_{21} is simply twice the ratio of V_{out} to V_s for a Z_0 source and a Z_0 load. This relationship provides a simple means of calculating S_{21} and S_{11} using the transistor equivalent circuit, and understanding the physical meanings of S_{21} and S_{11} in terms of impedance and voltage gain, which are familiar to analog designers. Another physical meaning of S_{21} is that $|S_{21}|^2$ gives the transducer gain for a Z_0 source and Z_0 load.

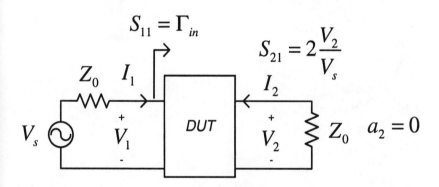

Figure 5.10 A simple method of calculating S_{11} and S_{21}. With a Z_0 drive and a Z_0 load, S_{11} is the input reflection coefficient looking into port 1, and S_{21} is twice the voltage gain V_2/V_s.

The measurements of S_{22} and S_{12} are similar. We terminate port 1 with a Z_0 load, and drive port 2 with a Z_0 source. S_{22} is essentially the output reflection coefficient looking back into the output port for a Z_0 source termination, S_{12} is the reverse gain, and $|S_{12}|^2$ is the reverse transducer gain for a Z_0 source and a Z_0 load.

Because of their intuitive relationship to the reflection coefficients, S_{11} and S_{22} are conveniently displayed on a Smith chart, while S_{21} and S_{12} are typically displayed on a polar plot. Figure 5.11(a) and (b) show an example of the S_{11} and S_{21} measured from 4 to 40 GHz for a SiGe HBT. Two collector currents of 1.26 mA and 25.0 mA are shown, with $V_{CB} = 1$ V. We see that the S_{11} for a bipolar transistor always moves clockwise as frequency increases on the Smith chart. The S_{11} data at higher I_C in general shows a smaller negative reactance, because of the higher EB diffusion capacitance. The S_{21} magnitude decreases with increasing frequency, as expected, because of decreasing forward transducer gain, while S_{21} is larger at higher I_C because of the higher f_T at that bias current. It follows from the above discussions that the S-parameters of a SiGe HBT will intimately depend on the transistor size, biasing condition, and operating frequency.

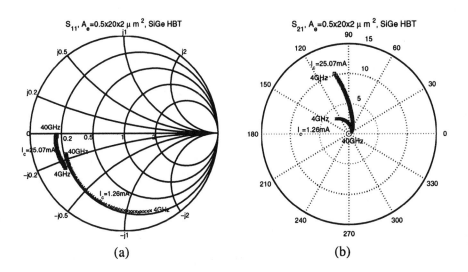

Figure 5.11 Example plots of (a) S_{11} and (b) S_{21} measured data for a SiGe HBT. Two traces are for $I_C = 1.26$ and 25 mA, with $V_{CB} = 1$ V. The frequency range is from 4 to 40 GHz, and $A_E = 0.5 \times 20 \times 2 \mu m^2$.

5.5 Stability, MAG, MSG, and Mason's U

5.5.1 Stability

One of the problems encountered in high frequency transistor amplifier design is oscillation, which can occur when the resistive input or output is *negative*. Negative resistance or conductance means that when a voltage is applied, the current flows from the lower potential end to the higher potential end, as in a battery. Negative resistance is indeed intentionally introduced in oscillator design, but is undesirable for amplifier design. Consider the general purpose transistor amplifier terminated with a $Y_s = G_s + jB_s$ source and a $Y_l = G_l + jB_l$ load in Figure 5.12(a). Using Y_s and Y_l as "boundary conditions" to $I = YV$, the input and output admittances are found to be

$$Y_{in} = Y_{11} - \frac{Y_{12}Y_{21}}{Y_l + Y_{22}}, \qquad (5.38)$$

$$Y_{out} = Y_{22} - \frac{Y_{21}Y_{12}}{Y_s + Y_{11}}. \qquad (5.39)$$

Notice that Y_{in} depends on Y_l, and in general Y_{out} depends on Y_s. These interactions are weak when Y_{12} is small (i.e., if $Y_{12} = 0$, $Y_{in} = Y_{11}$, and $Y_{out} = Y_{22}$). This condition is called the "unilateral assumption."

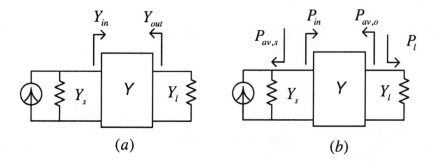

Figure 5.12 (a) Input and output admittance of an amplifier with Y_s source and Y_l load terminations. (b) Various power definitions of interest in amplifier design.

The amplifier is likely to oscillate when $\Re(Y_{in})$ or $\Re(Y_{out})$ is negative. The transistor is said to be "unconditionally" stable if both $\Re(Y_{in})$ and $\Re(Y_{out})$ are positive for *any positive* values of G_s and G_l. In other words, any passive load or source must produce a stable condition. Thus, the device is unconditionally stable if

$$K = \frac{2G_{11}G_{22} - \Re(Y_{12}Y_{21})}{|Y_{21}Y_{12}|} > 1, \qquad (5.40)$$

where G_{11} and G_{22} are the real parts of Y_{11} and Y_{22}. Here, K is the "Rollett stability factor" and is equal to the inverse of the "Linvil stability factor." When the 0 <

$K < 1$, amplifier stability can still be achieved by adding a parallel resistor with the output. The idea is to increase $\Re(Y_{out})$, which often increases $\Re(Y_{in})$ as well. Adding a resistor to the input works the same way, but degrades the overall noise figure, and is a less desirable approach. This K factor has direct implications on the various power gains, as detailed below.

5.5.2 Power Gain Definitions

A number of power gains can be defined to characterize the power transmission of a transistor amplifier. Consider the general purpose transistor amplifier shown in Figure 5.12(b) with a Y_s source and Y_l load. Four powers can be identified:

- $P_{av,s}$: power available from the source;
- P_{in}: power delivered to the amplifier input;
- $P_{av,o}$: power available from the amplifier output;
- P_l: power delivered to the load.

The power transmission gain is usually expressed by one of the following gains:

$$G_p = \frac{P_l}{P_{in}} \quad \text{operating power gain,} \quad (5.41)$$

$$G_a = \frac{P_{av,o}}{P_{av,s}} \quad \text{available gain,} \quad (5.42)$$

$$G_t = \frac{P_l}{P_{av,s}} \quad \text{transducer gain,} \quad (5.43)$$

where G_p is useful for power amplifiers, G_a is useful for low-noise amplifiers, and G_t is useful for cascaded circuits. All of these gains can be expressed as a function of the transistor Y-parameters Y_s and Y_l, through straightforward circuit analysis. For instance, G_t is given by

$$G_t = \frac{4G_s G_l |Y_{21}|^2}{|(Y_s + Y_{11})(Y_l + Y_{22}) - Y_{12}Y_{21}|^2}. \quad (5.44)$$

5.5.3 MAG and MSG

Of particular interest is simultaneous conjugate matching at both the input and the output

$$Y_s = Y_{in}^*, \quad (5.45)$$

$$Y_l = Y_{out}^*. \quad (5.46)$$

As a result, $P_{av,s} = P_{in}$, and $P_{av,o} = P_l$, and thus G_p, G_a, and G_t are equal to each other, and maximized. Therefore, "maximum available gain" (*MAG* or $G_{a,max}$), "maximum operating power gain" ($G_{p,max}$), and "maximum transducer gain" ($G_{t,max}$) are all identical when simultaneous conjugate matching exists. This fact justifies the use of $G_{p,max}$ in our previous derivation of f_{max}. Because of the interaction between input and output, only one particular Y_s and Y_l combination gives simultaneous conjugate matching. Solving (5.45) and (5.46) leads to

$$G_s = \frac{|Y_{12}Y_{21}|}{2G_{22}}\sqrt{K^2 - 1}, \qquad (5.47)$$

$$B_s = -B_{11} + \frac{\Im(Y_{12}Y_{21})}{2G_{22}}, \qquad (5.48)$$

$$G_l = \frac{|Y_{12}Y_{21}|}{2G_{11}}\sqrt{K^2 - 1}, \qquad (5.49)$$

$$B_l = -B_{22} + \frac{\Im(Y_{12}Y_{21})}{2G_{11}}, \qquad (5.50)$$

where G_{ij} and B_{ij} are real and imaginary parts of Y_{ij}, and $Y_{ij} = G_{ij} + jB_{ij}$, $i, j = 1, 2$, and K is the stability factor. Using the above expressions and (7.32), the maximum gain can be derived as

$$MAG = G_{a,max} = G_{p,max} = G_{t,max} = \frac{|Y_{21}|}{|Y_{12}|}\left(K - \sqrt{K^2 - 1}\right). \qquad (5.51)$$

Problems arise, however, when $K < 1$. In this case G_s and G_l, and hence *MAG*, are undefined for a negative argument to the square root. That is, simultaneous conjugate matching does not exist when $K < 1$. In such cases, the transistor must be stabilized by adding resistors at the input or output port, or by adding feedback.

In the absence of *MAG*, the first term in the *MAG* expression can be used as a figure-of-merit for the transistor. This figure-of-merit is called the *maximum stable gain* (*MSG*), and is equal to the *MAG* that would be obtained for $K = 1$, or

$$MSG = \frac{|Y_{21}|}{|Y_{12}|} = \frac{|S_{21}|}{|S_{12}|}. \qquad (5.52)$$

When K is equal to 1, *MSG*=*MAG*. When K is very large, a large margin of stability results, and the maximum available gain goes to zero. Thus, a fundamental tradeoff between stability and gain must be made in transistor circuit design. We note that *MSG* can be calculated even when $K > 1$ or *MAG* exists. Stablization can be accomplished by adding input and/or output shunt resistors. The idea is to increase the real parts of Y_{in} and Y_{out} without affecting *MSG*.

5.5.4 Mason's Unilateral Gain

In transistor amplifier design, a zero reverse transmission gain is often desired. This requires $S_{12} = 0$ or $Y_{12} = 0$ or $Z_{12} = 0$, which can be obtained by adding a feedback network around the transistor. Ideally, this feedback network should be loss-less. The transistor is then said to be "unilateralized." Inevitably, the Y-, Z-, or S-parameters of the resulting unilateralized transistor circuit are different from the two-port parameters of the original transistor. The MAG of the unilateralized transistor circuit that uses loss-less feedback is defined to be "Mason's unilateral gain" (U), and is related to the Y-parameters of the original transistor by [3]

$$U(\text{Mason}) = \frac{|Y_{21} - Y_{12}|^2}{4(G_{11}G_{22} - G_{12}G_{21})}. \qquad (5.53)$$

We note that the original derivation was made using Z-parameters. The expression has the same form as that shown above, except that all the Ys and Gs are replaced by Zs and Rs.

In practice, transistor maximum oscillation frequency f_{max} is often defined as the frequency at which the -20 dB/decade extrapolation of U drops to unity. Because U is the *MAG* of the unilateralized transistor circuit and hence contains a (loss-less) feedback network, we naively expect U to be different from the *MAG* of the transistor *itself*, as is often observed experimentally. The other impact of unilateralization is that the resulting transistor circuit is often unconditionally stable, even if the transistor itself is potentially unstable. As a result, *MAG* of the unilateralized transistor circuit, which is the U of the transistor by definition, exists, even if the *MAG* of the transistor itself does not exist. The loss-less feedback network needed for unilateralization is in general dependent on frequency, and is therefore narrow band. Clearly U is not necessarily the highest gain obtainable from the transistor. One can design networks around the transistor to produce gain higher than U, but with a correspondingly worse reverse isolation.

5.5.5 Which Gain Is Better?

A practical question presents itself. Which gain definition makes better sense, and should be used for experimental f_{max} extraction? This issue has emerged as a fairly contentious issue in the literature since some groups use U, while other groups use *MAG*, and the results are often quite different. From a practical standpoint, U is easier to extract (and is often less noisy), because U in general exists and can be measured at relatively lower frequencies. On the other hand, *MAG* often does not exist at lower frequencies, and requires measurements up to higher frequencies. For wireless designs targeted for a few gigahertz, there is a good chance that *MAG*

Dynamic Characteristics 163

does not exist. This is true, for instance, in SiGe HBTs with f_T above 100 GHz. In these situations, U is certainly more physical than MSG, which is the MAG for a stabilization that produces $K = 1$, the realization of which requires the use of input and/or output shunt resistors that are lossy. From a theoretical standpoint, U represents the MAG of a well-designed transistor circuit that has zero reverse transmission (i.e., perfect reverse isolation), while MAG is just the MAG of the transistor itself, as is, which may not even exist at the design frequency of interest.

Figure 5.13 shows the measured MAG, MSG, and Mason's U versus frequency for a SiGe HBT biased near its peak f_T. Observe that MAG exists only at higher frequencies where the K stability factor is larger than 1, while MSG and U exist at all frequencies measured. The measured values of MAG and U are fairly close over a large frequency range where both MAG and U decrease at a slope of -20 dB/decade. The f_{max} extracted from MAG and Mason's U are 96.2 and 113.4 GHz, respectively. [2]

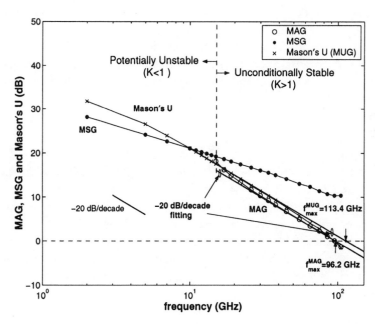

Figure 5.13 Measured MAG, MSG, and Mason's U versus frequency for a SiGe HBT biased near peak f_T.

---

[2]The f_{max} extracted from U data is almost always found to be higher than that extracted from MAG data, and hence has been logically selected by many groups as the favored approach! In light of the discussion above, however, reason would suggest that both numbers should be presented for any given technology.

Recall that we derived the classical f_{max} expression using MAG. We state here without proof that the use of Mason's U results in the same f_{max} expression. An important underlying assumption responsible for this "agreement" between the two definitions is that the simple transistor equivalent circuit used in the derivations continues to hold for both. The most significant assumption here is that all of the CB capacitances are placed between the intrinsic base and collector nodes. An extrinsic capacitance that appears between the extrinsic base and collector nodes can significantly degrade MAG, but not Masons's U. It does not affect Mason's U because any capacitances between the extrinsic terminals can be neutralized by inductances during unilateralization. This can also be readily understood from (5.53), because a capacitance between the external C and B nodes does not change the real parts of all the Y-parameters, and the changes to the imaginary parts of Y_{12} and Y_{21} are identical and cancel out.

We again emphasize that the source and load termination networks as well the feedback network required to achieve MAG and U are narrowband by definition. Therefore, these gains are not naturally good figures-of-merit for high-speed circuits used in optical communications systems with data rates above, say, 10 Gb/sec. For a given HBT technology, the degree of agreement between the f_{max} extrapolated from Mason's U and MAG depends on whether the CB capacitance is dominated by the intrinsic CB capacitance.

5.5.6 f_T Versus f_{max} Versus Digital Switching Speed

There is often heated debate concerning whether f_T or f_{max} is a better speed or bandwidth figure-of-merit. Device design can be tailored for either high f_T or high f_{max}. A careful inspection of the definition of f_T and f_{max}, however, can resolve this confusion. By definition, f_T is the extrapolated frequency at which the small-signal current gain with a shorted output termination is reduced to unity, while f_{max} is the extrapolated frequency at which the small-signal power gain with conjugate matching terminations at both input and output is reduced to unity. Therefore,

- Both f_T and f_{max} are narrowband small-signal parameters, and strongly depend on biasing current and voltage. Hence, they cannot directly predict the performance of large-signal circuits, such as digital circuits, or broadband circuits. They also require *specific* terminations and/or feedback networks.

- Ring oscillator speed or frequency divider data are more direct and accurate measures of digital switching speed. For instance, a record ECL ring oscillator gate delay of 4.3 psec was achieved using 210-GHz peak f_T SiGe HBT technology [4].

- Both f_T and f_{max} are *frequency domain* small-signal parameters. They do not contain any phase information for the input-output relations, or information regarding the time domain transient response, which is clearly important for many circuits (for ringing, etc.).

From an RF and analog circuit design standpoint, different circuits have different requirements on f_T and f_{max} [5]:

- For tuned ICs (e.g., RFICs or monolithic microwave ICs (MMICs)), f_{max} sets the gain and maximum operating frequency, but a low f_T/f_{max} ratio makes tuning difficult in most Si technologies.

- For lumped analog ICs such as transimpedance amplifiers found in optical receivers, we need high and comparable f_T and f_{max} values. Typically 1.5:1 f_{max}/f_T is considered good.

- For distributed amplifiers, f_{max} is in principle the performance limiting factor. A low f_T, however, makes design more difficult.

A popular practice today in SiGe HBTs is to tailor device design for similar values of f_T and f_{max}. This is a sound methodology since SiGe HBTs are being targeted for many different types of circuit applications.

5.6 Base and Emitter Transit Times

To understand the dynamic response of the SiGe HBT, and the role Ge plays in transistor frequency response, we must first formally relate the changes in the base transit time and emitter transit time to the physical variables of this problem. It is also instructive to carefully compare the *differences* between a comparably constructed SiGe HBT and a Si BJT. In the present analysis, the SiGe HBT and the Si BJT are taken to be of identical geometry, and it is assumed that the emitter, base, and collector doping profiles of the two devices are identical, apart from the Ge in the base of the SiGe HBT. For simplicity, a Ge profile that is linearly graded from the EB to CB junctions is assumed, as depicted in Figure 5.14.

5.6.1 τ_b in SiGe HBTs

The theoretical consequences of the Ge-induced bandgap changes to τ_b can be derived in closed-form for a constant base doping profile ($p_b(x) = N_{ab}^-(x) = N_{ab}^-$ = constant) by considering the generalized Moll-Ross transit time relation, which

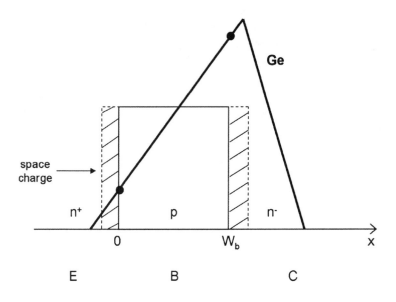

Figure 5.14 Schematic doping and Ge profiles used in the derivations.

holds for low-injection in the presence of both nonuniform base doping and nonuniform base bandgap at fixed V_{BE} and T [6]

$$\tau_b = \int_0^{W_b} \frac{n_{ib}^2(x)}{p_b(x)} \left[\int_x^{W_b} \frac{p_b(y)\,dy}{D_{nb}(y)\,n_{ib}^2(y)} \right] dx. \quad (5.54)$$

Following the analysis in Section 4.2, we can insert (4.3) into (4.2) to obtain (4.4), and substitute (4.4) into (5.54) to obtain

$$\tau_{b,SiGe} = \int_0^{W_b} \frac{n_{ib}^2(x)}{N_{ab}^-} \Biggl\{ \int_z^{W_b} \frac{N_b^-}{D_{nb}} \left[\frac{1}{\gamma n_{io}^2} e^{-\Delta E_{gb}^{app}/kT} e^{-\Delta E_{g,Ge}(0)/kT} \right. $$
$$\left. \cdot\, e^{-\Delta E_{g,Ge}(grade)y/W_b kT} \, dy \right] \Biggr\} dx. \quad (5.55)$$

Performing the first integration step yields,

$$\tau_{b,SiGe} = \int_0^{W_b} \frac{n_{ib}^2(x)}{N_{ab}^-} \Biggl\{ \frac{-N_{ab}^- W_b}{\widetilde{D}_{nb}\widetilde{\gamma} n_{io}^2} \frac{kT}{\Delta E_{g,Ge}(grade)} e^{-\Delta E_{gb}^{app}/kT} e^{-\Delta E_{g,Ge}(0)/kT}$$
$$\cdot \left[e^{-\Delta E_{g,Ge}(grade)/kT} - e^{-\Delta E_{g,Ge}(grade)x/W_b kT} \right] \Biggr\} dx, \quad (5.56)$$

Dynamic Characteristics

where we have accounted for the position dependence in both the mobility and the density-of-states product. Substitution of n_{ib}^2 from (4.4) into (5.56) and multiplying through gives

$$\tau_{b,SiGe} = \left\{ \frac{W_b \, kT \, \tilde{\gamma} \, n_{io}^2}{\widetilde{D}_{nb} \, \tilde{\gamma} \, n_{io}^2 \, \Delta E_{g,Ge}(grade)} \right\}$$

$$\cdot \int_0^{W_b} \left[1 - e^{\Delta E_{g,Ge}(grade) x / W_b \, kT} e^{-\Delta E_{g,Ge}(grade)/kT} \right] dx, \qquad (5.57)$$

which can be integrated and evaluated to obtain, finally [7, 8]

$$\tau_{b,SiGe} = \frac{W_b^2}{\widetilde{D}_{nb}} \frac{kT}{\Delta E_{g,Ge}(grade)}$$

$$\cdot \left\{ 1 - \frac{kT}{\Delta E_{g,Ge}(grade)} \left[1 - e^{-\Delta E_{g,Ge}(grade)/kT} \right] \right\}. \qquad (5.58)$$

As expected, we see that the base transit time in a SiGe HBT depends reciprocally on the amount of Ge-induced bandgap grading across the neutral base (i.e., for fixed base width, the band edge-induced drift field). It is instructive to compare τ_b in a SiGe HBT with that of a comparably designed Si BJT. In the case of a Si BJT (trivially derived from (5.54) for constant base doping and bandgap), we know that

$$\tau_{b,Si} = \frac{W_b^2}{2D_{nb}}, \qquad (5.59)$$

and hence can write

$$\frac{\tau_{b,SiGe}}{\tau_{b,Si}} = \frac{2}{\tilde{\eta}} \frac{kT}{\Delta E_{g,Ge}(grade)}$$

$$\cdot \left\{ 1 - \frac{kT}{\Delta E_{g,Ge}(grade)} \left[1 - e^{-\Delta E_{g,Ge}(grade)/kT} \right] \right\}, \qquad (5.60)$$

where we have used the ratio of electron diffusivities between SiGe and Si (4.11). Within the confines of our assumptions stated above, this can be considered an exact result. As expected from our intuitive discussion of the band diagram, observe that τ_b and hence f_T in a SiGe HBT depend reciprocally on the Ge-induced bandgap grading factor, and hence for finite Ge grading across the neutral base, τ_b is less than unity, and thus we expect enhancement in f_T for a SiGe HBT compared to a comparably constructed Si BJT. Figure 5.15 confirms this expectation experimentally. As can be seen in Figure 5.15, since f_T is increased across a

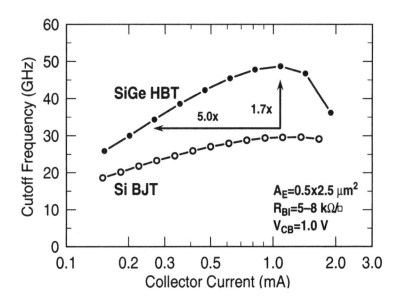

Figure 5.15 Measured comparison of unity gain cutoff frequency f_T as a function of bias current for a comparably constructed SiGe HBT and Si BJT.

large range of useful collector current, we can potentially gain dramatic savings in power dissipation for fixed frequency operation compared to a Si BJT. This power-for-performance trade-off can in practice be even more important than the sheer increase in frequency response, particularly for portable applications. In this case, if we decided, for instance, to operate the transistor at a fixed frequency of 30 GHz, we could reduce the supply current by a factor of 5×. Note as well, that as for the collector current density expression (4.15), the thermal energy (kT) plays a key role in (5.58), in this case residing in the numerator, and will thus have important favorable implications for SiGe HBT frequency response at cryogenic temperatures, as will be discussed in detail in Chapter 9.

Theoretical calculations using (5.60) as a function of Ge profile shape are shown in Figure 5.16 at 300 K and 77 K (the integrated Ge content is held fixed, and the Ge profile varies from a 10% triangular (linearly graded) to a 5% box (constant) Ge profile) [9].

5.6.2 Relevant Approximations

In similar manner as that for the collector current density (refer to Chapter 4), two physically relevant approximations can be made to obtain additional insight. First, we can assume that $\Delta E_{g,Ge}(grade) \gg kT$. This approximation can be termed the

Dynamic Characteristics

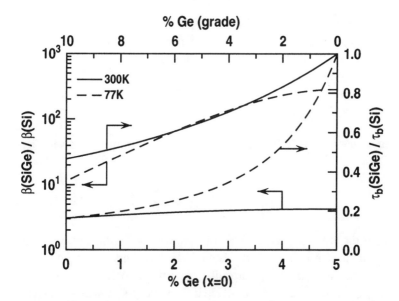

Figure 5.16 Theoretical calculations of base transit time ratio Γ as a function of Ge profile shape.

strong Ge grading scenario. In this case (5.58) reduces to

$$\tau_{b,SiGe} \simeq \frac{W_b^2}{2\widetilde{D}_{nb}} \frac{kT}{\Delta E_{g,Ge}(grade)}. \tag{5.61}$$

Note, however, that care should be exercised in applying this approximation. To check its validity for a realistic profile, assume that we have a 0% to 15% triangular Ge profile in a SiGe HBT operating at 300 K. Taking a band offset of roughly 75 meV per 10% Ge, we find that $\Delta E_{g,Ge}(grade)/kT = 4.3$ compared to unity, a reasonable but not overly compelling approximation. Clearly, however, as the temperature drops, the validity of this approximation improves rapidly as kT decreases. For example, in the case above, a 0% to 15% triangular Ge profile yields $\Delta E_{g,Ge}(grade)/kT = 17.0$ at 77 K, clearly $\gg 1$.

In addition to the strongly graded profile, we also can define an approximation for a weak Ge grading, that would be valid, for instance, in the case of a Ge box profile. In this case, $\Delta E_{g,Ge}(grade) \ll kT$. By expanding the exponential of the Ge grading factor in a Taylor's series, and canceling terms, we obtain

$$\tau_{b,SiGe} \simeq \frac{W_b^2}{2\widetilde{D}_{nb}}, \tag{5.62}$$

which is just the Si BJT result (5.59). As expected, we see that Ge grading is key to obtaining the desired frequency response improvement over a comparably constructed Si BJT, since the drift field that aids electron transport through the base is induced by compositional Ge grading.

5.6.3 τ_e in SiGe HBTs

The emitter charge storage time (τ_e) in a polysilicon-emitter contacted SiGe HBT can be written as [10]

$$\tau_{e,SiGe} \simeq \frac{1}{\beta_{ac}} \left(\frac{W_e}{S_{pe}} + \frac{W_e^2}{2D_{pe}} \right), \quad (5.63)$$

where S_{pe} is the hole surface recombination velocity at the the emitter contact, W_e is the neutral emitter width, and D_{pe} is the hole diffusivity in the emitter. We see, then, that τ_e in a SiGe HBT is reciprocally proportional to the *ac* current gain of the transistor (β_{ac}). For a graded-base SiGe HBT and a Si BJT with identical emitter contact technology (i.e., identical base currents), we can use (4.10) and (5.59) to write

$$\frac{\tau_{e,SiGe}}{\tau_{e,Si}} \simeq \frac{J_{C,Si}}{J_{C,SiGe}} = \frac{1 - e^{-\Delta E_{g,Ge}(grade)/kT}}{\widetilde{\gamma \eta} \frac{\Delta E_{g,Ge}(grade)}{kT} e^{\Delta E_{g,Ge}(0)/kT}}. \quad (5.64)$$

We thus see that τ_e depends much more strongly on the EB boundary value of the Ge-induced band offset than for τ_b. This can have important implications for technology scaling. In general, the required current gain of the transistor is determined by the given circuit application (e.g., $\beta = 100$), regardless of the technology generation, and thus τ_e's contribution to f_T is roughly fixed. On the other hand, because the base width naturally is thinned with technology evolution, τ_b will inherently decrease with vertical profile scaling. Logically, at some level of technology evolution, τ_b and τ_e will be of comparable magnitude. In this scenario is it far easier to tune τ_e than it is to tune τ_b, given the stronger dependence on the shape of the Ge profile. This is particularly true at decreased temperatures since the band offsets are thermally activated in the τ_e expression.

5.6.4 Other SiGe Profile Shapes

The analysis above holds for a range of Ge profiles between the triangular (linearly graded Ge) and box (constant Ge) profiles. There also exists, however, a class of technologically important Ge profiles that can be considered hybrid combinations of the triangular and box Ge profiles, which we will call Ge trapezoids, as depicted schematically in Figure 5.17. In this case, one takes a linearly graded profile and

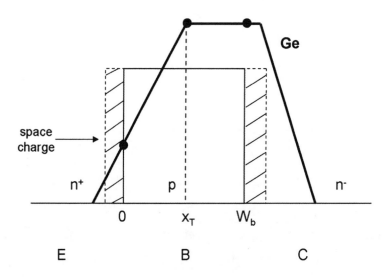

Figure 5.17 Schematic representation of the hybrid Ge trapezoidal profile.

truncates the grading at some intermediate position x_T in the neutral base, and the Ge content is held constant from x_T to W_b, and is then ramped down to zero as usual. At constant Ge stability and for fixed Ge profile width, this Ge trapezoidal profile design approach allows one to induce higher Ge grading across the more heavily doped EB end of the base, thereby maintaining good dynamic response, while using lower peak Ge content. The region of constant Ge in the neutral base, at least in principle, does not degrade *ac* performance since the CB side of the neutral base typically will have a doping-gradient-induced drift field in addition to the Ge-grading-induced drift field, which will in itself aid electron transport. Following the analysis in Chapter 4, for the Ge trapezoid, one can also derive base and emitter transit time expressions in the presence of constant base doping. In this case, the Ge-induced band offsets can be written as

$$\Delta E_{g,Ge}(x) = \begin{cases} \Delta E_{g,Ge}(0) + \Delta E_{g,Ge}(grade)\left(\frac{x}{x_T}\right) & ,0 \leq x \leq x_T \\ \Delta E_{g,Ge}(W_b) & ,x_T \leq x \leq W_b, \end{cases} \quad (5.65)$$

and the intrinsic carrier density is then [11]

$$n_{ib}^2(x) = \begin{cases} \gamma n_{io}^2 e^{\Delta E_{gb}^{app}/kT} e^{\Delta E_{g,Ge}(0)/kT} e^{[\Delta E_{g,Ge}(grade)x/x_T]/kT} & ,0 \leq x \leq x_T \\ \gamma n_{io}^2 e^{\Delta E_{gb}^{app}/kT} e^{\Delta E_{g,Ge}(W_b)/kT} & ,x_T \leq x \leq W_b. \end{cases} \quad (5.66)$$

In this case, the generalized Moll-Ross relation can be written as

$$\tau_b = \int_0^{x_T} \frac{n_{ib}^2(x)}{p_b(x)} \left\{ \int_x^{x_T} \frac{p_b(y)\,dy}{D_{nb}(y)\,n_{ib}^2(y)} + \int_{x_T}^{W_b} \frac{p_b(y)\,dy}{D_{nb}(y)\,n_{ib}^2(y)} \right\} dx$$
$$+ \int_{x_T}^{W_b} \frac{n_{ib}^2(x)}{p_b(x)} \left\{ \int_x^{W_b} \frac{p_b(y)\,dy}{D_{nb}(y)\,n_{ib}^2(y)} \right\} dx. \tag{5.67}$$

Substitution of (5.66) into (5.67) and evaluating yields [9],

$$\frac{\tau_{b,SiGe}}{\tau_{b,Si}} = \frac{2kT\,\xi^2}{\widetilde{\eta}\,\Delta E_{g,Ge}(grade)} \left\{ \left[\frac{1}{\xi} - 1 - \frac{kT}{\Delta E_{g,Ge}(grade)} \right] \right.$$
$$\left. \cdot \left[1 - e^{-\Delta E_{g,Ge}(grade)/kT} \right] + 1 + \frac{\Delta E_{g,Ge}(grade)}{2kT} \left[\frac{1}{\xi} - 1 \right]^2 \right\}, \tag{5.68}$$

where we have defined $\xi = x_T/W_b < 1$ to be the normalized trapezoidal intermediate boundary point. In this case, $\xi = 0$ corresponds to the pure Ge box profile, and $\xi = 1$ corresponds to the pure triangular Ge profile. We can evaluate the limits of both zero and infinite grading. If we define $\delta = \Delta E_{g,Ge}(grade)/kT$ then, for zero Ge grading,

$$\lim_{\delta \to 0} \left(\frac{\tau_{b,SiGe}}{\tau_{b,Si}} \right) = 1 \tag{5.69}$$

and we see that there is no improvement compared to a Si BJT, as expected. On the other hand, if δ becomes infinitely large (either by very high Ge grading or by operating at very low temperatures), then the ultimate performance improvement in τ_b is given by

$$\lim_{\delta \to \infty} \left(\frac{\tau_{b,SiGe}}{\tau_{b,Si}} \right) = \left(1 - \frac{1}{\xi} \right)^2. \tag{5.70}$$

We can plot the τ_b ratio as a function of reciprocal temperature for varying ξ values, as shown in Figure 5.18 (shown here for fixed Ge content). Expressions for τ_e for the Ge trapezoid can be easily obtained using the results for J_C in Chapter 4 to obtain

$$\frac{\tau_{e,SiGe}}{\tau_{e,Si}} = \frac{\frac{\xi kT}{\Delta E_{g,Ge}(grade)} + \left\{ 1 - \xi \left(1 + \frac{kT}{\Delta E_{g,Ge}(grade)} \right) \right\} e^{-\Delta E_{g,Ge}(grade)/kT}}{\widetilde{\gamma\eta}\, e^{\Delta E_{g,Ge}(0)/kT}} \tag{5.71}$$

and the functional dependence like that shown in Figure 4.10 ($\tau_e \sim 1/\beta$).

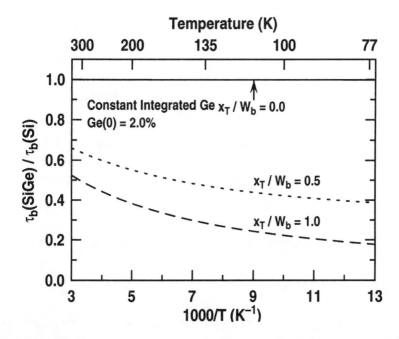

Figure 5.18 Current gain as a function of reciprocal temperature for varying values of ξ. Note that the integrated Ge content is not fixed in this case.

5.6.5 Implications and Optimization Issues for f_T

Based on the analysis above, we can make several observations regarding the effects of Ge on the frequency response of a SiGe HBT:

- For fixed bias current, the presence of Ge in the base region of a bipolar transistor affects its frequency response through the base and emitter transit times.

- The f_T enhancement for a SiGe HBT over a Si BJT depends reciprocally on the Ge grading across the base. This makes sense intuitively given the effects of the grading-induced drift field for the minority carrier transport. This observed dependence on Ge grading will play a role in understanding the best approach to profile optimization for a given application.

- For two Ge profiles of constant stability, a triangular Ge profile is better for cutoff frequency enhancement than a box Ge profile is, everything else being equal, *provided* τ_b is dominant over τ_e in determining f_T. While this is clearly the case in most first generation SiGe HBTs, it is nonetheless conceivable that for a τ_e dominated transistor, a more box-like Ge profile, which inherently favors β enhancement and hence τ_e improvement, might be a favored profile design for optimal frequency response. A compromise trapezoidal profile, which generally favors both τ_b and τ_e improvement, is a logical compromise profile design point. Such trade-offs are obviously technology generation dependent.

- Given that f_T is improved across the entire useful range of I_C, the f_T versus power dissipation trade-off offers important opportunities for portable applications, where power minimization is often a premium constraint.

- The Ge-induced f_T enhancement depends strongly on temperature, and for τ_b and τ_e, is functionally positioned in a manner that will produce a magnification of f_T enhancement with cooling, in stark contrast to a Si BJT.

5.7 ECL Gate Delay

The ECL gate represents the fundamental building block for modern high-speed bipolar-based digital systems. The historical origins of SiGe technology in the mid-late 1980s centered on developing a higher performance replacement of existing Si BJT ECL logic for mainframe computer systems. The ECL ring oscillator remains today a simple and powerful metric for assessing overall technology performance, since it provides more information than that captured by f_T and f_{max}, and yet is much simpler to design and test than a static or dynamic frequency divider. In addition, the frequently followed path in industry today is towards the realization of SiGe HBT BiCMOS technology as a "do-it-all" technology, for SiGe HBT analog, or RF, or digital circuits, integrated with on-chip digital CMOS where lower power is mandated over higher performance. Thus, the use of SiGe HBTs for ECL logic in communications systems remains widespread, and must be carefully understood and assessed. The relevant question in the present context is, how and why does SiGe affect the ECL power-delay performance?

The fundamental basis of the ECL gate is the differential amplifier, or from a digital viewpoint, more appropriately referred to as the "current switch." A current switch combined with emitter-follower output drivers forms the basic single-level ECL gate, as depicted in Figure 5.19 (multiple logic levels are often cascoded in modern ECL designs). The ECL gate is a low-logic-swing, nonsaturating logic family that thus provides high-speed switching, and also combines powerful logical functionality and efficient capacitive load driving capability. The logical outputs include both multi-input OR/NOR functions on a single gate and, in addition, emitter "dotting" of the emitter-follower outputs of multiple logical gates facili-

Dynamic Characteristics

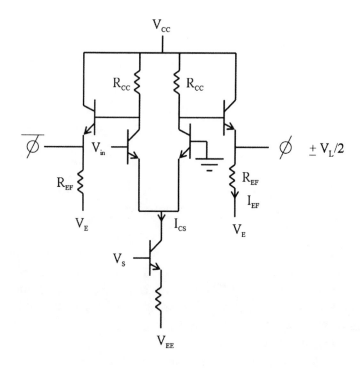

Figure 5.19 Circuit schematic for a generic ECL digital logic gate.

tates multi-input logical AND/NAND functionality. Typical ECL characteristics include: 400–800-mV logic swing, 1–10-mW power dissipation, 3.3–3.6-V supply, and sub-20-picosecond unloaded gate delay. ECL has been and remains today the workhorse high-speed bipolar logic family. Related higher-speed, and lower-power digital logic families such as current mode logic (CML), *ac*-coupled ECL (AC-ECL), and *ac*-coupled push-pull ECL (AC-PP-ECL) [12, 13], are closely related to the basic ECL gate.

5.7.1 ECL Design Equations

Simple first-order ECL design equations can be constructed based on circuit analysis of Figure 5.19, assuming, for instance, that the input is held low and looking at the in-phase output stage. If we imagine fixing V_{CC}, V_{EE}, V_S, and V_E, and the logic swing V_L (typically set by the system specifications: e.g., $V_{CC} = +1.4$ V, $V_{EE} = -1.2$ V, $V_S = -0.5$ V, $V_E = -0.7$ V, and $V_L = 500$ mV), then the requisite resistors, R_{CC}, R_{EF}, R_S, can be calculated according to the following procedure.

First, fix the desired total switch current (I_{CS}) and the ratio between the emitter-follower current (I_{EF}) and the current switch current such that $\overline{I_{EF}} = \kappa I_{CS}$ (typically $\kappa = 0.5 - 2.0$ depending on the size of the capacitive load to be driven). Using the fact that $R_{CC} \simeq V_L/I_{CS}$, we can solve for the required resistor values according to:

$$R_{CC} \simeq \frac{V_{CC} - V_{BE,EF} - I_{EF}R_E + V_L/2}{I_{CS}} \quad (5.72)$$

$$R_{EF} \simeq \frac{-V_L/2 - V_E}{I_{EF}} \quad (5.73)$$

$$R_S \simeq \frac{V_S - V_{BE,S} - I_{CS}R_E - V_{EE}}{I_{CS}}, \quad (5.74)$$

where $V_{BE,EF}$ and $V_{BE,S}$ are the base-emitter voltages at the desired current level, and can either be measured directly for the appropriate device geometry or simply obtained from a compact model. This process can be repeated for multiple current levels to span the desired power dissipation range (e.g., 1–10 mW).

To construct a simple ECL ring oscillator, we can link a string of ECL gates together and feed the last output back to the first input, such that an odd number of signal transitions occurs along the total delay path (to ensure instability). The gate chain should be lengthened sufficiently such that the total signal delay from stage 1 to stage n is long enough that it can be conveniently measured with existing probes and oscilloscope (several nanoseconds of total delay (i.e., < 1-GHz bandwidth) is usually sufficient). It is important in a practical design to configure the first two input stages such that the free-running logic swing can be checked and adjusted as needed to ensure a symmetric logic swing (e.g., ±250 mV). Given the excellent load-driving capability of the ECL, it is generally not necessary to buffer the output signal. If the output is fed directly into a 50-Ω scope input, and the total period of the voltage signal is measured, then the average ECL gate delay can be calculated by

$$\tau_{ECL} = \frac{1}{2} \frac{\text{measured waveform period}}{\text{total number of gates}}. \quad (5.75)$$

The extra factor of two is due to the fact that the signal must propagate twice through the ring to obtain one total period of the voltage waveform. Simple ECL gate delay measurements have consistently correlated well with both calibrated modeling results and direct measurements of the internal node-voltage waveforms, and can be viewed as a simple and very useful technology performance figure-of-merit.

5.7.2 ECL Power-Delay Characteristics

One of the most fundamental questions in digital logic design is to understand how transistor-level design couples to the so-called "power-delay characteristic" of the digital logic gate. The power-delay characteristic of a given logic gate is a fundamental measure of the amount of energy (i.e., ps × mW = fJ) dissipated in the $0 \rightarrow 1$ or $1 \rightarrow 0$ switching event. From a practical standpoint, we can design a series of ECL ring oscillators to span a practical range of switch currents (power dissipation), and experimentally trace out the power-delay performance of the ECL gate designed within a given technology. For ECL, we will inevitably obtain a power-delay curve similar to that depicted schematically in Figure 5.20. To

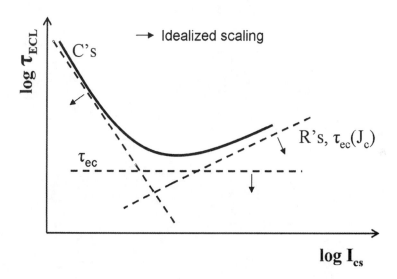

Figure 5.20 Schematic power-delay characteristics of a generic ECL digital logic gate.

understand why this characteristic shape arises, we can gain insight from a basic delay-equation analysis, which represents the switching delay of an ECL logic gate by a series of weighted RC time constants [14].

Using the basic current switch depicted in Figure 5.21, we can divide the overall power-delay performance into a "low-power" and a "high-power" regime. The low-power switching delay is dominated by the charging of the device capacitances through the pull-up current switch resistor

$$\tau_{low-power} \simeq R_{cc} \sum_{k=1}^{n} a_k C_k, \quad (5.76)$$

where a_k are the technology-dependent delay weighting factors. Using the fact that $R_{CC} \simeq V_L/I_{CS}$, we can write

$$\tau_{low-power} \simeq \left(\frac{V_L}{I_{CS}}\right) \{a_1 C_{cb} + a_2 C_{eb} + a_3 C_{cs} + a_4 C_W + \cdots\}. \quad (5.77)$$

or,

$$\tau_{low-power} \propto \frac{1}{I_{CS}} \propto \frac{1}{Power}. \quad (5.78)$$

In practice, C_{cb} is the dominant delay-limiting capacitance due to the Miller effect.

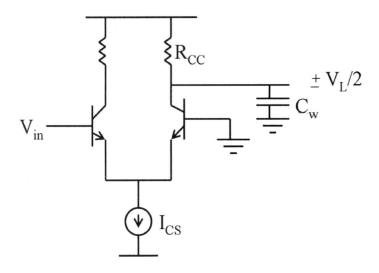

Figure 5.21 Circuit schematic of a generic current switch.

We can thus see that in the low-power regime, the ECL gate delay is reciprocally proportional to the power, and hence it is physically meaningful to plot the power-delay performance on log-log scales, since we expect a slope of -1 in the low-power regime, as depicted in Figure 5.20. This behavior makes intuitive sense given that if the transistor capacitances limit the switching delay, then supplying more current to charge those capacitances should speed up the process.

On the other hand, in the high-power regime, the delay is limited by the charging of the diffusion capacitance (C_{diff}) through the various device resistances, according to

$$\tau_{high-power} \simeq C_{diff} \sum_{k=1}^{n} b_k R_k, \qquad (5.79)$$

where b_k are again technology-dependent delay weighting factors. Since we can write

$$C_{diff} = \frac{q\tau_f I_C}{kT}, \qquad (5.80)$$

we see that ECL delay in the high-power domain can be expressed as

$$\tau_{high-power} \simeq \frac{q\tau_f I_C}{kT} \{b_1 R_{bi} + b_2 R_{bx} + b_3 R_e + b_4 R_c + \cdots\}, \qquad (5.81)$$

or

$$\tau_{high-power} \propto I_{CS} \propto Power. \qquad (5.82)$$

In practice, the intrinsic and extrinsic base resistances, R_{bi} and R_{bx}, are the delay-limiting device resistances. We can thus see that in the high-power regime the ECL gate delay degrades as the power increases, as reflected in Figure 5.20. In reality, in (5.81), both τ_{ec} and R_{bi} are current density dependent, but those dependences are opposite in sign and tend to cancel one another, such that (5.81) remains valid, and the gate delay degrades as the power continues to rise. This again makes intuitive sense, given the fact that if the resistances are dominant, supplying more current only serves to increase the voltage drop across them, slowing down the charging process. In between the low- and high-power domains, the minimum (optimum) ECL gate delay is reached, and is fundamentally limited by the total emitter-to-collector delay τ_{ec}, which can be considered a fundamental limit on ECL switching speed.

5.7.3 Impact of SiGe on ECL Power Delay

With this general analysis of the ECL power-delay characteristics in hand, we can make some intuitive predictions on the impact of SiGe on ECL gate delay. First, we know that introducing Ge in the base region of a bipolar transistor has three tangible consequences:

- β is increased for fixed R_b;
- V_A is increased for fixed R_b;
- f_T is increased for fixed R_b.

It can be easily shown via compact circuit modeling that ECL gate delay is very insensitive to transistor β or V_A (indeed, this is one of the advantages of ECL!), and thus two of the three transistor-level performance metrics affected by SiGe should not strongly influence the ECL performance. The f_T of the transistor, on the other hand, clearly matters to ECL delay, but only at minimum delay (i.e., τ_{ec}, and at high switch current levels (refer to (5.81)). In addition, to first order, SiGe has no impact on the transistor-level capacitances, and thus we would naively expect the SiGe HBT and Si BJT ECL gate delay to converge at sufficiently low current switch currents, since the device capacitances dominate the low-power delay regime (refer to (5.77)). That these assertions are valid in practice can be seen in Figure 5.22.

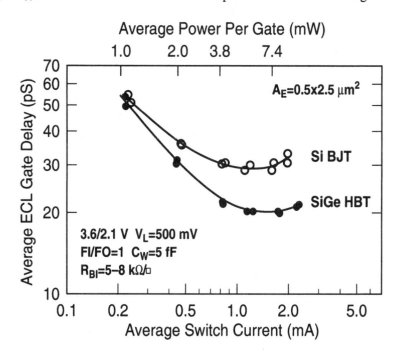

Figure 5.22 Measured power-delay performance of comparable SiGe HBT and Si BJT ECL ring oscillators.

From an "apples-to-apples" comparison (constant doping and processing conditions), we see that the SiGe HBT ECL circuit outperforms the Si BJT ECL gate by a substantial margin (nearly 10 psec), but that improvement does not translate uniformly across the entire power-delay range for practical applications. In addition, this substantial performance improvement at high currents is shown for an unloaded circuit. As circuit loading increases (as it will in any practical implementation), the improvement of SiGe over Si will diminish [15]. This observed limitation in ECL improvement with the addition of SiGe is important because with digital technology evolution, one generally wants to put more gates on a given wafer, and thus the inherent power constraints associated with bipolar digital logic will force switch currents to lower levels, further diminishing the performance delta between Si and SiGe ECL gates. This is not the whole story, of course, since one could in principle trade the higher β in the SiGe HBT for lower R_b and hence further improve the high current performance. Nevertheless, the case for SiGe as a driver of digital bipolar technologies is not overly compelling.

The conclusion here is not as bleak as one might naively anticipate, for two major reasons. First, the approach taken by most companies today is not to develop SiGe for pure digital applications, but rather to optimize SiGe technology for RF and analog applications, where the leverage over Si BJTs is much more compelling, and then, where needed, simply use the available SiGe device for ECL logic. That is, ECL is not the technology driver, but rather is relegated to the follower position. Given this, one cannot forget that 10-psec gate delay is still 10-psec gate delay! That is, one should not get overly mired in the question of merits of SiGe vs Si for ECL (or versus GaAs or InP for that matter), since what ultimately matters is whether the ECL gate switches at the required speed for the application. For 40-Gb/sec optical data links, for instance, one needs very high performance digital circuits. Period. If SiGe can provide that speed at lower cost than III-V it will be the favored technology. All of the Si-technology-compatible circuit records are currently held by SiGe, and will likely continue to be held by SiGe, whether they are optimized for ECL logic or not.

Finally, we note that there exist weighting-factor independent analytical formulations of the ECL gate delay that can be used in circuit sizings to quantify the impact of a given Ge profile change on ECL circuit performance. While these analytical formulations require approximations and are thus obviously not as accurate as full compact model solutions, they do provide very useful insight into the impact of Ge profile design changes on full circuit response. The first and simplest such formulation expresses the basic current switch gate delay as [16]

$$\tau_{CS} \simeq 1.7 \sqrt{\tau_{ec} \left(1 + 2\alpha R_b \frac{I_{CS}}{V_L}\right) \frac{V_L}{I_{CS}} (3C_{cb} + C_{cs})}. \qquad (5.83)$$

More complete (and complicated) versions for the full ECL gate can be found in [17].

References

[1] B. Jagannathan et al., "Self-Aligned SiGe NPN transistors with 285 GHz f_{max} and 207 GHz f_T in a manufacturable technology," *IEEE Elect. Dev. Lett.*, vol. 23, pp. 258-260, 2002.

[2] J.S. Rieh et al., "Measurement and modeling of thermal resistance of high speed SiGe heterojunction bipolar transistors," *Proc. IEEE Top. Workshop Si Mono. Int. Circ. RF Syst.*, pp. 110ij113, 2001.

[3] S.J. Mason, "Power gain in feedback amplifiers," *IRE Trans. Circ. Theory*, vol. CT-1, pp. 20-25, 1954.

[4] A.J. Joseph et al., "0.13 μm 210 GHz f_T SiGe HBTs – expanding the horizons of SiGe BiCMOS," *Tech. Dig. IEEE Int. Solid-State Circ. Conf.*, pp. 180-182, 2002.

[5] S. Long, "Basics of GaAs, InP and SiGe RFICs," *GaAs IC Symp. Short Course Notes*, 2001.

[6] H. Kroemer, "Two integral relations pertaining to electron transport through a bipolar transistor with a nonuniform energy gap in the base region," *Solid-State Elect.*, vol. 28, pp. 1101-1103, 1985.

[7] S.S. Iyer et al., "Heterojunction bipolar transistors using Si-Ge alloys," *IEEE Trans. Elect. Dev.*, vol. 36, pp. 2043-2064, 1989.

[8] D.L. Harame et al., "Si/SiGe epitaxial-base transistors – part II: process integration and analog applications," *IEEE Trans. Elect. Dev.*, vol. 42, pp. 469-482, 1995.

[9] A.J. Joseph, "The physics, optimization, and modeling of cryogenically operated silicon-germanium heterojunction bipolar transistors," Ph.D. Dissertation, Auburn University, 1997.

[10] D.J. Roulston, *Bipolar Semiconductor Devices*, McGraw-Hill, New York, NY, 1990.

[11] J.D. Cressler et al., "On the profile design and optimization of epitaxial Si- and SiGe-base bipolar technology for 77 K applications – part I: transistor dc design considerations," *IEEE Trans. Elect. Dev.*, vol. 40, pp. 525-541, 1993.

[12] C.T. Chuang, J.D. Cressler, and J.D. Warnock, "AC-coupled complementary push-pull ECL circuit with 34 fJ dower-delay product," *Elect. Lett.*, vol. 29, pp. 1938-1939, 1993.

[13] J.D. Cressler et al., "A high-speed complementary silicon bipolar technology with 12 fJ power-delay product," *IEEE Elect. Dev. Lett.*, vol. 14, pp. 523-526, 1993.

[14] P.M. Solomon and D.D. Tang, "Bipolar circuit scaling," *Tech. Dig. IEEE Int. Solid-State Circ. Conf.*, pp. 86-87, 1979.

[15] C.T. Chuang et al., "On the leverage of high-f_T transistors for advanced high-speed bipolar circuits," *IEEE J. Solid-State Circ.*, vol. 27, pp. 225-228, 1992.

[16] J.M.C. Stork, "Bipolar transistor scaling for minimum switching delay and energy dissipation," *Tech. Dig. IEEE Int. Elect. Dev. Meeting*, pp. 550-553, 1988.

[17] M.Y. Ghannam, R.P. Mertens, and R.J. van Overstraeten, "An analytical model for the determination of transient response of CML and ECL gates," *IEEE Trans. Elect. Dev.*, vol. 37, pp. 191-201, 1990.

Chapter 6

Second-Order Phenomena

In this chapter we examine in detail three important second-order phenomena that were not discussed in the *dc* and *ac* SiGe HBT device physics analysis presented in Chapters 4 and 5. While these second-order effects will always exist in SiGe HBTs, their specific impact on actual SiGe HBT circuits is both profile design and application dependent, and thus they must be carefully appreciated and kept in the back of the mind by practitioners of SiGe technology.

We first analyze the so-called "Ge grading effect" associated with the position dependence of the Ge content across the neutral base found in SiGe designs. The influence of Ge grading effect on SiGe HBT properties is physically tied to the movement of emitter-base space charge edge along the graded Ge profile with increasing base-emitter voltage. This Ge grading effect can present potential problems for circuit designs that require precise knowledge and control over the current dependence of both current gain and base-emitter voltage as a function of temperature. We then discuss the impact of neutral base recombination on SiGe HBT operation. A finite trap density necessarily exists in the base region of all bipolar transistors, and while the impact is usually assumed to be negligible in Si BJTs, it can become important in SiGe HBTs, particularly when they are operated across a wide temperature range. Neutral base recombination can strongly affect the output conductance (Early voltage) of SiGe HBTs, and is strongly dependent on the mode of base drive (i.e., whether the device is voltage or current driven), and hence the circuit application. Finally, we address high-injection heterojunction barrier effects in SiGe HBTs. Barrier effects associated with the collector-base heterojunction under high current density operation are inherent to SiGe HBTs, and if not carefully controlled, can strongly degrade both *dc* and *ac* performance at the large current densities which SiGe HBTs are often operated. We conclude each section with a brief discussion of the implications and potential problems imposed by these design constraints on both device and circuit designers (the bottom line).

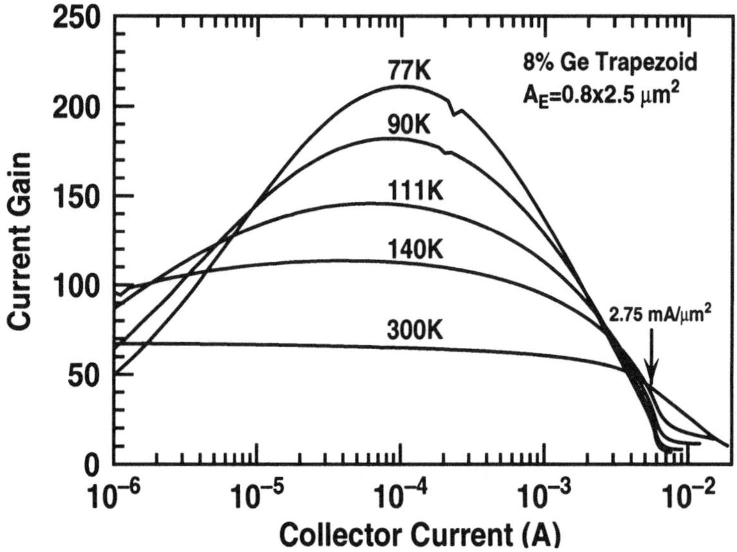

Figure 6.1 SiGe HBT current gain as a function of collector current at various temperatures, illustrating the Ge grading effect.

6.1 Ge Grading Effect

To ensure the applicability of SiGe HBTs to precision analog circuits, parameter stability over both temperature and bias must be ensured. Given the bandgap-engineered nature of the SiGe HBT, this can become an issue for concern, particularly for devices with non-constant (graded) Ge content across the base. Even a cursory examination of the bias current dependence of the current gain in a graded-base SiGe HBT, for instance, shows a profound functional difference from that of a Si BJT (compare Figure 6.1 to Figure 9.6). In particular, note that for a graded-base SiGe HBT, the current gain peaks at low-injection, and degrades significantly before the onset of high-injection effects. This medium-injection "collapse" of β is clearly enhanced by cooling, and thus can be logically inferred to be the result of a band-edge phenomenon.

To understand the physical origin of this bias-dependent behavior in the current gain in SiGe HBTs, consider Figure 6.2, which shows a schematic doping and Ge profile in a graded-base SiGe HBT. As derived in Chapter 4, the collector current at any bias of a graded-base SiGe HBT is exponentially dependent upon the amount of Ge at the edge of the emitter-base (EB) space-charge region. Physically, as the collector current density increases, the base-emitter voltage must also increase, and hence from charge balance considerations the EB space-charge width necessarily contracts, thereby reducing the EB boundary value of the amount of Ge ($\Delta E_{g,Ge}(0)$), and producing a bias and temperature dependence different from that of a Si BJT [1]. Since this Ge grading effect is the physical result of the modulation of the base width with increasing base-emitter voltage ($W_b(V_{BE})$), it can be logically associated with the so-called inverse Early effect (commonly known as the "late effect") [1] in SiGe HBTs.

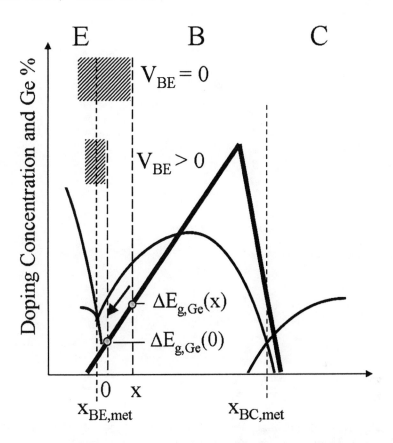

Figure 6.2 Schematic diagram of the base profile of a SiGe HBT, illustrating the physical origin of the Ge grading effect.

---

[1]Let it never be said that device engineers don't have a sense of humor! The use of "late" to describe the inverse Early effect is clearly a pun on Jim Early's name.

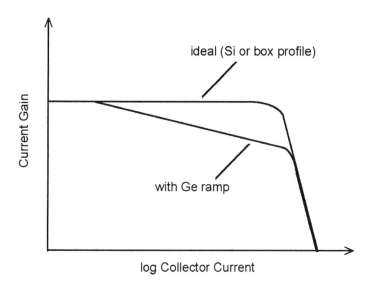

Figure 6.3 Illustration of the impact of Ge grading effects on SiGe HBT current gain as a function of collector current.

The dependence of the collector current density on the Ge profile shape in a SiGe HBT is given by

$$J_{C,SiGe}(V_{BE}, T) = \frac{q\, D_{nb}}{N_{ab}^{-} W_b} \left(e^{qV_{BE}/kT} - 1\right) n_{io}^2\, e^{\Delta E_{gb}^{app}/kT}$$
$$\cdot \left\{ \frac{\tilde{\gamma}\tilde{\eta}\, \Delta E_{g,Ge}(grade)/kT\, e^{\Delta E_{g,Ge}(0)/kT}}{1 - e^{-\Delta E_{g,Ge}(grade)/kT}} \right\}, \quad (6.1)$$

where $\Delta E_{g,Ge}(0)$ is the bandgap reduction due to Ge at the edge of the EB space-charge region at bias V_{BE}, and $\Delta E_{g,Ge}(grade)$ is the Ge grading across the neutral base ($\Delta E_{g,Ge}(W_b) - \Delta E_{g,Ge}(0)$) at bias V_{BE}. The result is an exponential degradation in J_C and hence β as the base-emitter bias increases (Figure 6.3). In addition, because J_C depends exponentially on the Ge-induced bandgap reduction at the EB junction divided by kT, this Ge-grading effect becomes much more pronounced at low temperatures [1]. For a uniformly doped base with a triangular Ge profile shape, the collector current density under low-injection is given approximately by

$$J_{C,SiGe} \simeq \frac{q D_{nb}}{N_{ab}^- W_b} \left(e^{qV_{BE}/kT} - 1 \right) n_{io}^2 e^{\Delta E_{gb}^{app}/kT}$$

$$\cdot \left\{ \tilde{\gamma}\tilde{\eta} \frac{\Delta E_{g,Ge}(grade)}{kT} e^{\Delta E_{g,Ge}(0)/kT} \right\}. \quad (6.2)$$

The relationship between J_C and Ge profile shape in (6.2) highlights the dependence of the collector current density on Ge profile design. Since $\Delta E_{g,Ge}(0)$ changes with increasing base-emitter bias, any changes in the amount of Ge seen by the device at that EB boundary will have a large impact due to the exponential relationship. Consequently, the more strongly graded the Ge profile, the more serious Ge grading effect can be expected to be.

Given that Ge grading effect in SiGe HBTs impacts the bias-current dependence of the current gain, a logical test-case circuit for examining the circuit-level influence of Ge grading effect is the ubiquitous bandgap reference circuit, since its functionality relies heavily on the identical dependence of V_{BE} (I_C) on temperature between transistors of differing size. Given two transistors with a (realistic) nonconstant base doping, and biased at the same collector current, two SiGe HBTs with a sufficiently strongly graded Ge profile might be expected to "feel" the Ge ramp effect differently, since the voltage-induced space-charge width changes would differ slightly between the two. The conceivable result would be a slight mismatch in V_{BE} over temperature between the two transistors, thereby degrading the output voltage stability of the bandgap reference circuit over temperature [2, 3].

6.1.1 Bandgap Reference Circuits

Since its introduction by Widlar in 1971 [4], the bandgap reference (BGR) circuit has been widely used as a voltage reference source in A/D and D/A converters, voltage regulators, and other precision analog circuits due to its good long-term stability and its ability to operate at low supply voltages. The BGR is able to provide temperature stability by summing voltages with positive and negative temperature coefficients. Modern BGRs can generate reference voltages with a temperature coefficient of better than 4 ppm/°C over a range of 0 to 125°C [5].

The negative temperature-coefficient in the BGR comes from the base-emitter voltage of a BJT (Figure 6.4). This voltage is added to the thermal voltage (kT/q), with its positive temperature coefficient multiplied by some constant set by the bandgap reference designer. For the purposes of BGR design, the collector current of a bipolar transistor can be written as

$$I_C(T) = B T^m e^{-E_{ge}/kT} e^{qV_{BE}/kT}, \quad (6.3)$$

190   Silicon-Germanium Heterojunction Bipolar Transistors

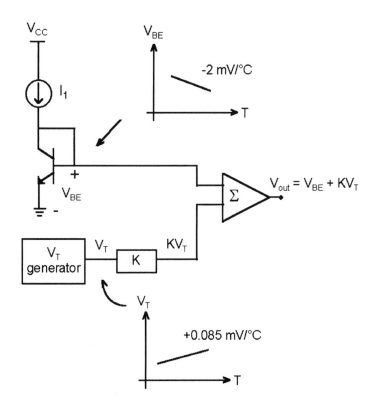

Figure 6.4 Illustration of the general principle underlying the bandgap reference circuit (after [6]).

where B represents lumped process-dependent parameters and E_{ge} is the bandgap energy in the presence of heavy doping ($E_{g0} - \Delta E_{gb}^{app}$). To arrive at an expression for the temperature-dependent base-emitter voltage, V_{BE} at a known temperature and collector current (their "reference" values) can be measured and substituted into (6.3). Solving for B yields

$$B = \frac{I_{C,R}}{T_R^m} e^{(E_{ge} - qV_{BE,R})/kT_R}. \tag{6.4}$$

Substituting for B in (6.3) and solving for V_{BE} gives

$$V_{BE} = \frac{E_{ge}}{q} + \frac{kT}{q} \ln \left\{ \frac{I_C}{I_{C,R} \left\{ \frac{T}{T_R} \right\}^m e^{(E_{ge} - qV_{BE,R})/kT_R}} \right\}. \tag{6.5}$$

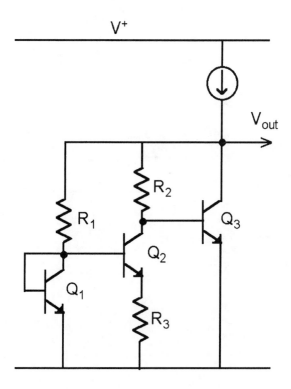

Figure 6.5 The Widlar bandgap reference circuit.

Rearranging (6.5) results in the familiar bandgap reference design equation [7].

$$V_{BE} = \frac{E_{ge}}{q} - \frac{T}{T_R}\left\{\frac{E_{ge}}{q} - V_{BE,R} - \frac{kT_R}{q}\ln\frac{I_C}{I_{C,R}} + m\frac{kT_R}{q}\ln\frac{T}{T_R}\right\}. \quad (6.6)$$

In (6.6), T_R, $I_{C,R}$, and $V_{BE,R}$ are the reference values of the respective parameters (typically referenced to 300 K). The negative temperature coefficient of V_{BE} can now be seen more clearly, as V_{BE} is the effective bandgap minus several temperature-dependent terms, the last of which controls the parabolic shape of V_{BE} on temperature.

Voltage references used Zener diodes as the reference element prior to Widlar's design [4]. These diodes typically had high breakdown voltages, limiting their broad applicability. Widlar proposed using the negative temperature coefficient of the emitter-base voltage of the BJT in conjunction with the positive temperature coefficient of the difference in emitter-base voltages of two transistors operating at

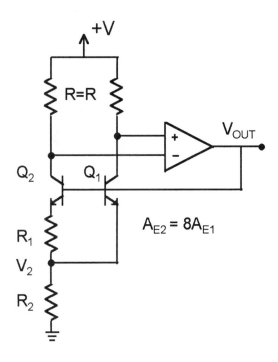

Figure 6.6 The Brokaw bandgap reference circuit.

different current densities. The combination of the positive and negative temperature coefficients would give a zero temperature coefficient reference voltage. In Figure 6.5, transistors Q_1 and Q_2 are biased at different current densities to produce temperature-proportional voltages across R_2 and R_3. Transistor Q_3 senses the output voltage through R_2, resulting in an output voltage of the V_{BE} of Q_3 plus the temperature-dependent voltage across R_2.

Brokaw improved upon Widlar's design by reducing the effects of the base currents flowing through R_1 and R_2 (Figure 6.6), as well as adding the capability to produce output voltages greater than the bandgap [7]. In Brokaw's circuit, the emitter area of transistor Q_2 is made larger than that of Q_1 (e.g., 8 times larger) in order to produce the difference in current densities. When the voltage at the common base is small, the voltage across R_1 is small, thus causing Q_2 to conduct more current through R_2. The imbalance in collector voltages then drives the op-amp to raise the base voltage until the two collector currents match. The difference in V_{BE} due to the difference in collector current densities appears across R_1, as given by

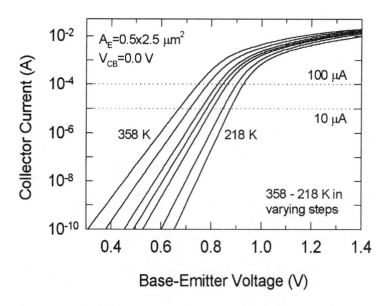

Figure 6.7 SiGe HBT collector current as a function of base-emitter voltage between 358 K and 218 K.

$$\Delta V_{BE} = V_{BE,1} - V_{BE,2} = \frac{kT}{q} \ln \frac{J_1}{J_2}. \quad (6.7)$$

Since the currents flowing through Q_1 and Q_2 are equal, the current in R_2 is twice that in R_1, resulting in

$$\frac{\Delta V_{BE}}{R_1} = \frac{V_2}{2 R_2}, \quad (6.8)$$

and a voltage across R_2 of

$$V_2 = \frac{2 R_2}{R_1} \frac{kT}{q} \ln \frac{J_1}{J_2}. \quad (6.9)$$

The resulting output voltage is the sum of the base-emitter voltage of Q_1 and the voltage across resistor R_2, yielding

$$V_{out} = V_{BE,1} + \frac{2 R_2}{R_1} \frac{kT}{q} \ln \frac{J_1}{J_2}. \quad (6.10)$$

The output voltage of the bandgap reference is the sum of a negative temperature-coefficient voltage, $V_{BE,1}$, and a positive temperature coefficient voltage, $k \Delta V_{BE}$.

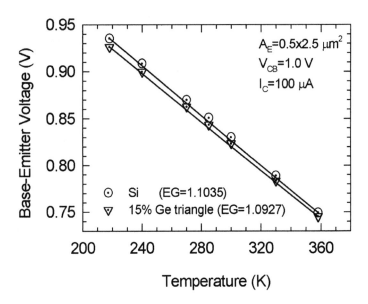

Figure 6.8 Base-emitter voltage as a function of temperature at fixed bias current for a Si BJT and a SiGe HBT. The lines represent a SPICE fit to the data using the *EG* fitting parameter.

In the context of SiGe HBT based BGR implementations, it is key that detailed knowledge of the impact of Ge profile shape on the temperature dependence of the base-emitter voltage exists. Clearly, V_{BE} in turn depends on the variation of I_C across the desired temperature range of interest (e.g., -55°C (218 K) to 85°C (358 K), as shown in Figure 6.7). As can be observed in Figure 6.8, the differences in $V_{BE}(T)$ between a Si BJT and a SiGe HBT are small, but clearly observable, and must be more carefully examined.

6.1.2 Theory

The Ge grading effect in SiGe HBTs is primarily determined by the "steepness" of the Ge profile through the EB space-charge region, and the magnitude and shape of $N_{ab}^-(x)$ at the space-charge to quasi-neutral base boundary. We can roughly estimate the variation on the SiGe-to-Si current gain ratio ($\Xi = \beta_{SiGe}/\beta_{Si}$) with V_{BE} for varying amounts of Ge grading by considering a linearly graded SiGe HBT with uniform doping levels in the emitter and base regions. From Chapter 4, we

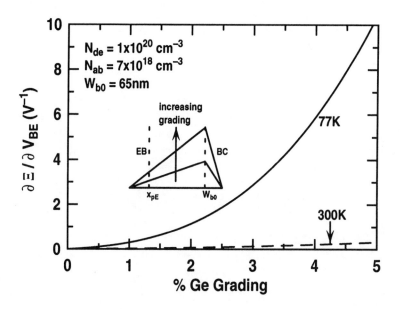

Figure 6.9 Theoretical dependence of SiGe-to-Si current gain enhancement factor on base-emitter voltage as a function of Ge grading at 300 K and 77 K.

have

$$\left.\frac{\beta_{SiGe}}{\beta_{si}}\right|_{V_{BE}} \equiv \Xi = \left\{\frac{\tilde{\gamma}\tilde{\eta}\Delta E_{g,Ge}(grade)/kT \, e^{\Delta E_{g,Ge}(0)/kT}}{1 - e^{-\Delta E_{g,Ge}(grade)/kT}}\right\}, \quad (6.11)$$

and can write

$$\frac{\partial \Xi}{\partial V_{BE}} = \left\{\frac{\partial \Xi}{\partial \Delta E_{g,Ge}(0)}\right\} \left\{\frac{\partial \Delta E_{g,Ge}(0)}{\partial x_{pE}}\right\} \left\{\frac{\partial x_{pE}}{\partial V_{BE}}\right\}, \quad (6.12)$$

to obtain [8]

$$\frac{\partial \Xi}{\partial V_{BE}} = \left\{\frac{-\Xi}{\phi_{bi,BE} - V_{BE}}\right\} \left\{\frac{x_{pE}}{2W_{b0}}\right\} \left[\frac{\Delta E_{g,Ge}(grade)/kT}{1 - e^{-\Delta E_{g,Ge}(grade)/kT}}\right], \quad (6.13)$$

where $\phi_{bi,BE}$ is the built-in potential of the EB junction, x_{pE} is the EB space-charge width on the base side of the junction, and W_{b0} is the neutral base width at zero-bias. As shown in Figure 6.9, as $\Delta E_{g,Ge}(grade)$ gets small (i.e., approaching a Ge box profile), the Ge grading effect becomes negligible, yielding a flat β versus I_C characteristic, as in a Si BJT. Equation (6.13) also predicts a weaker Ge grading effect in transistors with higher base doping, since x_{pE} becomes negligible with respect to W_{b0}. However, in practical SiGe HBT base profiles, which typically have a retrograded base doping level in the vicinity of the EB junction to reduce the EB electric field, the Ge grading effect is enhanced, since x_{pE} varies nonlinearly with V_{BE}. Finally, we note that due to the band-edge nature of Ge grading effect, its impact on device performance should be greatly magnified at reduced temperatures, as is clearly evident in Figure 6.9.

To determine the impact of the Ge grading effect on practical BGR circuits, we must recast the SiGe HBT collector current density into the familiar BGR design equation (6.6). First, the process-dependent parameters (B) and the Ge profile dependent terms (ξ) can be lumped together in I_C as

$$I_C(T) = \xi\, B\, T^m\, e^{-E_{g0}/kT}\, e^{E_{gb}^{app}/kT}\, e^{qV_{BE}/kT}, \qquad (6.14)$$

and rewritten in terms of the base-emitter voltage as

$$V_{BE} = \frac{E_{g0}}{q} - \frac{E_{gb}^{app}}{q} + \frac{kT}{q} \ln\left\{\frac{I_C}{\xi\, B\, T^m}\right\}. \qquad (6.15)$$

In practice, we can measure the base-emitter voltage at a reference temperature and collector current and solve for the lumped process parameters (B). Inserting the lumped parameters back into the original V_{BE} equation and simplifying yields the desired SiGe HBT result [9]

$$\begin{aligned}
V_{BE,SiGe} = & \frac{1}{q}\left\{E_{g0} - E_{gb}^{app} - \Delta E_{g,Ge}(0)\right\} - \frac{T}{qT_R}\left\{E_{g0} - E_{gb}^{app} - \Delta E_{g,Ge}(0)\right\} \\
& + \frac{T}{T_R}V_{BE,R} + \left\{\frac{kT}{q}\ln\frac{I_C}{I_{C,R}} - m\frac{kT}{q}\ln\frac{T}{T_R}\right\} \\
& - \left\{\frac{kT}{q}\ln\left(\frac{1 - e^{-\Delta E_{g,Ge}(grade)_R/kT_R}}{1 - e^{-\Delta E_{g,Ge}(grade)/kT}}\right)\right\} \\
& - \left\{\frac{kT}{q}\ln\left(\frac{T_R\, \Delta E_{g,Ge}(grade)}{T\, \Delta E_{g,Ge}(grade)_R}\right)\right\}. \qquad (6.16)
\end{aligned}$$

The effects of Ge on the base-emitter voltage of the transistor can be gleaned directly from this more generalized result. Observe that the effective bandgap at the emitter-base junction is simply the Si result in the presence of doping-induced bandgap narrowing ($E_{g0} - E_{gb}^{app}$), minus the bandgap reduction due to the amount of Ge at the EB junction ($\Delta E_{g,Ge}(0)$). In addition, the shape of V_{BE} versus temperature in a SiGe HBT is changed from that of a Si BJT due to the addition of Ge, as is apparent in the last two terms of the equation. The ratio T/T_R enhances this difference between Si BJTs and SiGe HBTs. For temperatures near the reference temperature, the last two terms of (6.16) have little effect on $V_{BE}(I_C, T)$, but as

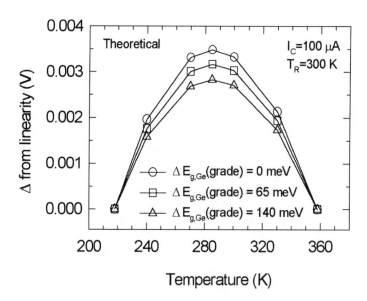

Figure 6.10 Theoretical dependence of the V_{BE} deviation from linearity as a function of temperature for various amounts of Ge grading.

the temperature decreases, these effects can become more pronounced. This result will later be used to compare the measurements of Si and SiGe devices.

Note that the effective bandgap parameters ($E_{g0} - E_{gb}^{app} - \Delta E_{g,Ge}(0)$) and m correspond to the SPICE modeling parameters EG and XTI, respectively. The amount of curvature in V_{BE} versus temperature is affected by the addition of Ge, as is apparent in the last two terms of (6.16). Assuming that the Ge grading ($\Delta E_{g,Ge}(grade)$) does not change significantly with temperature, the deviation from linearity of V_{BE} versus temperature (i.e., V_{BE} curvature) using (6.16) is actually reduced with increasing Ge grading across the base. In the curvature results presented, the deviation from linearity is calculated by drawing a line through the endpoints of V_{BE} across the relevant temperature range, and then subtracting the actual V_{BE} value from the value on the line at each temperature, according to

$$\Delta_{linearity}(T) = V_{BE}(T) - \left[V_{BE}(T_L) - \frac{V_{BE}(T_L) - V_{BE}(T_H)}{T_L - T_H}(T_L - T) \right], \quad (6.17)$$

where in this case $T_L = 218$ K (-55°C) and $T_H = 358$ K (85°C).

While this Ge-grading-induced V_{BE} curvature reduction might naively appear to be a good thing for BGR design, it in fact can worsen the performance of BGR circuits, as discussed below. Figure 6.10 shows the theoretical deviation from linearity that results from three different hypothetical Ge profiles: 1) no Ge grading; 2) 8.6% Ge grading; and 3) 18.6% Ge grading. Note that a box-shaped Ge profile (no Ge grading), in which the Ge concentration across the base is finite but constant, will have the same deviation from linearity as a Si BJT. Figure 6.11 shows the calculated percent reduction in peak deviation from linearity from 218 K to 358 K as a function of Ge grading in the base region of the device. Given sufficient Ge grading, it is clear that differences between Si BJTs and SiGe HBTs should be experimentally observable.

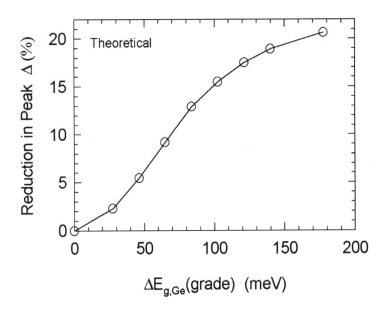

Figure 6.11 Theoretical dependence of the peak V_{BE} deviation from linearity as a function of Ge grading.

6.1.3 Measured Data and SPICE Modeling Results

In order to quantify the effects of the Ge profile shape on BGR operation, the experimental behavior of the base-emitter voltage as a function of bias current and temperature must be precisely known. Two experiments [10] were conducted to measure the base-emitter voltages for Si BJTs and SiGe HBTs of differing Ge profile shape and emitter area across temperature and bias. The results from these experiments were compared to both calibrated 1-D drift-diffusion simulations using SCORPIO, as well as SPICE simulations, to better understand the effect of the

Ge profile shape on BGR operation.

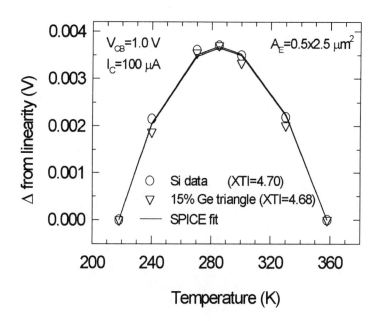

Figure 6.12 Measured V_{BE} deviation from linearity as a function of temperature for a 0.5 × 2.5 μm^2 Si BJT and SiGe HBT. The curves represent a SPICE fit to the data using the *XTI* fitting parameter.

Because proper design of BGR circuits requires extremely accurate knowledge of $V_{BE}(I_C, T)$, we might naively expect them to be sensitive to Ge grading effect. An additional important question is whether standard compact models (e.g., SPICE) can accurately capture $V_{BE}(I_C, T)$ in SiGe HBTs. To test this, the BGR design equation (6.6) was fit to the data to determine the values of the SPICE modeling parameters *EG* and *XTI*. A calibrated SPICE model was generated for each of the transistors measured. In this analysis, T_R, $I_{C,R}$, and $V_{BE,R}$ are the reference values of the respective parameters at 300 K, 50 μA, and $V_{BE}(I_{C,R}, T_R)$, respectively. Of interest is the inferred value of the SPICE parameter *EG*, which controls the slope and intercept of $V_{BE}(T)$, and the SPICE parameter *XTI*, which controls the deviation of $V_{BE}(T)$ from linearity (i.e., curvature). The base-emitter voltages at I_C = 10, 50, and 100 μA for Si BJTs of differing area and SiGe HBTs with a 15% triangular Ge profile were measured to six decimal place accuracy over a temperature range of -55 to 85°C. V_{BE} as a function of temperature at 100 μA for a 0.5 × 2.5 μm^2 transistor is shown in Figure 6.8, together with the fitted values of *EG* determined from SPICE. Note that the value of *EG* for the Ge profile is lower than that of the Si BJT, as expected. The measured deviation from linearity for both devices is shown in Figure 6.12, together with the fitted values of *XTI* determined from SPICE. The *XTI* value for the 0.5 × 2.5 μm^2 SiGe HBT is 4.68, slightly lower than the deviation from linearity shown for the Si BJT, whose *XTI* value is 4.70. Interestingly, however, if we compare the same set of data on a larger transistor, the variation between the Si BJT and SiGe HBT is measurably different (Figure 6.13). Observe that the measured deviation from linearity for the 0.5 × 2.5

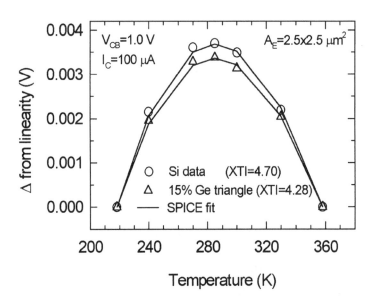

Figure 6.13 Measured V_{BE} deviation from linearity as a function of temperature for 2.5 × 2.5 μm^2 Si BJT and SiGe HBT. The curves represent a SPICE fit to the data using the *XTI* fitting parameter.

μm^2 Si BJT is approximately the same as that for the 2.5 × 2.5 μm^2 Si BJT, while the deviation from linearity for the 2.5 × 2.5 μm^2 SiGe HBT is substantially different from that for the 0.5 × 2.5 μm^2 SiGe HBT, resulting in a device mismatch. This result is expected for sufficient Ge grading (and nonconstant base doping) since at constant current, the two devices with differing emitter areas will feel the effects of the Ge-grading differently.

This $V_{BE}(T)$ mismatch between the SiGe HBTs can be expected to have an observable impact on BGR performance. To test this hypothesis, a simple BGR circuit (Figure 6.6) was modeled in SPICE using both Si BJT and SiGe HBT models fit to measured data. The output voltage for each circuit was simulated and

Second-Order Phenomena

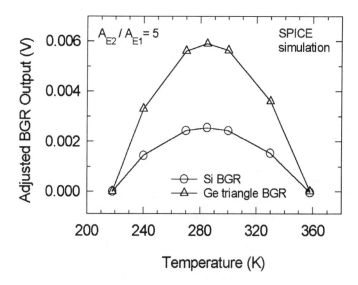

Figure 6.14 Calibrated SPICE modeling results for the adjusted BGR output voltage as a function of temperature for a Si BJT BGR and a SiGe HBT BGR.

adjusted by subtracting their minimum values for ease of comparison. The simulation results showing the shape of the BGR output voltage for the two different BGR circuits from 218 to 358 K appear in Figure 6.14. Observe that the change in output voltage across temperature is much worse for the SiGe HBT BGR circuit due to the difference in the shape of $V_{BE}(T)$ of the two different sized transistors, with a voltage stability of 31 ppm/°C compared to 15 ppm/°C for the Si BJT BGR. This increase in output voltage curvature of the SiGe circuit is directly related to the dissimilar *XTI* values used in the calibrated device models, and hence is reflective of the presence of the Ge grading effect.

6.1.4 The Bottom Line

When discussing any second-order effect in transistors, it is important to clearly understand both its physical origins and its potential implications for both device and circuit designers, so that it can be effectively "designed around." We can summarize these implications for Ge grading effect as follows:

- Ge grading effect is likely to be important only in precision analog circuits, not in digital or RF/microwave circuits. While the BGR circuit is a natural

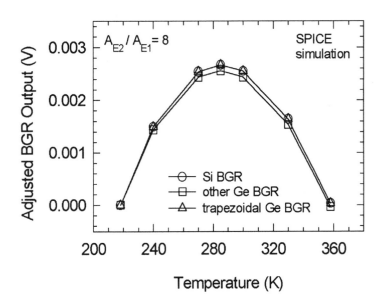

Figure 6.15 Calibrated SPICE modeling results for the adjusted BGR output voltage as a function of temperature for a Si BJT BGR and SiGe HBT BGRs constructed from two different SiGe profiles.

candidate for observing Ge grading effect, any analog circuit that depends strongly on current gain across a wide bias range, or that requires the matching of V_{BE} between multiple devices across both bias and temperature, could be potentially affected.

- While the Ge grading effect exists only in compositionally graded Ge profiles, these graded profile designs typically achieve the best dc and ac performance, and thus represent the vast majority of commercially relevant SiGe technologies. As such, Ge grading effect should never be discounted.

- The impact of the Ge grading effect is expected to be highly dependent on the specifics of the Ge profile shape, and thus will vary from technology to technology. The results presented above were shown for a 15% triangular Ge profile, and can be considered a worse case scenario for first generation SiGe HBT technology. As shown in Figure 6.15, a similar experiment conducted for more modest Ge content profiles (8% peak Ge for the "trapezoidal profile," and 10% peak for the "other Ge" profile), showed little effect on the output voltage of the SiGe HBT BGR. This observation is consis-

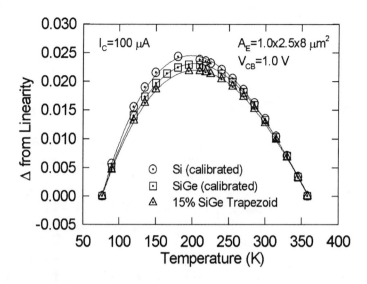

Figure 6.16 Calibrated SCORPIO simulations of the V_{BE} deviation from linearity down to 77 K for a Si BJT and two SiGe HBT profiles.

tent with the generally excellent performance reported for SiGe HBT BGRs fabricated from first generation SiGe HBT technology [11].

- Since the seriousness of the Ge grading effect depends on the Ge grading, it is a phenomenon that will generally worsen with technology scaling, since for constant strained layer stability, the peak Ge content in a SiGe HBT (and hence the grading across the neutral base) will naturally rise. This scaling-induced enhancement, however, will be at least partially offset by the natural increase in base doping with scaling.

- Conventional modeling methodologies employed in Gummel-Poon (SPICE) compact transistor models appear to adequately capture the Ge grading effect.

- Due to its thermally activated nature, cooling clearly exaggerates Ge grading effect, and thus is potentially important for precision analog circuits required to operate across a very wide temperature range. This can be easily seen in Figure 6.16, which compares calibrated SCORPIO simulations of a Si BJT, a 8% peak Ge content profile (labeled "SiGe (calibrated)"), and the 15% Ge triangle profile discussed above, down to 77 K. Observe that the curvature scales are much larger than those for the 218 K to 358 K temperature range.

6.2 Neutral Base Recombination

Neutral base recombination (NBR) in bipolar transistors involves the recombination of injected electrons transiting the neutral base with holes, via intermediate trap levels. Physically, NBR removes the desired injected electrons from the collector current via recombination (i.e., they don't exit the base), and increases the undesired hole density (required to support the recombination process), thereby degrading the base transport factor. Significant neutral base recombination thus leads to an increase in the base current and a simultaneous decrease in the collector current, thereby causing a substantial degradation in the current gain. Historically, NBR presented serious design constraints for achieving high gain in early bipolar transistors. Because NBR is determined by the presence of the various trapping centers and other defects in the neutral base region of the device, it is generally the quality of the bulk material after fabrication is complete that dictates the seriousness of NBR in a given transistor. One can show that it is thermodynamically impossible for a perfect (defect free) crystal to exist in nature, and thus NBR is an immutable fact-of-life in bipolar transistors. From a design point of view, the issue of the importance of NBR really becomes a matter of degree. Due to the fortuitous existence of the high purity Si bulk material for use in modern advanced transistors, the effectiveness of the annealing of implantation and processing-induced crystal damage, and the small base widths in high-speed devices, the impact of NBR in today's Si BJTs can generally be considered negligible.

For fixed trap density, the impact of NBR on transistor characteristics can be exaggerated, however, due to the presence of an increased total base minority carrier charge concentration (Q_{nb}) that participates with the trap recombination process. Because in a SiGe HBT the Ge-induced base bandgap reduction *exponentially* increases Q_{nb} compared to that in a comparably constructed Si BJT, one would naively expect that the NBR would be strongly enhanced in a SiGe HBT compared to a Si BJT, even at identical trap base density. This situation is also expected to become especially important as the temperature changes, due to the thermally activated nature of Q_{nb} in a SiGe HBT. It is essential, therefore, to understand the physical mechanism of NBR in SiGe HBTs, its impact on the transistor characteristics, and possible circuit implications. This section presents a comprehensive investigation of NBR in SiGe HBTs, and its influence on the temperature characteristics of V_A and βV_A. A direct consequence of NBR in SiGe HBTs is the degradation of V_A when transistors are operated with constant-current input (forced-I_B mode), as opposed to a constant-voltage input (forced-V_{BE} mode), the bias mode that is consistent with the first-order theory of output conductance presented in Chapter 4. In addition, experimental and theoretical evidence indicates that with cooling, V_A in SiGe HBTs degrade faster than in Si BJTs for forced-I_B mode of operation. The differences in V_A as a function of the input bias and temperature for SiGe HBTs can be accurately modeled using a modified version of SPICE. The performance of various practical SiGe HBT analog circuits are analyzed across temperature using this calibrated SPICE model, for situations with and without the presence of NBR. We then use 2-D device simulations to understand the physical location of the participating base trap levels. In light of these results, the implications of Ge profile design on NBR in SiGe HBTs are discussed, and we conclude with the bottom-line for SiGe HBT device and circuit designers.

6.2.1 Theory

The physical origin of NBR in bipolar transistors is the presence of traps in the base region, which facilitate the recombination of the injected minority carriers (electrons) and resident majority carriers (holes). Under forward bias, the electrons injected into the base region, while drift-diffusing through the neutral base region, will encounter base trap states, which in principle are distributed in some particular fashion in both physical and energy space, and recombine with the holes in the base region. The loss of charge in the neutral base due to NBR in a transistor will cause a decrease in the collector current and a corresponding increase in the base current, yielding a degraded β. The NBR current, therefore, is simply the rate at which the minority carriers recombine and can be written as the ratio between the excess minority carrier charge (Δn_b) and the minority carrier lifetime (τ_{nb}) integrated over the base region according to

$$J_{nbr} = q \int_0^{W_b} \frac{\Delta n_b(x)}{\tau_{nb}(x)} dx \qquad (6.18)$$

where Δn_b represents the spatial variation of the excess minority carrier density in the base region. In general, τ_{nb} is a function of position due to the random distribution of trap states in the base region. However, for mathematical convenience, τ_{nb} can be considered a constant by treating it as an effective minority carrier lifetime in the base (i.e., spatially averaged across the base). In general, Δn_b depends on various factors, including the applied bias (both V_{BE} and V_{CB}), temperature, and the specific shape of the base doping profile. Assuming: 1) steady-state conditions; 2) negligible carrier generation in the base; 3) a 1-D solution; 4) that Shockley-Read-Hall recombination dominates; 5) the Boltzmann approximation; 6) complete ionization of dopants; and 7) a constant position-averaged D_{nb}, we can determine Δn_b by solving the second-order differential equation which results from the current continuity and drift-diffusion transport equations [8]

$$\frac{d^2 \Delta n_b}{dx^2} + \frac{q\xi}{kT}\frac{d\Delta n_b}{dx} + \left[\frac{q}{kT}\frac{d\xi}{dx} - \frac{1}{L_{nb}^2}\right]\Delta n_b + \frac{d^2 n_{b0}}{dx^2} + \frac{q\xi}{kT}\frac{dn_{b0}}{dx} + \frac{qn_{b0}}{kT}\frac{d\xi}{dx} = 0, \quad (6.19)$$

where n_{b0}, ξ, and L_{nb} are the intrinsic minority carrier density in the base, the built-in base electric field, and the minority carrier diffusion length ($L_{nb} = \sqrt{(D_{nb}\tau_{nb})}$), respectively. This equation is especially useful for looking at situations having a constant built-in electric field (such as in a linearly graded base bandgap in a SiGe HBT or an exponentially varying base doping in a Si BJT). The standard EB and CB Shockley boundary conditions are used in solving (6.19).

For a Si BJT with a constant base doping profile, (6.19) reduces to the case from which J_{nbr} can be written as

$$J_{nbr}(Si) = \frac{q D_{nb,Si} n_{ib,Si}^2 e^{qV_{BE}/kT}}{N_{ab} L_{nb,Si}} \left\{\frac{\cosh \chi_{Si} - 1}{\sinh \chi_{Si}}\right\}, \quad (6.20)$$

where $\chi = W_b/L_{nb}$. The importance of J_{nbr} on the device operation clearly depends on the magnitude of χ. Ideally, when χ is much larger than unity (i.e., when τ_{nb} is very large), observe that (6.20) is equal to zero, as expected. Since for a Si BJT with constant doping we have

$$J_C(Si) = \frac{q D_{nb,Si} n_{ib,Si}^2}{W_b N_{ab}} e^{qV_{BE}/kT}, \quad (6.21)$$

J_{nbr} can also be expressed in terms of the ideal J_C by rewriting (6.20) using (6.21). The collector current in the presence of NBR can be determined by finding the slope of $\Delta n_b(x)$ at the CB space-charge region boundary of the quasi-neutral base region. The normalized variation in the collector current and the current gain in a Si BJT can then be determined as

$$\frac{\Delta J_C}{J_{C,ideal}}(\%) = \left\{\frac{\chi_{Si}}{\sinh \chi_{Si}} - 1\right\} \times 100, \quad (6.22)$$

and

$$\frac{\Delta \beta}{\beta_{ideal}}(\%) = \left\{\frac{1}{\cosh \chi_{Si} - 1 + \frac{\chi_{Si}}{\sinh \chi_{Si}}} - 1\right\} \times 100. \quad (6.23)$$

Observe that in the limit of $\chi_{Si} \to 0$ (the ideal case), (6.22) and (6.23) indicate that no change in either J_C or β should result. In the presence of strong NBR, however, (i.e., $\chi \to \infty$), both J_C and β fall rapidly to zero. Calculation of the percentage normalized change in both J_C and β for a Si BJT in the presence of NBR, as a function of τ_{nb}, clearly shows that the variation in J_C and β compared to the ideal case gets larger as τ_{nb} decreases (i.e., as the trap density in the base region increases), with an almost 10% decrease in J_C when τ_{nb} is about 10 psec. For the larger recombination lifetimes typically encountered in realistic Si BJTs base regions, however, the NBR component clearly has a negligible effect on both J_C and β. Solving (6.19) for a SiGe HBT requires the knowledge of ξ, which depends on both the shape of the base doping profile and the grading of the Ge profile across the neutral base. For a linearly graded Ge profile with a constant N_{ab}, there exists a constant Ge-grading-induced electric field across the base, which can be written as

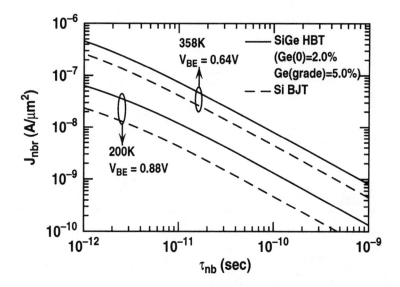

Figure 6.17 Theoretical variation in the NBR current density in both Si and SiGe transistors as a function of minority carrier lifetime in the base region, at both 358 K and 200 K.

$$\xi = \frac{kT}{q} \left\{ \frac{n_{ib}}{N_{ab}} \right\} \frac{d}{dx} \left\{ \frac{N_{ab}}{n_{ib}} \right\} = \frac{-\Delta E_{g,Ge}(grade)}{2q W_b}. \quad (6.24)$$

Using (6.19) and (6.24), one obtains a simplified differential equation for the SiGe

HBT [8],

$$\frac{d^2 \Delta n_b}{dx^2} - \left\{ \frac{\Delta E_{g,Ge}(grade)}{2W_b kT} \right\} \frac{d\Delta n_b}{dx} - \frac{\Delta n_b}{L_{nb}^2} + \frac{n_{bo}}{2} \left\{ \frac{\Delta E_{g,Ge}(grade)}{W_b kT} \right\}^2 = 0. \tag{6.25}$$

Solving (6.25) with the appropriate boundary conditions yields

$$\Delta n_b(x) = C_1 e^{m_1 x} + C_2 e^{m_2 x} - (1 + m_0) n_{b0}(x), \tag{6.26}$$

where the various terms are given by:

$$n_{b0}(x) = \frac{\gamma n_{ib}^2(Si)}{N_{ab}} e^{(\Delta E_{g,Ge}(grade) x/W_b + \Delta E_{g,Ge}(0))/kT}, \tag{6.27}$$

$$m_0(x) = \frac{2\chi_{SiGe}^2}{(\Delta E_{g,Ge}(grade)/kT)^2 - 2\chi_{SiGe}^2}, \tag{6.28}$$

$$m_{1(2)}(x) = \frac{\Delta E_{g,Ge}(grade)}{4W_b kT} \left\{ 1 + (-) \sqrt{1 + \left[\frac{4\chi_{SiGe}}{\Delta E_{g,Ge}(grade)/kT}\right]^2} \right\} \tag{6.29}$$

$$C_1 = \frac{n_{b0}(0) e^{m_2 W_b} \left\{ e^{qV_{BE}/kT} + m_0 \right\} - n_{b0}(W_b) \left\{ e^{qV_{BC}/kT} + m_0 \right\}}{e^{m_2 W_b} - e^{m_1 W_b}} \tag{6.30}$$

$$C_2 = \frac{n_{b0}(0) e^{m_1 W_b} \left\{ e^{qV_{BE}/kT} + m_0 \right\} - n_{b0}(W_b) \left\{ e^{qV_{BC}/kT} + m_0 \right\}}{e^{m_1 W_b} - e^{m_2 W_b}}. \tag{6.31}$$

Substituting (6.26)-(6.31) into (6.18) we obtain finally

$$J_{nbr}(SiGe) = \frac{q}{\tau_{nb,SiGe}} \left\{ \frac{C_1(e^{m_1 W_b} - 1)}{m_1} + \frac{C_2(e^{m_2 W_b} - 1)}{m_2} - \frac{(1 + m_0)[n_{b0}(W_b) - n_{b0}(0)]}{\Delta E_{g,Ge}(grade)/kT W_b} \right\}. \tag{6.32}$$

In general, τ_{nb} in a SiGe HBT can in principle be different from that in a Si BJT, due to the differences in the base profile. Using (6.18) and (6.32), one can roughly estimate the impact of τ_{nb}, Ge profile shape, and temperature on the NBR component in both Si and SiGe transistors, respectively. A comparison between the NBR in SiGe HBT and Si BJT can be made at fixed V_{BE}, which is modeled as a function of temperature to fit the data at $I_C = 5.0$ µA. Figure 6.17 shows the variation of NBR base current components in both a SiGe HBT and a Si BJT as a function of τ_{nb}, calculated at 358 K and 200 K, respectively. The SiGe HBT has a trapezoidal profile shape with a Ge content at $x = 0$ of 2% and Ge grading of 5%, with a base width of 65 nm and base doping of 3×10^{18} cm^{-3}. For simplicity, τ_{nb}, μ_{nb}, and the $N_C N_V$ product in the SiGe HBTs are assumed to be identical to that in the Si BJT. Observe that the NBR-induced base current component gets exponentially larger for decreasing τ_{nb} and gets comparable to I_C at very low τ_{nb} values (< 1 psec). It can also be clearly seen that J_{nbr} in the SiGe HBT is larger than that in a comparably constructed Si BJT and that this difference gets larger with cooling.

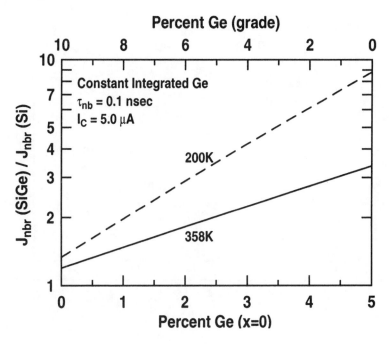

Figure 6.18 Theoretical variation of the ratio between the NBR currents in a SiGe HBT and Si BJT as a function of Ge profile shape, at both 358 K and 200 K.

The NBR component in a SiGe HBT will be a thermally activated function of the amount of Ge-induced bandgap reduction at the EB space-charge edge (that is, $\Delta E_{g,Ge}(0)$), since Q_{nb} depends strongly (exponentially) on this quantity. The Ge grading, however, is also expected to play a role in increasing the NBR component in a SiGe HBT since it affects the minority carrier base charge. Figure 6.18 shows the calculated ratio between the NBR component in a SiGe HBT and that in a comparably constructed Si BJT at 358 K and 200 K, for Ge profiles with varying Ge content at the EB boundary and a constant integrated Ge content (i.e., stability). As expected, the largest variation in NBR base current component is observed in a box-profile SiGe HBT. With cooling, this SiGe-to-Si ratio gets exponentially larger for all situations considered. Figure 6.19 shows the variation in the NBR ratio as a function of reciprocal temperature for three different situations: 1) a 0% Ge profile (i.e., Si BJT); 2) a 10% triangular Ge profile; and 3) a 7% trapezoidal profile (with 5% Ge grading). Clearly, the strongest increase in the NBR ratio is for a trapezoidal SiGe HBT, due to the presence of a large bandgap reduction at the EB space-charge boundary.

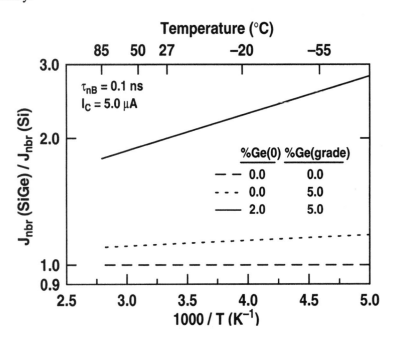

Figure 6.19 Theoretical variation of the ratio between the NBR currents in a SiGe HBT and Si BJT as a function of reciprocal temperature, for three different Ge profiles.

In real transistors, the shape of the base doping profile will also influence the J_{nbr} ratio due to the presence of a doping-induced, position dependent, built-in electric field in the base region. Solving (6.19) in such situations is difficult, and more often than not will not yield a closed-form solution. Instead, one can resort to more sophisticated numerical simulators to determine J_{nbr} under such situations. Note, however, that 1-D drift-diffusion simulators cannot easily model NBR because of the difficulty in modeling the base contact of a 1-D bipolar transistor (see Chapter 12). Since NBR is a bulk effect, a 2-D (or 3-D) simulation tool is required to properly account for this effect.

For an *npn* bipolar transistor with negligible EB space-charge region recombination, I_B under arbitrary forward-active bias is the sum of the hole current back-injected into the emitter, the hole current due to impact ionization in the collector-base region, and the NBR component under discussion. For small values of V_{CB}, the additional hole current due to impact-ionization is negligible and thus I_B is dominated by the other two components. As L_{nb} gets comparable to W_b, the NBR component of I_B becomes increasingly important. With negligible NBR (the ideal case), I_B will be independent of V_{CB} for any given V_{BE}. However, under nonnegligible NBR, any change in W_b with respect to L_{nb} will perturb the NBR component of I_B. Thus, an easy way to estimate the impact of NBR in a given transistor is to observe the rate of decrease in I_B with respect to varying V_{CB}, at a fixed V_{BE}. The base current in this case can be expressed as the sum of the drift-diffusion component and the NBR component as

$$J_B = J_{B,diff} + J_{nbr} = J_{b0}\, e^{qV_{BE}/kT} + J_{nbr,0}\, e^{qV_{BE}/kT}. \quad (6.33)$$

Here J_{b0} is assumed to be independent of V_{CB}, while $J_{nbr,0}$ is a function of V_{CB} because of χ_{Si} and is given by (from (6.20))

$$J_{nbr,0}(Si) = \frac{q\, D_{nb}\, n_{ib}^2}{N_{ab}\, L_{nb}} \left\{ \frac{\cosh \chi_{Si} - 1}{\sinh \chi_{Si}} \right\}. \quad (6.34)$$

Since the diffusion component of I_B is independent of V_{CB}, the change in I_B with V_{CB} will only be due to the variation in J_{nbr} through the variations in W_b. Therefore, in general, the input conductance of the transistor (g_μ) can be written as [8]

$$g_\mu = \left.\frac{\partial J_B}{\partial V_{CB}}\right|_{V_{BE}} = \left.\frac{\partial J_{nbr}}{\partial V_{CB}}\right|_{V_{BE}} = \left.\frac{\partial J_{nbr}}{\partial W_b}\right|_{V_{BE}} \left.\frac{\partial W_b}{\partial V_{CB}}\right|_{V_{BE}}. \quad (6.35)$$

Using (6.32),(6.33), and (6.34) we can determine the g_μ in a Si BJT

$$g_\mu(Si) = \frac{q\, D_{nb}\, n_{ib}^2}{N_{ab}\, L_{nb}} e^{qV_{BE}/kT} \left\{ \frac{\cosh \chi_{Si} - 1}{\sinh^2 \chi_{Si}} \right\} \frac{1}{L_{nb}} \left[\frac{\partial W_b}{\partial V_{CB}} \right]. \quad (6.36)$$

A more convenient way to compare the variations in J_{nbr} between devices and across temperature is to normalize g_μ to the base current at $V_{CB} = 0$ V (g'_μ).

Therefore, by rewriting (6.36) using the expressions for both J_C and V_A derived in Chapter 4, we finally obtain g'_μ in a Si BJT

$$g'_\mu(Si) = \frac{g_\mu(Si)}{I_{B,Si}(V_{BE}, V_{CB} = 0)} \approx \frac{-\beta_{Si}(V_{BE}, V_{CB} = 0) \chi^2_{Si}(\cosh \chi_{Si} - 1)}{V_{A,Si}(forced - V_{BE}) \sinh^2 \chi_{Si}},$$
(6.37)

where $\beta_{Si}(V_{BE}, V_{CB} = 0)$ represents the current gain measured at a given V_{BE} and $V_{CB} = 0$ V and $V_{A,Si}(forced - V_{BE})$ represents the Early voltage at a fixed V_{BE}. Observe that when χ_{Si} is zero (representing the ideal situation in the transistor with no NBR) that (6.37) predicts that g'_μ will be zero, as one would expect. On the other hand, when χ_{Si} becomes large (i.e., significant NBR is present), then (6.37) yields an increased value of g'_μ. Since β and V_A are weakly bias dependent in well-designed Si BJTs, it is expected that g'_μ will be relatively constant in the low-injection regime. Using (6.37) we can roughly estimate τ_{nb} from the knowledge of the experimentally determined values for β, V_A, and g'_μ in a Si BJT, for arbitrary bias.

Historically, the first experimental observation of NBR in SiGe HBTs was presented in [12], and a two lifetime region model was used to fit the simple first-order theory to the data. By assuming a low lifetime region near the CB space-charge region edge (associated with the SiGe growth interface), good agreement between measured data and theory was reported.

The g'_μ in a SiGe HBT with a trapezoidal Ge profile can in principle be obtained directly from (6.32), but is quite complicated mathematically. However, in order to qualitatively determine the device design parameters which strongly influence g'_μ, one can consider a simple box Ge profile. In this case, it is easily shown that g'_μ in a SiGe HBT is the same as (6.37), except for the differences in β. This is expected, since both J_{nbr} and β in a SiGe HBT are determined primarily by the amount of Ge-induced bandgap reduction at the EB space-charge edge (i.e., $\Delta E_{g,Ge}(0)$). In general, however, the NBR current component and hence g'_μ will be a function of the amount of Ge introduced into the base region of a SiGe HBT (that is, EB boundary value as well as Ge-grading). In addition, since the SiGe-to-Si J_{nbr} ratio is effectively amplified by cooling, it is expected that the SiGe-to-Si g'_μ ratio will also exponentially increase with decreasing temperature.

6.2.2 Experimental Results

In order to better understand the influence of NBR on SiGe HBT operation, we experimentally compared identically fabricated Si BJT and SiGe HBTs at various temperatures and biases in order to highlight the experimental observation of NBR,

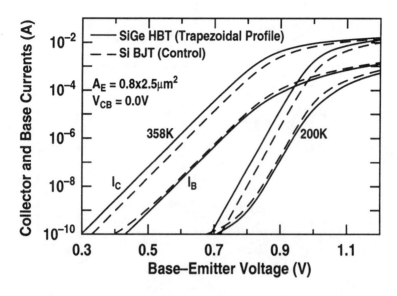

Figure 6.20 Measured Gummel characteristics for a comparably designed SiGe HBT and Si BJT, at both 358 K and 200 K.

the influence of NBR on V_A [13, 14], and the modeling of NBR in SiGe HBTs for circuit applications [15]. SiGe HBTs with two different Ge profiles were considered: a 7% trapezoidal Ge profile that has approximately 2% Ge at $x = 0$ and 5% Ge grading, and a 10% peak triangular Ge profile. Figure 6.20 shows the measured Gummel characteristics for the trapezoidal Ge profile SiGe HBT and the Si BJT, at 358 K and 200 K, respectively. The similarity of I_B at both temperature extremes is a good indication that the Si BJT is an excellent control for inferring Ge-induced differences in NBR. Figure 6.21 compares the variation in the normalized-I_B as a function of V_{CB} for both the Si and SiGe transistors at 358 K at 200 K, respectively. In this case, the transistors are biased in the low-injection region where their collector and base currents are ideal. One can clearly observe the decrease in I_B at low V_{CB} due to the modulation of the NBR current component for both transistors, at 358 K and 200 K, respectively. The strong decrease in I_B at larger values of V_{CB} is due to an increase in the impact-ionization base current component. By observing the variation in I_B with V_{CB} in the low-V_{CB} range, one can easily conclude that the Si BJT shows a weak NBR component ($\approx 0.5\%$ decrease in I_B), while the SiGe HBT shows not only a larger NBR base current component, but also an increase in the NBR with cooling, as anticipated from theory. It is important to note here that, although the NBR component in the SiGe HBT is clearly larger than that in the Si BJT, the magnitude of the NBR component is nevertheless still quite small ($\approx 3\%$

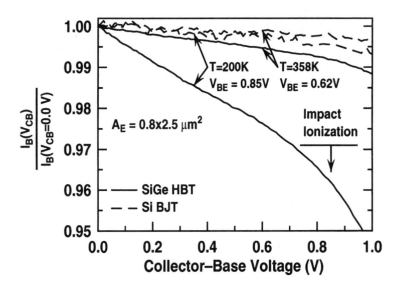

Figure 6.21 Measured normalized I_B as a function of V_{CB} for a SiGe HBT and a Si BJT, at both 358 K and 200 K.

of I_B at 200 K).

Using experimentally determined values for β, V_A, and g'_μ for the Si BJT, and approximate values for both W_b and μ_{nb} at 300 K, we estimate τ_{nb} to be about 0.2 nsec from (6.37) for this transistor. We expect that 0.2 nsec actually underestimates the real τ_{nb} value, because of the assumption of a constant doping profile in the base region used to derive (6.20). Previous studies [16, 17] on the estimation of minority-carrier lifetimes from experimental data suggest that a correction factor has to be used to properly account for the effect of the base doping gradient on the accurate estimation of τ_{nb} in real transistors. Therefore, by using an effective doping-induced base grading factor ($\eta = \ln(N_{ab}(0)/N_{ab}(W_b))$) of 4.6, which is reasonable in these transistors, we obtain a τ_{nb} of about 2 nsec for the Si BJTs. Due to the presence of large base doping concentration (peak $N_{ab} \approx 5 \times 10^{18}$ cm^{-3}) in these devices, τ_{nb} is expected to be dominated by Auger recombination rather than a pure SRH recombination process. Recent literature results [18] indicate that for acceptor doping levels close to 1×10^{19} cm^{-3}, τ_{nb} is approximately 55 nsec and that the Auger recombination component is at least an order of magnitude smaller than the SRH component. Indirect evidence for the presence of traps in the base region of various UHV/CVD Si and SiGe transistors has been demonstrated by performing liquid-helium temperature measurements of devices [19] (refer to Chapter 9).

Second-Order Phenomena 215

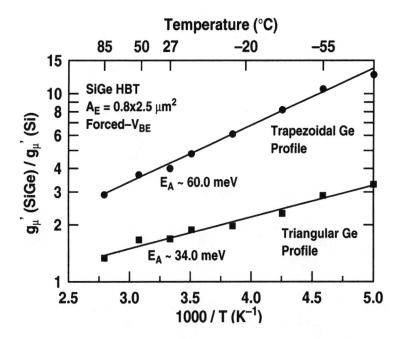

Figure 6.22 Measured ratio of g'_μ in a SiGe HBT and a Si BJT as a function of reciprocal temperature.

The measured g'_μ in SiGe HBTs is not only expected to be larger than for a comparably constructed Si BJT, but also thermally activated due to the presence of Ge band-offsets in the base region. Figure 6.22 confirms this expectation for both triangular and trapezoidal profile SiGe HBTs. Treating the presence of the graded Ge profile as equivalent to an increase in η, we can roughly estimate τ_{nb} for these SiGe HBTs by using (6.37). For instance, assuming that the 7% trapezoidal Ge profile is equivalent to a 10% increase in η compared to the Si BJT and by using the experimentally measured results for β, V_A, and g'_μ for the SiGe HBT, we obtain a value of τ_{nb} of approximately 1.5 nsec for these SiGe HBTs. As expected, the τ_{nb} values for both SiGe and Si transistors are comparable and the differences observed in the measured g'_μ are due primarily to the base bandgap differences induced by the Ge. It is important to note that this thermally activated relationship of the ratio between the g'_μ in a SiGe HBT and that in a Si BJT can only be explained by the variation in the base bandgap in SiGe HBTs and not by the small differences in τ_{nb} that might exist.

We have also experimentally observed that g'_μ in SiGe HBTs at low-injection levels shows only a weak V_{BE} dependence at all temperatures (Figure 6.23), mainly

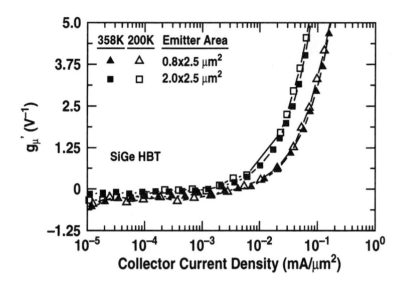

Figure 6.23 Measured g'_μ as a function of collector current density in a SiGe HBT with a triangular Ge profile. Two emitter areas are shown, at both 358 K and 200 K.

because of the cancellation of the bias dependence of β and V_A in (6.37). The NBR base current component in both the Si BJT and SiGe HBT showed only a weak dependence on the emitter size (Figure 6.23). This is not unexpected since the NBR component is primarily determined by the presence of traps in the bulk region of the intrinsic base region, independent of the emitter geometry.

6.2.3 Impact of NBR on Early Voltage

A direct consequence of NBR is a difference in the slope of the common-emitter output characteristics of a transistor depending on whether the device is biased using forced-I_B or forced-V_{BE} conditions [20]. This can be explained by comparing the dc characteristics for a transistor under an ideal situation (no NBR) with that in the presence of NBR (see Figure 6.24). Without NBR, the increase in I_C with V_{CB} is the same whether the transistor is biased under forced-V_{BE} or forced-I_B input drive, yielding the same V_A for both conditions. In the presence of NBR, however, V_A measured using both techniques will differ because of the decrease in I_B with V_{CB}. In a forced-I_B situation, V_{BE} is allowed to change in such a way as to maintain constant I_B. Due to the fact that I_B decreases with increasing V_{CB} in the presence of NBR, V_{BE} is forced to increase so as to maintain constant I_B.

Second-Order Phenomena 217

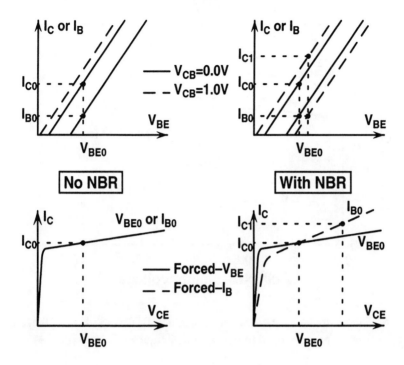

Figure 6.24 Illustration of the effects of NBR on the Gummel and output characteristics of a transistor.

This small increase in V_{BE} exponentially increases I_C, leading to a much smaller V_A. In a forced-V_{BE} situation, however, the I_C increase is due only to the decrease in W_b for an increase in V_{CB}, as one might expect in the ideal case. Thus, in the presence of NBR, $V_A(forced - I_B)$ will be smaller than $V_A(forced - V_{BE})$, and the two quantities are related through g'_μ, as detailed in the following derivation.

Under nonnegligible NBR, the forced-I_B operation of a transistor causes V_{BE} to increase for an increase in V_{CB}. From (6.33) one can determine

$$\left.\frac{\partial J_B}{\partial V_{CB}}\right|_{I_B} = 0 = J_B(0)\left[\frac{kT}{q}\left.\frac{\partial V_{BE}}{\partial V_{CB}}\right|_{I_B} + \frac{1}{J_{b0} + J_{nbr,0}}\left.\frac{\partial J_{nbr,0}}{\partial V_{CB}}\right|_{I_B}\right], \quad (6.38)$$

from which we have

$$\frac{1}{J_{b0} + J_{nbr,0}}\left.\frac{\partial J_{nbr,0}}{\partial V_{CB}}\right|_{I_B} = \frac{-q}{kT}\left.\frac{\partial V_{BE}}{\partial V_{CB}}\right|_{I_B}. \quad (6.39)$$

218 Silicon-Germanium Heterojunction Bipolar Transistors

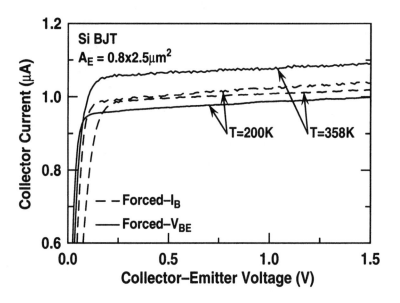

Figure 6.25 Measured output characteristics of a Si BJT at 358 K and 200 K, using both forced-I_B and forced-V_{BE} techniques.

In the case of forced-V_{BE} operation, we can define g'_μ as

$$g'_\mu = \frac{1}{J_B} \left.\frac{\partial J_B}{\partial V_{CB}}\right|_{V_{BE}} = \frac{1}{J_{b0} + J_{nbr,0}} \left.\frac{\partial J_{nbr,0}}{\partial V_{CB}}\right|_{V_{BE}}. \tag{6.40}$$

Note that (6.39) and (6.40) must be exactly equal because both equations represent the amount of NBR sampled in the transistor under the two different modes of operation. Therefore,

$$g'_\mu = \frac{-q}{kT} \left.\frac{\partial V_{BE}}{\partial V_{CB}}\right|_{I_B}. \tag{6.41}$$

In general, the Early voltage from forced-V_{BE} and forced-I_B can be derived from

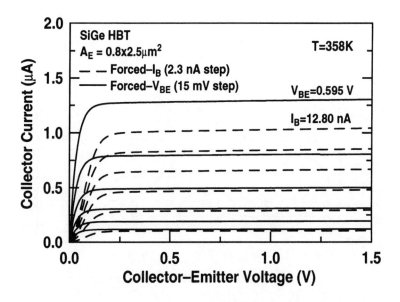

Figure 6.26 Measured output characteristics of a SiGe BJT at 358 K, using both forced-I_B and forced-V_{BE} techniques.

the Moll-Ross equation as (refer to Chapter 4)

$$V_A(forced - V_{BE}) = J_C(0) \left\{ \frac{\partial J_C}{\partial V_{CB}} \bigg|_{V_{BE}} \right\}^{-1}$$

$$\approx - \left[\frac{\frac{\partial}{\partial V_{CB}} \left\{ \int_0^{W_b} \frac{dx}{n_{ib}^2} \right\}}{\int_0^{W_b} \frac{dx}{n_{ib}^2}} \right]^{-1} \quad (6.42)$$

and

$$V_A(forced - I_B) = J_C(0) \left\{ \frac{\partial J_C}{\partial V_{CB}} \bigg|_{I_B} \right\}^{-1}$$

$$\approx \left[\frac{q}{kT} \frac{\partial V_{BE}}{\partial V_{CB}} \bigg|_{I_B} \frac{\frac{-\partial}{\partial V_{CB}} \left\{ \int_0^{W_b} \frac{dx}{n_{ib}^2} \right\}}{\int_0^{W_b} \frac{dx}{n_{ib}^2}} \right]^{-1} . \quad (6.43)$$

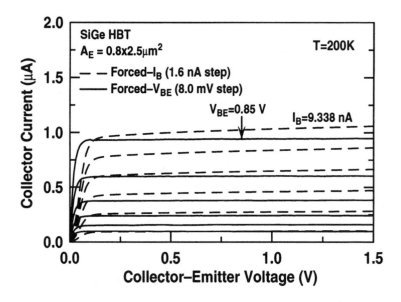

Figure 6.27 Measured output characteristics of a SiGe BJT at 200 K, using both forced-I_B and forced-V_{BE} techniques.

Using (6.41)–(6.43) and rearranging the terms give, finally [8]

$$g'_\mu \approx \frac{1}{V_A(forced - V_{BE})} - \frac{1}{V_A(forced - I_B)}. \qquad (6.44)$$

Since g'_μ is small in a well-made Si BJT, the difference between $V_A(forced - V_{BE})$ and $V_A(forced - I_B)$ is expected to be small, as is readily apparent from Figure 6.21. Equation (6.44) predicts that in SiGe HBTs, however, the difference between $V_A(forced - V_{BE})$ and $V_A(forced - I_B)$ will be greater because of the larger g'_μ compared to a Si BJT. Observe that using (6.44) one can also indirectly estimate the ratio of g'_μ between similarly constructed SiGe and Si transistors simply from the knowledge of measured V_A values. For the present case, using the measured V_A values for SiGe HBT (trapezoidal profile) and the Si BJT at 358 K and 200 K, we obtain $g'_\mu(SiGe)/g'_\mu(Si) \approx 3.0$ at 358 K and $g'_\mu(SiGe)/g'_\mu(Si) \approx 9.0$ at 200 K, which closely agrees with the results shown in Figure 6.22.

Recent transistor simulations have shown that $V_A(forced - I_B)$ and hence the βV_A of bipolar transistors is smaller in the presence of NBR when compared to an ideal situation [21, 22], and that variations in the base bandgap as well as temperature changes will also significantly affect the βV_A of transistors in the presence of NBR. Figure 6.25 shows typical common-emitter output characteristics of a Si

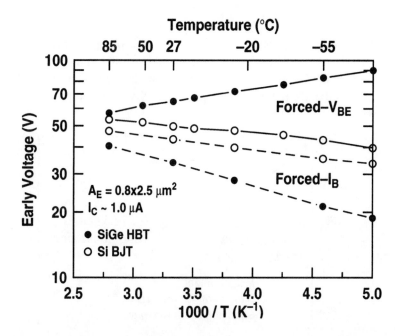

Figure 6.28 Measured Early voltage as a function of reciprocal temperature for a SiGe BJT and a Si BJT, using both forced-I_B and forced-V_{BE} techniques.

BJT measured using both forced-V_{BE} and forced-I_B input drive at 358 K and 200 K, respectively. Clearly, no noticeable difference exists between the two measurements because of the fact that the transistor has only a weak NBR component in the base current.

Figure 6.26 and Figure 6.27, however, show typical common-emitter output characteristics for the SiGe HBT, also obtained using forced-I_B and forced-V_{BE} at 358 K and 200 K, respectively. It is readily apparent from these figures that with cooling, the slope of I_C with respect to V_{CE} increases in a forced-I_B measurement, compared to a decrease in the same quantity for a forced-V_{BE} measurement. Figure 6.28 shows V_A obtained for both Si and SiGe transistors using forced-I_B and forced-V_{BE} conditions as a function of reciprocal temperature. Observe that the V_A in a Si BJT, obtained using both techniques, yields similar results, thus confirming the presence of only a weak NBR component in the base current of these transistors. In the SiGe HBTs, however, we can clearly observe a quasi-exponential degradation of $V_A(forced - I_B)$ compared to a quasi-exponential improvement in $V_A(forced - V_{BE})$ with cooling. While it is the bandgap grading in the SiGe HBT that increases the $V_A(forced - V_{BE})$, it is the amount of $\Delta E_{g,Ge}(x=0)$ that causes the exponential degradation in $V_A(forced - I_B)$ with cooling. From these experimental results, it is clear that such a strong temperature and input-bias dependent situation for SiGe HBTs could potentially have important consequences on the performance of SiGe HBT circuits which depend critically on the output conductance of the transistor.

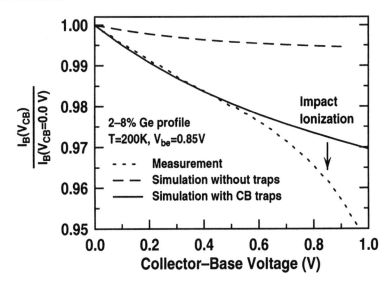

Figure 6.29 Measured and simulated normalized base current as a function of collector-base voltage.

6.2.4 Identifying the Physical Location of the NBR Traps

Given the experimentally determined presence of NBR in these SiGe HBTs, it is logical to wonder about the physical location of the responsible traps and how they may relate to the details of the device fabrication process. Calibrated 2-D device simulations (in this case with MEDICI [23]) are best suited to this task, and provide important insight into the physical mechanism of NBR in these SiGe HBTs. The MEDICI simulation-to-data calibration process is nontrivial, and information on proper meshing, parameter model choice, and the coefficient tuning methodology is discussed in [24], as well as in Chapter 12. Given the importance of temperature in experimentally assessing NBR in SiGe HBTs, simulation calibration across a wide temperature range (e.g., 300 K to 200 K) is important in the context of NBR. Note that the electron and hole lifetimes (i.e., trap densities) throughout the device

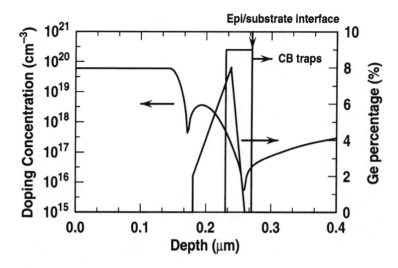

Figure 6.30 Doping and Ge profile for a SiGe HBT with finite trap density at the SiGe/Si growth interface.

are necessarily fixed during this calibration process.

Using these final calibrated parameters, we compared the measured and simulated normalized base current dependence on V_{CB} (Figure 6.29). Below the onset of impact ionization the simulations show only about 1.0% I_B reduction per volt V_{CB}, compared to the measured value (about 3.0% per volt V_{CB}). Note that introduction of a low lifetime region throughout the neutral base region changes the total I_B dramatically, and fails to accurately model the normalized I_B dependence on V_{CB}. This is not surprising given that: 1) the NBR-induced base current component is only a small fraction of the total I_B in these devices; and 2) the total base current of the SiGe HBT and Si BJT are nearly identical (meaning both have similar overall base trap density).

A logical explanation to this mismatch between data and simulation is that the additional traps responsible for the observed NBR are not uniformly distributed throughout the base, but rather are located in the vicinity of the CB junction. This hypothesis makes intuitive sense given that the original SiGe/Si growth interface is a plausible location for additional traps. As shown in Figure 6.30, a box-like, low-lifetime region (i.e., high trap density), located in the CB junction and centered on the SiGe growth interface, can explain the measured results (refer to the curve labeled "simulation with CB traps" in Figure 6.29). As can be seen in Figure 6.31, the hole recombination rate (R) is strongly modulated by the changing V_{CB}, and yet the total base current remains unperturbed, consistent with the data. Note that in

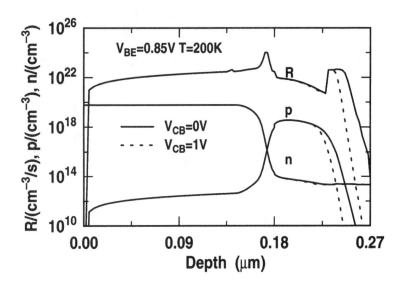

Figure 6.31 Simulated electron and hole densities, as well as the recombination rate profile for a SiGe HBT with collector-base traps.

this case, the recombination process is modulated by the majority carriers (holes), not the minority electrons one usually associates with NBR. One can estimate the value of the hole lifetime due to the traps by using

$$\Delta I_B \approx q R A_E \Delta W_{nbr} = \frac{q n A_E \Delta W_{nbr}}{\tau} = \frac{J_C A_E \Delta W_{nbr}}{v_s \tau} = \frac{I_C \Delta W_{nbr}}{v_s \tau}, \quad (6.45)$$

where ΔI_B is the I_B reduction due to the ΔV_{CB} increase, ΔW_{nbr} is NBR-induced positional shift in the point where $n = p$ as V_{CB} increases, v_s is the saturation velocity, and τ is the effective hole lifetime. The resulting normalized CB conductance becomes

$$g'_\mu = \frac{1}{I_B} \frac{\partial I_B}{\partial V_{CB}} \approx \frac{\beta}{v_s \tau} \frac{\partial W_{nbr}}{\partial V_{CB}}. \quad (6.46)$$

Using the lifetime determined in this manner, one finds that under the constraints of: 1) matching the total base current (hence β) at both 300 K and 200 K, while simultaneously 2) matching the CB conductance at 300 K and 200 K, that there is little flexibility in the trap lifetime and location if simulation-to-data calibration is to be maintained. Moving the traps closer to the EB junction, for instance, degrades β due to enhanced EB space-charge region recombination.

6.2.5 Circuit-Level Modeling Issues

Precision current sources (CSs), which are used extensively in many analog circuit applications, rely on the high V_A of the transistor to maintain a constant current output for large output voltage swings, and thus require very high output resistance (R_{out}). Given that in the presence of NBR there can be significant difference between the V_A of a SiGe HBT depending on whether it is voltage or current input driven, we can naively expect that the presence of NBR will have a strong impact on the performance of precision SiGe HBT CSs, particularly as the temperature changes. To assess the effects of NBR in precision SiGe HBT analog circuits, as well as shed light on the circuit-level modeling of NBR in SiGe HBTs, we have analyzed the behavior of various SiGe HBT current sources: the Wilson CS, the Cascode CS, the high-source-resistance CS, and the low-source-resistance CS, across the realistic operating range of -55°C to 85°C.

In order to quantify the impact of NBR on the temperature characteristics of these circuits, we first modified the conventional bipolar compact model (in this case, the Gummel-Poon PSPICE model, but the approach is easily extendable to any compact modeling tool) to accurately model the temperature dependence of V_A in SiGe HBTs across temperature [14]. The influence of NBR on the transistor characteristics is modeled in SPICE by considering a Miller resistance ($r_\mu = 1/g_\mu$) placed across the CB junction (Figure 6.32), whose values at various temperatures can be obtained experimentally from the inverse slope of the measured I_B versus V_{CB} characteristics. In a forced-V_{BE} input-drive situation, the variations in V_{CB} will force a current through r_μ that flows into the voltage source to decrease I_B by $\Delta i_\mu = V_{CB}/r_\mu$, as one would expect in the presence of NBR.

In this situation, the variation in I_C with V_{CB} is largely controlled by the SPICE VAF parameter. In the forced-I_B input-drive situation, however, Δi_μ is forced to flow into the device causing I_C to increase by $\beta \times i_\mu$, thereby degrading V_A. Figure 6.33 shows the close agreement between the appropriately modified SPICE model and the common-emitter output characteristics for the SiGe HBT at both 358 K and 200 K. In this case we have introduced a new parameter to control the temperature dependence of V_A (XTVAF).

Calibrated SPICE models were used to investigate and compare the temperature characteristics of R_{out} in CSs built with SiGe HBTs both with NBR and without NBR (i.e., the ideal case). Figure 6.34 shows the circuit schematic for both the Cascode and Wilson CSs, and Figure 6.35 shows the modeling results for R_{out} in these SiGe HBT CSs for situations both with and without NBR in the transistors, as a function of temperature. Under an ideal situation, both CSs show an increase in R_{out} because of the increase in both β and V_A with cooling. In the presence of NBR in the transistors, however, the R_{out} of both CSs is not only smaller (worse) com-

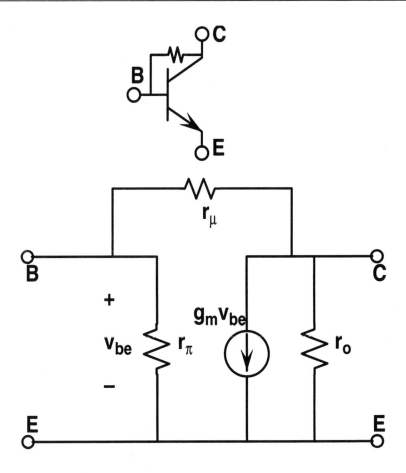

Figure 6.32 Schematic showing the approach to modeling NBR in SPICE using the collector-base resistance r_μ.

pared to the ideal case (no NBR) but also has an opposite temperature dependence. Observe that the default SPICE model does not accurately predict the temperature dependence of R_{out} in these CSs because of the lack of a parameter to account for the temperature dependence of V_A in the SiGe HBTs.

Other commonly used CS configurations include the high-source-resistance CS and low-source-resistance CS, shown in the inset of Figure 6.36. Unlike the Wilson, and Cascode CSs, these circuits correspond to the forced-I_B and forced-V_{BE} input-bias modes of operation, respectively. The modeling results for the ratio between the R_{out} of these circuits in situations with NBR to that without NBR, as a function of temperature are shown in Figure 6.36. The variable resistor value

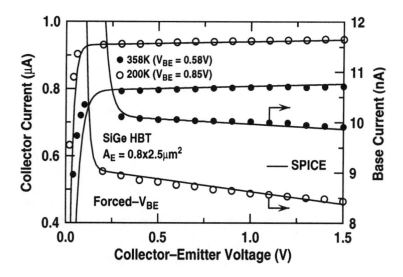

Figure 6.33 Measured data versus calibrated SPICE modeling results at 358 K and 200 K.

was chosen such that the transistor was biased at $I_C \approx 1.0$ μA at all temperatures. The low-source-resistance SiGe HBT CS shows near-ideal behavior of R_{out} across this temperature range because the transistor is biased in an almost pure forced-V_{BE} input-bias mode. The high-source-resistance SiGe HBT CS, however, shows a significant degradation in the R_{out} ratio with decreasing temperature because of the NBR-induced degradation in V_A resulting from operating the transistor in an almost pure forced-I_B input-bias mode. Finally, if we compare the performance of a SiGe HBT cascode and high-source-resistance CS with that of comparably designed Si BJT CSs, in the presence of NBR, we find that the use of SiGe HBT in the Cascode CS architecture gives substantial improvement in R_{out} to that obtained for a Si BJT-based Cascode CS ($R_{out}(SiGe)/R_{out}(Si) = 2.8$ at 300 K). As the circuits are biased closer to a forced-I_B input-bias mode situation, as in the high-source-resistance CS, however, one can clearly see the degrading effect of NBR in the SiGe HBT circuit. In this case, the performance of the SiGe HBT high-source-resistance CS and the Si BJT high-source-resistance CS are comparable ($R_{out}(SiGe)/R_{out}(Si) = 1.0$ at 300 K), and the Si BJT-based CS actually outperforms the SiGe HBT CS at lower temperatures ($R_{out}(SiGe)/R_{out}(Si) = 0.7$ at 200 K) [8].

While conventional Si BJT compact models can be appropriately modified to accurately capture NBR effects for circuit design, small-signal equivalent circuits of the transistor remain very useful for circuit analysis and offer additional insight

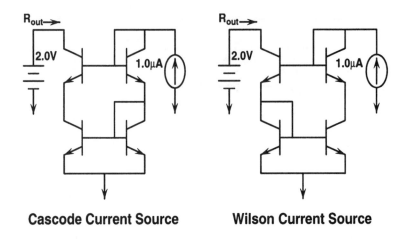

Cascode Current Source   **Wilson Current Source**

Figure 6.34 Circuit schematics of the Cascode and Wilson precision CSs.

into how device effects such as NBR couple to actual circuit operation. The first attempts to develop a small-signal model for the SiGe HBT were based on a linear superposition principle [25], but were shown to produce incorrect modeling of the emitter current. More recent work yielded a new hybrid-π small-signal model for the SiGe HBT, which properly accounts for the presence of significant NBR [26]. As depicted in Figure 6.37, the output resistance

$$r_o \equiv \frac{1}{\left.\frac{\partial I_C}{\partial V_{CB}}\right|_{V_{BE}}}, \tag{6.47}$$

and the collector-base resistance

$$r_\mu \equiv -\frac{1}{\left.\frac{\partial I_B}{\partial V_{CB}}\right|_{V_{BE}}} \tag{6.48}$$

are the required model parameters. From direct analysis of this new SiGe HBT hybrid-π model, we can determine that

$$r_\mu \approx \frac{\beta}{\left\{\frac{I_C}{V_A(\text{forced}-I_B)+V_{CE}} - \frac{I_C}{V_A(\text{forced}-V_{BE})+V_{CE}}\right\}} \tag{6.49}$$

and thus

$$g'_\mu \equiv \frac{g_\mu}{I_B} = \frac{1}{r_\mu I_B} \approx \frac{1}{V_A(\text{forced}-I_B)+V_{CE}} - \frac{I_C}{V_A(\text{forced}-V_{BE})+V_{CE}}, \tag{6.50}$$

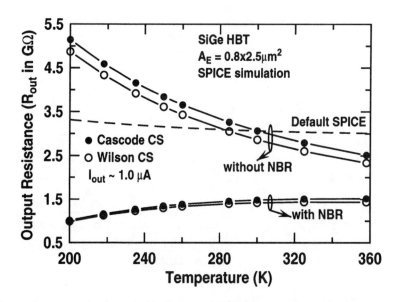

Figure 6.35 Comparison of calibrated SPICE modeling results both with NBR and without NBR for the output resistance of both Cascode and Wilson precision current sources as a function of temperature.

consistent with the result obtained by a more device-physics oriented transport formulation, as derived in (6.44). More details on the model extraction from experimental results, and its validity, are given in [26].

6.2.6 Device Design Implications

Experimental results, first-order theoretical calculations, and simulation results clearly show that the Ge-induced band offsets in the base region of a SiGe HBT can be used to exponentially enhance V_A and βV_A compared to that of a comparably constructed Si BJT. Unfortunately, however, these same Ge-induced band offsets will also amplify the otherwise weak NBR component in the transistors. Since the NBR current component in a SiGe HBT is a thermally activated function of the band offset at the EB space-charge region boundary, one would like to decrease the Ge-concentration at the EB boundary to weaken the impact of NBR on the device characteristics. This reduction in Ge, however, will also clearly reduce the β of the transistor, which is exponentially dependent on $\Delta E_{g,Ge}(x = 0)$. In practice, where one typically works under a constant thermodynamic stability criterion for the requisite SiGe strained layers, any reduction in the Ge-concentration at the EB-boundary will allow an increase in the Ge-grading across the base region. This is

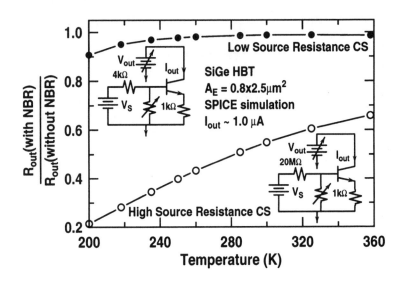

Figure 6.36 Calibrated SPICE modeling results for the ratio of the output resistance both with NBR and to that without NBR for the output resistance of both high- and low-source-resistance current sources.

advantageous for many SiGe HBT analog circuits, where one can trade β (typically not a limiting parameter) for larger V_A and higher f_{max}. Thus, a triangular-shaped Ge profile SiGe HBT is expected give the maximum benefit for V_A, βV_A, and f_{max}, and also result in a lower NBR-induced base current component. Experimental results, however, indicate that even in a triangular Ge profile SiGe HBT, the NBR component is significantly larger than in a comparably designed Si BJT (of equal neutral base trap density). Therefore, any introduction of Ge into the base region of a SiGe HBT, which is obviously intended to improve transistor performance, will naturally amplify the effects of any NBR base current component that may be present.

As discussed above, the presence of NBR causes I_B to be dependent on V_{CB}, hence affecting the relative $V_A(forced - V_{BE})$ versus $V_A(forced - I_B)$ values. A consequence of this difference is a much stronger-than-expected temperature dependence of the performance of SiGe HBT precision analog circuits, depending on the transistor input biasing condition of the circuit. For example, temperature characteristics consistent with simple SiGe theory (Chapter 4) for the SiGe HBT precision current sources are obtained when the transistors are biased using forced-V_{BE} input drive, whereas, a strongly degraded circuit performance at reduced temperatures is seen for a forced-I_B input biasing mode. Since in most analog circuit

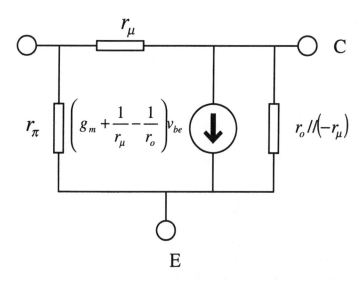

Figure 6.37 New hybrid-π equivalent circuit for the SiGe HBT which properly accounts for the dependence of terminal currents on V_{CB} in the presence of significant NBR.

designs the input-bias for a transistor is not purely V_{BE} driven, it is clear that the temperature performance of SiGe HBT precision circuits will be degraded due to the presence of NBR in the transistors. Therefore, in order to minimize the impact of NBR on the overall performance of SiGe HBT precision analog circuits, it is important that one carefully design the input-bias of those transistors from which large V_A is required to be closer to a forced-V_{BE} mode. The modeling methodology using a modified SPICE code that we have introduced should allow circuit designers to quantify the impact of NBR and thus gauge its importance in specific SiGe HBT circuit designs.

6.2.7 The Bottom Line

The presumption in this section is that significant NBR exists in the SiGe HBTs under consideration. In this situation we can say:

- An observable difference between $V_A(forced - V_{BE})$ and $V_A(forced - I_B)$ will exist in the SiGe HBT, and will be reflected in the output characteristics of the transistor.

- The measured $V_A(forced - V_{BE})$ value in the presence of NBR will be consistent with simple device theory (Chapter 4), but the $V_A(forced - I_B)$ will be degraded (lower) compared to simple theoretical expectations.

- This input-drive-dependent V_A difference will be amplified in the SiGe HBT compared to a comparably constructed Si BJT. That is, Ge-induced bandgap engineering will always act to enhance the effects of NBR.

- This input-drive-dependent V_A difference will get larger (worse) as the temperature decreases.

- Careful 2-D simulations can be used to identify the physical location of the traps responsible for the NBR component, and correlated with the fabrication process.

- Accurate compact modeling of SiGe HBTs for circuit design which includes NBR can be accomplished using existing Si BJT models, but may require an additional parameter to account for the inherently different temperature dependence in V_A between a SiGe HBT and a Si BJT. Such NBR-compatible models can provide a detailed assessment of the role of NBR-induced V_A changes on particular circuits.

NBR, while clearly inherent to bipolar transistor operation because finite trap densities necessarily exist in semiconductor crystals, does not necessarily strongly perturb the characteristics of modern SiGe HBTs. The experimental results presented in this chapter show a significant NBR base current component, and thus are instructive for understanding and modeling NBR in SiGe HBTs, but we have also measured devices from other SiGe technologies which do not show appreciable NBR-induced V_A changes. Thus, we do not consider NBR in any way to be a "show-stopper" for SiGe HBT deployment, but rather something to be carefully monitored and assessed during technology development and qualification. In this case, a simple bench-top measurement of $I_B(V_{CB})/I_B$ as a function of V_{CB} at two different temperatures (e.g., 300 K and 200 K) provides a simple and powerful tool for accurately assessing the presence of significant NBR in a given SiGe HBT technology generation. If present, appropriate steps can be taken to either try and correct the situation by process modification, or models can be developed which accurately account for the effect, thus ensuring that circuit designs are not negatively impacted.

6.3 Heterojunction Barrier Effects

In order to achieve maximum performance, SiGe HBTs must be biased at very high collector current densities (typically, above 1.0 mA/μm^2). High-injection heterojunction barrier effects (HBE), which occur in all HBTs, can cause severe degradation in key transistor metrics such as, β, g_m, V_A, f_T, and f_{max}, especially at reduced temperatures (see Chapter 9 for more detail on temperature effects). Careful transistor optimization is therefore required to delay the onset of the HBE to well above the current density levels required for normal circuit operation. Since the severity of the HBE is mainly determined by the amount of Ge-induced band offset at the SiGe/Si hetero-interface and the collector doping level, one needs to carefully design the CB junction of the HBT. In order to delay the onset of Kirk effect and hence HBE, one can easily increase the collector doping level (N_{dc}). Increasing N_{dc}, however, decreases f_{max} and BV_{CEO} due to the increase in C_{CB} and the CB electric field, respectively, presenting serious design constraints.

As will be seen, the shape and position of the Ge profile in the CB region of a SiGe are critical in determining the characteristics of the onset of HBE and the rate of degradation in HBT characteristics with increasing J_C. While large Ge grading is desirable for increasing V_A, f_T, and f_{max} of a SiGe HBT, the increased Ge concentration at the CB junction increases the induced barrier associated with HBE. To reduce the impact of the barrier on device performance, one can either gradually decrease the Ge at the CB region or place the SiGe-Si heterointerface deeper inside the collector region, instead of having an abrupt SiGe-Si transition at the interface. Obviously, these methods lead to an increase in the total Ge content of the film, which imposes film stability (and hence manufacturing) constraints on the fabrication process. These device design trade-offs clearly indicate that there exists no specific design solution to completely eliminate HBE. One can, however, tailor the CB design to suit the application at hand, and offers testament to the versatility that can be achieved with bandgap engineering.

We experimentally and theoretically examine the impact of Ge profile shape and the scaling of collector doping profile on high-injection HBE in SiGe HBTs operated over a wide temperature range of -73°C to 85°C [27, 28], as well as address compact modeling issues [29]. The results indicate that careful Ge profile design tailored with a proper collector profile design is required to push the barrier onset current density ($J_{C,barrier}$) to well beyond the typical circuit operating point.

We limit our discussion here to what we term "high-injection HBE," occurring at high-J_C in the device [8]. We note, however, for completeness, that there is another class of heterojunction barrier effects associated with the physical misplacement of the SiGe/Si hetero-interface with respect to the EB and CB junctions. As discussed in Chapter 4, as a general rule, the graded-base Ge profile should begin and end in the space-charge regions of the EB and CB junctions, outside the neutral base, since the transition from low bandgap (SiGe) to high bandgap (Si) is thereby effectively "buried" in the appropriate space-charge regions. If, instead,

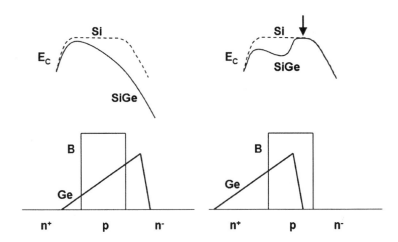

Figure 6.38 SiGe HBT band diagrams illustrating a Ge-misplacement-induced barrier.

the Ge profile is brought down *inside* the neutral base, parasitic conduction band barriers will result, as illustrated in Figure 6.38. These parasitic barriers can occur by an error in epitaxial growth conditions that places the boron base profile in an incorrect position with respect to the Ge profile, or more commonly, the base profile can out-diffuse past the Ge layer due to excessive thermal cycle, or some other enhanced diffusion process. In either case, due to barrier-induced minority carrier pile-up in the base, the electrical consequences of this Ge misplacement can be severe, resulting in a severe degradation in the transistor *dc* and *ac* performance, even at low-J_C values (it is the J_C dependence that differentiates this phenomenon from high-injection HBE). Such Ge-misplacement-induced HBE, however, are not fundamental to the operation of SiGe HBTs, and thus proper SiGe film growth and fabrication techniques can be used to easily eliminate them. For this reason, they will not be addressed in detail here, and the interested reader is referred to [30] for more detail.

6.3.1 High-Injection in SiGe HBTs

Although there are several ways to equivalently define high-injection in bipolar transistors, a state of high-injection in the device can be said to generally occur when the local minority carrier density in the emitter, base, or collector region, approaches and then exceeds the local ionized doping level. The onset of high-injection clearly depends on the construction and doping profiles of the device, but

Second-Order Phenomena 235

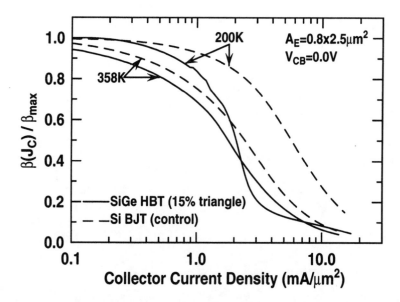

Figure 6.39 Normalized current gain as a function of collector current density for a Si BJT and SiGe HBT, at 358 K and 200 K.

as a general rule-of-thumb for modern Si-based bipolar transistors, current densities in excess of about 1.0 mA/μm^2 are usually sufficient to induce high-injection in the collector region. High-injection in the base and emitter are typically of lesser importance given the relatively higher doping densities in those regions.

In Si BJTs operated under high-injection in the collector, there are several phenomena that can cause the collector and base currents to deviate from their ideal low-injection behavior (i.e., $I_C, I_B \propto e^{qV_{BE}/kT}$), including Kirk effect [31], Webster-Rittner effect [32, 33], the *IR* drop associated with the base and emitter resistances, and quasi-saturation due to collector resistance. Among these, Kirk effect (or "base push-out") is usually the most important in practical Si BJTs (and SiGe HBTs). The physical basis of Kirk effect lies in the fact that the increased minority carrier concentration in the CB region, at high-injection, is sufficient to compensate for the doping-induced charge in the CB space-charge region, causing the space-charge region to first collapse, and then to be pushed deeper into the collector region as J_C (hence n_C) rises. The displacement of the CB space-charge region effectively increases the basewidth, which leads to a decrease in the collector current ($J_C \propto 1/W_b$), and an increase in the base transit-time ($\tau_b \propto 1/W_b^2$), thus causing a premature degradation in both β and f_T.

One can estimate the J_C at the onset of Kirk effect ($J_{C,Kirk}$) by considering the

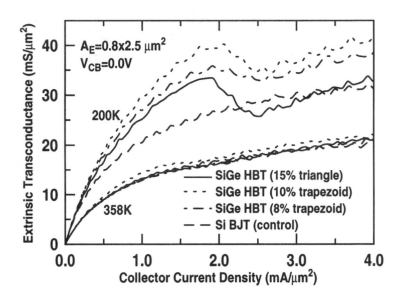

Figure 6.40 Extrinsic transconductance as a function of collector current density for a Si BJT and 3 SiGe HBT profiles, at 358 K and 200 K.

electron current density in the collector, which can be written generally as

$$J_C \approx q\, v_s\, n_C, \tag{6.51}$$

where v_s is the electron saturation velocity and n_C is the electron concentration in the collector. At the onset of Kirk effect, the built-in electric field at the original CB interface is reduced to zero and is moved (pushed) to a region deeper inside the collector. Using Poisson's equation, we can determine the electron concentration required to reduce the electric field to zero as

$$n_C \approx N_{dc} + \frac{2\epsilon(V_{CB} + \phi_{bi})}{q\, W_{epi}^2}, \tag{6.52}$$

where W_{epi} is the collector epi-layer thickness and ϕ_{bi} represents the CB built-in potential. Therefore, by combining (6.51) and (6.52) we obtain

$$J_{C,Kirk} \approx q\, v_s\, N_{dc}\left\{1 + \frac{2\epsilon(V_{CB} + \phi_{bi})}{q\, N_{dc}\, W_{epi}^2}\right\}. \tag{6.53}$$

The direct relationship between the onset of Kirk effect and the collector doping level is obvious. To get a feel for the numbers, if we assume realistic values for the

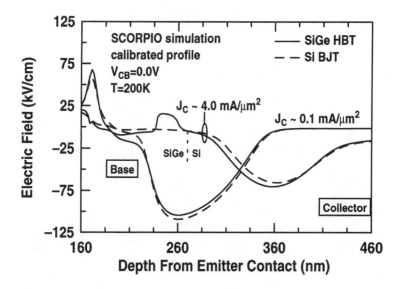

Figure 6.41 Simulated electric field distribution in a Si BJT and SiGe HBT at both low and high current density at 200 K.

uniformly doped collector (e.g., $N_{dc} = 1 \times 10^{17}$ cm^{-3}), and an epi-layer thickness of 0.5 μm, we thus expect from (6.53) that the onset of Kirk effect will occur at approximately 1.6 mA/μm^2.

Since maximum device performance is achieved at large current densities, one usually needs to increase N_{dc} to provide additional immunity to Kirk effect, thereby increasing the CB electric field, and decreasing the CB breakdown voltage. Thus, a fundamental trade-off exists in Si BJTs between device performance (i.e., peak f_T) and maximum operating voltage (i.e., BV_{CEO}), as reflected in the so-called "Johnson-limit" [34]. Note that at reduced temperatures, the slight increase in the saturation velocity with cooling will naturally increase $J_{C,Kirk}$ (refer to Chapter 9 for the design implications).

The story of high-injection is more interesting in SiGe HBTs. In SiGe HBTs, the transition from a narrow bandgap SiGe base layer to the larger bandgap Si collector layer introduces a valence band offset at the SiGe-Si heterointerface. Since this band offset is masked by the band bending in the CB space-charge region during low-injection operation, it has negligible effect on the device characteristics. At high-injection, however, the collapse of original CB electric field at the heterointerface exposes the offset, which opposes hole injection into the collector. The hole pile-up that occurs at the hetero-interface induces a conduction band barrier that then opposes the electron flow from base to collector, causing an increase in the

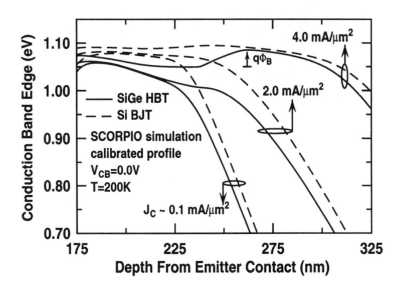

Figure 6.42 Simulated conduction band edge as a function of depth for a Si BJT and SiGe HBT, for various current densities at 200 K.

stored base charge which results in the sudden decrease in both f_T and f_{max}. The "pinning" of the collector current due to this induced conduction band barrier, and the simultaneous increase in the base current due to valence band offset, causes a rapid degradation in desirable characteristics of the SiGe HBT at a HBE onset current density, and can present serious device and circuit design issues. This effect was first reported in [35, 36], and later addressed by others authors [27, 37, 38]. In addition, since the transport currents are thermally activated functions of the barrier height, it is expected that the HBE will have a much more pronounced impact at reduced temperatures, raising important questions about operation over a wide temperature range [28, 39]. It is therefore essential that the collector profile and the Ge profile be designed properly to reduce the impact of HBE on circuit performance.

6.3.2 Experimental Results and Simulations

SiGe HBTs with three different Ge profiles were measured (an 15% Ge triangle, an 10% Ge trapezoid, and an 8% Ge trapezoid), along with a comparably designed Si BJT control. The collector profile was identical for all of the transistors and was selectively implanted to simultaneously optimize f_T (at high J_C) while maintaining an acceptable BV_{CEO} of about 3.3 V. The Gummel characteristics of all of

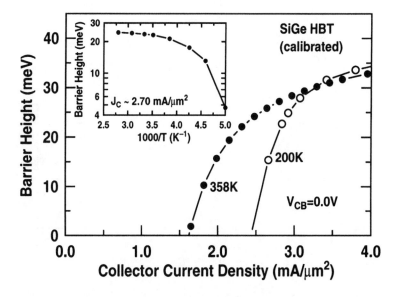

Figure 6.43 Simulated barrier height as a function of current density at 358 K and 200 K. Inset shows the temperature dependence of the barrier height at fixed current density.

the transistors are ideal across the measured temperature range of 200 K to 358 K. While the SiGe HBTs and Si BJTs have differing current gains, as expected, a normalization of β as a function of J_C shows that there is a clear difference in high-injection behavior for the SiGe and Si devices, particularly at reduced temperatures (Figure 6.39). As will be seen, this strong decrease in β at high current densities is the result of both J_C and J_B phenomena associated with HBE. A sensitive test for clearly observing high-injection HBE in SiGe HBTs is to extract the transconductance (g_m) at high J_C from the Gummel characteristics, at high and low temperatures. As shown in Figure 6.40, a clear dip in the g_m at 200 K at a J_C of about 2.0 mA/μm^2 can be clearly seen. By comparing g_m and β at $J_C = 2.0$ mA/μm^2 between the SiGe HBT and the Si BJT at 358 K and 200 K, respectively, one can easily deduce that the differences are associated with the Ge profile, and hence are a signature of high-injection HBE. In addition, Figure 6.40 suggests that the trapezoidal Ge profiles show a weaker degradation in g_m at 200 K compared to the triangular Ge profile, because of the presence of a smaller Ge band offset in the CB junction (15% Ge versus 8% and 10% Ge, respectively), indicating that the specific design of the Ge profile plays a role, as expected.

To shed light on both the physics of HBE in SiGe HBTs, as well as determine

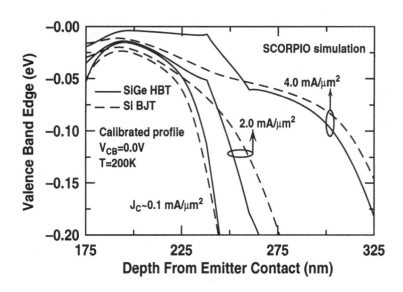

Figure 6.44 Simulated valence band edge as a function of depth for a Si BJT and SiGe HBT, for various current densities at 200 K.

the optimum doping and Ge profiles for scaled SiGe HBTs, we have used numerical device simulation (in this case, the 1-D drift-diffusion simulator SCORPIO [40]). Carefully calibrated simulations based on measured SIMS profiles show agreement with data for β, V_A, f_T, and f_{max} to within ±15% over -73°C to 85°C (and labeled as "calibrated profile" in subsequent figures). Figure 6.41 shows the electric field distribution in the base-collector region of both calibrated SiGe and Si transistors at low- and high-J_C. Observe that at low-J_C, the CB built-in electric field entirely covers the SiGe/Si heterointerface. At high-injection ($J_C = 4.0$ mA/μm^2, past peak f_T), however, the CB space-charge region is pushed deep into the collector region in both transistors due to Kirk effect and in the SiGe HBT a barrier is formed at the original SiGe/Si heterointerface, and can be clearly seen in the high-J_C field distribution.

Figure 6.42 shows the evolution of the induced conduction band barrier to electrons in the SiGe HBT as a function of J_C. Clearly, the electron barrier appears only at high injection and this can be correlated with the exposure of the SiGe/Si valence band offset (Figure 6.44). In addition, the magnitude of the induced conduction band barrier (ϕ_B) gets larger as the device is biased progressively into higher injection, while at very large current densities ϕ_B eventually saturates (Figure 6.43). Although ϕ_B at a fixed J_C decreases with cooling due to the shift in operating point with temperature (refer to Chapter 9), its impact will be much

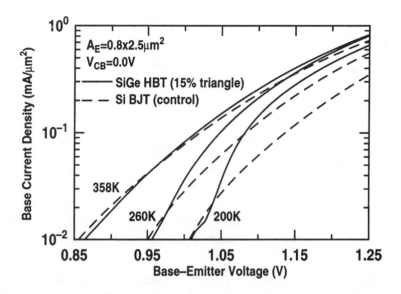

Figure 6.45 Measured base current density as a function of base-emitter voltage for a Si BJT and SiGe HBT, at three different temperatures.

greater at low temperatures due to its thermally activated nature as a band-edge phenomenon.

The sudden increase in J_B accompanying the barrier onset in a SiGe HBT (Figure 6.45) is the result of the accumulation of holes in the base region due to HBE (Figure 6.46). At low-injection one clearly sees that the hole concentration in the base is unperturbed compared to a Si BJT. At high-injection, however, not only is the hole profile pushed out into the collector region (Kirk effect) but also the presence of the barrier increases the hole concentration close to the CB junction. The calibrated simulations are clearly capable of quantitatively capturing the measured differences in $g_m(J_C)$, $\beta(J_C)$, and $f_T(J_C)$ between SiGe HBTs and Si BJTs operating across a wide temperature range [28].

6.3.3 Profile Optimization Issues

A fundamental trade-off in collector profile design exists between maximizing both BV_{CEO} and f_{max} in SiGe HBTs. RF and microwave power amplifiers require large BV_{CEO}, and therefore the collector doping must be reduced. Obviously, such a reduction in N_{dc} will adversely affect the large-signal performance due to the premature onset of HBE. We have investigated two hypothetical collector profiles for transistors suited for such high-power RF applications: 1) the calibrated profile

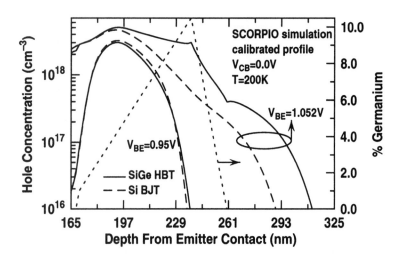

Figure 6.46 Simulated hole density profile in the base collector junction for a Si BJT and SiGe HBT at two different biases at 200 K.

with N_{dc} reduced by 0.2×; and 2) a flat 5×10^{16} cm^{-3} collector doping profile. The emitter, base, and the Ge profiles were left unchanged in order to investigate the performance differences solely due to N_{dc} scaling. In order to keep W_b constant while scaling the collector profile, we have abruptly dropped the N_{ab} at the CB junction. These profiles are expected to yield a BV_{CEO} in the range of 5–7 V.

Figure 6.47 shows the simulated f_T for these transistors compared to the calibrated SiGe HBT, at both 358 K and 200 K. Clearly, the reduced N_{dc} in both devices causes a much earlier onset of the barrier effect, and premature roll-off in f_T compared to the original profile. While both scaled-collector devices have comparable peak f_T, a more rapid decrease in the f_T of the SiGe HBT with constant collector doping profile is observed compared to the device with a 0.2× reduced N_{dc}. This is expected, since there is no retrograded doping to oppose the Kirk-effect/barrier effect process once it is triggered.

One can also, in principle, "tune" the barrier onset by properly adjusting the Ge retrograde profile shape. A higher Ge grading in the base region of a SiGe HBT provides better high-frequency performance throughout the temperature range. Increasing the Ge grading, however, necessarily increases the Ge content in the CB junction, which leads to a stronger barrier effect at high-injection. In order to reduce the impact of barrier effect in such cases, one can either more gradually decrease the Ge or push the Ge deeper into the collector. In either case, however, one is limited by the amount of Ge that can be added because of the stability constraints of the SiGe films. Various retrograde profile shapes that have varying stability have

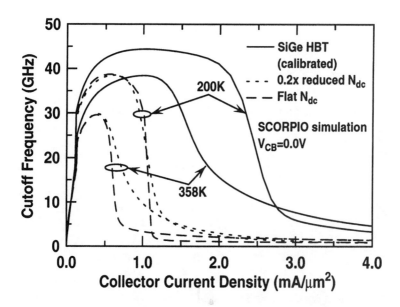

Figure 6.47 Simulated normalized cutoff frequency as a function of collector current density for different collector profiles at 358 K and 200 K.

been investigated to shed light on the optimum approach (Figure 6.48). Figure 6.49 shows the trade-offs in film stability with each SiGe retrograde design point. All three of these Ge profiles achieve similar β and peak f_T compared to the calibrated control profile, and thus shed light on the optimum approach to tailor the impact of HBE on high J_C performance.

Figure 6.50 compares the simulated f_T versus J_C for the various retrograded Ge profiles with that of the calibrated profile, at 358 K and 200 K. Observe that by increasing the retrograde slope one can simultaneously increase $J_{C,barrier}$ and reduce the f_T roll-off at high-J_C. The retrograde 2 profile is clearly not optimum, however, because only marginal improvement in $J_{C,barrier}$ is achieved at the cost of a significant lowering of the film stability. On the other hand, with an abrupt box-shaped Ge profile in the collector (retrograde 3), which has a stability point close to retrograde 1, we obtain similar peak f_T but with a much higher $J_{C,barrier}$. In this case, however, one faces a more rapid degradation in f_T beyond the barrier onset when compared to the other retrograde designs. These retrograde design tradeoffs are expected to hold for devices with more aggressive scaling of the base and Ge profiles, as discussed in detail in [28].

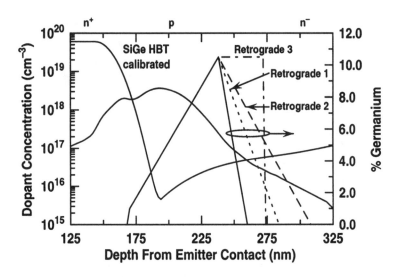

Figure 6.48 Doping profiles with varying Ge profile shapes used to investigate the effects of Ge retrograde on barrier effects in SiGe HBTs.

6.3.4 Compact Modeling

The successful insertion of SiGe HBTs into practical systems requires accurate compact circuit models for design. Because SiGe HBTs are typically modeled using Si BJT-based compact models (e.g., SPGP, VBIC, MEXTRAM, or HICUM), it is important to assess the accuracy of these models for capturing unique device phenomena such as high-injection HBE. Of interest in this context is how well existing compact models can fit $f_T - J_C$ and $\beta - J_C$ at very high current densities, particularly for SiGe HBTs optimized for high breakdown voltage (low N_{dc}).

As discussed above, the neutral-base charge, which dominates the extrapolated forward transit time (τ_f) at high current density, is subjected to two dominant high-injection effects in SiGe HBTs: Kirk effect and HBE. While semiempirical or empirical models of the neutral-base charge in SiGe HBTs have been presented (e.g., [41]), these models cannot accurately fit measured f_T data at very high J_C, especially for high-breakdown voltage devices ($BV_{CEO} > 5-7$ V). Figure 6.51 shows the measured and fitted cut-off frequency as a function of J_C for such a device. Observe that neither the VBIC or the HICUM models accurately capture the f_T-J_C data above 2.0 mA/μm^2. This modeling failure also causes discrepancies in other ac device parameters important in RF circuit design (e.g., C_{BE}).

In most compact models, the Kirk effect and HBE are lumped into a single function, assuming the Kirk effect and barrier effect occur simultaneously. This

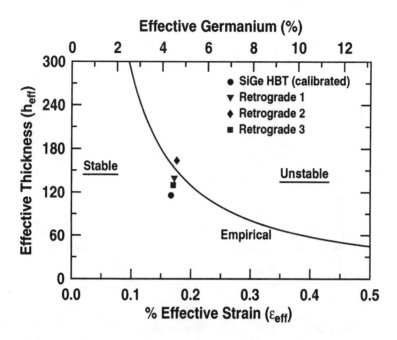

Figure 6.49 SiGe stability diagram showing the stability points of the simulated profiles.

assumption, however, is no longer valid when the SiGe/Si heterojunction is located either in the neutral base region or deeper in the epitaxial collector. The latter, for instance, might be found in SiGe HBTs optimized for high breakdown voltage. Figure 6.52 illustrates the impact of the Kirk effect and barrier effect on a SiGe HBT with a deep heterojunction. One finds that the neutral base is not subjected to the SiGe/Si barrier when the Kirk effect onset occurs. That is, the barrier effect is significant only when J_C is much higher than the critical onset current density of Kirk effect ($J_{C,barrier} > J_{C,Kirk}$).

The other effect that compact models fail to capture is that the functional form of the $f_T - J_C$ roll-off in SiGe HBTs is also determined by the depth of Si/SiGe heterojunction. Figure 6.53 shows simulated values of τ_f as a function of J_C for different SiGe/Si heterojunction locations ($x = 0$ implies the SiGe/Si heterointerface is at the metallurgical CB junction). For a shallow heterojunction location (i.e., 120 nm), barrier effect is significant when Kirk effect occurs (consistent with the experimental results presented in the previous section). For a deeper barrier location (i.e., 220 nm), however, the barrier effect is not significant, since the pushed-out base caused by Kirk effect has not yet reached the SiGe/Si heterointerface. A

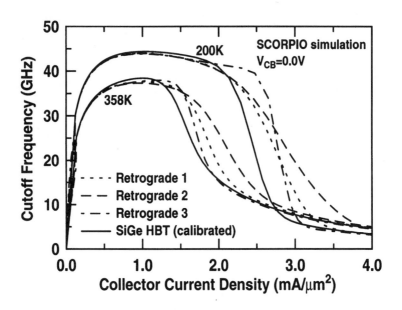

Figure 6.50 Simulated normalized cutoff frequency as a function of collector current density for different Ge profile shapes at 358 K and 200 K.

"kink" behavior in the transit time as a function of J_C results when the Kirk effect and barrier effect occur at different J_C, and can be seen in Figure 6.51 between 1-2 mA/μm^2. To accurately capture this phenomenon, a new transit time model that decouples the two effects is thus needed [29].

To obtain deeper insight into how the neutral-base charge density depends on J_C, and how high-injection in SiGe HBTs might best be modeled, we first consider calibrated 2-D MEDICI simulations [23]. Three different types of transistors were simulated: 1) a Si BJT; 2) a SiGe HBT with an infinitely deep SiGe/Si heterointerface (i.e., Ge extended into the subcollector); and 3) a SiGe HBT with 220-nm deep heterointerface (as measured from the metallurgical CB junction). The trapezoidal Ge profile used in the simulations has a peak Ge content of 10%, and is graded linearly across the neutral base, but is constant through the CB junction, as depicted in Figure 6.52.

Figure 6.54 shows the simulated energy band diagram as a function of depth at two high J_C values (1.96 and 5.40 mA/μm^2) for both SiGe HBT profiles. Both J_C values correspond to a large pushed-out base region (i.e., under Kirk effect). In this pushed-out base region, the electric field is low and the band is approximately flat, as expected. The electron mobility is maximized and is approximately constant across the neutral base region, and thus the assumption of charge neutrality in this

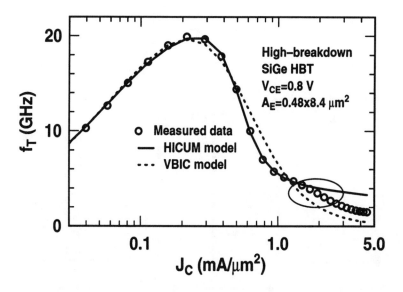

Figure 6.51 Comparison of measured $f_T - J_C$ characteristics with the HICUM and VBIC compact models.

region is valid.

At $J_C = 1.96$ mA/μm^2, the SiGe/Si heterointerface is still covered by the CB space-charge region, and no barrier is induced in E_C. At 5.4 mA/μm^2, however, the SiGe/Si heterointerface is exposed in the neutral base, and the heterojunction-induced electric field is suppressed by the piled-up holes, resulting in the formation of a barrier in the conduction band (Figure 6.54).

Considering a 1-D transistor, the forward transit time can be modeled as

$$\tau_f = \frac{q \int_E \Delta n(x)dx}{\Delta J_C} + \frac{q \int_B \Delta n(x)dx}{\Delta J_C} + \frac{q \int_C \Delta n(x)dx}{\Delta J_C} = \tau_e + \tau_b + \tau_c. \quad (6.54)$$

Hence, $\Delta n/\Delta J_C$ as a function of depth determines the dependence of τ_f on J_C. Figure 6.55 and Figure 6.56 show $\Delta n/\Delta J_C$ as a function of depth for the three transistors when the barrier effect *is not* significant (1.96 mA/μm^2), and *is* significant (5.4 mA/μm^2), respectively. The τ_e, τ_b, and τ_c can be obtained by the integration shown in (6.54). Since the barrier effect occurs when the heterojunction is located in the pushed-out base, it only changes τ_b. As shown in Figure 6.55 and Figure 6.56, when the SiGe HBT *is not* subjected to the barrier effect, the shape of $\Delta n/\Delta J_C$ in the neutral base is triangular. When the SiGe HBT *is* subjected to the barrier effect, however, the shape of $\Delta n/\Delta J_C$ in the neutral base is trapezoidal. Moreover, the slope of the trapezoid is approximately equal to the slope of the tri-

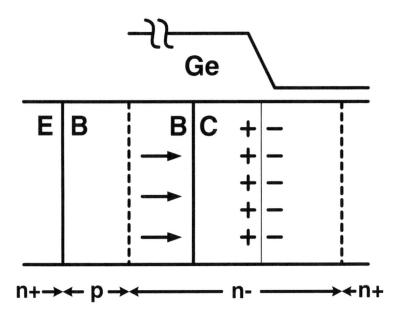

Figure 6.52 Illustration of Kirk effect and barrier effect in a SiGe HBT.

angle under the same bias conditions. A compact model of $\Delta\tau_b$ caused by the HBE can thus be derived based on these characteristics, as shown below.

Based on the simulations, the barrier effect only changes τ_b. Thus, the new transit time model can be written as $\tau_{f,new} = \tau_{f,old} + \Delta\tau_b$, where $\Delta\tau_b$ is the increase in the transit time due to barrier effect. An existing compact model (e.g., HICUM model) can be applied to model $\tau_{f,old}$. Assuming that $\beta \gg 1$, the relationship between the charge in the neutral base and the electron flow J_n can be written as [42]

$$-J_n \approx \mu_n n \frac{d(E_{Fn} - E_{Fp})}{dx} = \mu_n n \frac{d\Delta E_F}{dx}, \qquad (6.55)$$

where E_{Fn} and E_{Fp} are quasi-Fermi levels for the electrons and holes, respectively, μ_n is the mobility of the electrons, and x is the depth.

At high J_C, when Kirk effect occurs, the density of electrons (n) near the pushed-out base is much higher than local collector doping density (N_{dc}), suggesting that in this region $n \simeq (n - N_{dc}) \simeq p$. Figure 6.57 shows the simulated electron and hole distributions at different J_C, confirming that indeed $n \simeq p$ in the pushed-out base. Thus, n as a function of depth in the pushed-out base can be derived under the condition of high-injection as (assuming that $n \gg N_{dc}$)

$$n(n - N_{dc}(x)) \simeq n_{ib}^2(x) e^{\Delta E_F/kT} = n_i^2 e^{\Delta E_{g0}/kT} e^{(\Delta E_F + x\Delta G_{Eg,Ge})/kT} \qquad (6.56)$$

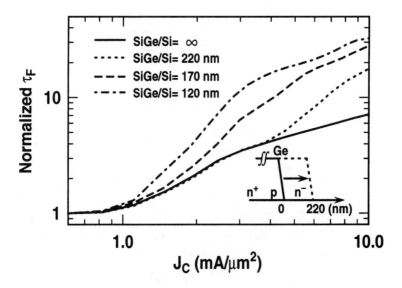

Figure 6.53 Transit time as a function of J_C with different barrier locations, simulated using MEDICI.

or

$$n \simeq n_i e^{\Delta E_{g0}/2kT} e^{(\Delta E_F + x\Delta G_{Eg,Ge})/2kT} \quad (6.57)$$

where $\Delta G_{Eg,Ge} = d[\Delta E_{g,Ge}(x)]/dx$ is the gradient of the Ge-induced bandgap narrowing (i.e., $\Delta G_{Eg,Ge}$ is constant assuming that the bandgap changes as a linear function of depth). Substituting n into (6.55) and using (6.57), one obtains

$$\frac{-J_n}{\mu_n} = n\left(\frac{2kT}{n}\frac{dn}{dx} - \Delta G_{Eg,Ge}\right)$$
$$\frac{-J_n}{2kT\mu_n} = \frac{dn}{dx} - \frac{\Delta G_{Eg,Ge}}{2kT}n$$
$$\frac{-1}{2kT\mu_n} = \frac{\partial}{\partial x}\left(\frac{\partial n}{\partial J_n}\right) - \frac{\Delta G_{Eg,Ge}}{2kT}\left(\frac{\partial n}{\partial J_n}\right). \quad (6.58)$$

In the epitaxial collector region, $\Delta G_{Eg,Ge} = 0$ since the Ge content is constant, and thus (6.58) can be solved as

$$\frac{\partial n}{\partial J_n} = Ax + B \quad (6.59)$$

where $A = -1/(2kT\mu_n)$, and B is a constant determined by the boundary conditions. Note that in the pushed-out base, μ_n can be shown to be approximately

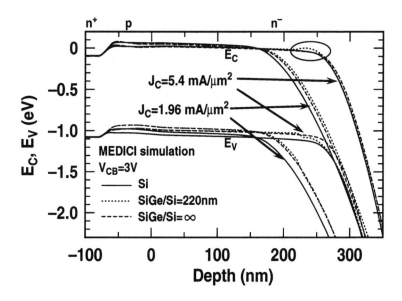

Figure 6.54 Simulated E_C, E_V as a function of depth for a Si BJT and two SiGe HBTs, at a J_C at which the barrier effect both is and is not significant.

constant, and thus the distribution of $\Delta n/\Delta J_C$ in the epitaxial collector is a linear function of the depth and the slope is approximately constant. When $\Delta G_{Eg,Ge}$ is nonzero, (6.58) can be solved as

$$\frac{\partial n}{\partial J_n} = Ce^{x\Delta G_{Eg,Ge}/2kT} + \frac{1}{\mu_n \Delta G_{Eg,Ge}} \quad (6.60)$$

where C is a constant determined by the boundary conditions.

At the interface between the pushed-out base and the BC junction, J_n is dominated by drift current, and the electric field is so high that the velocity of the electrons is almost saturated, and thus $J_n = qv_s n$. Hence, at the edge of the pushed-out base

$$\frac{\partial n(x_b)}{\partial J_n} = \frac{1}{qv_s}, \quad (6.61)$$

where x_b is the boundary at the edge of the pushed-out base. Near the barrier, the boundary condition on the electrons is similar to the space-charge region of a pn junction. Hence, one obtains

$$\Delta n(x_{bar}^-) = e^{\Delta E_{bar}/2kT} \times \Delta n(x_{bar}^+), \quad (6.62)$$

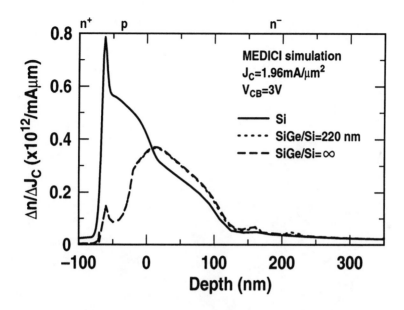

Figure 6.55 Simulated $\Delta n/\Delta J_C$ as a function of depth for a Si BJT and two SiGe HBTs, for a J_C at which barrier effects are negligible.

where x_{bar} is the SiGe/Si heterointerface, and ΔE_{bar} the induced barrier height at that J_C. Here x_{bar}^- is SiGe/Si heterointerface on the base side, while x_{bar}^+ is SiGe/Si heterointerface on the collector side (Figure 6.58). Thus

$$\frac{\partial n(x_{bar}^-)}{\partial J_n} = e^{\Delta E_{bar}/2kT} \times \frac{\partial n(x_{bar}^+)}{\partial J_n}, \tag{6.63}$$

and $\Delta n/\Delta J_n$ can be solved using the equations and conditions above. Figure 6.58 shows $\Delta n/\Delta J_n$ as a function of depth for a SiGe HBT both with and without barrier effect. The $\Delta \tau_b$ caused by barrier effect can then be calculated as

$$\Delta \tau_b \approx \begin{cases} 0 & , x_b \leq x_{bar} \\ qx_{bar}|A|(x_b - x_{bar})(e^{\Delta E_{bar}/2kT} - 1) & , x_b > x_{bar} \end{cases} \tag{6.64}$$

where $A = -1/(2kT\mu_n)$ is the constant defined above, x_b is a function of J_C due to the Kirk effect, and can be written as [41]

$$x_b(J_C) = \begin{cases} 0 & , J_C \leq J_{C,Kirk} \\ w_c(1 - J_{C,Kirk}/J_C) & , J_C > J_{C,Kirk} \end{cases} \tag{6.65}$$

where w_c is the length of epi-collector (Figure 6.58), and $J_{C,Kirk}$ the critical onset current for the Kirk effect. Substituting (6.65) into (6.64), one gets

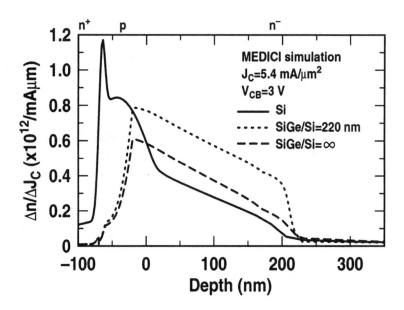

Figure 6.56 Simulated $\Delta n/\Delta J_C$ as a function of depth for a Si BJT and two SiGe HBTs, for a J_C at which barrier effects are significant.

$$\Delta \tau_b(J_C) \approx \begin{cases} 0 & , x_b \leq x_{bar} \\ \tau_{bb}(1 - J_{C,barrier}/J_C) & , x_b > x_{bar} \end{cases} \quad (6.66)$$

where

$$\tau_{bb} = q|A|(e^{\Delta E_{bar}/2kT} - 1)(w_c x_{bar} - x_{bar}^2), \quad (6.67)$$

and

$$J_{C,barrier} = J_{C,Kirk}/(1 - x_{bar}/w_c), \quad (6.68)$$

and the two additional parameters needed to account for the barrier effect are ΔE_{bar} and x_{bar}. In (6.66), τ_{bb} is the $\Delta \tau_b$ at $J_C = +\infty$, and $J_{C,barrier}$ is the critical onset current density for the barrier effect. Both are useful as model fitting parameters.

One can also define a smooth function to approximate (6.66) [41]

$$\Delta \tau_b = \tau_{bb} \frac{i_r + \sqrt{i_r^2 + a}}{1 + \sqrt{1 + a^2}} \quad (6.69)$$

where $i_r = 1 - J_{C,barrier}/J_C$, and a is an empirical parameter for smoothing the transition region. This new model for the change in $\Delta \tau_b$ caused by the barrier effect

Second-Order Phenomena

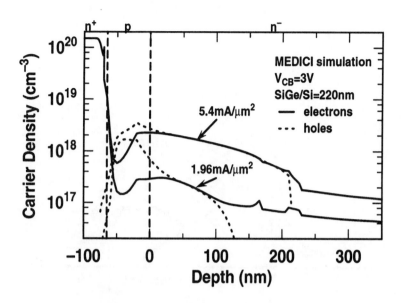

Figure 6.57 The simulated carrier density as a function of depth at different J_C values in a SiGe HBT.

Table 6.1 Key Parameters of the Present Model

Parameters	$V_{CE} = 0.8$ V	$V_{CE} = 3.0$ V
t_{ef0} (psec)	2.0	2.0
t_{hcs} (psec)	37	37
a_{hc}	0.5	0.5
$J_{C,Kirk}$ (mA/μm^2)	0.8	1.9
τ_{bb} (psec)	110	110
$J_{C,barrier}$ (mA/μm^2)	2.3	4.7

in SiGe HBTs can be easily added to existing compact models (e.g., the HICUM model). Figure 6.59 compares f_T as a function of J_C using the present transit time model implemented in HICUM, and the original HICUM result versus measured data. Table 6.1 shows the key fitting parameters used in the present model (note that $x_{bar} = 260$ nm and $\Delta E_{bar} = 70$ meV in this case). The first four parameters are identical to that used in the HICUM model, the last two are parameters derived above (note that for a 1-D model, $J_{C,Kirk}$ is identical to $I_{CK}/Area$ in HICUM). The comparison shows that the present compact model accurately captures the cut-off

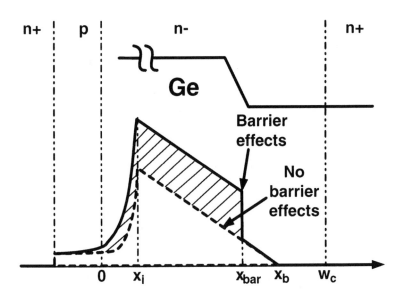

Figure 6.58 Illustration of $\Delta n / \Delta J_n$ as a function of depth for a SiGe HBT both with and without barrier effect present.

frequency and transit time as a function of J_C in the presence of HBE.

6.3.5 The Bottom Line

Due to the presence of SiGe/Si heterojunctions in SiGe HBTs, heterojunction barrier effects are inherent in SiGe HBT design and operation, and thus in some sense can be considered the most serious of the three second-order phenomena considered in this chapter. Given this situation, HBE must always be carefully "designed around." This is not overly difficult for low-BV_{CEO} transistors where the collector doping is relatively high, effectively retarding Kirk effect. For applications requiring higher breakdown voltage devices (e.g., power amplifiers), however, care must be taken to ensure that HBE do not adversely impact circuit designs, and that they are accurately modeled. For HBE in SiGe HBTs we can state the following:

- Heterojunction barrier effects fall into two general categories: 1) induced barriers due to Ge misplacement; and 2) high-injection-induced barriers. The former can be corrected with proper growth and fabrication techniques, and are thus not inherent to a given SiGe technology. The latter, however, can be considered fundamental to the operation of SiGe HBTs, and must be carefully accounted for and accurately modeled by designers.

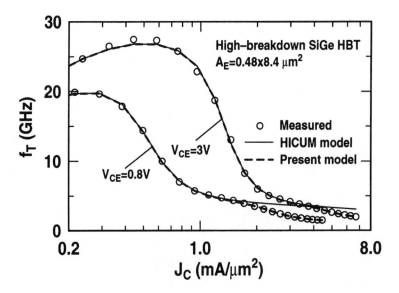

Figure 6.59 The measured and modeled cut-off frequency as a function of J_C using both the present model, and the default HICUM model.

- High-J_C HBE causes a rapid degradation in β, g_m, and f_T once the barrier is induced. The critical onset current density for HBE ($J_{C,barrier}$) is thus a key device design parameter.

- HBE are induced in the conduction band when the hole density in the pushed-out base under high-J_C is effectively blocked from moving into the collector by the SiGe/Si heterojunction. Both J_C and J_B are strongly affected.

- For low-breakdown voltage devices, HBE and Kirk effect generally occur at similar current densities. In higher-breakdown voltage devices, or devices with deep SiGe/Si heterojunctions, however, the two effects can occur at very different current densities, producing unusual structure in the f_T-J_C characteristics, which first-order compact models do not accurately capture.

- Changes to the Ge retrograde can be used to effectively retard the onset of HBE, but at the expense of reduced film stability.

- Changes to the collector doping profile can be used to effectively retard the onset of HBE, but at the expense of increased CB capacitance and reduced breakdown voltage.

- The impact of HBE on device and circuit performance will rapidly worsen as the temperature decreases, because they are band edge phenomena.

For any SiGe HBT technology generation, it is a prudent exercise to carefully characterize the transistors and assess the significance of HBE on the overall device response, and determine $J_{C,barrier}$. This is easily accomplished by plotting linear g_m on linear J_C at two temperatures (e.g., 300 K and 200 K), and this knowledge can then be communicated to circuit designers. If $J_{C,barrier}$ is low enough for practical concerns, then Ge and/or collector profile modifications can be implemented to alleviate any problems. When moving to a new technology generation with different Ge and doping profiles, HBE should always be revisited.

References

[1] E.F. Crabbé et al., "Current gain rolloff in graded-base SiGe heterojunction bipolar transistors," *IEEE Elect. Dev. Lett.*, vol. 14, pp. 193-195, 1993.

[2] M.S. Latham et al., "The impact of Ge grading on the bias and temperature characteristics of SiGe HBT precision voltage references," *J. de Physique IV*, vol. 6, pp. 113-118, 1996.

[3] S.L. Salmon et al., "The impact of Ge profile shape on the operation of SiGe HBT precision voltage references," *Proc. IEEE Bipolar/BiCMOS Circ. Tech. Meeting*, pp. 100-103, 1997.

[4] R.J. Widlar, "New developments in IC voltage regulators," *IEEE J. Solid-State Circ.*, vol. 6, pp. 2-7, 1971.

[5] M. Gunawan et al., "A curvature-corrected low-voltage bandgap reference," *IEEE J. Solid-State Circ.*, vol. 28, pp. 667-670, 1993.

[6] P.R. Gray and R.G. Meyer, *Analysis and Design of Analog Integrated Circuits*, New York: Wiley, 1977.

[7] A.P. Brokaw, "A simple three-terminal IC bandgap reference," *IEEE J. Solid-State Circ.*, vol. 9, pp. 388-393, 1974.

[8] A.J. Joseph, "The physics, optimization, and modeling of cryogenically operated silicon-germanium heterojunction bipolar transistors," Ph.D. Dissertation, Auburn University, 1997.

[9] S.L. Salmon, "The impact of germanium profile shape on the operation of SiGe HBT precision voltage references," M.S. Thesis, Auburn University, 1998.

[10] S.L. Salmon et al., "The impact of Ge profile shape on the operation of SiGe HBT precision voltage references," *IEEE Trans. Elect. Dev.*, vol. 47, pp. 292-298, 2000.

[11] H.A. Ainspan and C.S. Webster, "Measured results on bandgap references in SiGe BiCMOS," *Elect. Lett.*, vol. 34, pp. 1441-1442, 1998.

[12] Z.A. Shafi et al., "The importance of neutral base recombination in compromising the gain of SiGe/Si heterojunction bipolar transistors," *IEEE Trans. Elect. Dev.*, vol. 38, pp. 1973-1976, 1991.

[13] A.J. Joseph et al., "Neutral base recombination in advanced SiGe HBTs and its impact on the temperature characteristics of precision analog circuits," *Tech. Dig. IEEE Int. Elect. Dev. Meeting*, pp. 755-758, 1995.

[14] A.J. Joseph et al., "Neutral base recombination and its impact on the temperature dependence of Early voltage and current gain-Early voltage product in UHV/CVD SiGe heterojunction bipolar transistors," *IEEE Trans. Elect. Dev.*, vol. 44, pp. 404-413, 1997.

[15] A.J. Joseph et al., "Optimization of Early voltage for cooled SiGe HBT precision current sources," *J. de Physique IV*, vol. 6, pp. 125-130, 1996.

[16] J.P. Downing and R.J. Whittier, "Simple determination of base transport factor in bipolar transistors," *Solid-State Elect.*, vol. 14, pp. 221-225, 1971.

[17] M.S. Birrittella, A. Neugroschel, and F.A. Lindholm, "Determination of the minority-base base lifetime of junction transistors by measurements of basewidth-modulation conductances," *IEEE Trans. Elect. Dev.*, vol. 26, pp. 1361-1363, 1979.

[18] D.B.M. Klaassen, "A unified mobility model for device simulation – II. temperature dependence of carrier mobility and lifetime," *Solid-State Elect.*, vol. 35, pp. 961-967, 1992.

[19] A.J. Joseph, J.D. Cressler, and D.M. Richey, "Operation of SiGe heterojunction bipolar transistors in the liquid-helium temperature regime," *IEEE Elect. Dev. Lett.*, vol. 16, pp. 268-270, 1995.

[20] R.C. Jaeger and A.J. Broderson, "Self-consistent bipolar transistor models for computer simulation," *Solid-State Elect.*, vol. 21, pp. 1269-1272, 1978.

[21] J.M. McGregor et al., "Output conductance of bipolar transistors with large neutral-base recombination current," *IEEE Trans. Elect. Dev.*, vol. 39, pp. 2569-2575, 1992.

[22] S. Mohammadi and C.R. Selvakumar, "Analysis of BJTs, pseudo-HBTs, and HBTs by including the effect of neutral base recombination," *IEEE Trans. Elect. Dev.*, vol. 41, pp. 1708-1715, 1994.

[23] MEDICI, 2-D Semiconductor Device Simulator, Avant!, Fremont, CA, 1997.

[24] G. Niu, J.D. Cressler, and A.J. Joseph, "Quantifying neutral base recombination and the effects of collector-base junction traps in UHV/CVD SiGe HBTs," *IEEE Trans. Elect. Dev.*, vol. 45, pp. 2499-2504, 1998.

[25] J.S. Hamel, R.J. Alison, and R.J. Blaikie, "Experimental method to extract *ac* collector-base resistance from SiGe HBTs," *IEEE Trans. Elect. Dev.*, vol. 44, pp. 1944-1948, 1997.

[26] G. Niu et al., "A new common-emitter hybrid-π small-signal equivalent circuit for bipolar transistors with significant neutral base recombination," *IEEE Trans. Elect. Dev.*, vol. 46, pp. 1166-1173, 1999.

[27] A.J. Joseph et al., "Impact of profile scaling on high-injection barrier effects in advanced UHV/CVD SiGe HBTs," *Tech. Dig. IEEE Int. Elect. Dev. Meeting*, pp. 253-256, 1996.

[28] A.J. Joseph et al., "Optimization of SiGe HBTs for operation at high current densities," *IEEE Trans. Elect. Dev.*, vol. 46, pp. 1347-1356, 1999.

[29] Q. Liang et al., "A physics-based, high-injection transit-time model applied to barrier effects in SiGe HBTs," *IEEE Transactions on Electron Devices*, vol. 49, pp. 1807-1813, 2002.

[30] J.W. Slotboom et al., "Parasitic energy barriers in SiGe HBTs," *IEEE Elect. Dev. Lett.*, vol. 12, pp. 486-488, 1991.

[31] C.T. Kirk, "Theory of transistor cutoff frequency falloff at high current densities," *IRE Trans. Elect. Dev.*, vol. 3, pp. 164-170, 1964.

[32] W.M. Webster, "On the variation of junction-transistor current-amplification factor with emitter current," *Proc. IRE*, vol. 42, pp. 914-916, 1954.

[33] E.S. Rittner, "Extension of the theory of the junction transistor," *Phys. Rev.*, vol. 94, pp. 1161-1171, 1954.

[34] E.O. Johnson, "Physical limitations on frequency and power parameters of transistors," *RCA Review*, pp. 163-177, 1965.

[35] S. Tiwari, "A new effect at high currents in heterostructure bipolar transistors," *IEEE Elect. Dev. Lett.*, vol. 9, pp. 142-144, 1988.

[36] S. Tiwari, "Analysis of the operation of GaAlAs/GaAs HBTs," *IEEE Trans. Elect. Dev.*, vol. 36, pp. 2105-2121, 1989.

[37] P.E. Cottrell and Z. Yu, "Velocity saturation in the collector of Si/Ge$_x$Si$_{1-x}$ HBTs," *IEEE Elect. Dev. Lett.*, vol. 11, pp. 431-433, 1990.

[38] Z.P. Yu, P.E. Cottrell, and R.W. Dutton, "Modeling and simulation of high-level injection behavior in double heterojunction transistors," *Proc. IEEE Bipolar/BiCMOS Circ. Tech. Meeting*, pp. 192-194, 1990.

[39] J.D. Cressler et al., "High-injection barrier effects in SiGe HBTs operating at cryogenic temperatures," *J. de Physique IV*, vol. 4, pp. C6/117-C6/122, 1994.

[40] D.M. Richey, J.D. Cressler, and A.J. Joseph, "Scaling issues and Ge profile optimization in advanced UHV/CVD SiGe HBTs," *IEEE Trans. Elect. Dev.*, vol. 44, pp. 431-440, 1997.

[41] M. Schröter, "Physics-based minority charge and transit time modeling of bipolar transistors," *IEEE Trans. Elect. Dev.*, vol. 46, pp. 288-300, 1999.

[42] H. Kroemer, "Two integral relations pertaining to the electron transport through a bipolar transistor with nonuniform energy gap in the base region," *Solid-State Elect.*, vol. 28, pp. 1101-1103, 1985.

Chapter 7

Noise

One of the most desirable attributes of SiGe HBTs is their low noise capability. To illustrate the importance of noise, consider a mobile receiver. The lower limit of dynamic range is set by the noise, while the upper limit is set by the linearity. The first stage of the receiver, typically an LNA, must amplify signals as low as -100 dBm, while maintaining adequate signal-to-noise ratio. The amount of noise added by the LNA (as measured by the noise figure) must be sufficiently low. The receiver must also be highly linear in order to handle signal levels which fluctuate between 20–40 dB from the mean value as a function of time [1]. The predominant contributor to the LNA noise figure is the RF noise of the transistor. Another key concern in RF circuits is "phase noise." The transistor low-frequency noise is upconverted to RF frequencies through the device nonlinearities, producing phase noise in local oscillator (LO) reference signals. The LO phase noise mixes with the RF signal, thus broadening the signal in the intermediate frequency band. The minimum channel spacing of the receiver is thus limited by the LO phase noise. In this chapter, we describe the physical origins of RF noise, low-frequency noise, and phase noise in SiGe HBTs, and we offer a device-level generic noise modeling and optimization methodology. Applications of this methodology to SiGe HBT profile design for low noise, as well as optimal transistor sizing and biasing for circuit design are presented.

To describe transistor RF noise performance, we first introduce the "linear noisy two-port theory" [2]. Based on this theoretical formalism, a generic modeling methodology is then developed, which enables the estimation of all the noise parameters and the associated gain using either measured or simulated Y-parameters. Analytical equations are derived to offer intuitive insight into how noise performance is related to the key transistor parameters, and used for optimal transistor sizing and biasing in circuit design. The ability to simultaneously achieve high current gain (β), high cutoff frequency (f_T), and low base resistance (r_b) is the fundamental underlying reason for the superior noise performance of SiGe HBTs. SiGe profile design for low noise is then illustrated using first generation SiGe HBT technology [3, 4]. We finally discuss the measurement and modeling of $1/f$ noise as well as the upconversion of low-frequency noise to phase noise at RF frequencies.

7.1 Fundamental Noise Characteristics

7.1.1 Thermal Noise

Imagine that we measure the voltage across a resistor using an ideal voltmeter. We will inevitably measure random voltage fluctuations. These fluctuations occur even if we do not apply any external power supply, and are referred to as the "thermal noise" of a resistor. We cannot use the average value of the voltage as a figure-of-merit, because the time-averaged value is always zero. Instead, we use its mean square value

$$\overline{v_n^2} \equiv \overline{(v - \overline{v})^2} = \lim_{T \to \infty} \int_0^T (v - \overline{v})^2 dt. \tag{7.1}$$

The use of the square makes it possible to take a meaningful average of the fluctuation. Thermal noise has microscopic origins and results from the random (Brownian) motion of the transporting electrons, which causes random current flow even if there is no external bias or electric field.[1] The random current flow leads to random voltage fluctuations across the resistor. This noise increases with temperature, as expected.

In practice, the amount of noise measured depends on the bandwidth of the measurement system. A common measurement involves a very narrow bandwidth Δf, centered on a frequency f. The noise voltage spectral components within this bandwidth have a mean square value. The ratio of this mean square value with respect to Δf, as Δf approaches zero, gives the "power spectral density" (PSD) of the voltage noise, denoted by $S_v(f)$, where $S_v(f)$ has units of V^2/Hz. A more rigorous definition of power spectral density for voltage and current noise is given below. For a resistor of value R,

$$S_v = 4kTR, \tag{7.2}$$

---

[1]The microscopic origin of electrical noise remains an active research field, and an entire body of literature is associated with it (quantum fluctuation theory). Noise processes of various types are a fundamental aspect of the universe, and noise is thus far more general than just in the field of electrical engineering where it is usually formally treated.

and
$$\overline{v_n^2} = 4kTR\Delta f. \quad (7.3)$$

Using Thevenin's theorem, we can also describe the thermal noise using a current source with a mean square value of

$$\overline{i_n^2} = 4kTg\Delta f = 4kT\frac{1}{R}\Delta f. \quad (7.4)$$

We have assumed that the noise spectrum density has no frequency dependence (from $f = 0$ to ∞). This condition is also referred to as "white noise," an analog to "white light" (i.e., a uniform spectral density across all wavelengths). Strictly speaking, this is clearly an unphysical condition. However, we state here without proof that it is generally safe to assume white noise behavior for the frequency range of practical interest in most electrical engineering problems.

7.1.2 Shot Noise in a *pn* Junction

"Shot noise" refers to the fluctuations associated with the *dc* current flow across a potential barrier, which naturally occurs, for instance, under forward bias in a *pn* junction. The passage of carriers across this barrier is a random event, and the resulting current consists of a large number of independent current pulses. The average of these current pulses is the *dc* diode current. However, if we can measure the current using an ideal ammeter (i.e., with zero input impedance), the current also has fluctuations. The mean square of the current fluctuations is proportional to the average diode current I_{dc}

$$\overline{i_n^2} = 2qI_{dc}\Delta f. \quad (7.5)$$

Again, the shot noise spectral density does not have frequency dependence in the frequency range of practical interest to most circuits.

7.1.3 Shot Noise in Bipolar Transistors

From fundamental bipolar transistor theory, the essence of bipolar action is to use an additional junction to separate the electron and hole diffusion currents in the EB junction. Assuming negligible phase delay in the CB junction (an assumption that can clearly be violated at very high frequencies [5, 6]), we have

$$\overline{i_b^2} = 2qI_B\Delta f, \quad (7.6)$$

$$\overline{i_c^2} = 2qI_C\Delta f. \quad (7.7)$$

The base current shot noise is amplified by the transistor, producing a noise much stronger than the collector current shot noise. For the same I_C, a SiGe HBT has a lower I_B than a Si BJT, and hence a lower $\overline{i_b^2}$ than a Si BJT, because of the inherently higher β. Furthermore, a SiGe HBT can be operated at a lower I_C than a Si BJT for the same RF gain, because of the higher f_T and f_{max}, further reducing $\overline{i_b^2}$. A summary of existing models for the correlation between the base and collector shot noises can be found in [6].

7.2 Linear Noisy Two-Port Network Theory

A linear noisy electrical network can be equivalently represented by its noise-free counterpart with added external noise generators. The magnitude of the external generator is expressed either by an equivalent noise resistance or an equivalent noise temperature. If the network has four terminals or two (input and output) ports, the inherent noisiness of the network is expressed by the "noise figure."

7.2.1 Two-Port Network

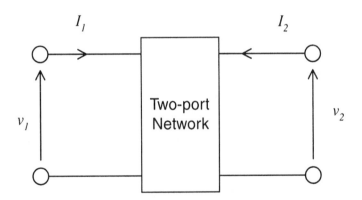

Figure 7.1 A noiseless linear two-port.

Figure 7.1 shows a noiseless linear two-port network. The currents and voltages at the input and output are related to each other by the impedance matrix Z or the admittance matrix Y

$$\begin{pmatrix} V_1 \\ V_2 \end{pmatrix} = \begin{pmatrix} Z_{11} & Z_{12} \\ Z_{21} & Z_{22} \end{pmatrix} \begin{pmatrix} I_1 \\ I_2 \end{pmatrix}, \qquad (7.8)$$

or

$$\begin{pmatrix} I_1 \\ I_2 \end{pmatrix} = \begin{pmatrix} Y_{11} & Y_{12} \\ Y_{21} & Y_{22} \end{pmatrix} \begin{pmatrix} V_1 \\ V_2 \end{pmatrix}. \tag{7.9}$$

The subscripts 1 and 2 refer to the input and output ports, respectively. Currents flowing into the network are defined as positive. The upper case letters I and V indicate the Fourier transforms of the current and voltage, which are frequency dependent [2].

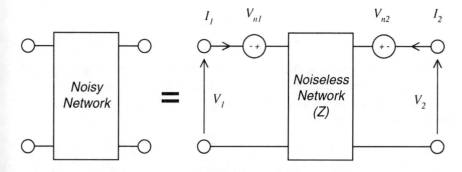

Figure 7.2 Thevenin equivalent circuit with two external series voltage noise generators.

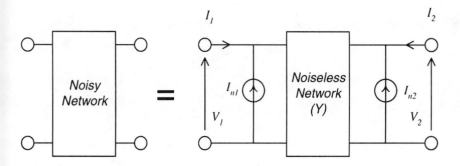

Figure 7.3 Norton equivalent circuit with two external parallel current noise generators.

A noisy two-port can be equivalently represented by an extension of Thevenin's theorem. In Figure 7.2, a series noise voltage generator appears at each port of the noiseless two-port. They generate the same amount of noise to the outside circuit as the actual noisy network would. Some degree of correlation in general exists between these generators because the same physical mechanism may be responsible for the open-circuit voltage fluctuations. Two external parallel current generators are convenient if the two-port is described using the Y matrix, as shown in Figure 7.3. In this case, the $I - V$ relation becomes

$$\begin{pmatrix} V_1 + V_{n1} \\ V_2 + V_{n2} \end{pmatrix} = \begin{pmatrix} Z_{11} & Z_{12} \\ Z_{21} & Z_{22} \end{pmatrix} \begin{pmatrix} I_1 \\ I_2 \end{pmatrix}, \quad (7.10)$$

or

$$\begin{pmatrix} I_1 + I_{n1} \\ I_2 + I_{n2} \end{pmatrix} = \begin{pmatrix} Y_{11} & Y_{12} \\ Y_{21} & Y_{22} \end{pmatrix} \begin{pmatrix} V_1 \\ V_2 \end{pmatrix}. \quad (7.11)$$

The conventional Fourier transform cannot be applied to a noise voltage or current source, because these are random fluctuations extending over all time and thus have infinite energy content. Instead, the Fourier transforms for noise voltage and current are defined by [2]

$$v_n(f) = \int_{-\tau/2}^{\tau/2} v_n(t) e^{-j2\pi f t} dt, \quad (7.12)$$

$$i_n(f) = \int_{-\tau/2}^{\tau/2} i_n(t) e^{-j2\pi f t} dt. \quad (7.13)$$

The corresponding power spectral densities (PSDs) are defined by

$$S_{v_n}(f) = \lim_{\tau \to \infty} \frac{2\overline{|v_n(f)|^2}}{\tau}, \quad (7.14)$$

$$S_{i_n}(f) = \lim_{\tau \to \infty} \frac{2\overline{|i_n(f)|^2}}{\tau}. \quad (7.15)$$

A more convenient representation is to use a series voltage generator and a parallel current generator, both at the input, as shown in Figure 7.4. The $I - V$ relations then become

$$\begin{pmatrix} I_1 + I_{na} \\ I_2 \end{pmatrix} = \begin{pmatrix} Y_{11} & Y_{12} \\ Y_{21} & Y_{22} \end{pmatrix} \begin{pmatrix} V_1 + V_{na} \\ V_2 \end{pmatrix}. \quad (7.16)$$

The new input voltage and current noise generators V_{na} and I_{na} can be related to I_{n1} and I_{n2} by comparing the $I - V$ relationships

$$V_{na} = -\frac{I_{n2}}{Y_{21}}, \quad (7.17)$$

and

$$I_{na} = I_{n1} - \frac{Y_{11}}{Y_{21}}I_{n2} = I_{n1} - \frac{I_{n2}}{H_{21}}. \quad (7.18)$$

This circuit arrangement is also known as the "chain representation." The choice of the name "chain" becomes obvious if we consider the series connection of several networks as a chain (i.e., a cascade). The chain representation is extremely convenient in noise figure calculations. Note that the input noise current I_{na} in the chain representation is different from the input noise current I_{n1} shown in Figure 7.3.

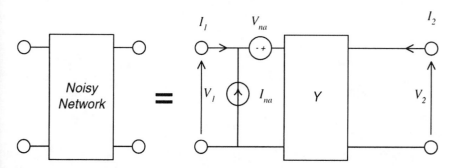

Figure 7.4 The chain representation of a linear noisy two-port with an input noise current generator and an input voltage noise generator.

7.2.2 Input Noise Voltage and Current in BJTs

Consider the shot noise in an ideal 1-D bipolar transistor. The base current shot noise and the collector current shot noise correspond to I_{n1} and I_{n2}, as illustrated in Figure 7.5. We can convert Figure 7.5 into the chain representation using (7.17) and (7.18)

$$V_{na} = -\frac{i_c}{Y_{21}}, \quad (7.19)$$

$$I_{na} = i_b - \frac{Y_{11}}{Y_{21}}i_c = i_b - \frac{i_c}{H_{21}}. \quad (7.20)$$

Now we can solve for the spectral densities of V_{na}, I_{na}, and $V_{na}I_{na}^*$

$$S_{v_n} = \frac{V_{na}V_{na}^*}{\Delta f} = \frac{S_{ic}}{|Y_{21}|^2} = \frac{2qI_C}{|Y_{21}|^2}, \quad (7.21)$$

$$S_{i_n} = \frac{I_{na}I_{na}^*}{\Delta f} = S_{ib} + \frac{S_{ic}}{|H_{21}|^2} = 2qI_B + \frac{2qI_C}{|H_{21}|^2}, \quad (7.22)$$

where (7.19) and (7.20) were used. For a given operating I_C, a higher β is desired to reduce I_B and hence S_{i_n}. A higher f_T is also desired to increase H_{21} and hence reduce S_{i_n}. Fundamentally, the inherently high β and high f_T of SiGe HBTs will thus lead to low input noise current. Finally, we have

$$S_{v_n i_n^*} = \frac{1}{\Delta f} \frac{i_c}{Y_{21}} \left(\frac{Y_{11}}{Y_{21}} i_c \right)^*$$

$$= S_{i_c} \frac{Y_{11}^*}{|Y_{21}|^2} = 2qI_C \frac{Y_{11}^*}{|Y_{21}|^2}, \qquad (7.23)$$

which is equivalent to

$$S_{i_n v_n^*} = \left[S_{v_n i_n^*} \right]^* = 2qI_C \frac{Y_{11}}{|Y_{21}|^2}. \qquad (7.24)$$

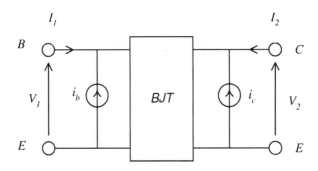

Figure 7.5 Shot noise in an ideal 1-D bipolar transistor.

In reality, however, the current flow in bipolar transistor is 2-D (or even 3-D). The base current must be supplied laterally, which creates a nonuniform emitter-base junction voltage along the current path. This is the well-known "current crowding effect." This two-dimensional phenomenon is often described by adding a fictitious base resistance at the base of an ideal 1-D bipolar transistor. The noise implications of the current crowding effect, for both the shot noise and the thermal noise related to the base resistance, are rarely discussed. It is common practice instead to simply perform noise analysis using the base resistance and a 1-D transistor equivalent circuit, without realizing where this equivalence comes from.

Figure 7.6 shows an equivalent circuit for the noise sources in a bipolar transistor used in this work. It differs from conventional practice in the manner in which the base current shot noise ($2qI_B$) is accounted for. In this case, the base current shot noise is directly tied between the emitter and the base, as opposed to between the emitter and an internal base node connected to the base through the base resistance [7]–[10]. Strictly speaking, the distributive nature of current flow requires a distributive description of the associated shot noise and base resistance thermal noise. One approach would be to split the total base current shot noise between the internal base node and the external base node by a bias and frequency dependent factor. In real devices, however, such a separation makes little difference at frequencies below the cutoff frequency, because the "lumped" base resistance is far smaller than both the transistor input impedance and the optimum source impedance for minimum noise figure [6]. Connecting the base shot noise directly between the external base and the emitter considerably simplifies analytical noise analysis. The assumption used in [10] to simplify the noise figure equation is indeed equivalent to directly connecting the base current shot noise between the external base and the emitter.

Similarly, the lumped intrinsic base resistance is a fictitious equivalent resistance used to account for the distributive 2-D current flow using a lumped resistance. Strictly speaking, the equivalent thermal noise for this fictitious resistance is not simply $4kTr_b$, which is often assumed without justification based on the lumped base resistance concept. Instead, the equivalent thermal noise is a function of current crowding [11, 12]. Fortunately, for low-noise transistors, and particularly at low current density, current crowding effects are not significant. This is particularly true in SiGe HBTs, since the base regions are quite heavily doped. Thus the widely used $4kTr_b$ thermal noise description for the base resistance does not introduce significant errors. The thermal noise is uncorrelated with either the base current shot noise or the collector current shot noise. Therefore, it does not contribute to $S_{i_n v_n^*}$, and only S_{v_n} is affected by the base resistance

$$S_{v_n} = S_{v_n}^{\text{shot}} + S_{v_n}^{\text{thermal}} = \frac{2qI_C}{|Y_{21}|^2} + 4kTr_b. \quad (7.25)$$

We note that for the same β, the additional bandgap engineering leverage in SiGe HBTs allows one to use higher base doping than in conventional implanted-base Si BJTs, which in turn reduces the input noise voltage. The base resistance is typically extracted from S-parameters using various impedance circle methods [13, 14]. In our experience, all of these extraction techniques give similar results. The S-parameters can be either obtained from measurement or device simulation. Spectral densities of the input noise current, voltage, and their cross-correlation are then readily obtained using (7.22), (7.25), and (7.24), respectively. Analytical expressions of all the Y-parameters can also be used to derive analytical expressions for S_{i_n}, S_{v_n}, and $S_{i_n v_n^*}$.

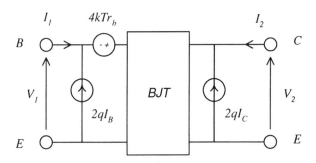

Figure 7.6 A complete description of major noise sources in a BJT including the thermal noise and shot noise.

7.2.3 Noise Figure of a Linear Two-Port

Noise figure is defined to be the signal-to-noise ratio at the input divided by the signal-to-noise ratio at the output. This is equivalent to the ratio of the total noise power to the noise power arising from the source. The concept is straightforward when we consider the linear two-port as an amplifier, as shown in Figure 7.7. Both useful signals and unwanted noise at the input are amplified by the same factor, and in addition, the amplifier adds its own noise. The signal-to-noise ratio thus becomes smaller (worse) after amplification. Noise figure is therefore a useful measure of the amount of noise added by the amplifier, and an indicator of the minimum detectable signal. A perfect amplifier would have 0-dB noise figure. [2]

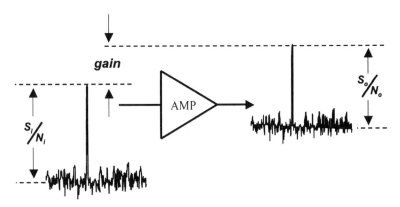

Figure 7.7 Illustration of the definition of noise figure for an amplifier.

---

[2]Throughout the discussions in this chapter, the terms noise figure (NF) and noise factor (F) are used interchangeably, but to be precise, the noise figure is formally defined to be $NF = 10 \log F$.

The noise figure of a transistor circuit is determined by the source termination admittance $Y_s = G_s + jB_s$, and the noise parameters of the circuit, including the minimum noise figure NF_{min}, the optimum source admittance $Y_{s,opt}$, and the noise resistance R_n, through [2]

$$NF = NF_{min} + \frac{R_n}{G_s}\left| Y_s - Y_{s,opt} \right|^2, \qquad (7.26)$$

where G_s is the real part of Y_s. Noise figure reaches its minimum when $Y_s = Y_{s,opt}$, while R_n determines the sensitivity of noise figure to deviations from $Y_{s,opt}$. In measurements, the optimum reflection coefficient Γ_{opt} is often used, and Γ_{opt} is related to Y_{opt} by

$$\Gamma_{opt} = \frac{1 - Y_{opt} Z_0}{1 + Y_{opt} Z_0}, \qquad (7.27)$$

where Z_0 is typically 50 Ω. The smallest possible NF_{min} is obviously desired. A small R_n is also desired when the source is intentionally terminated at a value different from $Y_{s,opt}$ in order to have a higher gain than at $Y_{s,opt}$. In general, the optimum Y_s for minimum noise figure (i.e., at "noise matching") differs from the optimum Y_s for maximum power transfer (i.e., at "gain matching"). The proximity of noise matching and gain matching conditions determines the "associated gain" (G_a), and is defined as the maximum available gain for a noise matching source termination ($Y_s = Y_{s,opt}$).

The noise parameters NF_{min}, $Y_{s,opt} = G_{s,opt} + jB_{s,opt}$, and R_n are functions of the input noise current i_{na}, the input noise voltage v_{na}, and their cross-correlation $v_{na}i_{na}^*$, all of which can be expressed using the physical noise sources and the Y-parameters, as detailed below. Denoting the spectral densities of i_{na}, v_{na}, and $i_{na}v_{na}^*$ as S_{i_n}, S_{v_n}, and $S_{i_n v_n^*}$, one obtains the following equations for R_n, $G_{s,opt}$, $B_{s,opt}$, and NF_{min} from linear noisy two-port theory [2]

$$R_n = \frac{S_{v_n}}{4kT}, \qquad (7.28)$$

$$G_{s,opt} = \sqrt{\frac{S_{i_n}}{S_{v_n}} - \left[\frac{\Im(S_{i_n v_n^*})}{S_{v_n}}\right]^2}, \qquad (7.29)$$

$$B_{s,opt} = -\frac{\Im(S_{i_n v_n^*})}{S_{v_n}}, \qquad (7.30)$$

$$NF_{min} = 1 + 2R_n \left(G_{s,opt} + \frac{\Re(S_{i_n v_n^*})}{S_{v_n}} \right), \qquad (7.31)$$

where $\Re$ and $\Im$ stand for the real and imaginary parts of the various factors, respectively.

7.2.4 Associated Gain

In addition to the noise parameters, the associated gain G_A^{ass}, defined as the available gain G_A at noise matching ($Y_S = Y_{s,opt}$), is also an important figure-of-merit. It is a measure of the maximum output power achievable when the input is noise-matched in order to minimize noise. An amplifier with low noise figure is useless if it does not provide sufficient gain. The available gain is defined as the transducer gain G_t for conjugate matching at the output (i.e., $Y_L = Y_{out}^*$). The transducer gain G_t is given by [15]

$$G_t = \frac{\text{power delivered to the load}}{\text{power available from the source}}$$

$$= \frac{4G_L G_S |Y_{21}|^2}{|Y_{11} + Y_S|^2 |Y_{out} + Y_L|^2}. \quad (7.32)$$

The available gain G_A is obtained as the G_t for $Y_L = Y_{out}^*$

$$G_A = G_t \text{ for conjugate matching at output}$$

$$= \left|\frac{Y_{21}}{Y_{11} + Y_S}\right|^2 \frac{G_S}{G_{out}}. \quad (7.33)$$

The associated gain is derived from the available gain by noise matching the source admittance $Y_S = Y_{s,opt}$

$$G_A^{ass} = G_A \text{ for noise matching at the input}$$

$$= \left|\frac{Y_{21}}{Y_{11} + Y_{s,opt}}\right|^2 \frac{G_{s,opt}}{G_{out}}, \quad (7.34)$$

where

$$G_{out} = \Re(Y_{out}), \quad (7.35)$$

and

$$Y_{out} = Y_{22} - \frac{Y_{12} Y_{21}}{Y_{11} + Y_{s,opt}}. \quad (7.36)$$

We see then that the associated gain is derived from the transducer gain G_t by conjugate matching the load admittance (i.e., $Y_L = Y_{out}^*$) and noise matching the source admittance. Noise matching at the source is in general *not* conjugate matching. Ideally, we would like the noise matching and conjugate matching to be as close as possible to simultaneously achieve low noise and high power output. In practice, the "closeness" of noise and conjugate matching can be readily checked by plotting the noise matching impedance and the conjugate matching impedance on the same Smith chart and then comparing them. A device design that offers *both* low noise and high gain is certainly desirable. Achieving this goal, however, is often difficult in practice, at both the device level and at the circuit level.

7.2.5 Y-Parameter Based Modeling

We established above that the input noise current and voltage, as well as their correlation, can be calculated from measured or simulated Y-parameters using (7.22), (7.24), and (7.25), respectively. The noise parameters, including NF_{min}, R_n, and $Y_{s,opt}$, can then be calculated using (7.28)–(7.31). Associated gain is then readily calculated using (7.34). This methodology facilitates the estimation of noise parameters and associated gain from S-parameter measurements or even device simulation, which are useful alternatives to direct noise measurement for noise optimization.

The above procedures can be easily automated on a computer. The calculation of noise parameters is then straightforward once the program is written. It also has the utility of allowing one to easily compare different noise models by simply modifying the noise sources. The following is a sample program in MATLAB for doing this:

```
v_rn = rn(v_vn2);    %    1/G_va
v_gn = gn(v_in2);    %    corresponding to g_ia
c = corr(v_vn_inc);  %    cross correlation between Va and Ia
v_fmin = fmin(c, v_rn, v_gn);
v_gopt = gopt(c, v_rn, v_gn);
v_bopt = bopt(c, v_rn);
y_opt = v_gopt + j*v_bopt;
gamma_opt = (1-y_opt*z0)./(1+y_opt*z0);
mag_opt = abs(gamma_opt);
angle_opt = angle(gamma_opt)*180/pi;
nfmin = 10*log10(v_fmin);
ga_associated = gass(y_opt);
```

Figure 7.8 shows NF_{min} and R_n as a function of frequency obtained using the Y-parameter based noise modeling approach for a SiGe HBT, together with measured noise data. Figure 7.9 shows the magnitude and angle of Γ_{opt}, and Figure 7.10 shows the associated gain. The emitter area in this case is $0.4 \times 12 \times 2$ μm^2, and the collector current is 10.0 mA. This simple modeling approach gives a reasonably good agreement with the data. The only parameter that needs to be extracted independently is the base resistance r_b, but this can be obtained from S-parameters using circle impedance techniques. At high frequencies (relative to f_T), NF_{min} is overestimated, primarily because the base and collector current shot noises were assumed to be independent of each other. An additional parameter τ_n, which has dimensions of time, can be introduced to describe the correlation between the base and collector current shot noises [6]. In our experience, adjustment of τ_n can al-

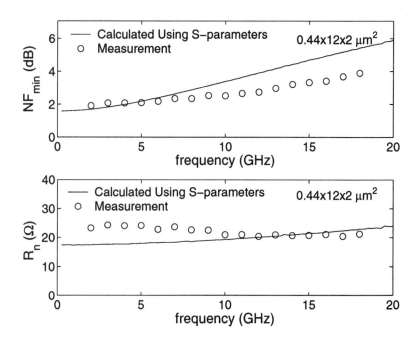

Figure 7.8 NF_{min} and R_n as a function of frequency. The symbols are measurement data, and the lines are modeling results.

ways lead to much better agreement with measured noise data for all of the requisite parameters, including NF_{min}, R_n, Y_{opt} (Γ_{opt}), and G_A^{ass}. The correlation between base and collector shot noises is reduced to zero by setting $\tau_n = 0$. Such an approach is suitable for noise modeling in circuit simulators, but is not suitable for device design, because τ_n has to be extracted from measured noise data.

7.3 Analytical Modeling

To gain additional intuitive insight into device optimization for noise, analytical expressions for NF_{min}, R_n, Y_{opt} and G_A^{ass} are desirable. This can be accomplished using analytical Y-parameter equations. Accuracy must be balanced against simplicity of functional form in order to make such analytical expressions useful. Recall that the power spectral densities of the input noise current (S_{i_n}), the input noise

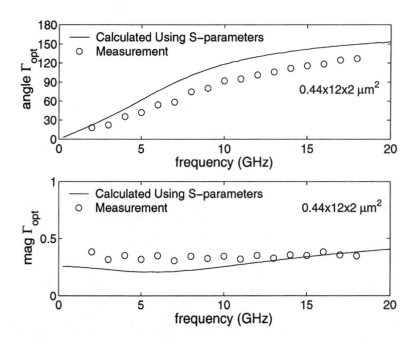

Figure 7.9 Magnitude and angle of Γ_{opt} as a function of frequency. The symbols are measurement data, and the lines are modeling results.

voltage (S_{v_n}), and their cross-correlation ($S_{i_n v_n^*}$) are given by

$$S_{i_n} = 2q\frac{I_C}{\beta} + \frac{2qI_C}{\left|\frac{Y_{21}}{Y_{11}}\right|^2}, \tag{7.37}$$

$$S_{v_n} = 4kTr_B + \frac{2qI_C}{|Y_{21}|^2}, \tag{7.38}$$

$$S_{i_n v_n^*} = \frac{2qI_C Y_{11}}{|Y_{21}|^2}. \tag{7.39}$$

Next, we express the Y-parameters in terms of fundamental device parameters, such as β, g_m, and f_T. A simplified small-signal equivalent circuit is shown in Figure 7.11. At frequencies smaller than f_T, the base resistance is not important for the input impedance, and thus can be neglected for simplicity, even though it is significant as a noise voltage generator. The $I - V$ relation thus becomes

$$\begin{pmatrix} i_1 \\ i_2 \end{pmatrix} = \begin{pmatrix} g_{be} + j\omega(C_{be} + C_{bc}) & -j\omega C_{bc} \\ g_m - j\omega C_{bc} & j\omega C_{bc} \end{pmatrix} \begin{pmatrix} v_1 \\ v_2 \end{pmatrix}. \tag{7.40}$$

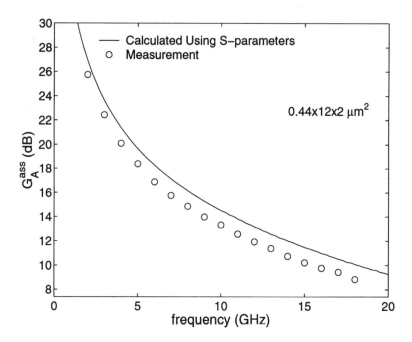

Figure 7.10 Associated gain G_A^{ass} as a function of frequency. The symbols are measurement data, and the lines are modeling results.

The Y-parameters can then be obtained

$$Y_{11} = \frac{g_m}{\beta} + j\omega C_i, \quad (7.41)$$

$$Y_{12} = -j\omega C_{bc}, \quad (7.42)$$

$$y_{21} \approx g_m, \quad (7.43)$$

$$y_{22} = j\omega C_{bc}, \quad (7.44)$$

where $g_m = qkT/I_C$, $C_i = C_{be} + C_{bc}$. Here, C_{be} consists of the EB depletion capacitance C_{te} and the EB diffusion capacitance $g_m\tau$ ($C_{be} = C_{te} + g_m\tau$), with τ being the transit time, and C_i is related to f_T through

$$f_T = g_m/2\pi C_i. \quad (7.45)$$

Finally, S_{i_n}, S_{v_n}, and $S_{i_n v_n^*}$ can then be expressed in terms of I_C (or g_m), β, C_i, and

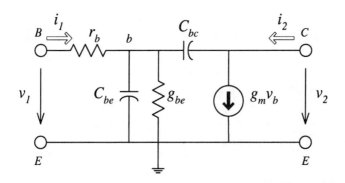

Figure 7.11 Equivalent circuit for the Y-parameter derivation used in analytical noise modeling.

r_b by substituting (7.41)–(7.44) into (7.37)–(7.39) to yield

$$S_{i_n} = 2qI_C \left[\frac{1}{\beta} + \frac{1}{\beta^2} + \left(\frac{\omega C_i}{g_m} \right)^2 \right]$$

$$\approx 2qI_C \left[\frac{1}{\beta} + \left(\frac{\omega C_i}{g_m} \right)^2 \right], \quad (7.46)$$

$$S_{v_n} = 4kT \left(r_b + \frac{1}{2g_m} \right), \quad (7.47)$$

$$S_{i_n v_n^*} = 2qI_C \frac{Y_{11}}{|Y_{21}|^2}$$

$$= 2qg_m \frac{kT}{q} \frac{1}{g_m^2} \left(\frac{g_m}{\beta} + j\omega C_i \right)$$

$$= 2kT \left(\frac{1}{\beta} + \frac{j\omega C_i}{g_m} \right), \quad (7.48)$$

where $I_C = g_m kT/q$ was used. With the help of (7.46)–(7.48), we can determine R_n, $G_{s,opt}$, $B_{s,opt}$, and NF_{min} from (7.28)–(7.31).

7.3.1 Noise Resistance

Substituting (7.47) into (7.28) leads to

$$R_n = \frac{S_{vn}}{4kT} = r_b + \frac{1}{2g_m}. \quad (7.49)$$

Equation (7.49) indicates that R_n is directly proportional to the base resistance, and is thus independent of frequency at a given biasing current. Note that R_n is the first parameter that needs to be derived in order to obtain F_{min}, even though it is typically the least familiar among all the noise parameters. In the bias range where $r_b \gg \frac{1}{2g_m}$, a condition that is often met in practical circuits, $R_n \approx r_b$. Since R_n determines the sensitivity of NF to deviations of source termination from $Y_{s,opt}$, a smaller r_b helps in keeping NF close to NF_{min} when the source is impedance-matched instead of noise-matched.

7.3.2 Optimum Source Admittance

Substituting (7.46)–(7.48) into (7.29) and (7.30), one has [16]

$$G_{s,opt} = \sqrt{\frac{S_{i_n}}{S_{v_n}} - \left[\frac{\Im(S_{i_n v_n^*})}{S_{v_n}}\right]^2}$$

$$= \sqrt{\frac{1}{R_n}\frac{1}{2}g_m\left[\frac{1}{\beta} + \left(\frac{\omega C_i}{g_m}\right)^2\right] - \left(\frac{\omega C_i}{2g_m R_n}\right)^2}$$

$$= \sqrt{\frac{g_m}{2R_n}\frac{1}{\beta} + \frac{(\omega C_i)^2}{2g_m R_n}\left(1 - \frac{1}{2g_m R_n}\right)}, \qquad (7.50)$$

$$B_{s,opt} = -\frac{\Im(S_{i_n v_n^*})}{S_{v_n}} = -\frac{\omega C_i}{2g_m R_n}. \qquad (7.51)$$

Equations (7.50) and (7.51) indicate that the imaginary part of the optimum noise matching admittance is negative. A series inductor at the input is thus needed for noise matching of the imaginary part.

7.3.3 Minimum Noise Figure

The NF_{min} is obtained by substituting (7.46)–(7.48) into (7.31) to yield [16]

$$NF_{min} = 1 + 2R_n\left[\sqrt{\frac{g_m}{2R_n}\frac{1}{\beta} + \frac{(\omega C_i)^2}{2g_m R_n}\left(1 - \frac{1}{2g_m R_n}\right)} + \frac{1}{2\beta R_n}\right]$$

$$= 1 + \frac{1}{\beta} + \sqrt{\frac{2g_m R_n}{\beta} + \frac{2R_n(\omega C_i)^2}{g_m}\left(1 - \frac{1}{2g_m R_n}\right)}. \qquad (7.52)$$

Equation (7.52) can be used to identify the frequency and bias current dependence of NF_{min}. Observe that NF_{min} monotonically increases with frequency. The current dependence is more complicated. We need to point out that the capacitance term is I_C dependent for bipolar transistors. However, it can be easily shown that at relatively high current, F_{min} increases with I_C monotonically through the g_m term. Equation (7.52) suggests that NF_{min} decreases (improves) with increasing β, decreasing C_i (transit time), and decreasing r_b. In circuit applications where $g_m r_b \gg 1/2$, (7.52) can be further simplified to

$$NF_{min} = 1 + \frac{1}{\beta} + \sqrt{2 g_m r_b} \sqrt{\frac{1}{\beta} + \left(\frac{f}{f_T}\right)^2}. \tag{7.53}$$

The two terms inside the second square root become equal at $f = f_T/\sqrt{\beta}$, which defines a transition of NF_{min} from a white noise behavior (independent of frequency) to a 10 dB/decade increase as the frequency increases. A smaller r_b in the transistor reduces not only the sensitivity of NF to deviations from noise matching, but also the minimum noise figure NF_{min}. Equation (7.53) also indicates that a low r_b is the key to reducing NF_{min} when $f > f_T/\sqrt{\beta}$.

7.3.4 Associated Gain

Substituting (7.50) and (7.51) into (7.34)–(7.36), we can analytically express associated gain as a function of β, C_{bc}, C_i, g_m, and r_b as [16]

$$G_A^{ass} = \frac{1}{\omega^2 C_{bc} C_i r_b} \sqrt{\frac{g_m r_b + 1/2}{2} \frac{g_m^2}{\beta} + \frac{(\omega C_i)^2}{2} g_m r_b}, \tag{7.54}$$

where (7.49) has been used. The derivation of (7.54) is mathematically complicated but conceptually straightforward. A number of observations can be made from (7.54):

- G_A^{ass} increases with I_C through the g_m term. Thus, we need a certain amount of I_C to achieve sufficient associated gain.
- G_A^{ass} decreases with increasing β through the first term inside the square root. This suggests that any reduction of noise figure due to β increase must necessarily result in a reduction of associated gain. The impact of β on G_A^{ass} is important at low frequencies, but becomes less important at high frequencies.
- G_A^{ass} decreases with increasing frequency. At relatively low frequencies, the first term inside the square root dominates, making the frequency dependence of G_A^{ass} close to $1/\omega^2$ (i.e., -20 dB/decade).

280   Silicon-Germanium Heterojunction Bipolar Transistors

- At high frequencies, the second term inside the square root dominates, making the frequency dependence close to $1/\omega$ (i.e., -10 dB/decade). An inspection of (7.54) shows that the two terms inside the square root become equal at $f = f_T/\sqrt{\beta}$.

- A smaller transit time (i.e., higher f_T) is desired to increase G_A^{ass} because of smaller C_i, and is consistent with conventional device design approaches for noise reduction.

- A smaller CB capacitance C_{bc} is desired to increase G_A^{ass}. This is consistent with conventional design approaches for obtaining higher f_{max}.

7.4 Optimal Sizing and Biasing for LNA Design

We have derived transistor-level models for understanding and added insight into the various noise parameters of a single SiGe HBT. For circuit designers, however, additional relevant questions exist, including: what geometry and bias current are optimum for a given low noise application? The geometry of a bipolar transistor is defined by the emitter stripe width (W_E), the emitter length (L_E), or the number of unit cells of a given single emitter stripe. We now examine the considerations involved in the judicious choice of transistor geometry for low noise SiGe HBT applications.

7.4.1 Emitter Width Scaling at Fixed J_C

For simplicity, let us assume that r_b is dominated by the intrinsic base resistance, and then compare the NF_{min} for two emitter widths ranging from W_{E0} to $W_{Es} = M \times W_{E0}$, where M is a scaling factor which has a value from 0 to 1 (e.g., M=0.5). The subscripts "0" and "s" denote the reference device and scaled device, respectively. At the same V_{BE}, the two devices operate at the same collector current density J_C, and thus have the same f_T. The NF_{min} for the reference device, $NF_{min,0}$, is obtained from (7.53) as follows

$$NF_{min,0} = 1 + \frac{1}{\beta} + \sqrt{2g_{m,0}r_{b,0}}\sqrt{\frac{1}{\beta} + \left(\frac{f}{f_T}\right)^2}. \qquad (7.55)$$

For the scaled device, $g_{m,s} = M \times g_{m,0}$, $r_{b,s} = M \times r_{b,0}$, and thus $NF_{min,s}$ is given by

$$NF_{min,s} = 1 + \frac{1}{\beta} + M\sqrt{2g_{m,0}r_{b,0}}\sqrt{\frac{1}{\beta} + \left(\frac{f}{f_T}\right)^2}. \qquad (7.56)$$

Because $0 < M < 1$, $NF_{min,s} < NF_{min,0}$. A smaller emitter width thus reduces NF_{min}. The amount of improvement, ΔNF_{min}, is given by

$$\Delta NF_{min} = (1 - M)\sqrt{2g_{m,0}r_{b,0}}\sqrt{\frac{1}{\beta} + \left(\frac{f}{f_T}\right)^2}. \quad (7.57)$$

Equation (7.57) suggests that the NF_{min} improvement resulting from emitter width reduction is more pronounced at higher frequencies. The emitter width in a SiGe HBT BiCMOS process is typically proportional to the minimum feature size of the parent CMOS process, and decreases as the CMOS technology advances. The noise performance of SiGe HBTs consequently will improve as a direct result of lateral scaling. The situation is quite different for CMOS noise performance, on the other hand, because the gate resistance generally *increases* with lateral scaling, since the gate current flows along the channel width direction. For a circuit designer concerned mostly about broadband noise, the appropriate choice of emitter width is an easy one: the minimum allowable feature size should be used for improving NF_{min}.

7.4.2 Emitter Length Scaling at Fixed J_C

Consider a reference emitter length L_{E0} and a scaled emitter length $L_{Es} = N \times L_{E0}$. The base resistance is reduced by the scaling factor N ($r_{b,s} = r_{b,0}/N$), but the transconductance g_m and all the capacitances increase by the same scaling factor under the same V_{BE} or J_C ($g_{m,s} = N \times g_{m,0}$). The cutoff frequency thus remains the same. An inspection of (7.53) immediately shows that $NF_{min,s} = NF_{min,0}$. The emitter length does not affect the achievable NF_{min} and hence the noise performance for a given technology generation. This is the case because NF_{min} is fundamentally dependent on the operating current density J_C, not the bias current I_C.

The appropriate choice of emitter length is clear if the lowest achievable NF_{min} is the only concern: a circuit designer should use the smallest emitter length realizable in order to minimize I_C and hence power dissipation. Clearly J_C need only be sufficient in order to achieve adequate f_T. The problem, however, is that the optimum source impedance required for $NF = NF_{min}$, for a small device is too far away from the driving impedance of the RF source, typically 50 Ω. The optimum source impedance $Z_{s,opt} = R_{s,opt} + jX_{s,opt}$ is related to the optimum source admittance by

$Z_{s,opt} = 1/Y_{s,opt}$. Using (7.50) and (7.51), one obtains

$$R_{s,opt} = \frac{G_{s,opt}}{G_{s,opt}^2 + B_{s,opt}^2}$$

$$= \sqrt{\frac{g_m}{2R_n}\frac{1}{\beta} + \frac{(\omega C_i)^2}{2g_m R_n}\left(1 - \frac{1}{2g_m R_n}\right)} \bigg/ \left[\frac{g_m}{2R_n}\frac{1}{\beta} + \frac{(\omega C_i)^2}{2g_m R_n}\right] \quad (7.58)$$

$$X_{s,opt} = \frac{-B_{s,opt}}{G_{s,opt}^2 + B_{s,opt}^2}$$

$$= \frac{\omega C_i}{2g_m R_n} \bigg/ \left[\frac{g_m}{2R_n}\frac{1}{\beta} + \frac{(\omega C_i)^2}{2g_m R_n}\right]. \quad (7.59)$$

Observe that $R_{s,opt}$ and $X_{s,opt}$ scale with the emitter length in a similar manner to r_b, which also scales with the emitter length. Consider increasing the emitter length by a factor of N. Thus, $g_m \rightarrow Ng_m$, $R_n \rightarrow R_n/N$, $C_i \rightarrow NC_i$, and therefore $R_{s,opt} \rightarrow R_{s,opt}/N$ and $X_{s,opt} \rightarrow X_{s,opt}/N$.

To achieve $NF = NF_{min}$, the source driving impedance $R_s = 50\ \Omega$ must be transformed to $Z_{s,opt}$ using a passive matching network. In fact, this is how low-noise amplifiers are designed in traditional RF circuits that employ discrete transistors. For a minimum size device (e.g., a $0.25 \times 1.0\ \mu m^2$ SiGe HBT), $Z_{s,opt} \gg 50\ \Omega$, making impedance matching difficult. The resulting passive network has losses and can significantly degrade both noise figure and gain. In addition, the bandwidth is very narrow for a large impedance transform ratio.

For RFIC design, the designer has the freedom of choosing the emitter length such that $R_{s,opt}$ is equal to 50 Ω [10]. Noise matching of the resistive component for $NF = NF_{min}$ is thus achieved by transistor sizing without incurring losses, because the transistor is an active element. The reactive matching can then be achieved via an inductor in series with R_s. The $R_{s,opt}$ of a test device of known emitter length, $R_{s,opt}^0$, is obtained from either noise measurement, device-level simulation, or ADS/SPICE modeling at a particular J_C of $J_{C,0}$. The J_C only needs to be sufficient to meet the f_T requirement. The scaling factor N_{nm} required to produce $R_{s,opt} = 50\ \Omega$ is simply given by

$$N_{nm} = \frac{R_{s,opt}^0}{50\ \Omega}. \quad (7.60)$$

For many practical I_C and r_b values, $g_m \gg 1/2r_b$, and $g_m R_n \gg 1/2$. If $f \gg$

$f_T/\sqrt{\beta}$, (7.58) can be further simplified to

$$R_{s,opt} = \frac{f_T}{f}\sqrt{\frac{2r_b}{g_m}} = \frac{f_T}{f}\sqrt{\frac{2r_b kT}{qI_C}}$$

$$= \frac{f_T}{f}\frac{1}{L_E}\sqrt{\frac{2(r_b \times L_E)}{J_C \times W_E}\frac{kT}{q}}, \qquad (7.61)$$

where W_E and L_E are the emitter width and length, respectively, and $r_b \times L_E$ is the base resistance normalized by the emitter length. Because r_b itself is inversely proportional to L_E, the normalized base resistance $r_b \times L_E$ becomes independent of the emitter length, thus facilitating analysis of emitter length scaling. Similarly, $R_{s,opt}$ becomes independent of the emitter width W_E in this case if r_b is dominated by the intrinsic base resistance, because the intrinsic base resistance is proportional to W_E.

Equation (7.61) can be used to estimate $R_{s,opt}^0$ of the test device when these assumptions are satisfied. The bias current of the scaled device is also determined by

$$I_{C,s} = J_{C,0} \times N_{nm} \times A_{E,0}, \qquad (7.62)$$

where $A_{E,0}$ is the emitter area of the test device. Note that the $R_{s,opt}$ has been adjusted to R_S. Recall that $X_{s,opt}$ is positive, and thus an inductor can be added in series with R_S to produce a source reactance of $X_{s,opt}$. The resulting circuit is shown in Figure 7.12. In practice, an emitter inductor instead of a base inductor is often used for noise matching [17]. The operation of that circuit configuration will become apparent after we address simultaneous impedance and noise matching.

The circuit in Figure 7.12 is now noise-matched, and has a noise figure of NF_{min} at the chosen J_C. The input impedance, however, will not in general equal 50 Ω. An input impedance of 50 Ω is required for optimum performance of the filter preceding the LNA as well as for minimum RF power reflection. The transistor input impedance, however, is mainly capacitive at RF frequencies. The question is then how to produce a resistive component of the input impedance without the increasing noise figure. A popular technique is to use an emitter inductor to produce an input resistance, and then use a base inductor for simultaneous noise matching and impedance matching, as discussed below.

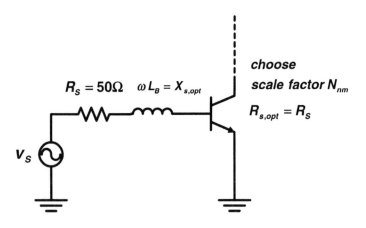

Figure 7.12 Noise matching of a single transistor amplifier using emitter length scaling.

7.4.3 Simultaneous Impedance and Noise Matching

The input impedance looking into the base of a SiGe HBT with an emitter inductance L_e is given by

$$Z_{in} = \frac{1}{j\omega C_{be}} + (1 + \beta_{RF}) j\omega L_e. \qquad (7.63)$$

Assuming that the RF operating frequency is far above f_β (where the ac β begins to decrease mainly due to C_{be}), and that Miller effect is negligible, the equivalent circuit can be simplified to the form shown in Figure 7.13. The β_{RF} is given by

$$\beta_{RF} = -j\frac{g_m}{\omega C_{be}} = -j\frac{\omega_T}{\omega}. \qquad (7.64)$$

Substituting (7.64) into (7.63) leads to

$$Z_{in} = \frac{1}{j\omega C_{be}} + j\omega L_e + \omega_T L_e. \qquad (7.65)$$

A resistive component is thus produced by using an emitter inductor. The value of L_e needed to match a 50-Ω RF source impedance is obtained from (7.65) using $\Re(Z_{in}) = R_s = 50\ \Omega$

$$L_e = \frac{50\ \Omega}{2\pi f_T}. \qquad (7.66)$$

Now the resistive component of the input impedance is matched to the source resistance R_s, but the reactive component is not matched to the source as yet. Another

Noise

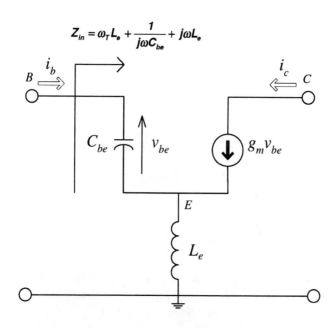

Figure 7.13 Simplified equivalent circuit for a transistor with an emitter inductor. A resistive component of $\omega_T L_e$ is produced.

question to consider is whether or not the source impedance required for noise matching is close to that required for impedance matching. The ideal case would be that the source impedance required for noise matching is identical to the source impedance required for impedance matching for the transistor with emitter inductive degeneration. This is indeed the case in some situations, as shown below.

Since an ideal inductor does not introduce noise, the NF_{min} of the transistor with L_e present is nearly identical to the NF_{min} of the transistor itself. Using the noisy two-port circuit analysis techniques described in [18], the noise parameters for the combination of the transistor and the emitter inductor can be related to the noise parameters for the transistor alone

$$NF_{min}^c = NF_{min} \qquad (7.67)$$

$$R_{s,opt}^c = R_{s,opt} \qquad (7.68)$$

$$X_{s,opt}^c = X_{s,opt} - \omega L_e, \qquad (7.69)$$

where the superscript "c" stands for the combination of the transistor and the emitter inductors. Note that $R_{s,opt}$ and NF_{min} are not changed by the emitter inductor, while $X_{s,opt}$ is reduced by ωL_e. When we have $f \gg f_T/\sqrt{\beta}$, $X_{s,opt}$ in (7.59) can be simplified to

$$X_{s,opt} = \frac{1}{\omega C_i} \approx \frac{1}{\omega C_{be}}, \qquad (7.70)$$

and finally $X^c_{s,opt}$ is obtained

$$X^c_{s,opt} = \frac{1}{\omega C_{be}} - \omega L_e. \qquad (7.71)$$

In addition, L_b, the base inductance required for noise matching of the imaginary part $X^c_{s,opt}$, is also obtained

$$L_b = \frac{X^c_{s,opt}}{\omega} = \frac{1}{\omega^2 C_{be}} - L_e. \qquad (7.72)$$

Here L_b is typically much larger than L_e, and is implemented off-chip, while L_e can be implemented as a combination of on-chip and bond-wire inductance. An inspection of (7.65) and (7.71) immediately shows that

$$X^c_{s,opt} = -\Im(Z_{in}). \qquad (7.73)$$

Therefore, the source impedance $R^c_{s,opt} + jX^c_{s,opt}$ produces both noise matching and impedance matching to the circuit consisting of the transistor and the emitter inductor. The resulting circuit is shown in Figure 7.14.

Noise matching and impedance matching are simultaneously achieved using this circuit topology when the underlying assumptions are satisfied. In practice, however, the assumptions are not always well satisfied (for instance, the Miller effect is often not negligible), and hence L_b and L_e must be adjusted to simultaneously achieve both low noise figure and good input impedance matching. Another often-violated assumption is $f \gg f_T/\sqrt{\beta}$. For example, for $f_T = 20$ GHz, and $\beta = 100$, $f_T/\sqrt{\beta} = 2$ GHz, which is comparable to a practical RF operating frequency of 2.4 GHz. In that case, the noise figure achieved is slightly higher than NF_{min}. The input impedance, however, remains matched to the 50-Ω source. The r_b of the final device can be subtracted from 50 Ω in (7.66) for better estimation of L_e.

Equations (7.69) and (7.71) also explain how a single emitter inductor can be used for noise matching, a common design practice. In this case, L_e can be adjusted such that $L_b = 0$. The resulting input impedance, however, will not equal 50 Ω, in general, and a compromise between noise matching and impedance matching must be made [17]. An emitter inductor also helps in improving linearity by decreasing v_{be} for a given v_s, and produces a resistive input. Since both the base and emitter

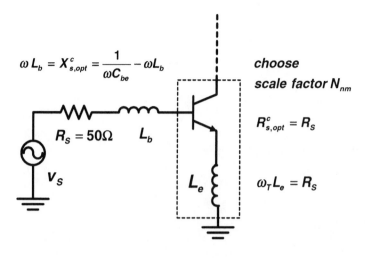

Figure 7.14 Simultaneous noise and impedance matching with L_b and L_e. The transistor is scaled to produce $R_{s,opt} = R_S$, and $L_e = R_S/\omega_T$ produces resistive impedance matching without increasing NF_{min} and $R_{s,opt}$. The change in $X_{s,opt}$ is compensated for by $L_b = 1/\omega^2 C_{be} - L_e$, which also produces reactive impedance matching under certain conditions.

naturally have bond-wire inductance, the use of both L_b and L_e for simultaneous noise and impedance matching is a logical choice. Experimental LNA results using this simultaneous noise and impedance matching topology have recently been reported for various SiGe HBT technologies [19]–[21].

7.4.4 Current Density Selection

For a source voltage v_s, the input current is simply $v_s/2R_s$ because of input impedance matching. The output current is given by

$$i_c = \beta_{RF} \times i_b = -j\frac{f_T}{f}\frac{v_s}{2R_s}, \qquad (7.74)$$

where $\beta_{RF} \simeq h_{21}$ was assumed. This leads to an overall transconductance (G_m) of

$$G_m = \frac{f_T}{f}\frac{1}{2R_s} = \frac{f_T}{f}\frac{1}{100\,\Omega}. \qquad (7.75)$$

Note that G_m is only a function of the f_T/f ratio. Since f_T increases with increasing J_C, the transconductance requirement (related to the gain requirement) sets the lower limit of usable circuit-level J_C. The noise figure requirement and power consumption constraints set the upper limit on usable J_C. In the J_C range that satisfies both power consumption and noise figure requirements, the J_C that produces the best linearity (highest $IP3$) should be chosen.

7.4.5 A Design Example

We now give an example of how to choose J_C and the corresponding noise matching scaling factor N_{nm}. We first simulate NF_{min} and h_{21} as a function of I_C for a reference device of known emitter length. This can be accomplished using parameterized S-parameter and dc sweep in ADS. Figure 7.15 gives an example setup designed specifically for this purpose. The design frequency here is 5 GHz, and the device used is typical of first generation SiGe HBTs featuring a 50-GHz peak f_T. Figure 7.16 shows the simulated NF_{min} and h_{21} as a function of V_{BE} (equivalent to J_C). For an 18-dB requirement of h_{21}, V_{BE} needs to be 0.860 V, which sets NF_{min} to 1.5 dB, and $R_{s,opt} = 121\,\Omega$ at $V_{BE} = 0.86$ V (shown by the markers). To produce a 50-Ω $R_{s,opt}$, the emitter length thus needs to be scaled by a factor $N_{nm} = 2.42$. The bias current I_C for the scaled device is determined to be N_{nm} times of the I_C of the reference device, or 6.7 mA.

7.4.6 Frequency Scalable Design

A robust methodology for scaling an existing LNA design for operation in another frequency band is also highly desirable. Typical dual-band front-ends in use today, for instance, have two radios on chip. To develop such a scaling method, we first examine the equations of $R^c_{s,opt}$, L_e, $X^c_{s,opt}$, NF_{min}, and the overall G_m using (7.61), (7.66), (7.71), (7.53), and (7.75):

$$R^c_{s,opt} = \frac{f_T}{f}\frac{1}{L_E}\sqrt{\frac{2(r_b \times L_E)}{J_C \times W_E}\frac{kT}{q}} = R_s = 50\,\Omega, \qquad (7.76)$$

$$L_e = \frac{50\,\Omega}{2\pi f_T}, \qquad (7.77)$$

Noise

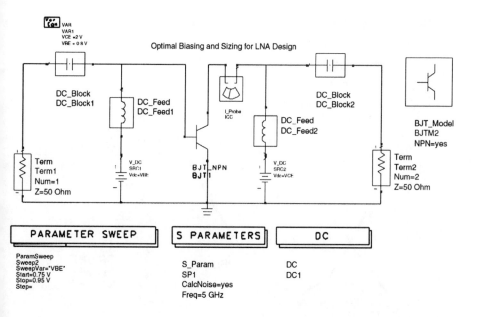

Figure 7.15 ADS simulation setup for optimal transistor sizing and biasing.

$$X^c_{s,opt} = \frac{1}{\omega C_{be}} - \omega L_e, \tag{7.78}$$

$$NF_{min} = 1 + \frac{1}{\beta} + \sqrt{2g_m r_b} \sqrt{\frac{1}{\beta} + \left(\frac{f}{f_T}\right)^2}, \tag{7.79}$$

$$G_m = \frac{f_T}{f} \frac{1}{100\,\Omega}. \tag{7.80}$$

The cutoff frequency f_T is related to J_C through

$$\frac{1}{2\pi f_T} = \tau_f + \frac{kT}{qJ_C} C_t, \tag{7.81}$$

where C_t is the depletion capacitance per unit area, and τ_f is the forward transit time. One approach is to keep the f_T/f ratio constant as the frequency increases, thus keeping the overall G_m of the scaled design the same as in the existing design. Denoting the frequencies as f_1 and $f_2 = Kf_1$, the $J_{C,2}$ required to increase f_T to

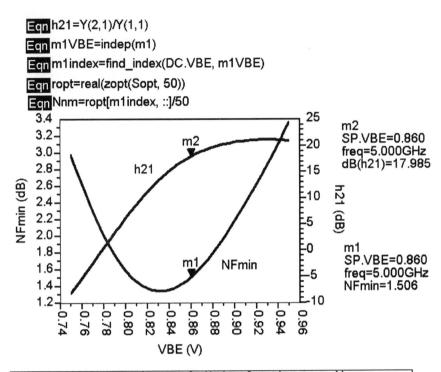

Figure 7.16 ADS data display setup for determining optimal transistor sizing and biasing.

Kf_T can be related to $J_{C,1}$ using (7.81) as follows

$$R_{J_C} \triangleq \frac{J_{C,2}}{J_{C,1}} = \frac{1/\omega_{T,1} - \tau_f}{1/K\omega_{T,1} - \tau_f}. \quad (7.82)$$

The ratio R_{J_C} is less than the frequency scaling factor K because of the τ_f term. Note that $R^c_{s,opt}$ needs to be kept at R_s (50 Ω). Since f_T/f is held constant, and the normalized base resistance $r_b \times L_E$ is independent of L_E, the emitter length L_E scaling factor can be obtained from (7.76) as

$$R_{L_E} \triangleq \frac{L_{E2}}{L_{E1}} = \sqrt{\frac{1}{R_{jc}}} = \sqrt{\frac{1/K\omega_{T,1} - \tau_f}{1/\omega_{T,1} - \tau_f}}. \quad (7.83)$$

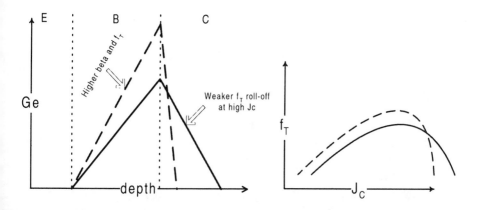

Figure 7.17 Schematic of the SiGe profile design trade-offs encountered when improving f_T and β at low-injection and suppressing the f_T roll-off at high-injection.

The bias current scaling factor is given by

$$R_{I_C} \triangleq \frac{I_{C,2}}{I_{C,1}} = R_{L_E} \times R_{J_C} = \sqrt{\frac{1/\omega_{T,1} - \tau_f}{1/K\omega_{T,1} - \tau_f}}. \qquad (7.84)$$

Here, R_{L_E} and R_{I_C} determine the transistor emitter length and bias current at $f_2 = Kf_1$. From this, C_{be} can then be determined, and L_e and L_b are then found using (7.77) and (7.78). The NF_{min} is finally obtained from (7.79). The noise figure is equal to NF_{min} because of noise matching, and the input impedance remains matched to 50 Ω in this case.

7.5 SiGe Profile Design Trade-offs

The noise performance of SiGe HBTs is largely determined by the device-level layout and profile design, which determines β, r_b, and f_T. As we will show, a higher β also leads to lower (better) low-frequency ($1/f$) noise and hence improved phase noise capability. At a given SiGe technology generation, the emitter width and base sheet resistance are fixed, and thus r_b is fixed. However, the SiGe profile itself can be optimized to increase β and f_T [22] and represents a nice example of using bandgap engineering for optimization of RF circuit response.

A need for higher β and a higher f_T dictate the use of more Ge content and a larger Ge gradient across the neutral base (refer to Chapters 4 and 5). The total integrated Ge content, however, is limited by the SiGe film stability (refer to Chapter 2). For constant film stability, a higher f_T and higher β can only be realized in practice by pushing the edge of the Ge retrograde in the collector significantly closer to the surface [22, 23]. The additional Ge can then be used to reduce the effective Gummel number in the neutral base for higher β, and increase the Ge grading for higher f_T. A large Ge grading in the neutral base necessarily creates a large retrograding in the collector-base space charge region. This retrograding of Ge does not have an impact on device operation at low injection, because of the carrier depletion in the CB space-charge region.

Problems arise, however, at high-injection levels. At high J_C, the minority carrier charge is sufficient to compensate the ionized depletion charge in the CB space-charge region. At sufficiently high J_C, the neutral base pushes out (Kirk effect), exposing the SiGe-Si heterojunction, which induces a conduction band barrier, and thereby strongly degrades both β and f_T [24, 25] (refer to Chapter 6). In SiGe HBTs, we are thus forced at a fundamental level to trade high-J_C f_T performance for improved NF_{min}.

These trade-offs are further illustrated in Figure 7.17. The solid profile and the dashed profile have the same integrated Ge content. The dashed profile has a larger Ge content and a higher Ge gradient in the neutral base, and therefore higher β and f_T, and hence lower noise. The solid profile has a deeper and less abrupt Ge retrograding into the collector, and therefore a better (weaker) f_T roll-off at high injection. The associated risk for using the dashed profile is premature f_T roll-off. If the f_T rolls off too early, for instance, the high peak f_T potential offered by the larger Ge grading in the neutral base may never be realized. The resulting peak f_T may even be lower than the peak f_T for the solid profile. The key is to achieve noise performance improvements while minimizing the f_T degradation at high injection. A realistic goal that can be achieved with careful optimization is to maintain the peak f_T and the f_{max} to values comparable with the original control profile design point.

Experimental results illustrating this noise optimization strategy are given below for first generation SiGe HBT technology. This methodology is easily extendable to other SiGe HBT technology generations. Our goal here is to optimize the SiGe profile for higher β, higher f_T, and lower NF_{min}. The first step towards optimization is to identify the limiting factors for the input noise current and voltage for the technology in question.

7.5.1 Input Noise Current Limitations

The spectral density of the input noise current, S_{ia}, is readily obtained from the measured Y-parameters using (7.24), as derived above. The Y-parameters cannot be directly measured at RF frequencies, and are thus converted from measured S-parameters. Figure 7.18 shows S_{i_n} versus I_C at 2 GHz for a 0.5 (emitter width) ×20 (emitter length) ×2 (stripe number) μm^2 SiGe HBT. The individual contributions from the base and collector shot noises are plotted separately, together with the total input noise current.

Observe that the $2qI_B$ contribution dominates for most of the I_C bias range. Hence, we must improve the current gain to effectively reduce the input noise current. For an ideal transistor with infinite current gain and infinite f_T (h_{21}), S_{i_n} would be zero. This is fundamentally responsible for the increase in NF_{min} with rising I_C. An easy way to get around is to reduce the biasing current, but this is not a favorable approach, since it also means a reduction of gain, since f_T and f_{max} can become unacceptably low at very low I_C. As a result, functionally, the NF_{min} in a SiGe HBT decreases with I_C at very low I_C, reaches a minimum (as much as 10× lower in bias current than for peak f_T), and then increases at higher I_C.[3]

The bias dependence of $2qI_C$ to S_{i_n} (i.e., $2qI_C/|h_{21}|^2$) has two contributions. First, $|h_{21}|$ appears in the denominator and increases with increasing I_C because of increasing f_T. Such an increase, however, saturates when f_T becomes much higher than the frequency under question (2 GHz in this case). Typically $|h_{21}|$ is constant at low frequencies, and then decreases at higher frequencies, at a slope of -20 dB/decade. In addition, $2qI_C$, which appears in the numerator, increases monotonically with increasing I_C. As a result, the ratio $2qI_C/|h_{21}|^2$ decreases with increasing I_C at first when the increase of $|h_{21}|^2$ dominates over the increase of I_C. At higher currents, however, when the increase of I_C dominates over the increase of $|h_{21}|^2$, the ratio $2qI_C/|h_{21}|^2$ starts to increase again.

7.5.2 Input Noise Voltage Limitations

The input noise voltage has two contributions from the base resistance, as well as the collector current shot noise, as shown in (7.25). For the frequency range of interest, y_{21} can be approximated by qI_C/kT. The collector current shot noise contribution then becomes

$$S_{vn}^{2qIc} = \frac{2qI_C}{|y_{21}|^2} = \frac{2(kT)^2}{qI_C}. \quad (7.85)$$

---

[3]The fundamental functional form of the bias dependence of broadband noise in a bipolar transistor is quite different from that in a MOSFET, in which noise figure typically decreases monotonically with increasing drain current. Hence, ironically, if the only consideration is broadband noise, the SiGe HBT effectively offers much lower power dissipation than MOSFETs.

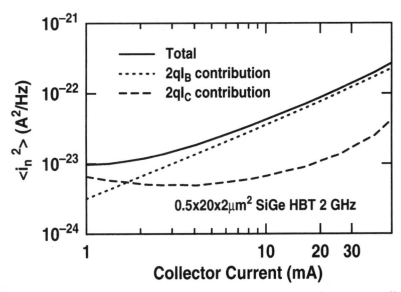

Figure 7.18 Spectral density of the equivalent input current noise versus collector current at 2 GHz calculated using measured Y-parameters.

Note that (7.85) sets the fundamental limit for the noise performance of a SiGe HBT at a given I_C for zero r_b, infinite β, and infinite f_T (h_{21}). This fundamental noise limit depends only on the bias current and temperature, and is independent of device technology (III-V or Si or SiGe). The $1/I_C$ functional form results from the exponential $I_C - V_{BE}$ relation that underlies the $g_m = qI_C/kT$ translinear relation, and is fundamental to all bipolar transistors.

Figure 7.19 shows S_{v_n} versus I_C for the same SiGe HBT at 2 GHz. The data is calculated from measured Y-parameters using (7.25). Before the high-injection f_T roll-off, $|y_{21}|$ is independent of frequency in the range of interest, and can be approximated by qI_C/kT. Therefore, the contribution of $2qI_C$ to S_{va} is solely determined by $1/I_C$ prior to the f_T roll-off I_C, and is thus independent of any other transistor parameters. This theoretical $1/I_C$ dependence is corroborated by the data prior to high injection in Figure 7.19. For the SiGe HBT technology generation under discussion (i.e., first generation), the contribution of r_b dominates over most of the bias current range, as can be seen from Figure 7.19. Therefore, significant improvement of noise performance can be expected by increasing the base doping and decreasing the emitter width in subsequent technology scaling. Because S_{v_n} is dominated by the thermal noise $4kTr_b$ and S_{i_n} is dominated by the base current shot noise $2qI_B$, the cross-correlation term $S_{v_n i_n^*}$ can be neglected in these devices at this frequency.

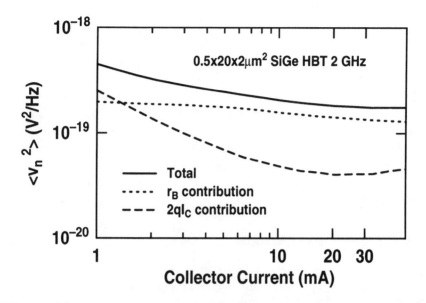

Figure 7.19 Spectral density of the equivalent input noise voltage versus collector current at 2 GHz calculated from measured Y-parameters.

7.5.3 Approaches to Noise Improvement

To improve SiGe HBT noise performance, we must reduce either S_{va} or S_{ia}, or ideally both. According to the above analysis, the base resistance r_b needs to be reduced. For meaningful comparisons between devices, the r_b needs to be normalized by the emitter length when devices with differing emitter lengths are used. For similar reasons, noise figure comparisons should be made at the same current density or the same V_{BE}. We emphasize that (contrary to popular opinion) while a simple increase of the emitter length indeed reduces the r_b of the device, it does not improve the noise capability of the transistor, because all of the capacitances increase by the same factor. However, the emitter length (or the number of unit cells) can be optimized to simplify noise matching, as described above.

At a given SiGe technology generation, the minimum emitter width is determined by the minimum feature size. The base sheet resistance is determined by the amount of boron dopants that can be kept in place during processing, and is obviously limited by the overall thermal cycle. Therefore, r_b and S_{va} are basically fixed. That is, the input noise voltage can only be reduced by lateral and vertical scaling, and the reduction of process thermal cycle, the addition of carbon doping, etc. There is no room for further reduction of S_{v_n} at a given technology generation. The input noise current S_{ia}, however, can be reduced by increasing β (to reduce I_B) and increasing f_T (to increase h_{21}), according to (7.37). In particular, at relatively high currents where the RF power gain is large, $2qI_B$ dominates S_{ia} in these devices. Therefore, significant noise improvement can only be achieved through an increase of β at relatively high bias currents. As we will show below, a high β is also desired in order to reduce the $1/f$ noise corner frequency and hence the phase noise in amplifiers and oscillators. The underlying approach to improving broadband noise performance at a given SiGe technology node is clear: the SiGe profiles must to be optimized for higher β and f_T under the fundamental constraint of maintaining overall SiGe film stability.

7.5.4 SiGe Profile Optimization

To illustrate this noise optimization methodology, we use calibrated 2-D simulations to determine the optimum Ge profiles for achieving the best noise performance. The simulator calibration process can be found in [26], as well as Chapter 12. The transistor Y-parameters were simulated directly by using MEDICI. The general noise modeling methodology developed above is used here to obtain NF_{min}. The PSDs of the input noise current and voltage, as well as their cross correlations, are then calculated from (7.22), (7.25), and (7.23), respectively. The f_T and f_{max} are also extracted from simulated Y-parameters using standard techniques.

Figure 7.20 shows two optimized low-noise Ge profiles obtained using this optimization approach ("LN1" and "LN2"). They maintain the overall SiGe film stability, as well as the peak f_T and peak f_{max} of the SiGe control profile, but have a simulated NF_{min} that is substantially lower than the control (starting) Ge profile (by 0.2 dB). They also have much lower $1/f$ noise corner frequency than the SiGe control profile, as shown below. All of the SiGe profiles here are unconditionally stable to defect generation, as shown in the stability diagram (Figure 7.21). Compared to the SiGe control profile, the two low-noise profiles have a smaller effective thickness and larger effective strain because of the higher Ge content in the neutral base.

7.5.5 Experimental Results

The two low-noise profiles, together with a Si BJT and a SiGe control, were fabricated in the same wafer lot under identical processing conditions using first generation SiGe HBT technology. Figure 7.22 shows the doping and Ge profiles for the 18% peak Ge low-noise profile LN2, as measured by SIMS. To within the

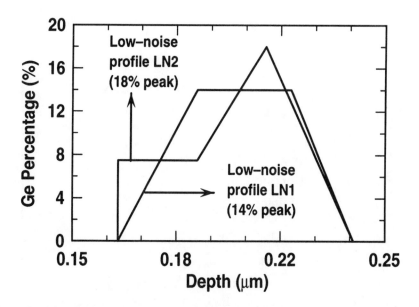

Figure 7.20 Schematic of the two optimized low-noise Ge profiles. Both are unconditionally stable.

SIMS resolution limit, the designed Ge shape is basically reproduced. Table 7.1 summarizes the measured device parameters of the four fabricated profiles. The penalty in BV_{CEO} for LN1 and LN2 is due to the higher β, but should only have a small impact on LNA designs, which see a finite source impedance (i.e., not a true "open").

Figure 7.23 shows the measured Gummel characteristics for a 0.5 × 20 × 2 μm^2 unit cell. At the same V_{BE}, the I_B is the same for all of the profiles because of identical emitter structure, as expected. The I_C is the highest in LN1 and LN2, because of the higher Ge content and the larger Ge gradient across the neutral base, again as expected. The β improvement is more easily seen on the $\beta - I_C$ curves shown in Figure 7.24.

The $f_T - I_C$ and the $f_{max} - I_C$ characteristics are shown in Figures 7.25 and 7.26, respectively. In addition to a much higher β, a modest increase in f_T is achieved in the two low-noise designs, primarily due to increased Ge gradient in the neutral base. The f_{max} of the two low-noise designs are comparable to that of the SiGe control, indicating that the high power gain in the SiGe control design point is retained.

The high-injection design trade-off is clearly confirmed by the experimental results. The two low-noise profiles have a higher β and f_T at low J_C because of

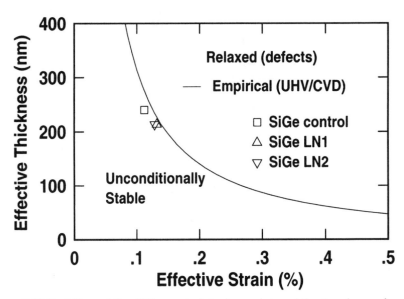

Figure 7.21 Stability of the SiGe control design point and the two low noise profiles. All of these profiles are unconditionally stable.

Table 7.1 Transistor Parameters for the Noise Optimization Experiment

Parameter	Si BJT	SiGe Control	SiGe LN1	SiGe LN2
β at $V_{BE}=0.7$V	67	114	350	261
V_A(V)	19	60	58	113
BV_{CEO} (V)	3.5	3.2	2.7	2.7
R_{Bi} (kΩ/□)	12.8	9.8	10.3	10.7
Peak f_T (GHz)	38	52	52	57
Peak f_{max} (GHz)	57	64	62	67

their higher total Ge content and larger Ge gradient across the neutral base. In contrast, the SiGe control and the Si BJT have a weaker (better) f_T roll-off at high J_C because of deeper Ge retrograding in the collector. For RF circuit applications, the upper bias current limit of these devices is set by the minimum f_T and f_{max} requirements. For instance, the upper collector current limit for meeting a requirement of $f_{max} > 30$ GHz is 50 mA for the two low-noise profiles. We have also achieved the stated goal of maintaining the peak f_T and peak f_{max} of the SiGe control design point, despite the enhanced (worse) high-injection barrier effects in

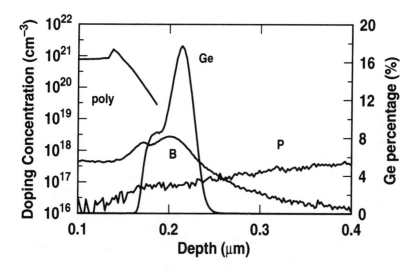

Figure 7.22 Measured SIMS data for the doping and Ge profiles of the fabricated SiGe low noise profile "LN2."

the two low noise profiles. In fact, the peak f_T and peak f_{max} values in the LN2 device are higher than for the SiGe control.

The NF_{min} data measured at 2 GHz is shown in Figure 7.27. The improvements of β and f_T translate into a clear improvement of the NF_{min} over the Si BJT and SiGe control profiles. The NF_{min} of both the LN1 and LN2 SiGe HBTs are 0.2 dB lower at 2 GHz than the NF_{min} of the SiGe control profile, which is consistent with the simulations.[4] Similar improvement is achieved at higher frequencies, where, for instance, a 0.3 dB NF_{min} improvement is obtained at $f = 10$ GHz and $I_C = 5$ mA. Figure 7.28 shows the measured associated gain as a function of I_C. The associated gain at noise matching is above 13 dB at NF_{min} for all of the profiles. Interestingly, the Si BJT has the highest associated gain, despite its poorer noise figure. This observed associated gain – noise figure trade-off can be explained by the smaller β in the Si BJT using previously derived associated gain expression (7.54). Importantly, these device-level noise figure improvements translate directly into measurable circuit-level performance improvements in test circuits fabricated on the same wafers [27]. The circuits were actually designed for the SiGe control devices, and were used as is without adjustments for the new low-noise profiles, and consequently can be considered unoptimized.

---

[4]While to the uninitiated 0.2-dB improvement may appear to be a small change, in a sub-1.0-dB NF_{min} transistor, it isn't!

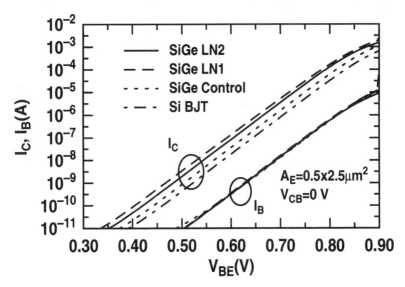

Figure 7.23 Measured I_C and I_B versus V_{BE} for the Si BJT, the SiGe control, and the two low-noise optimized SiGe HBTs.

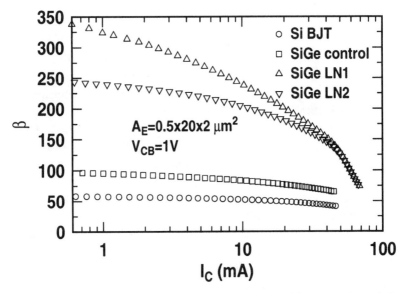

Figure 7.24 Measured β versus I_C for the Si BJT, the SiGe control, and the two low-noise SiGe HBTs.

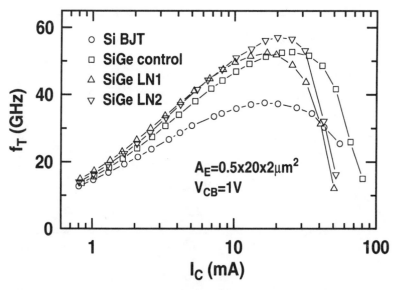

Figure 7.25 Measured f_T versus I_C for the fabricated Si BJT, the SiGe control, and the two low noise SiGe profiles.

7.6 Low-Frequency Noise

One of the advantages of SiGe HBTs over GaAs HBTs is their superior low $1/f$ noise at low frequencies [28], thus making them excellent choices for low-noise amplifiers, oscillators [29], and power amplifiers [30]. One might wonder why low-frequency noise ($<$ 10 kHz) is so important for circuits operating at (multi-GHz) RF frequencies? First, low-frequency noise is upconverted to RF frequencies through the nonlinear $I-V$ and $C-V$ relationships inherent in the transistor to produce transistor phase noise (parasitic sidebands on the carrier frequency that fundamentally limit spectral purity of the system). Second, low-frequency noise is clearly important for emerging wireless receivers utilizing zero intermediate frequency (IF) architectures.

7.6.1 Upconversion to Phase Noise

Consider applying an RF signal, f_{RF}, and a low-frequency signal, f_{LF}, to the base of a SiGe HBT. Spectral components of frequencies $f_{RF} \pm f_{LF}$ are generated in

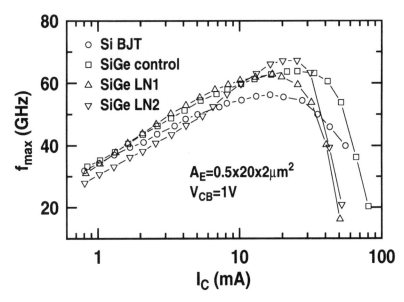

Figure 7.26 Measured f_{max} versus I_C for the fabricated Si BJT, the SiGe control, and the two low noise SiGe profiles.

the collector current because of the strongly nonlinear $I_C - V_{BE}$ relation. The low-frequency signal is thus upconverted to RF by frequency-mixing through the nonlinear circuit elements in the transistor.

Similarly, the low-frequency $1/f$ noise in a RF transistor amplifier is up-converted to RF frequencies by mixing with the incoming RF signal. This produces both amplitude and phase noise at the output, thus degrading spectral purity. Alternatively, one can describe this situation as a fluctuation of the dc bias voltage V_{BE} in the RF amplifier, which is caused by the $1/f$ noise. The magnitude and phase of the amplifier gain at RF frequencies thus fluctuate with V_{BE}, resulting in amplitude and phase noise at the output.

Both amplitude noise and phase noise are important for low-noise amplifiers and power amplifiers. The phase noise measured for an open loop amplifier is also referred to as the "residual phase noise" [31]. Residual phase noise is an important concern for direct conversion receivers [32]. When the amplifier is used to build an oscillator, the amplitude noise is not important because of inherent amplitude stabilization. The amplifier phase noise, however, is directly translated into oscillation frequency noise (phase noise), because the total phase shift is zero (or 360 degrees) on a loop turn [33, 34]. The oscillation frequency fluctuates in order to compensate for the amplifier phase fluctuations. In a heterodyne RF receiver, the local oscillator (LO) phase noise results in a broadening of the downconverted sig-

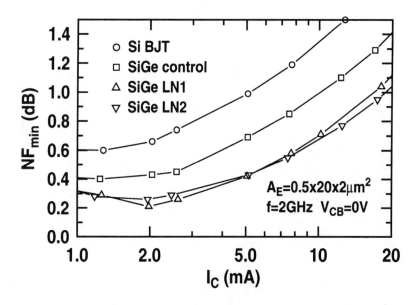

Figure 7.27 Measured minimum noise figure (NF_{min}) versus collector current at 2 GHz for the fabricated Si BJT, SiGe control, and the two SiGe low noise profiles.

nals at the intermediate frequency (IF), thus limiting the minimum channel spacing between adjacent channels. For a 900-MHz transceiver, for instance, a 30-kHz channel spacing typically requires less than -100 dBc/Hz phase noise at 100-kHz offset from the carrier.

7.6.2 Measurement Methods

Experimentally, it is well established that the major low-frequency noise source in typical SiGe HBTs is the base current noise, as it is in typical polysilicon emitter Si BJTs. The low-frequency noise behavior can be described using a noise current source placed between the base and emitter. The noise can be measured either indirectly at the collector terminal or directly at the base terminal. Figure 7.29 shows the diagram of the indirect measurement method. The *dc* biasing at the base and collector is adjusted through potentiometers P_B and P_C, respectively. Batteries are typically used as power supplies in order to minimize spurious noise. The thermal noises from P_B and P_C are dynamically short-circuited by two large capacitors C_B and C_C, respectively, and C_B and C_C also provide an *ac* ground to the $1/f$ noise. We note that C_B and C_C are sometimes left out for simplicity, but care must

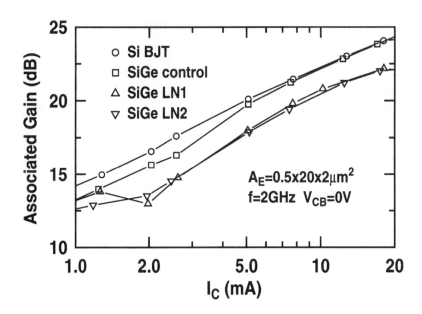

Figure 7.28 Measured associated gain versus collector current at 2 GHz for the fabricated Si BJT, SiGe control, and the two SiGe low noise profiles.

be exercised in determining the effective source and load resistances seen by the transistor.

The base bias resistor R_B is chosen such that $R_B \gg r_\pi$, with r_π being the transistor input impedance. The base noise current thus flows into the transistor instead of R_B, and is amplified by the transistor current gain β. The voltage noise at R_C is further amplified by a low-noise preamp, and detected by a dynamic signal analyzer (DSA). The spectral density of the base current noise (S_{I_B}) is obtained from the spectral density of the voltage noise measured at R_C (S_{V_C}) as

$$S_{I_B} = \frac{S_{V_C}}{(R_C \beta)^2}. \tag{7.86}$$

The transistor output impedance is assumed to be much higher than R_C, which can always be satisfied by proper choice of R_C. If capacitor C_C is not used, the effective dynamic load resistance seen by the collector node needs to be calculated and used in place of R_C in (7.86). This method, though indirect, is quite popular because of its simplicity.[5] Strictly speaking, the small-signal β should be used in

---

[5] We note that "simple" is not a word generally associated with $1/f$ noise measurements by those who have actually made the measurements! Obtaining robust and clean noise data can be very time consuming.

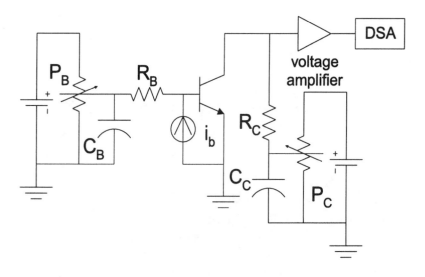

Figure 7.29 Test circuit used to indirectly measure the base current $1/f$ noise from collector voltage fluctuations.

(7.86). The base bias resistance R_B typically ranges from 50 kΩ to 10 MΩ, and the collector sampling resistance R_C is on the order of 2 kΩ. Because of the $R_B \gg r_\pi$ requirement, very large R_B is needed for measurement at low I_B values, which often presents difficulties in practice when attempting to obtain data across a large bias range.

At low I_B, low-frequency noise can be directly measured at the base using a current amplifier connected in series with the base biasing network, as shown in Figure 7.30. A large bypass capacitance C_B short-circuits the noise from the base biasing network, and creates a low impedance path for the base current $1/f$ noise. As long as the input impedance of the current amplifier is much lower than the transistor input impedance (r_π), all of the base noise current flows into the current amplifier. The *dc* base current is also directly read off the current amplifier's *dc* output voltage. No assumptions need to be made here for the transistor equivalent circuit. The spectral density of the current amplifier output voltage is proportional to the base noise current, and the gain of the current amplifier has units of V/A. In our experience, at least for a medium current range, the two measurement methods give identical results.

Figure 7.31 shows a typical low-frequency base current noise spectrum (S_{I_B}) for a first generation SiGe HBT. The noise spectrum shows a clear $1/f$ component as well as the $2qI_B$ shot noise level. The corner frequency f_C is determined from the intercept of the $1/f$ component and the $2qI_B$ shot noise level. The roll-off seen

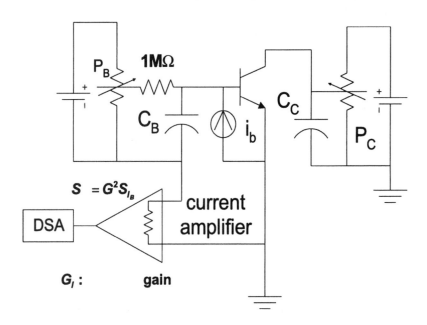

Figure 7.30 Test circuit used to directly measure the base current $1/f$ noise using a current amplifier connected in series with the base biasing network.

above 10 kHz is due to the bandwidth limitation of the preamp, and at higher I_B values, the $2qI_B$ shot noise level cannot be directly observed for this reason. The calculated $2qI_B$ value can be used to determine f_C in this case.

7.6.3 Bias Current Dependence

For a given device geometry, S_{I_B} is a function of I_B and is modeled by

$$S_{I_B} = K_F \frac{I_B^\alpha}{f}, \tag{7.87}$$

where K_F and α correspond to the KF and AF model parameters used in SPICE. The α value provides information on the physical origins of the $1/f$ noise. First-order theory predicts $\alpha = 1$ for carrier mobility fluctuations, and $\alpha = 2$ for carrier number fluctuations [36]–[42]. The α for typical SiGe HBTs is close to 2, and varies only slightly with SiGe profile and collector doping profile (2 ± 0.2).

The observed $S_{I_B} - I_B$ dependence for the SiGe HBTs is approximately the same as for comparably fabricated Si BJTs, and independent of the SiGe profile, as shown in Figure 7.32. The same devices described in Section 7.5 are used here,

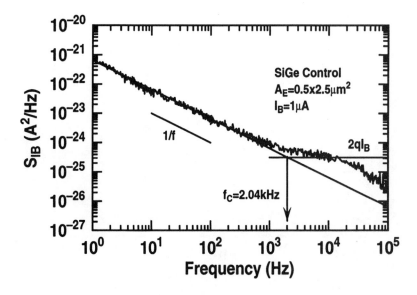

Figure 7.31 A typical low-frequency noise spectrum of a first generation SiGe HBT ($A_E = 0.5 \times 2.5$ μm^2, and $I_B = 1$ μA).

including a 10% peak Ge profile control, a 14% peak Ge low-noise design (LN1), an 18% peak Ge low-noise design (LN2), and an epi-base Si BJT (all fabricated in the same wafer lot). At a given V_{BE}, the I_B is the same for the SiGe HBTs and the Si BJT because of the identical emitter structure (Figure 7.23). Assuming that the $1/f$ noise is solely a function of the number of minority carriers injected into the emitter, we intuitively expect the same $1/f$ noise at a given V_{BE} (I_B).

The fact that SiGe HBTs have the same S_{I_B} as a comparably constructed Si BJT for a given I_B is clearly good news for RF circuit designers. If we compare S_{I_B} at the same I_C, however, S_{I_B} is significantly *lower* (better) in SiGe HBTs than in Si BJTs, because of the lower I_B (higher β) found in SiGe HBTs, all else being equal. The relevant question is which comparison (constant I_B or constant I_C) makes better sense? A constant I_C comparison is more meaningful in the context of RFIC design, because many RF figures-of-merit fundamentally depend on I_C instead of I_B (e.g., f_T and f_{max}). In addition, NF_{min}, though dependent on I_B, is often compared at the same operating I_C as well. Since $S_{I_B} \propto I_C^2/\beta^2$, the S_{I_B} for the LN1 and LN2 SiGe HBTs should be naturally lower than for the SiGe control and Si BJT because of their higher β. This is corroborated by the measured data shown in Figure 7.33.

The most meaningful question with respect to low-frequency noise is, for a given situation, how much V_{BE} fluctuation is induced, and how much phase noise

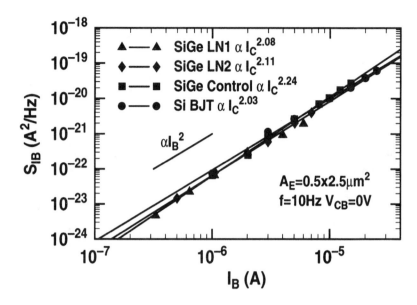

Figure 7.32 Measured S_{I_B} at 10 Hz as a function of I_B for the Si BJT, the SiGe control, and the two low-noise SiGe HBTs.

is generated for a given amount of $1/f$ noise? Recall that in the direct method of noise measurement, the base bias resistance must be sufficiently high compared to transistor input impedance to force all the noise current flow into transistor base. In that case, the V_{BE} fluctuation is the same for different SiGe HBTs biased at the same I_C. The S_{I_B} differences are exactly canceled by the β differences. Thus, SiGe HBTs would appear to have no advantage at all over Si BJTs.

The biasing network in practical RF circuits, however, is quite *different* from that used in the indirect $1/f$ noise measurement circuit. In fact, the *dc* biasing network in RF oscillators is often optimized to minimize the impact of $1/f$ noise on the transistor nonlinearity control voltages [31]. The low-frequency impedance presented by the *dc* biasing network to the base current noise source is in general smaller (sometimes much smaller) than r_π. The V_{BE} fluctuation is then determined by S_{I_B} and the low-frequency impedance of the biasing network (independent of r_π). The V_{BE} fluctuation is hence smaller for the high β SiGe HBTs, which have lower S_{I_B}. We conclude, therefore, that the $1/f$ noise of bipolar transistors (Si or SiGe or III-V) should be compared at the same I_C as opposed to I_B, at least for RF circuits. The two low-noise optimized SiGe HBTs have the lowest $1/f$ noise at a given I_C, and are thus expected to have better overall phase noise performance.

Figure 7.34 compares the residual phase noise measured on single transistor

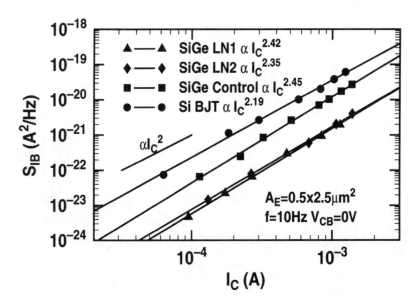

Figure 7.33 Measured S_{I_B} at 10 Hz as a function of I_C for the Si BJT, the SiGe control, and the two low-noise SiGe HBTs.

amplifiers with different SiGe profile designs. The measured results confirm our theoretical expectations. The 14% peak Ge LN1 SiGe HBT shows the lowest residual phase noise. The carrier frequency in these measurements is 10 GHz, the input power is 0 dBm, and both the source and load were terminated at 50 Ω. The residual phase noise level in these SiGe HBTs is excellent compared to competing technologies (of any type), with values as low as -165 dBrad/Hz at 10 kHz offset from the carrier. The measured residual phase noise improvement in the two low-noise SiGe designs is consistent with ADS phase noise simulations [43].

Because of the extremely low levels of phase noise in SiGe HBTs, care must be exercised to reduce the phase noise floor in the measurement setup. A diagram of the experimental setup used here is shown in Figure 7.35 [34]. A microwave carrier from a high-spectral-purity oscillator is injected into both the DUT as well as a line. The carrier is modulated by the device low-frequency noise through the device nonlinearities. The noisy carrier at the DUT output is demodulated through a mixer in quadrature mode (phase detector) using the unmodulated carrier in the line as a reference signal. The demodulated baseband signal is input into a Fast Fourier Transform (FFT) spectrum analyzer after lowpass filtering and low noise amplification. The phase noise of the reference oscillator is canceled due to a small electrical delay between the two arms of the phase detector. Two low-noise mixers

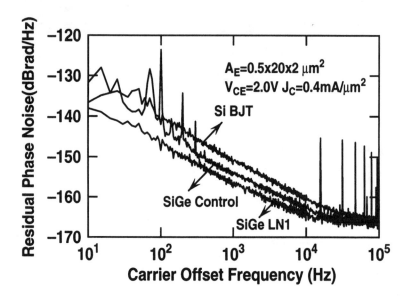

Figure 7.34 Measured residual phase noise spectra for the Si BJT, the SiGe control, and the LN1 low-noise SiGe HBT ($I_C = 8$ mA, $A_E = 0.5 \times 40$ μm^2, P_{in}=0 dBm, and the carrier frequency is 10 GHz). Both the input and the output are terminated at 50 Ω.

are then employed to reduce the noise floor of the measurement system (as low as -175 dBc/Hz at 10 kHz offset from the carrier). The final data is obtained as a cross-spectrum of the outputs of the two mixers by the spectrum analyzer controlled by a computer through an HP-IB bus, and the setup is placed inside a Faraday shield to avoid external spurious signals.

7.6.4 Geometry Dependence

The $1/f$ noise amplitude, as measured by the K_F factor, scales inversely with the total number of carriers in the noise-generating elements, according to the Hooge's theory [35]. The $1/f$ noise generated by sources in the EB spacer oxide at the device periphery is inversely proportional to the emitter perimeter $P_E = W_E + L_E$, while the $1/f$ noise generated by sources located at the intrinsic EB interface (i.e., the emitter polysilicon-silicon interface) across the emitter window is inversely proportional to the emitter area $A_E = W_E \times L_E$. The K_F factor is often examined as a function of the emitter area, the emitter perimeter, or the perimeter-to-area ratio as a means of locating the contributing $1/f$ noise sources [38]–[42]. For instance, for fixed frequency, the combination of a $1/A_E$ dependence with an I_B^2

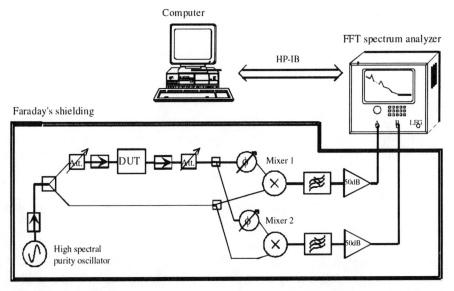

Figure 7.35 A residual phase noise measurement setup [34]. The noise floor is -175 dBc/Hz at 10 kHz offset from carrier.

bias dependence for S_{I_B} is consistent with a uniform areal distribution of noise-generating traps across the emitter region. In practice, caution must be exercised in interpreting P_E or A_E scaling data, because test devices are often designed with the emitter width equal to the minimum feature size, and with an emitter length much larger than the emitter width. As a result, such data tend to scale with the emitter perimeter and area in a similar manner, making interpretation difficult. A wide distribution of device sizes and P_E/A_E ratios thus needs to be used when designing test structures for noise scaling studies in order to make a clear distinction between P_E and A_E scaling in SiGe HBTs. For all of the SiGe HBTs described in Section 7.5, the $1/f$ noise K_F factor is inversely proportional to A_E. Equation (7.87) can thus be rewritten as

$$S_{I_B} = \frac{K}{A_E} \frac{I_B^2}{f} = \frac{K}{\beta^2} \frac{1}{A_E} \frac{I_C^2}{f}, \qquad (7.88)$$

where K is a factor independent of the emitter area and is defined as $K = K_F \times A_E$, where $\alpha = 2$ is assumed. Equation (7.88) is written as a function of I_C to facilitate technology comparisons for RFIC circuit design, for reasons discussed above. Because the K factor for low-frequency noise is approximately independent of base profile design, a higher β SiGe HBT has a lower S_{I_B}, and hence generates lower phase noise when used in RF amplifiers and oscillators. For a given operating current, a larger device can clearly be used to reduce S_{I_B}. This tactic, however, reduces f_T because of the lower J_C. The maximum device size one can use is usually limited by this f_T requirement. Optimum transistor sizing is thus important not only for reducing NF_{min}, but also for reducing phase noise [43].

7.6.5 $1/f$ Noise Figures of Merit

Traditionally, $1/f$ noise performance is characterized by the corner frequency (f_C) figure-of-merit, defined to be the frequency at which the $1/f$ noise equals the shot noise level $2qI_B$. Equating (7.88) with $2qI_B$ leads to

$$f_C = \frac{KI_B}{2qA_E} = \frac{KJ_C}{2q\beta}, \qquad (7.89)$$

where J_C is the collector current density, and β is the dc β. Equation (7.89) suggests that f_C is proportional to J_C and K, and inversely proportional to β. We note that this conclusion differs from that derived in [44]. The derivation in [44] showed that f_C is independent of bias current density, because $\alpha = 1$ was assumed (i.e., according to mobility fluctuation theory). This dependence of α, however, is not the case in typical SiGe HBTs, which show an α close to 2. Figure 7.36 shows the measured and modeled $f_C - J_C$ dependence for the devices described in Section 7.5. As expected, f_C is the lowest in the two low-noise SiGe HBTs, LN1 and LN2, and highest for the Si BJT. The modeling results calculated using (7.89) fit the measured data well.

The corner frequency alone, however, does not take into account transistor frequency response, and is thus not suitable for adequately assessing transistor capability for applications such as oscillators. For instance, Si BJTs typically also have low (good) f_C, but do not have sufficient gain to sustain oscillation at RF and microwave frequencies because of their limited f_T. GaAs HBTs, on the other hand, have high f_T, but typically also have high f_C and hence generate larger phase noise when used in oscillators. SiGe HBTs provide f_T comparable to GaAs HBTs, and lower f_C than Si BJTs (as shown below), making them a very attractive choice for ultra-low phase noise oscillators. A better figure-of-merit to gauge transistor $1/f$ noise performance for oscillator applications is the f_C/f_T ratio recently defined in [44], since it also takes into account transistor frequency response via f_T.

The cutoff frequency f_T is related to J_C by (7.81) (rewritten for convenience)

$$\frac{1}{2\pi f_T} \approx \tau_f + \frac{1}{g_m}C_t$$
$$= \tau_f + \frac{kT}{qJ_C}C_t, \qquad (7.90)$$

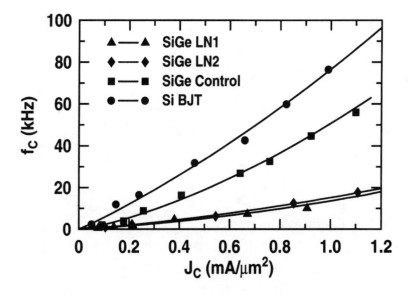

Figure 7.36 Measured and modeled f_C versus J_C for the standard breakdown voltage Si BJT, the SiGe control, and the two low-noise SiGe HBTs.

where τ_f is the forward transit time, $g_m = qJ_C/kT$ is the transconductance per unit area, and C_t is the total junction depletion capacitance per unit area. Prior to f_T roll-off at high J_C, τ_f and C_t are nearly constant. The f_C/f_T ratio can thus be obtained by combining (7.89) and (7.90)

$$\frac{f_C}{f_T} = K\frac{\pi}{q}\frac{J_C}{\beta}\left(\tau_f + kT\frac{C_t}{qJ_C}\right)$$

$$= \frac{K\pi}{\beta q}\left(\tau_f J_C + kT/qC_t\right). \quad (7.91)$$

This model thus suggests a *linear* increase of the f_C/f_T ratio with operating collector current density J_C, provided that β and τ_f are constants. This is in contrast to the prediction of a J_C independent f_C/f_T ratio found in [44], which assumed $\alpha = 1$. At larger values of J_C where f_T is high, $\tau_f J_C \gg kT/qC_t$, and $f_C/f_T \approx K\pi\tau_f J_C/\beta q$. The f_C/f_T ratio is thus determined by the $K\tau_f/\beta$ term at higher J_C values. A smaller τ_f, a higher β, and a smaller K factor are desired in order to reduce (improve) f_C/f_T. A smaller f_C/f_T indicates better phase noise performance at high frequencies. Figure 7.37 shows the measured and modeled f_C/f_T-J_C dependence for a first generation SiGe HBT. The agreement between data and model is quite good. The two low-noise SiGe HBTs show the best (lowest) f_C/f_T because of highest f_T and the lowest f_C, as expected.

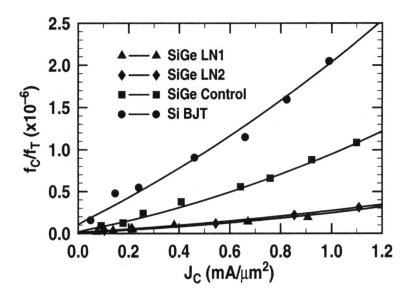

Figure 7.37 Measured and modeled f_C/f_T ratio as a function of J_C for the standard breakdown voltage Si BJT, the SiGe control, and the two low-noise SiGe HBTs.

These results confirm that SiGe profiles optimized for high β and high f_T have better phase noise performance for the same operating frequency. To achieve the same RF gain, a higher f_T transistor can operate at a lower J_C, thus reducing f_C/f_T, which further improves (lowers) f_C. As can be seen from (7.91), the τ_f/β ratio can be used as a figure-of-merit for SiGe profile optimization, because f_C/f_T is proportional to $K\tau_f/\beta$ according to (7.91). The K factor is primarily determined by the emitter structure, and is independent of the SiGe profile used in the base as well the collector doping profile, as evidenced by the experimental data. A SiGe profile producing the lowest τ_f/β ratio leads to the best f_C/f_T ratio, and hence should have the best phase noise performance at RF/microwave frequencies.

7.7 Substrate and Cross-Talk Noise

An important issue for SiGe HBT RF/microwave technology is the use of a low resistivity p-type silicon substrate (typically 10–20 Ωcm for a SiGe BiCMOS process). The low resistivity (conductive) substrate not only contributes thermal noise, but also acts as a coupling medium between various dc and ac sources on the IC, leading to cross-talk noise. RF losses to the low resistivity substrate also increase amplifier noise figure and degrade gain. These substrate noise effects can be significant and must be accounted for, particularly in mixed-signal design where noisy digital functions are integrated with sensitive analog/RF functions on the same die. Three methods can be used to minimize substrate and cross-talk noise in SiGe technologies, including: 1) the use of a sufficient number of substrate contacts to control the potential of the substrate around sensitive nodes; 2) the use of the naturally available n^+ subcollector as a ground plane; and 3) simply by maintaining appropriate spacing between noisy components. A combination of these approaches is often used, since neither method works for all situations.

7.7.1 Noise Grounding Using Substrate Contacts

Noise currents are induced in the substrate by supply voltage glitches arising from the inductive nature of bond wires and package pins, as well as via signal fluctuations on interconnects, which can be capacitively coupled into the substrate through the isolation oxides. To minimize the distribution of substrate noise current, one can reduce the effective resistance between the *ac* ground and the substrate, as shown in Figure 7.38. Deep trenches, if available, can be used to laterally isolate cross-talk from other devices. Substrate contacts clearly occupy space, and thus should only be used in and around critical devices or circuits. The noise injection paths can be shunted to the *ac* ground by placing as many substrate contacts as possible around sensitive circuits.

7.7.2 Noise Grounding Using n^+ Buried Layers

An alternative to substrate contacting is to use the naturally available n^+ collector buried layer found in SiGe HBT processes as an RF ground plane. An area of buried layer that is connected to *ac* ground can be placed near critical devices and underneath critical signal lines in order to provide a low resistive return path to *ac* ground, as shown in Figure 7.39. The low resistivity n^+ buried layer effectively shunts the substrate noise signals to *ac* ground. This approach is effective for substrate noise under the metal interconnects, but not as effective for substrate noise generated underneath a SiGe HBT, because the collector-to-substrate junction capacitance is in series with the n^+ buried layer return path. A combination of n^+ buried layer and top substrate contacts can be used to achieve the best noise immunity. The n^+ buried layer ground planes can also be used underneath bondpads for shielding from the substrate.

One recent promising technique to remove substrate and cross-talk noise is the micromachined etching of substrates below the noisy components [45]. Because of cost concerns, however, this technique is not suitable for standard processing

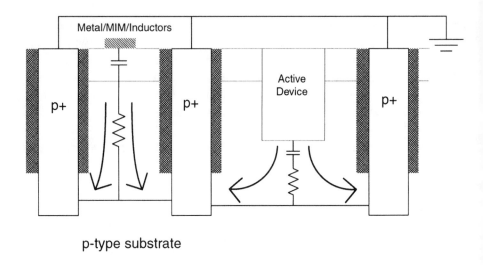

Figure 7.38 Grounding of substrate and cross-talk noise using substrate contacts around critical devices.

as yet, even though it has been successfully applied to producing high-Q on-chip inductors [46]. Another solution, of course, is to simply use an silicon-on-insulator (SOI) or high-resistivity substrate. This approach may in fact be needed for microwave and mm-wave applications of SiGe HBTs, but the additional process complexity (and hence cost) presents a substantial barrier for its mainstream use. Other potential problems exist for SOI, including increased thermal resistance because of the buried oxide. Self-heating as well as thermal-coupling between devices can cause additional problems at the circuit level.

References

[1] S. Martin et al., "Device noise in silicon RF technologies," *Bell Labs Tech. J.*, pp. 30-45, 1997.

[2] H.A. Haus et al., "Representation of noise in linear twoports," *Proc. IRE*, vol. 48, pp. 69-74, 1960.

[3] D.C. Ahlgren et al., "Manufacturability demonstration of an integrated SiGe HBT technology for the analog and wireless marketplace," *Tech. Dig. IEEE Int. Elect. Dev. Meeting*, pp. 859-862, 1996.

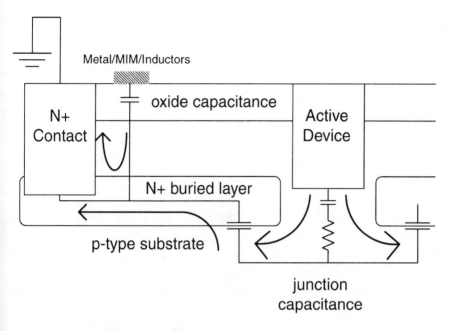

Figure 7.39 Illustration of the grounding of substrate and cross-talk noise by using the n^+ buried layer.

[4] S. Subbanna et al., "Integration and Design Issues in Combining Very-High-Speed Silicon-Germanium Bipolar Transistors and ULSI CMOS for System-on-a-Chip Applications," *Tech. Dig. IEEE Int. Elect. Dev. Meeting*, pp. 845-848, 1999.

[5] A. van der Ziel, "Noise in solid-state devices and lasers," *Proc. IEEE*, vol. 58, pp. 1178-1206, 1970.

[6] G. Niu et al., "A unified approach to RF and microwave noise parameter modeling in bipolar transistors," *IEEE Trans. Elect. Dev.*, vol. 48, pp. 2568-2574, 2001.

[7] E.G. Nielson, "Behavior of noise figure in junction transistors," *Proc. IRE*, vol. 45, pp. 957-963, 1957.

[8] R.J. Hawkins, "Limitations of Nielsen's and related noise equations applied to microwave bipolar transistors, and a new expression for the frequency and current dependent noise figure," *Solid-State Elect.*, vol. 20, p. 191-196, 1977.

[9] H. Fukui, *Low-Noise Microwave Transistors and Amplifiers*, New York, IEEE Press, 1981.

[10] S.P. Voinigescu et al., "A scalable high-frequency noise model for bipolar transistors with application to optimal transistor sizing for low-noise amplifier design," *IEEE J. Solid-State Circ.*, vol. 32, pp. 1430-1438, 1997.

[11] J.A. Pals, "On the noise of a transistor with DC current crowding," *Philips Res. Rep.*, vol. 26, p. 91, 1971.

[12] H.C. de Graaff and F.M. Klassen, *Compact Transistor Modeling for Circuit Design*, Springer-Verlag, New York, 1990.

[13] T. Nakadai and K. Hashimoto, "Measuring the base resistance of bipolar transistors," *Proc. IEEE Bipolar/BiCMOS Circ. Tech. Meeting*, pp. 200-203, 1991.

[14] W. Kloosterman, J. Paasschens, and D. Klaassen, "Improved extraction of base and emitter resistance from small signal high frequency admittance measurements," *Proc. IEEE Bipolar/BiCMOS Circ. Tech. Meeting*, pp. 93-96, 1999.

[15] R.S. Carson, *High Frequency Amplifiers*, Wiley, New York, 1975.

[16] G. Niu et al., "Noise-gain tradeoff in RF SiGe HBTs," *Solid-State Elect.*, vol. 46, pp. 1445-1451, 2002.

[17] H. Shumacher et al., "Low-noise, low-power wireless frontend MMICs using SiGe HBTs," *IEICE Trans. Elect.*, vol. E82-C, pp. 1943-1950, 1999.

[18] H. Hillbrand and P. Russer, "An efficient method for computer aided noise analysis of linear networks," *IEEE Trans. Circ. Syst.*, vol. 23, pp. 235-238, 1976.

[19] O. Shana'a, I. Linscott, and L. Tyler, "Frequency-scalable SiGe bipolar RFIC front-end design," *Proc. IEEE Cust. Int. Circ. Conf.*, pp. 183-186, 2000.

[20] J. Sadowy et al., "Low noise, high linearity, wide bandwidth amplifier using a 0.35 μm SiGe BiCMOS for WLAN applications," *Proc. RFIC Symp.*, pp. 217-220, 2002.

[21] D.A. Johnson, "5-6 GHz RFIC front-end components in silicon-germanium HBT Technology," M.S. Thesis, Virginia Polytechnic Institute, 2001.

[22] G. Niu et al., "Noise modeling and SiGe profile design tradeoffs for RF applications," *IEEE Trans. Elect. Dev.*, vol. 47, pp. 2037-2044, 2000.

[23] W.E. Ansley, J.D. Cressler, and D.M. Richey, "Base profile optimization for minimum noise figure in advanced UHV/CVD SiGe HBT," *IEEE Trans. Micro. Theory Tech.*, vol. 46, pp. 653-660, 1998.

[24] A.J. Joseph et al., "Optimization of SiGe HBT's for operation at high current densities," *IEEE Trans. Elect. Dev.*, vol. 46, pp. 1347-1354, 1999.

[25] G. Niu and J.D. Cressler, "The impact of bandgap offset distribution between conduction and valence bands in Si-based graded bandgap HBT's," *Solid-State Elect.*, vol. 43, pp. 2225-2230, 1999.

[26] G. Niu et al., "Noise parameter optimization of UHV/CVD SiGe HBT's for RF and microwave applications," *IEEE Trans. Elect. Dev.*, vol. 46, pp. 1347-1354, 1999.

[27] L. Sheng, J. Jensen, and L. Larson, "A Si/SiGe HBT sub-harmonic mixer/downconverter," *Proc. IEEE Bipolar/BiCMOS Circ. Tech. Meeting*, pp. 71-74, 1999.

[28] L. Vempati et al., Low-frequency noise in UHV/CVD epitaxial Si and SiGe bipolar transistors," *IEEE J. Solid-State Circ.*, vol. 31, pp. 1458-1467, 1996.

[29] B. Van Haaren et al., "Low-frequency noise properties of SiGe HBT's and application to ultra-low phase-noise oscillators," *IEEE Trans. Micro. Theory Tech.*, vol. 46, pp. 647-652, 1998.

[30] D. Harame et al., "Current status and future trends of SiGe BiCMOS Technology," *IEEE Trans. Elect. Dev.*, vol. 48, pp. 2575-2594, 2001.

[31] O. Llopis et al., "Low phase noise SiGe based microwave oscillators," *Tech. Dig. Workshop Euro. Micro. Conf.*, 2001.

[32] A. Abidi, "Low-power radio frequency IC's for portable communications," in *RF/Microwave Circuit Design for Wireless Communications*, L. Larson, Editor, Artech House, Norwood, MA, 1996.

[33] D.B. Leeson, "A simple model of feedback oscillator noise spectrum," *Proc. IEEE*, vol. 54, pp. 329-330, 1966.

[34] O. Llopis et al., "Phase noise in cryogenic microwave HEMT and MESFET oscillators," *IEEE Trans. Micro. Theory Tech.*, vol. 41, pp. 369-373, 1993.

[35] A. Van der Ziel, *Noise in Solid State Devices and Circuits*, Wiley, New York, 1986.

[36] L.K.J. Vandamme, "Noise as a diagnostic tool for quality and reliability of electronic devices," *IEEE Trans. Elect. Dev.*, vol. 41, pp. 2174-2187, 1994.

[37] A. Mounib et al., "Low-frequency noise sources in polysilicon emitter BJT's: influence of hot-electron-induced degradation and post-stress recovery," *IEEE Trans. Elect. Dev.*, vol. 42, pp. 1647-1652, 1995.

[38] M.J. Deen et al., "Low frequency noise in polysilicon-emitter bipolar junction transistors," *J. Appl. Phys.*, vol. 77, pp. 6278-6285, 1995.

[39] M. Koolen and J. C. J. Aerts, "The influence of non-ideal base current on 1/f noise behaviour of bipolar transistors," *Proc. IEEE Bipolar/BiCMOS Circ. Tech. Meeting*, pp. 232-235, 1990.

[40] H. A. W. Markus and T. G. M. Kleinpenning, "Low-frequency noise in polysilicon emitter bipolar transistors," *IEEE Trans. Elect. Dev.*, vol. 42, pp. 720-727, 1995.

[41] P. Llinares, D. Celi, and O. Roux-dit-Buisson, "Dimensional scaling of 1/f noise in the base current of quasi self-aligned polysilicon emitter bipolar junction transistors," *J. Appl. Phys.*, vol. 82, pp. 2671-2675, 1997.

[42] M.J. Deen, S. L. Rumyantsev, and M. Schroter, "On the origin of 1/f noise in polysilicon emitter bipolar transistors," *J. Appl. Phys.*, vol. 85, pp. 1192-1195, 1999.

[43] G. Niu et al., "Transistor noise in SiGe HBT RF technology," *IEEE J. Solid-State Circ.*, vol. 36, pp. 1424-1427, 2001.

[44] L.K.J. Vandamme and G. Trefan, "Review of low-frequency noise in bipolar transistors over the last decade," *Proc. IEEE Bipolar/BiCMOS Circ. Tech. Meeting*, pp. 68-73, 2001.

[45] A. Yeh et al., "Copper-encapsulated silicon micromachined structures," *J. Microelectromechanical Syst.*, vol. 9, pp. 281-287, 2000.

[46] H. Jiang, A. Yeh, and N.C. Tien, "New fabrication method for high-Q on-chip spiral inductor," *Proc. SPIE*, vol. 3876, pp. 153-159, 1999.

Chapter 8

Linearity

"Linearity" is the counterpart of "distortion," or "nonlinearity," and in the present context refers to the ability of a device, circuit, or system to amplify input signals in a linear fashion. The term "linearity" often has different meanings for different types of circuits, and must be examined in the context of a given circuit application. SiGe HBTs, like other semiconductor devices, are in general nonlinear elements. For instance, it has an exponential $I_C - V_{BE}$ nonlinearity common to all bipolar transistors, and in fact represents the strongest nonlinearity found in nature! This exponential $I_C - V_{BE}$ nonlinearity actually underlies the "translinear principle," which enables a large variety of linear and nonlinear functions to be realized using bipolar transistors. Despite our intuition, the distortion in translinear circuits is *not* *caused* by the exponential $I_C - V_{BE}$ relationship, but rather is due to the *departure* from it, by various means (i.e., series resistance, high-level injection, impact ionization, Early effect, and inverse Early effect).

SiGe HBTs can be used to build both "nonlinear" and "linear" circuits depending on the required application and the circuit topology used. While unavoidable nonlinearities in transistors might be naively viewed to always be a bad thing, this is clearly not the case. In fact, transistor nonlinearity is both a blessing and a curse. Nonlinearity can be a blessing because:

- We need nonlinearity to translate frequency from baseband to RF for signal transmission (upconversion), and translate frequency from RF to baseband for signal reception (downconversion).

- Nonlinearity is necessary to realize frequency multiplication, which is used in frequency synthesis.

- Nonlinearity is required to build an oscillator.

Nonlinearity can also be a curse, however, because it creates distortion in the various signals we are interested in preserving, amplifying, or transmitting. For instance,

- Nonlinearity causes intermodulation of two adjacent strongly interfering signals at the input of a receiver, which can corrupt the nearby (desired) weak signal we are trying to receive.

- In the transmit path, nonlinearity in power amplifiers clips the large amplitude input. Digitally modulated signals used in modern wireless communications systems typically have several dB of variation in instantaneous power as a function of time, and thus require highly linear amplifiers. Clipping of the large amplitude input causes a spreading of the output spectrum to adjacent channels (i.e., adjacent channel power ratio (ACPR)). The power "leaked" to adjacent channels interferes with other users.

Perhaps surprisingly, SiGe HBTs exhibit excellent linearity in both small-signal (e.g., LNA) and large-signal (e.g., PA) RF circuits, despite their strong $I - V$ and $C - V$ nonlinearities. Clearly, the overall circuit linearity strongly depends on the interaction (and potential cancellation) between the various $I - V$ and $C - V$ nonlinearities, the linear elements in the device, as well as the source termination, the load termination, and any feedback present. This issues will be illustrated using Volterra series [1]–[3], a powerful formalism for analysis of nonlinear systems.

8.1 Nonlinearity Concepts

When the input signal is sufficiently weak, the operation of a transistor circuit is linear and dynamic. The response of a linear and dynamic circuit is characterized by an impulse response function in the time domain and a linear transfer function in the frequency domain. Strictly speaking, for a bipolar transistor, the validity of the small-signal requires V_{BE} variation to be much smaller than the thermal voltage kT/q.

For larger input signals, an active transistor circuit becomes a nonlinear dynamic system. Its impulse response can be approximated with a "Volterra series." We will postpone the discussion of Volterra series for the moment because of the complexity involved. Many of the nonlinearity concepts, however, can be illustrated using simple power series, a concept which only applies to a memory-less circuit. In practice, even a linear circuit has memory elements (e.g., capacitors). Nevertheless, the use of power series simplifies the illustration of many commonly used linearity figures of merit.

Linearity

Under small-signal input, the output voltage $y(t)$ of a memoryless nonlinear circuit can be related to its input voltage $x(t)$ by a power series

$$y(t) = k_1 x(t) + k_2 x^2(t) + k_3 x^3(t), \quad (8.1)$$

where for simplicity we have truncated the series at third-order. The effect of storage elements such as capacitors and higher-order nonlinear terms are not accounted for. The concepts of "harmonics," "intermodulation," and "gain compression" are introduced below using (8.1).

8.1.1 Harmonics

If we let the input voltage $x(t) = A \cos \omega t$, the output voltage $y(t)$ then becomes

$$\begin{aligned} y(t) =& k_1 A \cos \omega t + k_2 A^2 \cos^2 \omega t + k_3 A^3 \cos^3 \omega t \\ =& \frac{k_2 A^2}{2} & \text{dc shift} \\ &+ \left(k_1 A + \frac{3 k_3 A^3}{4} \right) \cos \omega t & \text{fundamental} \\ &+ \frac{k_2 A^2}{2} \cos 2\omega t & \text{second harmonic} \\ &+ \frac{k_3 A^3}{4} \cos 3\omega t & \text{third harmonic.} \quad (8.2) \end{aligned}$$

Equation (8.2) has a *dc* shift, a "fundamental output" at ω, a "second-order harmonic term" at 2ω, and a "third-order harmonic term" at 3ω. Thus, an "nth-order harmonic term" is proportional to A^n. In practice, the *relative* level of a given harmonic with respect to the fundamental output is of great interest. The relative "second harmonic distortion" (HD_2) is obtained from (8.2) as

$$HD_2 = \frac{k_2 A^2}{2} \Big/ (k_1 A) = \frac{1}{2} \frac{k_2}{k_1} A, \quad (8.3)$$

where the $3 k_3 A^3 / 4$ term added to $k_1 A$ is neglected. Therefore,

- For small A, the output behaves in a weakly nonlinear manner. The fundamental output at ω grows with A, while the second harmonic at 2ω grows with A^2.

- The *extrapolation* of the output at 2ω and ω intersect at a certain input level defined as $IHD2$. $IHD2$ is obtained from (8.3) by letting $HD2 = 1$

$$IHD2 = 2 \frac{k_1}{k_2}. \quad (8.4)$$

Note that $IHD2$ is independent of the input signal level. Once $IHD2$ is known, one can calculate $HD2$ for any desired small-signal input A using

$$HD2 = \frac{A}{IHD2}. \qquad (8.5)$$

- The output level at the intercept point, $OHD2$, is simply the product of the small-signal gain G and $IHD2$

$$OHD2 = G \cdot IHD2$$

$$= 2\frac{k_1^2}{k_2}. \qquad (8.6)$$

The preferred use of the input or output number for the intercept point is immaterial. There is no solid reason to use one preferentially over the other, because usually both gain and one of the two intercept numbers are specified. In a similar manner, the third-order harmonic distortion $HD3$, the input and output intercept of the third harmonic distortion $IHD3$, and $OHD3$ can be defined.

8.1.2 Gain Compression and Expansion

The small-signal gain is obtained by neglecting the harmonics. In (8.2), the small-signal gain is k_1 when we neglect the nonlinearity-induced term $3k_3 A^3/4$. However, as the signal amplitude A grows, $3k_3 A^3/4$ becomes comparable to or even larger than $k_1 A$. The gain thus changes with the input. This variation of gain with input signal level is a fundamental manifestation of nonlinearity.

If $k_3 < 0$, then $3k_3 A^3/4 < 0$. That is, the gain decreases with increasing input level (A), and eventually diminishes to zero. This phenomenon is referred to as "gain compression" in many (though not all) RF circuits, and is often quantified by the "1 dB compression point," or P_{1dB}. In real circuits, three terms of the power series are not usually sufficient to describe the nonlinear behavior at the 1dB compression point, because of large-signal operation. Once again, either the input or the output value can be used for characterization purposes. The input value is the input magnitude at which the gain drops by 1 dB. Signal power as opposed to voltage is often used in RF circuits. The transformation between voltage and power involves a reference impedance, usually 50 Ω. Typical RF front-end amplifiers require -20 to -25-dBm input power at the 1dB compression point.

8.1.3 Intermodulation

Consider a two-tone input voltage $x(t) = A\cos\omega_1 t + A\cos\omega_2 t$. The output has not only harmonics of ω_1 and ω_2, but also "intermodulation products" at $2\omega_1 - \omega_2$

Linearity 325

and $2\omega_2 - \omega_1$. A full expansion of (8.1) using $x(t) = A\cos\omega_1 t + A\cos\omega_2 t$ shows that the output contains signals at ω_1, ω_2, $2\omega_1$, $2\omega_2$, $3\omega_1$, $3\omega_2$, $\omega_1 + \omega_2$, $\omega_1 - \omega_2$, $2\omega_2 - \omega_1$, $2\omega_2 + \omega_1$, $2\omega_1 - \omega_2$, and $2\omega_1 + \omega_2$.

When ω_1 and ω_2 are closely spaced, the third-order intermodulation products at $2\omega_2 - \omega_1$ and $2\omega_1 - \omega_2$ are the major concerns, because they are close in frequency to ω_1 and ω_2, and thus within the amplifier bandwidth, and inaccessible to filtering. Consider a weak desired signal channel, and two nearby strong interferers at the input. One intermodulation product falls in band, and corrupts the desired component, as illustrated in Figure 8.1. The fundamental signal and intermodulation

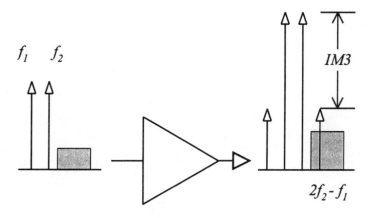

Figure 8.1 Illustration of the corruption of desired signals by the intermodulation product of two strong interferers.

products in the output are given by

$$y(t) = \left(k_1 A + \frac{3k_3 A^3}{4} + \frac{3k_3 A^3}{2}\right)\cos\omega_1 t + \cdots \quad \text{fundamental}$$

$$+ \frac{3k_3 A^3}{4}\cos(2\omega_2 - \omega_1)t + \cdots \quad \text{intermodulation.} \quad (8.7)$$

The ratio of the amplitude of the *IM* product to the amplitude of the fundamental output is defined as the "third-order intermodulation distortion" (*IM3*), similar to the definition of HD3. Neglecting the higher order terms added to $k_1 A$, one has

$$IM3 = \frac{3k_3 A^3}{4} \bigg/ k_1 A = \frac{3}{4}\frac{k_3}{k_1}A^2. \quad (8.8)$$

A few general observations can now be made:

- For small input signals (A), the fundamental output at ω_1 grows *linearly* with A, while the *IM* product at $2\omega_2 - \omega_1$ grows as A^3. A 1-dB increase in the input results in a 1-dB increase of fundamental output, but a 3-dB increase of *IM* product!

- The extrapolation of the fundamental output and the *IM3* versus the input intersect at a given input level. That level is defined to be the "input third-order intercept point" (*IIP3*). This condition corresponds to $IM3 = 1$.

- The *IIP3* is obtained from (8.8) by letting $IM3 = 1$

$$IIP3 = \sqrt{\frac{4}{3}\frac{k_1}{k_3}}. \tag{8.9}$$

Clearly *IIP3* is a more useful figure of merit than *IM3* because it does not depend on the input signal level. Given *IIP3*, *IM3* can be calculated for any desired small input A

$$IM3 = \frac{A^2}{IIP3}. \tag{8.10}$$

Because *IM3* grows with A^2 (8.8), *IIP3* can be measured at a single input level A_0,

$$IIP3^2 = \frac{A_0^2}{IM3_0}, \tag{8.11}$$

where $IM3_0$ is the measured relative intermodulation distortion. Note that *IIP3* and A_0 are voltages, and thus $IIP3^2$ and A_0^2 are measures of power. Taking $10 \log$ on both sides, one has

$$20 \log IIP3 = 20 \log A_0 - 10 \log IM3_0. \tag{8.12}$$

Here, $20 \log IIP3$ is the power expressed in dB at the intercept point, and $20 \log A_0$ is now the input power level expressed in dB. The reference power level does not enter into the equation. Now (8.12) can be rewritten in terms of power

$$P_{IIP3} = P_{in} + \frac{1}{2}\left(P_{o,1st} - P_{o,3rd}\right), \tag{8.13}$$

and

$$P_{OIP3} = P_{o,1st} + \frac{1}{2}\left(P_{o,1st} - P_{o,3rd}\right). \tag{8.14}$$

This is indeed how *IP3* data in commercial load-pull systems and CAD tools (e.g., ADS) is defined for each input power level. The following is the sample output of a load-pull measurement on a SiGe HBT amplifier. The two tones are at 2.000 and 2.001 GHz (i.e., 1-MHz spacing).

Pin	Pout,1st	Gain	Pout,3rd	P(OIP3)
dBm	dBm	dB	dBm	dBm
-30.00	-11.72	18.28	-74.48	19.65
-29.00	-10.75	18.25	-72.68	20.20
-28.00	-9.74	18.26	-69.91	20.35
-27.00	-8.74	18.26	-67.24	20.51
-26.00	-7.72	18.28	-64.89	20.87
-25.00	-6.77	18.23	-62.28	20.98
-24.00	-5.74	18.26	-59.57	21.18
-23.00	-4.73	18.27	-57.15	21.47
-22.00	-3.75	18.25	-54.66	21.71
-21.00	-2.74	18.26	-52.07	21.92
-20.00	-1.72	18.28	-49.63	22.37
-19.00	-.73	18.27	-47.31	22.69
-18.00	.30	18.30	-45.20	23.05
-17.00	1.28	18.28	-43.13	23.49
-16.00	2.32	18.32	-41.19	24.07
-15.00	3.34	18.34	-39.16	24.59
-14.00	4.34	18.34	-36.82	24.93
-13.00	5.38	18.38	-33.63	24.88

. . . .

```
Frequency :     2.00 GHz
Source State :  1   #377
Source Imp :    .02  -170.1
Load State :    1   #550
Load Imp :      .70   14.9
Vc           3.001 V
Vb            .816 V
Date       :30 Apr 1998
Time       :16:22:43
```

The reader is encouraged to calculate *OIP3* from the data in columns 1 to 4 above using (8.14), and then compare the results with the data given in column 5.

The measured fundamental and third-order *IM* product power versus input power data for the above SiGe HBT amplifier are plotted in Figure 8.2, along with the gain. The measured slope of the *IM* product curve deviates from 3 : 1, because of the "high" input power level used in the measurement. As a result, the *IP3* numbers measured at different input powers are different, as can be seen from the data output. One would obtain an *OIP3* of 35 dBm by simply extrapolating the linear portions of the measured $P_{out,1st}$ and $P_{out,3rd}$ data. The *OIP3* based on a theoretical 3 : 1 slope at $P_{in} = -30$ dBm is only 20 dBm, however, and therefore, caution must be exercised in interpreting the *IP3* numbers. The gain compression at very high input power level can also be clearly seen here.

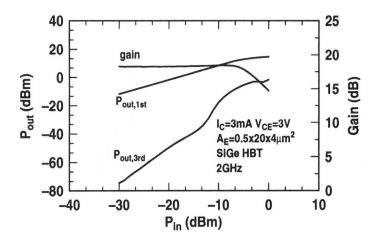

Figure 8.2 A typical P_{out} versus P_{in} curve for a first generation SiGe HBT ($I_C = 3$ mA and $V_{CE} = 3$ V). The input power at the 1-dB compression point is -3 dBm.

Clearly *IIP3* is an important figure of merit for front-end RF/microwave low-noise amplifiers, because they must contend with a variety of signals coming from the antenna. The interfering signals are often much stronger than the desired signal, thus generating strong intermodulation products that can corrupt the weak but desired signal. To some extent, *IIP3* is a measure of the ability of a handset, for instance, not to "drop" a phone call in a crowded environment. For many LNA applications, *IIP3* is just as important (if not more so) as the noise figure. The *dc* power consumption must also be kept very low because the LNA is likely to be continuously listening for transmitted signals of interest and hence continuously draining power. The power consumption aspect is taken into account by another figure of merit, the *linearity efficiency*, which is defined as $IIP3/P_{dc}$, where P_{dc} is the *dc* power dissipation. First generation SiGe HBTs typically exhibit excellent linearity efficiencies above 10, which is competitive with III-V technologies. We note, however, that $IIP3/P_{dc}$ is not adequate for describing the Class *AB* operating mode for transistors in the driver and output stage of power amplifiers [4].

8.2 Physical Nonlinearities

Figure 8.3 depicts a typical transistor equivalent circuit that includes the dominant physical nonlinearities in a SiGe HBT:

- I_{CE} represents the collector current transported from the emitter, and is a nonlinear function of the controlling voltage V_{BE}. The $I_{CE} - V_{BE}$ nonlinearity is a nonlinear transconductance.

- I_{BE} represents the hole injection into the emitter, and is also a nonlinear function of V_{BE}.

- I_{CB} represents the avalanche multiplication current, and is a strong nonlinear function of both V_{BE} and V_{CB}. The nonlinear current source I_{CB} has a 2-D nonlinearity because it has two controlling voltages.

- C_{BE} is the EB junction capacitance, and includes the diffusion capacitance and depletion capacitance. C_{BE} is a strong nonlinear function of V_{BE} when the diffusion capacitance dominates, because diffusion charge is proportional to the transport current I_{CE}.

- C_{BC} is the CB junction capacitance.

We examine now the details of various $I - V$ and $C - V$ nonlinearities.

8.2.1 The I_{CE} Nonlinearity

To first order, the transport current I_{CE} is controlled by V_{BE}. This results in a nonlinear transconductance in weakly nonlinear circuit analysis. Denoting the nonlinear current $i(t)$ as a nonlinear function (f) of the controlling voltage $v_C(t)$, one has

$$i(t) = f(v_C(t)) = f(V_C + v_c(t)) \quad (8.15)$$

$$= f(V_C) + \sum_{k=1}^{\infty} \frac{1}{k!} \left. \frac{\partial^k f(v(t))}{\partial v^k} \right|_{v=V_C} \times v_c^k(t), \quad (8.16)$$

where $i(t)$ is the sum of the dc and ac currents, $v_c(t)$ is the ac voltage which controls the conductance, and V_C is the dc controlling (bias) voltage.

For small $v_c(t)$, considering the first three terms of the power series is usually sufficient. Mathematically speaking, the definition of *small* clearly depends on the magnitude of the derivatives. A number of nonlinearity coefficients can be defined to characterize an $I - V$ nonlinearity, including:

$$g = \left. \frac{\partial f(v)}{\partial v} \right|_{v=V_C}, \quad (8.17)$$

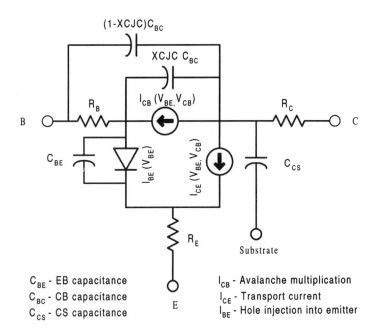

Figure 8.3 Equivalent circuit of the SiGe HBT used for Volterra series simulations.

$$K_{2g} = \frac{1}{2!}\frac{\partial^2 f(v)}{\partial v^2}\bigg|_{v=V_C}, \quad (8.18)$$

$$K_{3g} = \frac{1}{3!}\frac{\partial^3 f(v)}{\partial v^3}\bigg|_{v=V_C}, \quad (8.19)$$

and

$$K_{ng} = \frac{1}{n!}\frac{\partial^n f(v)}{\partial v^n}\bigg|_{v=V_C}. \quad (8.20)$$

The ac current-voltage relation can be rewritten as

$$i_{ac}(t) = g \cdot v_c(t) + K_{2g} \cdot v_c^2(t) + K_{3g} \cdot v_c^3(t) + \cdots, \quad (8.21)$$

where g is essentially the small-signal (trans)conductance of the linearized element, and K_{2g} and K_{3g} are the second-order and third-order nonlinearity coefficients. The subscript g for K_2 and K_3 denotes that these two coefficients are associated with the linearized (trans)conductance g. For an ideal SiGe HBT, I_{CE} increases exponentially with V_{BE}

$$I_{CE} = I_S \exp\left(\frac{qV_{BE}}{kT}\right). \quad (8.22)$$

Linearity

The nonlinearity coefficients are thus

$$g_m = \frac{qI_{CE}}{kT}, \qquad (8.23)$$

$$K_{2g_m} = \frac{1}{2!}\frac{q^2 I_{CE}}{(kT)^2}, \qquad (8.24)$$

$$K_{3g_m} = \frac{1}{3!}\frac{q^3 I_{CE}}{(kT)^3}, \qquad (8.25)$$

and

$$K_{ng_m} = \frac{1}{n!}\frac{q^n I_{CE}}{(kT)^n}. \qquad (8.26)$$

For a dc bias current of 1.0 mA, at 300 K, we have $kT/q = 26$ mV, $g_m = 0.0387$ A/V, $K_{2g_m} = 0.751$ A/V^2, and $K_{3g_m} = 9.70$ A/V^3. When measured $I_{CE} - V_{BE}$ data is used in numerically evaluating the derivatives, care must be exercised because of the strong nonlinearity. For instance, the first-order derivative can be numerically calculated using $I_{C0}d(\ln I_{C0})/dV_{BE}$, which gives much more accurate results than using dI_{C0}/dV_{BE} directly. At a given operating I_{CE}, the $I_{CE} - V_{BE}$ nonlinearity coefficients are identical for a SiGe HBT and a Si BJT.

From a circuit point of view, the consequence of the $I_{CE} - V_{CE}$ nonlinearity is to make the effective transconductance a function of v_{be} (as opposed to a constant in a linear circuit)

$$g_{m,\mathit{eff}} = \frac{i_c}{v_{be}} = g_m(1 + \overbrace{\frac{1}{2}\frac{qv_{be}}{kT} + \frac{1}{6}\frac{q^2 v_{be}^2}{(kT)^2} + \cdots}^{\text{nonlinear contributions}}). \qquad (8.27)$$

Equation (8.27) indicates that the nonlinear contributions to $g_{m,\mathit{eff}}$ increase with the voltage drop across the EB junction v_{be}. In typical bipolar amplifiers, v_{be} decreases with increasing biasing current, making $g_{m,\mathit{eff}}$ closer to a constant, as it is in a linear circuit. The linearity of SiGe HBT circuits can therefore generally be improved by increasing the biasing current, although at a cost of increased power consumption. Emitter resistance or inductance also helps in improving linearity by decreasing v_{be}, although at the expense of gain.

8.2.2 The I_{BE} Nonlinearity

For an ideal SiGe HBT with a constant current gain β, the base current tracks the collector current,

$$I_{BE} = \frac{I_{CE}}{\beta}. \qquad (8.28)$$

Therefore, the nonlinearity coefficients of the nonlinear EB conductance simply track the nonlinearity coefficients of the nonlinear transconductance

$$g_{be} = \frac{g_m}{\beta}, \tag{8.29}$$

$$K_{2g_{be}} = \frac{K_{2g_m}}{\beta}, \tag{8.30}$$

$$K_{3g_{be}} = \frac{K_{3g_m}}{\beta}, \tag{8.31}$$

and

$$K_{ng_{be}} = \frac{K_{ng_m}}{\beta}. \tag{8.32}$$

For better accuracy, measured $I_{BE} - V_{BE}$ data can be directly used in determining the nonlinearity coefficients.

8.2.3 The I_{CB} Nonlinearity

The I_{CB} term represents the impact ionization (avalanche multiplication) current

$$\begin{aligned} I_{CB} &= I_{CE}(M - 1) \\ &= I_{C0}(V_{BE})F_{Early}(M - 1), \end{aligned} \tag{8.33}$$

where $I_{C0}(V_{BE})$ is I_C measured at zero V_{CB}, M is the avalanche multiplication factor, and F_{Early} is the Early effect factor, and can all be measured [5] (refer to Chapter 4). It is important to distinguish Early effect and avalanche multiplication here, although they both contribute to an increase of I_C with increasing V_{CB}, because of their fundamentally different impact on the base and emitter currents [5].

At low current density, the avalanche multiplication factor M in SiGe HBT can be successfully modeled as a function of V_{CB} using the empirical "Miller equation"

$$M = \frac{1}{1 - (V_{CB}/V_{CBO})^m}, \tag{8.34}$$

where V_{CBO} and m are two fitting parameters. A sample $M - 1$ plot measured at $V_{BE} = 0.6$ V for a typical SiGe HBT is shown in Figure 8.4 (reproduced from Chapter 4 for convenience). The measured data can be accurately captured using (8.34).

This avalanche multiplication model, however, is of little use in practical linearity modeling, because avalanche multiplication also strongly depends on operating

Linearity

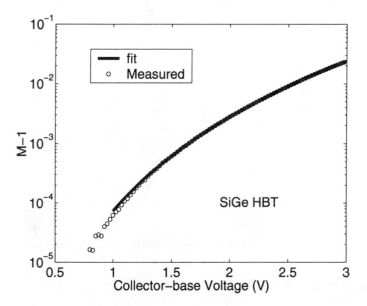

Figure 8.4 Measured $M - 1$ versus V_{CB} for a typical SiGe HBT operating at low current density.

current density J_C. At a given V_{CB}, the avalanche multiplication rate is constant only at low J_C where f_T and f_{max} are very low. At the higher J_C of practical interest, the avalanche multiplication factor decreases with increasing J_C, because of decreasing peak electric field in the CB junction (Kirk effect). Figure 8.5 shows measured M-1 versus J_C data, together with a model fit using [6]

$$M - 1 = \frac{V_{CB}}{\alpha V_{CBO}} \exp\left(-\frac{m\alpha^{-1/3}}{V_{CB}^{2/3}}\right), \tag{8.35}$$

$$\alpha = 1 - \tanh\left[\frac{I_C}{I_{CO}} \exp\left(\frac{V_{CB}}{V_R}\right)\right], \tag{8.36}$$

where m, V_{CBO}, I_{CO}, and V_R are fitting parameters. The measured M-1 data is accurately captured as a function of J_C. As J_C increases, the mobile carrier charge $(-J_C/v)$ compensates the depletion charge (qN_{dc}), which effectively reduces the net charge density on the collector side of the CB junction, decreasing the peak electric field and hence M-1. As expected, the measured decrease of M-1 with J_C also varies with V_{CB}.

Another closely related high-injection effect is the cut-off frequency (f_T) roll-off at high J_C. At sufficiently high J_C, the net charge density reduces to zero,

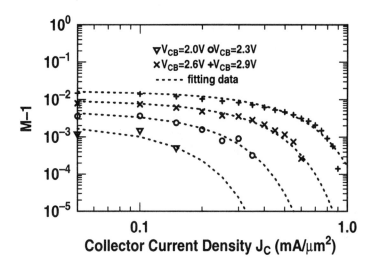

Figure 8.5 Avalanche multiplication factor ($M-1$) as a function of J_C for different V_{CB}.

and base push-out commences, resulting in the roll-off of f_T. In SiGe HBTs, however, the situation is further worsened by the retrograding of SiGe profile into the collector (refer to Chapter 6). Unlike $M-1$, which is strongly dependent on V_{CB}, f_T and f_{max} are weakly dependent on V_{CB} for $V_{CB} > 0$ prior to the f_T roll-off. Figure 8.6 shows f_T and f_{max} as a function of J_C measured at $V_{CB} = 1.0$ V for the same device used in Figure 8.5. The f_T and f_{max} peaks occur near a J_C of 1.0–2.0 mA/μm^2 (Figure 8.6), while $M-1$ starts to decrease at much smaller J_C values (see Figure 8.5). When base push-out occurs the electric field becomes nearly constant in the collector, resulting in a low peak electric field, and hence a low $M-1$. Past the f_T peak, the electrical CB junction and hence the peak electric field position shifts from the metallurgical CB junction to the n-epi/n^+ buried layer interface. The peak electric field and $M-1$ thus increase with very high J_C (well past the f_T peak). This subsequent increase of $M-1$ with J_C is not included in this new $M-1$ model. For the purposes of this chapter, the biasing current is limited to below peak f_T, which is the case for most RF circuits.

The avalanche current I_{CB} is controlled by two voltages, V_{BE} (through J_C) and V_{CB}, and thus needs to be described by a 2-D power series. This series can be split into three subseries i_u, i_v and i_{uv}. The first two series, i_u and i_v, contain powers of one single voltage as for a 1-D nonlinear transconductance

$$i_u = g_u \cdot u_c + K_{2g_u} \cdot u_c^2 + K_{3g_u} \cdot u_c^3 + \cdots, \quad (8.37)$$

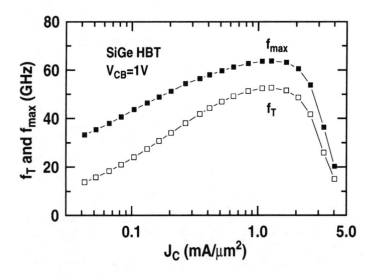

Figure 8.6 Measured f_T and f_{max} as a function of J_C for a SiGe HBT.

and

$$i_v = g_v \cdot v_c + K_{2g_v} \cdot v_c^2 + K_{3g_v} \cdot v_c^3 + \cdots, \quad (8.38)$$

where u_c and v_c are the controlling voltages. The cross-term contribution, i_{uv}, is given by

$$i_{uv} = K_{2g_u \& g_v} \cdot u_c \cdot v_c + K_{32g_u \& g_v} \cdot u_c^2 \cdot v_c + K_{3g_u \& 2g_v} \cdot u_c \cdot v_c^2. \quad (8.39)$$

The nonlinearity coefficient $K_{mjg_u \&(m-j)g_v}$ with $m > j$ is defined as

$$K_{mjg_u \&(m-j)g_v} = \frac{1}{j!} \frac{1}{(m-j)!} \frac{\partial^m f(u,v)}{\partial u^j \partial v^{m-j}} \bigg|_{u=U_C, v=V_C}. \quad (8.40)$$

If j or $m - j$ is equal to unity, then it is omitted as a subscript, such as in $K_{2g_u \& g_v}$.

8.2.4 The C_{BE} and C_{BC} Nonlinearities

In both bipolar and field-effect transistors, current is controlled through voltage modulation of carrier densities (charges), which effectively produces a capacitive effect. Like the terminal currents, these capacitances are nonlinear functions of the terminal voltages. For small-signal distortion, the charge storage associated with a nonlinear capacitor can be described as

$$Q(t) = f(v_C(t)) = f(V_C + v_c(t)) \quad (8.41)$$

$$= f(V_C) + \sum_{k=1}^{\infty} \frac{1}{k!} \frac{\partial^k f(v(t))}{\partial v^k} \bigg|_{v=V_C} \times v_c^k(t), \quad (8.42)$$

where the power series represents the *ac* part of the stored charge. The first-order, second-order, and third-order nonlinearity coefficients are defined as

$$C = \frac{\partial f(v)}{\partial v}\bigg|_{v=V_C}, \tag{8.43}$$

$$K_{2C} = \frac{1}{2!}\frac{\partial^2 f(v)}{\partial v^2}\bigg|_{v=V_C}, \tag{8.44}$$

and

$$K_{3C} = \frac{1}{3!}\frac{\partial^3 f(v)}{\partial v^3}\bigg|_{v=V_C}. \tag{8.45}$$

The *ac* charge can then be written as

$$q_{ac}(t) = C \cdot v_c(t) + K_{2C} \cdot v_c^2(t) + K_{3C} \cdot v_c^3(t) + \cdots, \tag{8.46}$$

where C is essentially the small-signal linear capacitance. The nonlinearity coefficient of order n is defined by the nth order derivative of the charge Q (*not* the capacitance C) with respect to V.

The excess minority carrier charge Q_D in a SiGe HBT is proportional to the transport current I_{CE} through the transit time τ_f

$$Q_D = \tau_f I_{CE} = \tau_f I_S \exp\left(\frac{qV_{BE}}{kT}\right), \tag{8.47}$$

and thus,

$$C_D = \tau_f \cdot g_m = \tau_f \frac{qI_{CE}}{kT} = \tau_f \frac{qI_{CE}}{kT}, \tag{8.48}$$

$$K_{2C_D} = \tau_f \cdot K_{2g_m} = \tau_f \frac{q^2 I_{CE}}{2(kT)^2}, \tag{8.49}$$

$$K_{3C_D} = \tau_f K_{3g_m} = \tau_f \frac{q^3 I_{CE}}{6(kT)^3}, \tag{8.50}$$

and

$$K_{nC_D} = \tau_f K_{ng_m} = \frac{\tau_f q^n I_{CE}}{n!(kT)^n}. \tag{8.51}$$

Here, C_D, K_{2C_D}, and K_{3C_D} are proportional to g_m, K_{2g_m}, and K_{3g_m}, respectively, through a constant τ_f. The effective capacitance becomes a function of v_{be}, as opposed to a constant in a linear circuit,

$$C_{D,\text{eff}} = \frac{q_D}{v_{be}} = C_D(1 + \overbrace{\frac{1}{2}\frac{qv_{be}}{kT} + \frac{1}{6}\frac{q^2 v_{be}^2}{(kT)^2} + \cdots}^{\text{nonlinear contributions}}). \tag{8.52}$$

Similar to the effective transconductance, the "effective" diffusion capacitance can be made more linear (i.e., closer to a constant) by increasing the biasing current at the expense of power consumption. A higher I_C leads to a larger C_D, and hence a smaller v_{be}. A larger value of C_D itself generally helps in improving circuit linearity.

The EB and CB junction depletion capacitances are often modeled by

$$C_{dep}(V_f) = \frac{C_0}{\left(1 - \frac{V_f}{V_j}\right)^{m_j}}, \qquad (8.53)$$

where C_0, V_j and m_j are known model parameters for a given junction. The nonlinearity coefficients can be analytically evaluated in this case. In practice, more complicated models such as those used in MEXTRAM can be used, and the nonlinearity coefficients are then evaluated numerically.

The CB depletion capacitance is in general much smaller than the EB depletion capacitance, because the CB junction is under reverse bias for normal (forward-active) operation. However, the CB depletion capacitance is important in determining linearity, because of its feedback function, as we will show below using Volterra series. Both the absolute value and the derivative of C_{CB} with respect to V_{CB} are important.

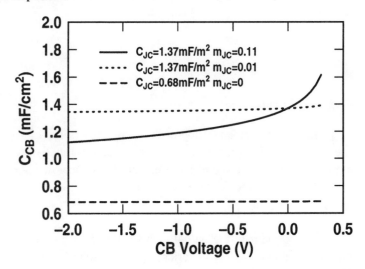

Figure 8.7 C_{CB} versus CB junction voltage for different m_j values.

Figure 8.7 shows the $C - V$ characteristics of the CB junction of a first generation SiGe HBT (here $V_j = 0.7$ V, and $m_j = 0.11$, $m_j = 0.01$, and $m_j = 0$ are compared). Observe that m_j affects not only $\partial C/\partial V$, but also the absolute value of C. At a reverse bias of 2 V ($V_f = -2$ V), a decrease of m_j from 0.11 to 0.01 not only makes the capacitance more linear (smaller derivative), but also increases the absolute value of C_{CB}. Therefore, caution must be exercised in identifying whether the absolute value or the derivative is dominant in determining the transistor overall linearity [8]. A higher C_{CB} value in general improves linearity because of increased feedback, and an associated penalty in gain.

Having discussed the individual physical $I - V$ and $C - V$ nonlinearities in a SiGe HBT, we are now in a position to examine how they affect the linearity of a SiGe HBT amplifier, and to then address an often-asked, and highly relevant question: Which nonlinearity is dominant in the device? The analysis method used is Volterra series, which applies to small-signal distortion such as that found in front-end low-noise amplifiers and mixers. Compared to other distortion analysis methods, Volterra series allow us to easily identify the contribution of various individual nonlinearities, as well as identify the interaction between individual nonlinearities.

The mathematical derivation of Volterra series is quite complicated, and has been well treated in [1] and [2]. Applying Volterra series to circuit linearity analysis, however, is straightforward, and can be readily performed with the help of a matrix solver such as that found in MATLAB. The fundamental concepts and analysis procedures are reviewed below to the extent that should allow interested readers to repeat our analysis for their own SiGe HBTs.

8.3 Volterra Series

8.3.1 Fundamental Concepts

Volterra series is a general mathematical approach for solving systems of nonlinear integral and integral-differential equations, such as the equations found in nonlinear circuits. Volterra series can be viewed as an extension of the theory of linear systems to weakly nonlinear systems. The essentials of Volterra series can be briefly summarized as:

- Volterra series approximate the output of a nonlinear system in a manner similar to the more familiar Taylor series approximation of analytical functions. Similarly, the analysis is applicable only to weak nonlinearities (in our case, only small-signal inputs).

- The response of a nonlinear system to an input $x(t)$ is equal to the sum of the responses of a series of transfer functions of different orders (H_1, H_2, ...,

Linearity

H_n), as shown in Figure 8.8

$$Y = H_1(x) + H_2(x) + H_3(x) + \cdots. \tag{8.54}$$

- In the time domain, H_n is described as an impulse response $h_i(\tau_1, \tau_2, \cdots, \tau_n)$. As in linear circuit analysis, frequency domain representation is often more convenient, and thus $H_n(s_1, \cdots, s_n)$, the nth-order transfer function in the frequency domain, is obtained through a multidimensional Laplace transform of the time domain impulse response

$$\overbrace{H_n(s_1, \cdots, s_n)}^{\text{frequency domain}} = \tag{8.55}$$

$$\overbrace{\int_{-\infty}^{+\infty} \cdots \int_{-\infty}^{+\infty}}^{\text{Laplace transform}} \overbrace{h_n(\tau_1, \tau_2, \cdots, \tau_n)}^{\text{time domain}} e^{-(s_1\tau_1 + s_2\tau_2 + \cdots + s_n\tau_n)} d\tau_1 \cdots d\tau_n.$$

Here, H_n takes n frequencies as the input, from $s_1 = j\omega_1$ to $s_n = j\omega_n$.

- The first-order transfer function $H_1(s)$ is essentially the transfer function of the small-signal linear circuit at dc bias. Higher-order transfer functions represent higher-order phenomena.

- Solving the output of a nonlinear circuit is equivalent to solving the Volterra series $H_1(s)$, $H_2(s_1, s_2)$, and $H_3(s_1, s_2, s_3)$, $\cdots$

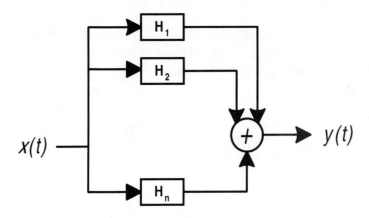

Figure 8.8 An illustration of Volterra series.

For a very small input, the output can be accurately described using only the linear (i.e., first-order) operator $H_1(s)$, the transfer function of the linearized circuit. As the input signal increases, however, a substantial part of the output is generated by the nonlinearities. For sufficiently small inputs, these nonlinear effects can be described accurately using only the second- and the third-order transfer functions $H_2(s_1, s_2)$ and $H_3(s_1, s_2, s_3)$.

As for $H_1(s)$, $H_2(s_1, s_2)$ and $H_3(s_1, s_2, s_3)$ are determined by the linear and nonlinear properties of all of the circuit elements, as well as by the way in which the different elements are connected. To solve $H_1(s)$, the nonlinear circuit is first linearized and solved at $s = j\omega$. This solution requires first-order derivatives of the nonlinear circuit elements. The solution of $H_2(s_1, s_2)$ and $H_3(s_1, s_2, s_3)$ also need the second-order and third-order nonlinearity coefficients of the nonlinear circuit elements.

Before discussing how to solve $H_2(s_1, s_2)$ and $H_3(s_1, s_2, s_3)$, let us first see how they can be used in determining the response of a nonlinear circuit to a two-tone excitation $A_1 \cos \omega_1 t + A_2 \cos \omega_2 t$. The output consists of 18 responses at 13 distinct frequencies, including the fundamental signal, the various harmonics, intermodulation products, desensitizations, and compressions. The intermodulation products, harmonics, and the dc shift are listed in Table 8.1.

The solution of Volterra series in nonlinear electrical circuits is fortunately a straightforward case. The transfer functions can be solved in increasing order by repeatedly solving the same linear circuit using different excitations at each order. It is for this reason that both the absolute values and derivatives of nonlinear conductances and capacitances are important in determining the overall circuit linearity.

To illustrate the analysis procedure, consider a bipolar transistor amplifier with an RC source and an RL load. Figure 8.9 shows the linearized circuit. For simplicity, we neglect all of the nonlinear capacitances in the transistor, the base and emitter resistance, as well as the avalanche multiplication current. After mastering the analysis procedures, one can easily add all of these elements in a straightforward manner with the aid of a matrix calculator such as found in MATLAB. As in linear circuit analysis, $H_n(s_1, \cdots, s_n)$ can be defined for each nodal voltage. The nth-order transfer functions for all of the nodal voltages are denoted by a vector $\vec{H}_n(s_1, \cdots, s_n)$.

8.3.2 First-Order Transfer Functions

We now number the base node as "1," and the collector node as "2." Applying nodal analysis leads to

$$Y(s) \cdot \vec{H}_1(s) = \vec{I}_1, \qquad (8.56)$$

Linearity

Table 8.1 Responses of a Nonlinear System to a Two-Tone Excitation Expressed in Terms of Volterra Series (the Input Here is $A_1 \cos \omega_1 t + A_2 \cos \omega_2 t$)

Order	Output Freq	Amplitude of Output	Type of Response
1	ω_1	$A_1\|H_1(j\omega_1)\|$	Linear
1	ω_2	$A_2\|H_1(j\omega_2)\|$	Linear
2	$\omega_1 + \omega_2$	$A_1 A_2\|H_2(j\omega_1, j\omega_2)\|$	Intermodulation
2	$\|\omega_1 - \omega_2\|$	$A_1 A_2\|H_2(j\omega_1, -j\omega_2)\|$	Intermodulation
2	$2\omega_1$	$\frac{1}{2}A_1^2\|H_2(j\omega_1, j\omega_1)\|$	Harmonic
2	$2\omega_2$	$\frac{1}{2}A_2^2\|H_2(j\omega_1, j\omega_1)\|$	Harmonic
2	0	$\frac{1}{2}A_1^2\|H_2(j\omega_1, -j\omega_1)\|$	dc shift
2	0	$\frac{1}{2}A_2^2\|H_2(j\omega_2, -j\omega_2)\|$	dc shift
3	$2\omega_1 + \omega_2$	$\frac{3}{4}A_1^2 A_2\|H_3(j\omega_1, j\omega_1, j\omega_2)\|$	Intermodulation
3	$\|2\omega_1 - \omega_2\|$	$\frac{3}{4}A_1^2 A_2\|H_3(j\omega_1, j\omega_1, -j\omega_2)\|$	Intermodulation
3	$\omega_1 + 2\omega_2$	$\frac{3}{4}A_1 A_2^2\|H_3(j\omega_1, j\omega_2, j\omega_2)\|$	Intermodulation
3	$\|\omega_1 - 2\omega_2\|$	$\frac{3}{4}A_1 A_2^2\|H_3(j\omega_1, -j\omega_2, -j\omega_2)\|$	Intermodulation
3	$3\omega_1$	$\frac{1}{4}A_1^3\|H_3(j\omega_1, j\omega_1, j\omega_1)\|$	Harmonic
3	$3\omega_2$	$\frac{1}{4}A_2^3\|H_3(j\omega_2, j\omega_2, j\omega_2)\|$	Harmonic

where $Y(s)$ is the admittance matrix evaluated at frequency s, $\vec{H}_1(s)$ is the vector of the first-order transfer functions, and $\vec{I}_1$ is a vector of excitations. We cannot directly apply nodal analysis because the current flowing through the zero-impedance voltage source is unknown. This voltage source, however, is readily eliminated by converting the voltage source into a current source, a technique known as compact modified nodal analysis (CMNA) [9]. This conversion results in the equivalent circuit shown in Figure 8.10. Applying Kirchoff's current law at node 1 yields

$$Y_S(V_1 - V_S) + g_{be}V_1 = 0, \qquad (8.57)$$

where

$$Y_S(s) = \frac{1}{Z_S(s)} = \frac{1}{R_S + \frac{1}{j\omega C_S}}, \qquad (8.58)$$

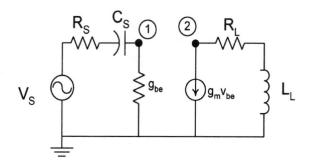

Figure 8.9 Linearized circuit of a SiGe HBT amplifier terminated with an *RC* source and an *RL* load. The capacitances inside the transistor are neglected here for simplicity.

and

$$Y_L(s) = \frac{1}{Z_L(s)} = \frac{1}{R_L + j\omega L_L}. \qquad (8.59)$$

Similarly, for node 2

$$g_m V_1 + Y_L V_2 = 0. \qquad (8.60)$$

The corresponding matrix representation is thus

$$\begin{bmatrix} Y_S + g_{be} & 0 \\ g_m & Y_L \end{bmatrix} \cdot \begin{bmatrix} V_1 \\ V_2 \end{bmatrix} = \begin{bmatrix} Y_S V_S \\ 0 \end{bmatrix}. \qquad (8.61)$$

The 2×2 matrix is the CMNA "admittance matrix." In it, V_1 and V_2 become the transfer functions of the voltages at node 1 and node 2 for an input voltage of unity (i.e., $V_S = 1$ V). We denote the two linear transfer functions as $H_{11}(s)$ and $H_{12}(s)$. The first subscript represents the order of the transfer function, and the second subscript represents the node number

$$\begin{bmatrix} Y_S + g_{be} & 0 \\ g_m & Y_L \end{bmatrix} \cdot \begin{bmatrix} H_{11}(s) \\ H_{12}(s) \end{bmatrix} = \begin{bmatrix} Y_S \\ 0 \end{bmatrix}. \qquad (8.62)$$

Solving the above matrix gives the first-order transfer functions of the nodal voltages.

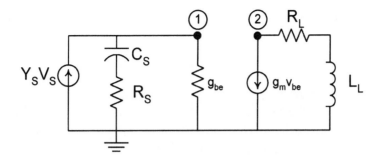

Figure 8.10 CMNA equivalent of the circuit shown in Figure 8.9.

8.3.3 Second-Order Transfer Functions

The second-order transfer functions, $\vec{H}_2(s_1, s_2)$, are then obtained by solving the same linear circuit function, but now with different excitations. The real citation V_s is not set to zero. Instead, the so-called second-order "virtual nonlinear current sources" are applied to excite the circuit. The circuit responses (nodal voltages) under these *virtual* excitations are the second-order transfer functions.

Each nonlinear element in the original circuit contributes a virtual current source excitation. The current source is placed in parallel with the corresponding linearized element. Its orientation is identical to that of the original large-signal nonlinear current source. Like $\vec{H}_2(s_1, s_2)$, the virtual current source is defined for two input frequencies, s_1 and s_2, representing the second-order corrections to the linear response, and is determined by:

- The second-order nonlinearity coefficients of the specific $I - V$ or $C - V$ nonlinearity in question.

- The first-order transfer functions of the controlling voltage(s).

The second-order virtual current source for a $I - V$ nonlinearity, i_{NL2g}, is given by

$$i_{NL2g}(u) = K_{2g}(u) H_{1u}(s_1) H_{1u}(s_2), \qquad (8.63)$$

where $H_{1u}(s)$ is the first-order transfer function of the controlling voltage, u, evaluated at frequency s, and $K_{2g}(u)$ is the second-order nonlinearity coefficient that determines the second-order response of i to u. The subscript "u" in H_{1u} is not necessarily a node number. Instead, it refers to the controlling voltage, u. For instance, if the base node is numbered "1," and the emitter node is numbered "2," H_{1u} refers to the transfer function of the controlling voltage, $u = v_{be}$. That is, $H_u = H_{11} - H_{12}$.

Similarly, the second-order virtual current source for a $C - V$ nonlinearity, i_{NL2C}, is given by

$$i_{NL2C}(u) = (s_1 + s_2)K_{2C}(u)H_{1u}(s_1)H_{1u}(s_2), \qquad (8.64)$$

where $K_{2C}(u)$ is the second-order nonlinearity coefficient of the nonlinearity capacitance C, and u is the controlling voltage. The virtual current source flows from the positive terminal to the negative terminal of the corresponding capacitance.

For a 2-D nonlinearity (i.e., a nonlinearity with two controlling voltages), such as the avalanche multiplication current, I_{CB}, the second-order virtual current source consists of two virtual current sources for each controlling voltage, as well as a cross-term

$$i_{NL2g}(u, v) = i_{NL2g_u}(u) + i_{NL2g_v}(v)$$
$$+ \frac{1}{2}K_{2g_u \& g_v} [H_{1u}(s_1)H_{1v}(s_2) + H_{1v}(s_1)H_{1u}(s_2)], \qquad (8.65)$$

where $H_{1u}(s)$ is the first-order transfer function of the first controlling voltage, u, evaluated at frequency s, and $H_{1v}(s)$ is the first-order transfer function of the second controlling voltage, v, evaluated at frequency s.

Using $\vec{H}_1$ and the second-order nonlinearity coefficients, a second-order virtual current source is calculated for each nonlinear circuit element. For the example circuit, the two virtual current sources corresponding to g_{be} and g_m are given by

$$i_{NL2g_{be}} = K_{2g_{be}}H_{11}(s_1)H_{11}(s_2), \qquad (8.66)$$

and

$$i_{NL2g_m} = K_{2g_m}H_{11}(s_1)H_{11}(s_2). \qquad (8.67)$$

The controlling voltage v_{be} is equal to the voltage at node "1," because the emitter is grounded.

The real excitation, V_S, is now shorted. The virtual current sources are then used to excite the same linearized circuit, but at a frequency of $s_1 + s_2$. A circuit schematic is shown in Figure 8.11. The nodal voltages obtained using these "virtual" excitations are the second-order transfer functions. The solution is straightforward, and

$$Y \cdot \vec{H}_2(s_1, s_2) = \vec{I}_2, \qquad (8.68)$$

where

- Y is the same CMNA admittance matrix used in (8.56), but now evaluated at a frequency of $s_1 + s_2$,

Linearity 345

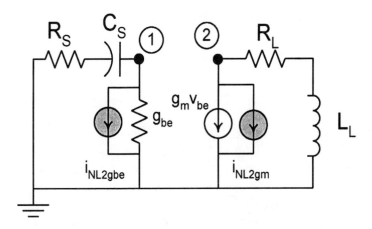

Figure 8.11 Network for solving the second-order transfer functions of the SiGe HBT amplifier example.

- $\vec{H}_2(s_1, s_2)$ is the second-order transfer function vector, and

- $\vec{I}_2$ is a linear combination of all the second-order nonlinear current sources, and can be obtained by applying Kirchoff's current law at each node.

For our example, the linearized circuit already exists, and thus all we need to do is excite the circuit with the two virtual current sources $i_{NL2g_{be}}$, and i_{NL2g_m}. The admittance matrix is evaluated at the frequency $s_1 + s_2$. Applying Kirchoff's current law at nodes 1 and 2 gives

$$\begin{bmatrix} Y_S + g_{be} & 0 \\ g_m & Y_L \end{bmatrix} \cdot \begin{bmatrix} H_{21}(s_1, s_2) \\ H_{22}(s_1, s_2) \end{bmatrix} = \begin{bmatrix} -i_{NL2g_{be}} \\ -i_{NL2g_m} \end{bmatrix}. \quad (8.69)$$

The admittance matrix remains the same, except for the evaluation frequency.

8.3.4 Third-Order Transfer Functions

In a similar manner, the third-order transfer functions, $\vec{H}_3(s_1, s_2, s_3)$, can be solved as a response to virtual excitations determined by the previously solved first- and second-order transfer functions and nonlinearity coefficients

$$Y \cdot \vec{H}_3(s_1, s_2, s_3) = \vec{I}_3. \quad (8.70)$$

The third-order virtual current source for a $I - V$ nonlinearity, i_{NL3g}, is given by

$$i_{NL3g}(u) = K_{3g}(u)H_{1u}(s_1)H_{1u}(s_2)H_{1u}(s_3)$$
$$+ \frac{2}{3}K_{2g}[H_{1u}(s_1)H_{2u}(s_2, s_3) + H_{1u}(s_2)H_{2u}(s_1, s_3)$$
$$+ H_{1u}(s_3)H_{2u}(s_1, s_2)], \qquad (8.71)$$

where u is the controlling voltage, $H_{1u}(s)$ is the first-order transfer function of u evaluated at frequency s, $K_{2g}(u)$ is the second-order nonlinearity coefficient that determines the second-order response of i to the controlling voltage u, and $K_{3g}(u)$ is the third-order nonlinearity coefficient that determines the third-order response of i to the controlling voltage u.

The third-order virtual current source for a $C - V$ nonlinearity, i_{NL3C}, is given by

$$i_{NL3C}(u) = (s_1 + s_2 + s_3)K_{3C}(u)H_{1u}(s_1)H_{1u}(s_2)H_{1u}(s_3)$$
$$+ (s_1 + s_2 + s_3)\frac{2}{3}K_{2C}[H_{1u}(s_1)H_{2u}(s_2, s_3)$$
$$+ H_{1u}(s_2)H_{2u}(s_1, s_3) + H_{1u}(s_3)H_{2u}(s_1, s_2)], \qquad (8.72)$$

where $H_{2u}(s_1, s_2)$ is the second-order transfer function of the controlling voltage u, evaluated at the frequency pair (s_1, s_2).

For a $I-V$ nonlinearity with two controlling voltages, the virtual current source excitation consists of two virtual current sources for each controlling voltage, u and v, and cross-terms

$$i_{NL3g}(u, v) = i_{NL3g_u}(u) + i_{NL3g_v}(v)$$
$$+ i_{NL3C1g}(u, v) + i_{NL3C2g}(u, v) + i_{NL3C3g}(u, v), \qquad (8.73)$$

where

$$i_{NL3C1g}(u, v) = \frac{1}{3}K_{2g_u \& g_v} \cdot \Big[H_{1u}(s_2)H_{2v}(s_1, s_3) + H_{1u}(s_1)H_{2v}(s_2, s_3)$$
$$+ H_{1u}(s_3)H_{2v}(s_1, s_2) + H_{1v}(s_2)H_{2u}(s_1, s_3)$$
$$+ H_{1v}(s_1)H_{2u}(s_2, s_3) + H_{1v}(s_3)H_{2u}(s_1, s_2) \Big], \qquad (8.74)$$

$$i_{NL3C2g}(u, v) = \frac{1}{3}K_{32g_u \& g_v} \cdot \Big[H_{1u}(s_1)H_{1u}(s_2)H_{1v}(s_3) + H_{1u}(s_1)H_{1u}(s_3)H_{1v}(s_2)$$
$$+ H_{1u}(s_2)H_{1u}(s_3)H_{1v}(s_1) \Big], \qquad (8.75)$$

$$i_{NL3C3g}(u, v) = \frac{1}{3}K_{3g_u\&2g_v} \cdot \Big[H_{1u}(s_1)H_{1v}(s_2)H_{1v}(s_3) + H_{1u}(s_1)H_{1v}(s_3)H_{1v}(s_2)$$
$$+ H_{1u}(s_2)H_{1v}(s_3)H_{1v}(s_1)\Big]. \tag{8.76}$$

In our example SiGe HBT amplifier circuit, the third-order virtual current sources corresponding to g_{be} and g_m, which are $i_{NL3g_{be}}$ and i_{NL3g_m}, are given by

$$i_{NL3g_{be}} = K_{3g_{be}}(u)H_{11}(s_1)H_{11}(s_2)H_{11}(s_3)$$
$$+ \frac{2}{3}K_{2g_{be}}[H_{11}(s_1)H_{21}(s_2, s_3) + H_{11}(s_2)H_{21}(s_1, s_3)$$
$$+ H_{11}(s_3)H_{21}(s_1, s_2)], \tag{8.77}$$

$$i_{NL3g_m} = K_{3g_m}(u)H_{11}(s_1)H_{11}(s_2)H_{11}(s_3)$$
$$+ \frac{2}{3}K_{2g_m}[H_{11}(s_1)H_{21}(s_2, s_3) + H_{11}(s_2)H_{21}(s_1, s_3)$$
$$+ H_{11}(s_3)H_{21}(s_1, s_2)]. \tag{8.78}$$

where H_{ij} stands for the ith-order transfer function of the voltage at node j. In both cases, the controlling voltage is equal to the voltage of node "1," because the emitter is grounded. The third-order transfer function vector is then solved from Figure 8.12

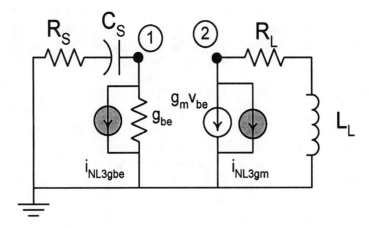

Figure 8.12 Network for solving the third-order transfer functions of a simple SiGe HBT amplifier.

$$\begin{bmatrix} Y_S + g_{be} & 0 \\ g_m & Y_L \end{bmatrix} \cdot \begin{bmatrix} H_{31}(s_1, s_2, s_3)) \\ H_{32}(s_1, s_2, s_3) \end{bmatrix} = \begin{bmatrix} -i_{NL3g_{be}} \\ -i_{NL3g_m} \end{bmatrix}. \tag{8.79}$$

The admittance matrix now needs to be evaluated at a frequency of $s_1 + s_2 + s_3$.

From the above analysis, it is clear that the nonlinear behavior of the overall circuit is determined by:

- The linearized circuit, including linearized elements as well as purely linear elements. This explains why the absolute values of nonlinear capacitances are important in determining overall circuit linearity.

- The nonlinearity coefficients of all the nonlinear elements. These coefficients directly affect the magnitude of the virtual current source excitations at each order.

- Circuit topology, including the placement of both linear and nonlinear elements.

- Source and load terminations, at both the fundamental frequencies and the harmonic frequencies, since they enter into the evaluation of the admittance matrix for different orders.

The ability afforded by Volterra series for predicting nonlinear circuit response by repeatedly solving the same linear circuit has important practical implications. Insights gained from linear circuit analysis can now be used in understanding nonlinear circuit behavior. Furthermore, one can selectively turn on and turn off the nonlinear current sources for each individual nonlinearity when solving for Volterra series, facilitating identification of the dominant nonlinearity for a given situation.

8.4 Single SiGe HBT Amplifier Linearity

We now analyze the linearity of a single SiGe HBT common-emitter transistor amplifier using the methodology described above. Figure 8.13 shows the circuit schematic. The SiGe HBT is biased through RF chokes at both the input and the output, and is buffered from the RF source and load by capacitors. The transistor has four 0.5×20 μm^2 emitter fingers. The load is tuned to maximize power gain at $I_C = 3$ mA and $V_{CE} = 3$ V.

Here we use the large-signal equivalent circuit shown in Figure 8.3. For accurate evaluation of the required derivatives, we use table-lookup from measured device data for $I_{C0}(V_{BE})$, $I_{B0}(V_{BE})$, $F_{Early}(V_{CB})$, C_{BC}, C_{BE}, C_{CS}, and their derivatives. The avalanche model takes into account both the current and voltage dependence of $M - 1$. The elements values were extracted using measured dc data and measured S-parameters up to 40 GHz.

Linearity 349

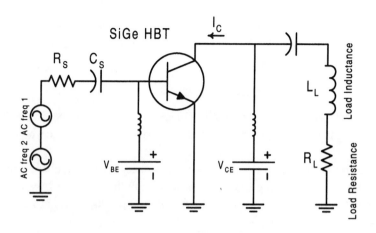

Figure 8.13 A single SiGe HBT amplifier used for Volterra series analysis ($A_E = 0.5 \times 20 \times 4$ μm^2). The default parameter values are $I_C = 3$ mA, $V_{CE} = 3$ V, $R_S = 50$ Ω, $C_S = 300$ pF, $R_L = 186$ Ω, and $L_L = 9$ nH.

8.4.1 Circuit Analysis

The first solution step is to linearize the large-signal equivalent circuit in Figure 8.3 at the *dc* bias point, and solve the resulting linear circuit using CMNA [9]

$$Y(s) \cdot \vec{H}_1(s) = \vec{I}_1, \tag{8.80}$$

where $Y(s)$ is the CMNA [9] admittance matrix at frequency $s(j\omega)$, $\vec{H}_1(s)$ is the vector of first-order transfer functions of the node voltages, $\vec{I}_1$ is the vector of the node excitations with a unity input voltage, and Y and $\vec{I}_1$ are obtained by applying the Kirchoff's current law at every circuit node. The voltages that control the various nonlinearities can be expressed as a linear combination of the nodal voltages (i.e., elements of $\vec{H}_1(s)$).

With $\vec{H}_1(s)$, we now excite the same circuit using the second-order virtual excitation vector $\vec{I}_2$. Here $\vec{I}_2$ is determined by the first-order voltages that control the individual nonlinearities, and the second-order nonlinearity coefficients of all the $I - V$ and $C - V$ nonlinearities. The node voltages under such an excitation are contained in the second-order transfer function vector $\vec{H}_2(s_1, s_2)$

$$Y(s_1 + s_2) \cdot \vec{H}_2(s_1, s_2) = \vec{I}_2, \tag{8.81}$$

where $Y(s_1 + s_2)$ is the same CMNA admittance matrix used in Eq. (8.80), but evaluated now at the frequency $s_1 + s_2$.

350 Silicon-Germanium Heterojunction Bipolar Transistors

In a similar manner, we proceed to solve the third-order transfer function vector $\vec{H}_3$ using excitations specified in terms of the previously determined first- and second-order transfer functions

$$Y(s_1 + s_2 + s_3) \cdot \vec{H}_2(s_1, s_2, s_3) = \vec{I}_3. \qquad (8.82)$$

Now P_{out} vs P_{in}, the third-order input intercept (*IIP3*) at which the first-order and third-order signals have equal power, and the (power) gain can then be obtained from $\vec{H}_3$ and $\vec{H}_1$.

With a multitone input, the node voltages at each mixed frequency for each node can be expressed using the Volterra series solutions. For a two-tone input $A(cos\omega_1 t + cos\omega_2 t)$, the third-order intermodulation product (*IM*) at $2\omega_1 - \omega_2$ is given by

$$V_{\text{IM3}} = \frac{3}{4} A^3 H_{3o}(j\omega_1, j\omega_1, -j\omega_2), \qquad (8.83)$$

where H_{3o} is the third-order transfer function of the voltage at the load resistance node (the output node element of $\vec{H}_3$). The subscript "o" here means "output." Similarly, the third-order *IM* product at $2\omega_2 - \omega_1$ can be calculated. The frequency difference between $2\omega_1 - \omega_2$ and $2\omega_2 - \omega_1$ is typically so small that the corresponding *IM* results have equal magnitude.

The output voltage at the fundamental frequency ω_1 is given by

$$V_{\text{fundamental}} = A H_{1o}(j\omega_1), \qquad (8.84)$$

where $H_{1o}(j\omega_1)$ is the output node element of $\vec{H}_1(j\omega_1)$. Here, $H_{1o}(j\omega_1)$ is essentially the voltage gain, from which the power gain can be calculated together with the source and load impedance values. The input voltage magnitude A_{IP3}, at which the extrapolated fundamental output voltage equals the intermodulation output voltage, can be obtained by equating (8.83) and (8.84)

$$A_{IP3} = \sqrt{\frac{4}{3} \cdot \frac{|H_{1o}(j\omega_1)|}{|H_{3o}(j\omega_1, j\omega_1, -j\omega_2)|}}. \qquad (8.85)$$

Accordingly, *IIP3*, the input power at which the fundamental output power equals the intermodulation output power, is

$$\begin{aligned} IIP3 &= \frac{A_{IP3}^2}{8R_S} \\ &= \frac{1}{6R_S} \cdot \frac{|H_{1o}(j\omega_1)|}{|H_{3o}(j\omega_1, j\omega_1, -j\omega_2)|}, \end{aligned} \qquad (8.86)$$

where R_S is the source resistance. Note that *IIP3* is often expressed in dBm using $IIP3_{dBm} = 10 \log \left(10^3 \cdot IIP3 \right)$.

8.4.2 Distinguishing Individual Nonlinearities

It is often highly relevant to determine *which* physical nonlinearity dominates for a given device technology and circuit topology. To answer this question, we need a means of distinguishing the various individual nonlinearities. Traditionally, individual nonlinearities are distinguished from each other by separating $\vec{H}_3$ into components related to each individual nonlinearity [2]. The overall $\vec{H}_3$ is the sum of the individual $\vec{H}_3$. This approach does not *completely* distinguish individual nonlinearities, however, because the solution of $\vec{H}_2$ involves *all* of the nonlinearities, and $\vec{H}_2$ was used in the calculation of $\vec{I}_3$.

A complete separation can be made if we include *only* the virtual excitation related to the specific nonlinearity in question in solving both $\vec{H}_2$ and $\vec{H}_3$. An individual *IIP3* is thus obtained for each nonlinearity. The value that gives the lowest *IIP3* (i.e., the highest distortion) can be identified as the dominant nonlinearity. The overall circuit *IIP3* can then be calculated by including all of the nonlinearities in solving both $\vec{H}_2$ and $\vec{H}_3$. The overall $\vec{H}_3$, however, is no longer equal to the sum of individual $\vec{H}_3$s.

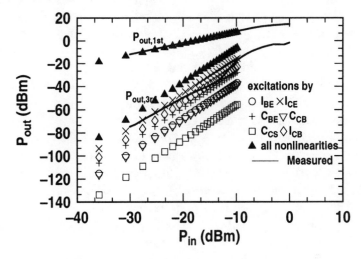

Figure 8.14 P_{out} versus P_{in} for both the fundamental signal and third-order intermodulation signal at 2 GHz. The tone spacing is 1 MHz.

Figure 8.14 shows the calculated fundamental and third-order intermodulation power as a function of input power. The biasing conditions and elements values are the same as in Figure 8.13. The solid symbol (denoted by the ▲) is the overall *IM* distortion power, and is higher than any of the individual distortion powers. In this case, the I_{CE} nonlinearity yields the highest individual *IM* distortion power, and is thus the dominant nonlinearity. As we will see below, however, other nonlinearities can dominate for other biasing conditions and circuit configurations.

Strictly speaking, Volterra series is valid only at low input power levels, such as for the signals found in mobile receivers, which are often as low as -100 dBm. We note that *IIP3* measurements, however, are often made at much higher P_{in}, for practical reasons. Nevertheless, a data-to-model comparison provides an approximate test of the Volterra series model accuracy. The comparison was made here at 2 GHz with a 1-MHz tone spacing. The agreement is excellent for the fundamental signal, and is within 5 dB for the intermodulation signal at $P_{in} = -30$ dBm. This is an acceptable agreement, considering that -30 dBm, the lowest P_{in} used in the measurement, is reasonably large for the small-signal distortion requirement.

8.4.3 Collector Current Dependence

Experimentally, it is well established that *IIP3* depends on the operating current. Figure 8.15 shows *IIP3* and gain as a function of I_C up to 60 mA, where the f_T and f_{max} peaks occur ($V_{CE} = 3$ V). At low I_C (< 5 mA), the exponential $I_{CE} - V_{BE}$ nonlinearity (denoted by ×) yields the lowest individual *IIP3*, and hence is the dominant nonlinearity. For 5 mA $< I_C <$ 25 mA, the I_{CB} nonlinearity due to avalanche multiplication (denoted by ◊) dominates. For $I_C > 25$ mA, the C_{CB} nonlinearity due to the CB capacitance (denoted by ▽) dominates. Interestingly, the overall *IIP3* obtained by including all of the nonlinearities is close to the lowest individual *IIP3* for all the I_C in this case, and indicates a weak interaction between the various individual nonlinearities.

The overall *IIP3* increases with I_C for $I_C < 5$ mA when the exponential I_{CE} nonlinearity dominates. For $I_C > 5$ mA, where the avalanche current (I_{CB}) nonlinearity dominates, the I_C dependence of the overall *IIP3* has two determining factors: 1) the initial current for avalanche I_{CE} increases with I_C; and 2) the avalanche multiplication factor ($M - 1$) decreases with I_C. While the detailed structure of the simulated overall *IIP3* curve cannot be easily explained, clearly the increase of the avalanche *IIP3* and hence the overall *IIP3* for $I_C > 17$ mA can be readily understood to be a result of the decrease of $M - 1$ with increasing J_C. For $I_C > 25$ mA, the overall *IIP3* becomes limited by the C_{CB} nonlinearity, and is approximately independent of I_C. The optimum biasing current is therefore $I_C = 25$ mA for this case ($V_{CE} = 3$ V). The use of a higher I_C only increases power consumption, and does not improve the linearity. The decrease of $M - 1$ with increasing J_C is therefore quite beneficial to the linearity of these SiGe HBTs. To our knowledge, this is the only beneficial effect of the charge compensation by mobile carriers in the CB junction space charge region of a SiGe HBT, since the other high-injection phenomena degrade the transistor response. These Volterra

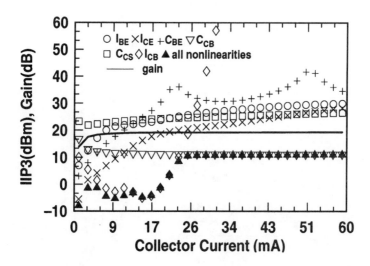

Figure 8.15 *IIP3* and gain as a function of I_C. The circuit parameters are the same as in Figure 8.13.

series results indicate the importance of modeling the J_C dependence of $M - 1$ for linearity analysis of SiGe HBT circuits.

In LNA design, these SiGe HBTs are typically biased at a J_C of 0.1–0.2 mA/μm^2 to minimize broadband noise, which corresponds to a I_C of 4–8 mA in Figure 8.15. In this I_C range, *IIP3* is limited by avalanche multiplication for the circuit configuration in Figure 8.13. For further improvement of *IIP3*, a lower collector doping is thus desired, provided that the noise performance is not inadvertently degraded. In Chapter 7 we showed that the noise figure is relatively independent of the collector doping as long as Kirk effect does not occur at the J_C of interest [7]. Thus, there must exist an optimum collector doping profile for simultaneously producing low-noise SiGe HBTs with the best linearity.

8.4.4 Collector Voltage Dependence

The collector voltage mainly affects two nonlinearities: the C_{CB} nonlinearity and the I_{CB} nonlinearity. A higher value of V_{CE} leads to a larger reverse CB junction bias, thus making C_{CB} more linear. On the other hand, a higher V_{CE} also leads to more avalanche multiplication and a stronger I_{CB} nonlinearity. One can therefore intuitively expect that an optimum V_{CE} for maximum *IIP3* exists. Figure 8.16 shows the overall *IIP3* as a function of V_{CE} up to 3.3 V, the BV_{CEO} of this SiGe HBT. A peak in *IIP3* generally exists as V_{CE} increases. At $I_C = 10$ mA, the optimum V_{CE} is 2.4 V, yielding an *IIP3* of 9 dBm. Observe that this is 11 dB higher than the *IIP3* at V_{CE} = 3 V (-2 dBm), and illustrates the importance of proper biasing in linearity optimization.

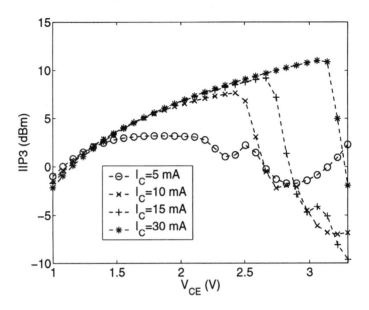

Figure 8.16 *IIP3* as a function of V_{CE} for different I_C. The circuit parameters are the same as in Figure 8.13.

To better illustrate the importance of biasing current and voltage selection, *IIP3* is shown as a function of I_C for different V_{CE} in Figure 8.17. As discussed above, at sufficiently high I_C, *IIP3* approaches a value that depends on V_{CE}. The threshold I_C where *IIP3* reaches its maximum is higher for a higher V_{CE}. For a given V_{CE}, I_C must be above the threshold to achieve good *IIP3*. On the other hand, the use of an I_C well above the threshold does not further increase *IIP3*, and only increases power consumption. The optimum I_C is thus at the threshold value, which is 10 mA for V_{CE} = 2 V, in this case. Figure 8.18 shows contours of *IIP3* as a function of I_C and V_{CE}, which can be used for selection of the appropriate biasing current and voltage.

8.4.5 Load Dependence

Figure 8.19 shows the simulated *IIP3* for the individual nonlinearities as well as for all of the nonlinearities, as a function of load resistance. The gain clearly varies with load, and peaks when the load is closest to conjugate matching, as expected.

Linearity

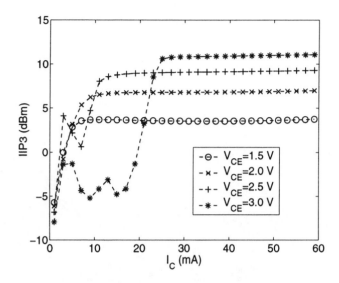

Figure 8.17 *IIP3* as a function of I_C for different V_{CE}. The circuit parameters are the same as in Figure 8.13.

Observe that *IIP3*, however, is much more sensitive to load variations. Physically, the load dependence results from the CB feedback, primarily due to the CB capacitance C_{CB} and the avalanche multiplication current I_{CB}. Both feedback mechanisms are nonlinear, although the load dependence would still exist for a linear CB feedback [8] (for instance, using an externally connected linear CB capacitance). Another feedback mechanism is the emitter resistance, which is negligible in these devices because of low emitter contact resistance, but which can be important for some technologies. The Early effect contribution to the load dependence is also small, because of the large Early voltage in these SiGe HBTs.

Another contributor to the linearity load dependence is the collector-substrate capacitance (C_{CS}) nonlinearity, since the nonlinearity controlling voltage V_{CS} is a function of the load condition. The contribution of C_{CS} nonlinearity to the overall *IIP3*, however, is generally negligible because of its placement in the equivalent circuit. For verification, we now repeat the simulation with $C_{CB} = 0$ and $I_{CB} = 0$, and using experimentally extracted F_{Early} and R_E values. The results are shown in Figure 8.20. Note that *IIP3* becomes virtually independent of load condition for all of the nonlinearities except for the C_{CS} nonlinearity. Closed-form analysis using Volterra series for a transistor with zero C_{CB}, zero I_{CB}, and zero R_e indeed proves that *IIP3* becomes independent of the load conditions when these feedback mechanisms are removed. The underlying physics is straightforward if we examine the

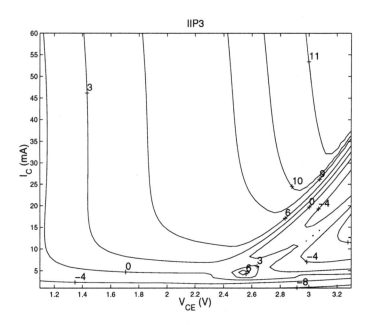

Figure 8.18 Contours of *IIP3* in dBm on the I_C–V_{CE} plane. The circuit parameters are the same as in Figure 8.13.

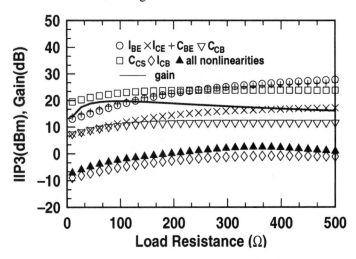

Figure 8.19 *IIP3* and gain as a function of load resistance at $I_C = 13$ mA and $V_{CE} = 3$ V. The circuit parameters are the same as in Figure 8.13.

Linearity

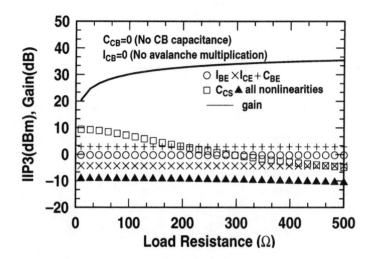

Figure 8.20 *IIP3* and gain versus load resistance at $I_C = 13$ mA and $V_{CE} = 3$ V ($C_{CB} = 0$ and $I_{CB} = 0$).

equivalent circuit shown in Figure 8.21. The nonlinearity controlling voltage V_{BE} that controls the $I_{BE} - V_{BE}$, $C_{BE} - V_{BE}$, and $I_{CE} - V_{BE}$ nonlinearities becomes independent of the load condition when $C_{CB} = 0$, $I_{CB} = 0$, and $R_E = 0$. Because of the weak Early effect in SiGe HBTs, the nonlinear current I_{CE} is solely determined by V_{BE}, and therefore is also independent of the load condition. Consequently, *IIP3* is independent of the load condition, consistent with the simulations.

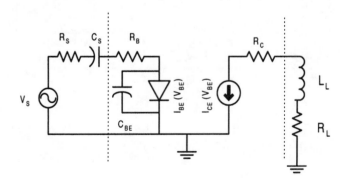

Figure 8.21 Equivalent circuit of a single transistor amplifier with $C_{CB} = 0$, $I_{CB} = 0$, and $F_{Early} = 1$.

8.4.6 Dominant Nonlinearity Versus Bias

Figure 8.22 shows the dominant nonlinearity factor on the $I_C - V_{CE}$ plane. The source and load conditions are the same as in Figure 8.13. The upper limit of I_C is where f_T reaches its peak value. The avalanche multiplication and C_{CB} nonlinearities are the dominant factors for most of the bias currents and voltages. From basic device physics, both avalanche multiplication and the C_{CB} nonlinearities can be decreased by reducing the collector doping. This process, however, fundamentally conflicts with the need for high collector doping to suppress Kirk effect and heterojunction barrier effects in SiGe HBTs. Fortunately, typical SiGe HBT processes provide at least two collector doping profiles through selective ion implantation to achieve different breakdown voltages, and thus the higher breakdown voltage devices can be utilized in circuit designs that require higher *IIP3*.

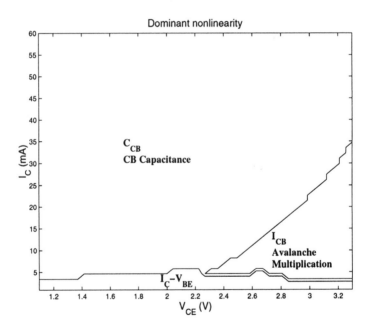

Figure 8.22 Dominant nonlinearity factor on the $I_C - V_{CE}$ plane. The circuit parameters are the same as in Figure 8.13.

8.4.7 Nonlinearity Cancellation

From the Volterra series analysis methodology, one can expect interaction between the virtual excitations due to different physical nonlinearities. In some cases, the overall *IIP3* obtained by including all of the nonlinearities can be larger (better) than an individual *IIP3* because of cancellation between the various individual nonlinearities. This cancellation process can be a complicated function of circuit topology and operating conditions. Using Volterra series, a comparison of the individual *IIP3* and the overall *IIP3* can be easily made to illuminate the interaction between the various individual nonlinearities.

Nonlinearity cancellation can be clearly observed in Figure 8.19, where the *IIP3* with all nonlinearities turned on (denoted by ▲) is noticeably higher than the *IIP3* with the avalanche current (I_{CB}) nonlinearity alone (denoted by ◊). In this case, the two dominant nonlinearities, as evidenced by their lowest individual *IIP3* contributions, are the I_{CB} and C_{CB} nonlinearities. The cancellation between the I_{CB} and C_{CB} nonlinearities leads to an overall *IIP3* value that is higher (better) than the *IIP3* obtained using the I_{CB} nonlinearity alone. The degree of cancellation depends on the biasing, as well as the source and load conditions, as expected from the Volterra series analysis. The widely cited cancellation phenomenon between the I_{CE} and C_{BE} analyzed in [10] is not important in this case, but obviously can become important in other situations (e.g., in a cascode circuit), and highlights the point that there is no general description of cancellation effects in SiGe HBTs that is applicable to all practical circuit scenarios.

Thus far we have discussed linearity of a single transistor amplifier without considering other figures of merit, such as input impedance matching or noise figure. Circuits such as LNAs, however, often simultaneously require high linearity, low noise, good impedance matching, and low power consumption. The relevant question then is: how can we make the best trade-off between various performance metrics? We describe below an approach to optimum design that balances all of the design considerations for an LNA using the commonly used "cascode" architecture.

8.5 Cascode LNA Linearity

Figure 8.23 shows a simplified schematic of a cascode amplifier (without the associated biasing network). With proper transistor sizing, the common-emitter stage produces simultaneous impedance matching and noise matching to the 50-Ω source impedance, as described in Chapter 7. The common-base stage provides additional gain, and improves isolation between the output and the input. The disadvantage of this scheme is that it requires a higher power supply to turn on both Q_1 and Q_2, which in this case is about 1.6 V. The first stage can be viewed as a transconductance stage that converts the incoming RF voltage to an RF current, which is subsequently reproduced in the load by the common-base stage with a near-unity

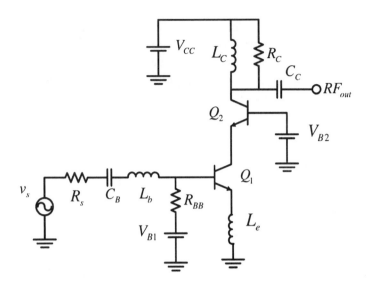

Figure 8.23 An inductively degenerated cascode LNA. Here, the L_e and L_b are used to match the input impedance to 50 Ω.

current gain. Overall nonlinearity is mainly introduced in the first stage. The second stage process is fairly linear, similar to that in a current mirror. The sizing and biasing of the common-emitter stage are thus the keys to design optimization. The output impedance is matched to a 50-Ω load using a shunt-L/series-C L-match. R_C is used to improve isolation and stability, and has little effect on the power gain when properly chosen. It does affect output impedance matching, however, which can be compensated for by the L-match consisting of L_c and C_C. C_C also serves as a dc blocking capacitor. An alternative output matching scheme is to adjust the size of Q_2 to produce a 50-Ω output resistance, and then use a series inductor to cancel the reactance [11]. Because of the low V_{CB} across both transistors, the avalanche multiplication nonlinearity is negligible in this problem. Because of the low voltage gain of the first stage, the C_{CB} nonlinearity is also much weaker than in the single transistor amplifier discussed above.

8.5.1 Optimization Approach

To find the optimum transistor size and biasing current, we first need to determine the design goals in the 2-D design space of emitter length L_E and I_C. We can narrow down our design goals to noise figure (*NF*), linearity (*IIP3*), and gain, by first optimizing L_b and L_e for impedance matching, and then limiting the I_C to a given power constraint. For a given L_E and I_C we should:

1. Adjust L_e and L_b to match the input impedance to 50 Ω using either analytical equations or optimization in a CAD tool such as ADS.

2. Calculate NF and gain using a small-signal analysis (or S-parameter analysis).

3. Calculate $IIP3$ using harmonic balance (in ADS) or by using Volterra series.

The results can then be plotted as contours in the design space, from which an optimum L_E and I_C are chosen. The above design process is illustrated below using a 2-GHz LNA design. A custom program written in MATLAB is used instead of ADS for more efficient design as well as to facilitate better insight into the linearity results. The program calculates noise figure and gain using standard linear circuit analysis and calculates $IIP3$ using Volterra series. Here, L_b and L_e are determined using numerical optimization algorithms.

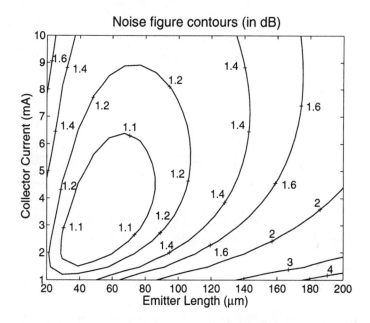

Figure 8.24 Noise figure contours as a function of emitter length and collector current. The input impedance is matched to R_S at each combination of emitter length and collector current.

Figure 8.24 shows NF contours in I_C versus L_E space. We emphasize here that NF differs from NF_{min}, and refers to the actual noise figure achieved when L_b and

L_e are adjusted for input impedance matching. We find that NF is minimum at $I_C = 4$ mA and $L_E = 60$ μm. This design point, though optimum for noise, is not necessarily the best choice when $IIP3$ is considered. Interestingly, the variation of NF with I_C and L_e is quite small across a wide range of I_C and L_E near the NF valley. An inspection of the $NF = 1.1$ dB and $NF = 1.2$ dB contours clearly shows this. This slowly varying nature of NF near the valley of NF is beneficial for circuit design in SiGe HBTs. For instance, a different (L_E, I_C) can be chosen for better linearity or lower power consumption with little increase in noise figure.

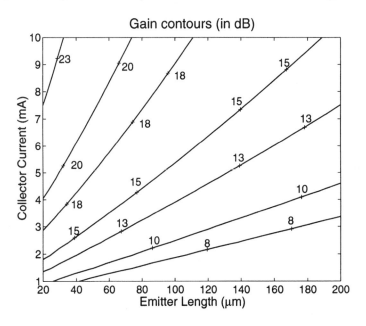

Figure 8.25 Gain contours as a function of emitter length and collector current for 2-GHz LNA. The input impedance is matched at all the design points.

Figure 8.25 shows gain contours in the design space. For a given gain requirement (e.g., 15 dB), the design point should be located above the corresponding gain contour. The contours are nearly straight lines for the following reasons. Because the input impedance is matched to the 50-Ω source, the input current is always $i_b = v_s/2R_s$. The output current is $h_{21} \times i_b$, and H_{21} increases with f_T, which itself increases with J_C. Hence, the output current (and hence gain) increases with J_C. This occurs as L_E decreases for a fixed I_C, or as I_C increases for a fixed L_E.

Figure 8.26 shows $IIP3$ contours in the design space. The variation of $IIP3$ with L_E and I_C is more complicated than the variation of noise figure and gain. Also shown are the design space for $NF \leq 1.2$ dB (within the dashed line), and the

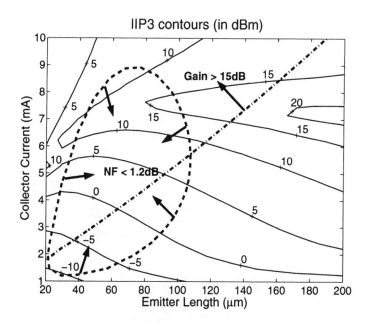

Figure 8.26 *IIP3* contours as a function of emitter length and collector current for input impedance matched LNA design.

design space for gain $\geq$ 15 dB (above the dashed-dotted line). Within a common design space of $NF \leq 1.2$ dB and gain $\geq$ 15 dB, *IIP3* changes from approximately -5 to 15 dBm. The optimum design point for *IIP3* is $L_E = 80$ μm and $I_C = 7.5$ mA. The resulting *IIP3* is above 15 dBm. The noise figure obtained is 1.15 dB, because of the slow variation of *NF* near the design point that minimizes *NF*. It is worth noting that *IIP3* varies much more rapidly with both I_C and L_E than *NF*, and hence is a more sensitive metric to optimize.

If one chooses the design point for minimum *NF* ($I_C = 4$ mA, $L_E = 60$ μm), the *IIP3* obtained is only 0 dBm. The optimum design point for *IIP3* is thus a better choice, because noise figure is still near its minimum, while *IIP3* is significantly higher (by 15 dB). For the optimized design of $L_E = 80$ μm and $I_C = 7.5$ mA, *IIP3* = 15.8 dBm, gain = 18 dB, *NF* = 1.15 dB, and $|s11| < -20$ dB. If the power constraint is tightened to $I_C \leq 5.5$ mA, an *IIP3* of 5 dBm can be obtained at $I_C = 5.5$ mA, $L_E = 50$ μm, with a near minimum noise figure of 1.08 dB.

The I_C dependence of *IIP3* can be better understood from Figure 8.27. A peak of *IIP3* can be observed for each emitter length. An *IIP3* peak occurs at $I_{ce} \simeq 6$ mA for $L_E = 40$ μm. Past this point, a further increase in collector current degrades *IIP3*. The peaks are physically a result of the maximization of the cancellation

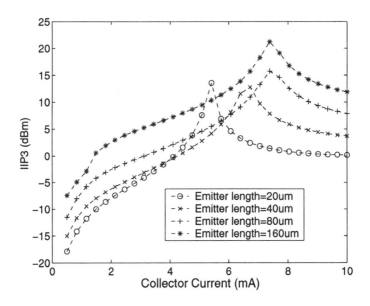

Figure 8.27 *IIP3* as a function of I_C in 0.5 × 20, 0.5 × 40, 0.5 × 80, and 0.5 × 160 μm^2 SiGe HBTs at 2 GHz. The input impedance is matched as I_C varies.

between the various individual nonlinearities, as can be seen from Figure 8.28 for $L_E = 40$ μm, where the *IIP3* for each individual nonlinearity is shown.

Across most of the I_C range used, the overall *IIP3* (∇) is higher than at least one of the individual *IIP3* contributions due to: 1) the g_m (denoted by ×) or the $I_{CE} - V_{BE}$ nonlinearity; 2) the g_{be} (o) or $I_{BE} - V_{BE}$ nonlinearity; and 3) the c_{be} (+) or $C_{BE} - V_{BE}$ nonlinearity. Individual nonlinearities partially cancel each other, producing an overall *IIP3* that is better than the individual *IIP3*. The cancellation here is much stronger than observed in the single transistor amplifier discussed above. Interestingly, no peak is observed for the individual nonlinearity *IIP3* contributions, while a clear peak is observed for the overall *IIP3*. This can only be explained by a maximization of the cancellation effects between the various nonlinearities. Increasing the collector current has two effects on *IIP3*: 1) it monotonically decreases the impedance in the EB junction, thereby reducing the first-order (linear) *ac* voltage on the EB junction, which in turn controls the g_m, g_{be}, and c_{be} nonlinearities. As a result, the *IIP3* calculated by turning on these individual nonlinearities increases monotonically with I_C. And 2), it changes the cancellation between the individual nonlinearities, which is maximized at a given collector current. The resulting increase in *IIP3* leads to a peak in the overall *IIP3*.

Linearity

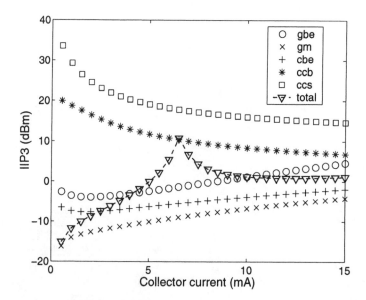

Figure 8.28 Individual contributions and overall *IIP3* as a function of I_C ($L_E = 40$ μm). The input impedance is always matched as I_C varies.

The distortions generated by each of the individual nonlinear sources partially cancel, making the overall distortion lower than for the individual distortions. The observed nonmonotonic behaviour of *IIP3* is in contrast to the conventional wisdom that *IIP3* improves monotonically with increasing collector current [12]. This is because L_e and L_b are adjusted for input impedance matching as the collector current increases. In a similar fashion, *IIP3* peaks as L_E varies, as shown in Figure 8.29.

8.5.2 Design Equations

The design optimization presented above can also be performed using a set of analytical equations [13]. Such an analytical scheme provides additional intuitive insight into the design trade-offs, and can be used to reduce simulation time. The input to those equations are β, f_T, τ_f, etc., which can be obtained from either measured data or from a device model.

A number of assumptions were made in deriving the traditional L_b and L_e equations for impedance matching, including $\beta \cdot f \gg f_T$ and negligible Miller effect. In a high f_T SiGe HBT technology, however, $\beta \cdot f \gg f_T$ is often not valid. The following equations produce much better agreement with numerical simulation

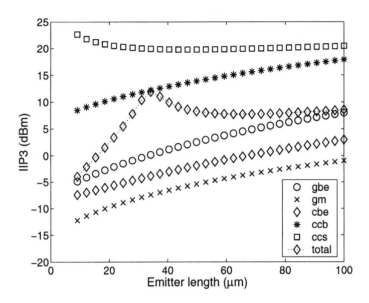

Figure 8.29 *IIP3* as a function of L_E ($I_C = 6$ mA). The input impedance is always matched.

results (e.g., using ADS), and can be used either as an initial guess for simulation or even as a replacement for simulation

$$L_e = \frac{R_s}{2\pi f_T} - \frac{2\pi f_T(1/g_{be} - R_s)}{\omega^2 \beta^2}, \tag{8.87}$$

and

$$L_b = \frac{2\pi f_T(1/g_{be} - R_s)}{\omega^2 \beta} - L_e, \tag{8.88}$$

where $\omega = 2\pi f$ and $g_{be} = g_m/\beta$.

A simplified noise figure equation can be derived by assuming: 1) the base resistance, the CB capacitance, and the Early effect are all negligible as far as impedance is concerned; and 2) the cascode stage's noise contribution remains negligible. The only impact of r_b is as a thermal noise source. The total noise figure has three contributions from the base current shot noise $\bar{i}_b$, collector current shot noise $\bar{i}_c$, and the base thermal noise $\bar{v}_b$, and is given by

$$NF = 10\log_{10}(1 + n_{ib} + n_{ic} + n_{vb}), \tag{8.89}$$

where n_{ib}, n_{ic}, and n_{vb} are the ratios of the noise power produced by $\bar{i}_b$, $\bar{i}_c$, and $\bar{v}_b$ to the noise power due to the source $\bar{v}_s$, and

$$n_{ib} = \frac{(g_{be}R_s)^2 + [B(1 - g_{be}R_s)]^2}{2g_{be}R_s},$$

$$n_{ic} = \frac{4(g_{be}R_s)^2 + [g_{be}R_s/B + B(1 - g_{be}R_s)]^2}{2g_m R_s},$$

$$n_{vb} = \frac{r_b}{R_s}, \tag{8.90}$$

where $B = 2\pi f_T/\omega \beta$ [14]. Note that the L_b and L_e values needed for input impedance matching are used in deriving (8.90). For a typical LNA design, I_C is less than 10 mA, and $g_{be}R_s \ll 1$ for a 50-Ω R_s. Equation (8.90) can be further simplified to

$$n_{ib} = \frac{(g_{be}R_s)^2 + B^2}{2g_{be}R_s},$$

$$n_{ic} = \frac{4(g_{be}R_s)^2 + [g_{be}R_s/B + B]^2}{2g_m R_s},$$

$$n_{vb} = \frac{r_b}{R_s}. \tag{8.91}$$

At low injection, and for $f = 2$ GHz, $B \ll 1$, but B is an increasing function of J_C, g_{be} and g_m are increasing functions of I_C, and r_b is a decreasing function of emitter length. For a fixed L_E, the n_{ic} term decreases with increasing I_C, while the n_{ib} term increases as with increasing I_C, thus producing a valley. In a similar fashion, with increasing L_E, NF first decreases, then reaches a minimum, and finally increases again. Therefore, the noise figure contours are closed in shape, with the lowest NF at the minimum, as can be seen from Figure 8.24.

Assuming that the Miller effect is negligible for the common-emitter stage, and that the current gain of the common-base stage is close to unity, an analytical gain equation can be derived. Since the input impedance is matched to R_s, the input current i_b is equal to $v_s/2R_s$. Thus, the current gain of the circuit is equal to

$$\frac{\beta g_{be}}{g_{be} + j\omega C_{be}} = \frac{\beta}{1 + j\omega \beta/2\pi f_T},$$

and the transducer power gain can be written as

$$G = \frac{\beta^2 R_L}{[1 + (\omega \beta/2\pi f_T)]^2 R_s}, \tag{8.92}$$

where R_L is the load resistance. According to (8.92), as J_C increases, f_T increases, and thus the gain increases. Therefore, the gain increases with increasing I_C for fixed emitter length, and decreases with increasing emitter length for a fixed I_C.

Simulation results have shown that the nonlinearities due to collector-substrate capacitance, impact ionization, and CB capacitance are negligible for a cascode SiGe HBT LNA. We can therefore neglect these nonlinearities here and derive an analytical equation for $IM3$, which relates to $IIP3$ by $IIP3 = P_{in} - IM3/2$. Even though the C_{CB} nonlinearity is not significant, the role of C_{CB} as a linear circuit element cannot always be neglected, particularly at low J_C. We therefore include C_{CB} as a linear circuit element in the derivation. Denoting the two frequencies by $s_1 = j\omega_1$ and $s_2 = j\omega_2$, and their difference by $\Delta s = s_1 - s_2$, $IM3$ is obtained as

$$|IM3| \simeq \left| \frac{3}{4} \cdot C(s_1, s_2) \cdot L(2s_1 - s_2) \cdot (1 - G(2s_1) - 2G(\Delta s)) \right|, \tag{8.93}$$

where

$$C(s_1, s_2) = \frac{q^2}{6(kT)^2} \cdot v_{be,1}(s_1) \cdot v_{be,1}(-s_2)$$

$$\simeq \frac{q^2}{6(kT)^2} \cdot |v_{be,1}^2(s_1)|. \tag{8.94}$$

Here, $v_{be,1}(s)$ is the first-order voltage drop across the EB junction (v_{be}) at frequency s

$$v_{be,1}(s) = \frac{v_s \cdot kT/q}{B'(s) + A'(S) \cdot I_C}, \tag{8.95}$$

$$A'(s) = A(s) + E(s) \cdot Z_b, \tag{8.96}$$

$$B'(s) = B(s) + F(s) \cdot Z_b \cdot kT/q, \tag{8.97}$$

$$A(s) = \left(s \cdot \tau_f + \frac{1}{\beta} \right) \cdot (Z_b + Z_e) + Z_e, \tag{8.98}$$

$$B(s) = kT/q \cdot [1 + s \cdot C_{te}(Z_b + Z_e)], \tag{8.99}$$

$$E(s) = \frac{[(1 + 1/\beta + s \cdot \tau_f) \cdot Z_e + r_e] \cdot s \cdot C_{cb}}{1 + s \cdot C_{cb} \cdot r_e}, \tag{8.100}$$

$$F(s) = \frac{(1 + s \cdot C_{te} \cdot Z_e) \cdot s \cdot C_{cb}}{1 + s \cdot C_{cb} \cdot r_e}, \tag{8.101}$$

$$Z_b(s) = s \cdot L_b + R_s, \tag{8.102}$$

$$Z_e(s) = s \cdot L_e, \tag{8.103}$$

$$r_e = 1/g_m = qI_C/kT, \tag{8.104}$$

where v_s is the RF source voltage, R_s is the source resistance, τ_f is the transit time, C_{te} is the EB depletion capacitance, C_{cb} is the CB capacitance, and

$$L(s) = \frac{(kT/q) \cdot K(s) \cdot G(s)}{(1 - E(s)) \cdot I_C + F(s) \cdot kT/q}, \quad (8.105)$$

$$G(s) = \frac{A'(s) \cdot I_C}{B'(s) + A'(s) \cdot I_C}, \quad (8.106)$$

where

$$K(s) = (1 - E(s)) \cdot \frac{B'(s)}{A'(s) \cdot kT/q} + F(s). \quad (8.107)$$

Here, $C(s_1, s_2)$ and $L(2s_1 - s_2)$ are monotonic functions of emitter length and I_C, and *IIP3* peaks with varying emitter length or I_C as the cancellation term $|(1 - G(2s_1) - 2G(\Delta s))|$ is minimized. Figure 8.30 shows the three terms: $|C(s_1, s_2)|$, $|L(2s_1 - s_2)|$, and $|(1 - G(2s_1) - 2G(\Delta s))|$ together with the total *IM3* for $v_s = 1$ V, and *IIP3* for the input-impedance matched amplifier. The minimum of the cancellation term and the minimum of *IM3* (or the maximum of *IIP3*) occurs at the same value of I_C, suggesting that the overall *IM3* variation is dominated by the variation of the cancellation term (denoted by *).

References

[1] D.D. Weiner and J.E. Spina, *Sinusoidal Analysis and Modeling of Weakly Nonlinear Circuits*, Van Nostrand Reinhold, 1980.

[2] P. Wambacq and W. Sansen, *Distortion Analysis of Analog Integrated Circuits*, Kluwer Academic, New York, 1998.

[3] S. Narayanan, "Transistor distortion analysis using volterra series representation," *Bell System Technical Journal*, vol. 46, no. 3, pp. 991-1024, May-June 1967.

[4] H. Jos, "Technology developments driving evolution of celluar phone power amplifiers to integrated RF front-end modules," *IEEE J. Solid-State Circ.*, vol. 36, pp. 1382-1389, 2001.

[5] G. Niu et al., "Measurement of collector-base junction avalanche multiplication effect in advanced UHV/CVD SiGe HBT's," *IEEE Trans. Elect. Dev.*, vol. 46, pp. 1007-1015, 1999.

[6] G. Niu et al., "RF linearity characteristics of SiGe HBTs," *IEEE Trans. Micro. Theory Tech.*, vol. 49, pp. 1558-1565, 2001.

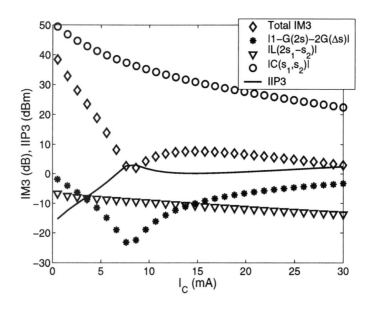

Figure 8.30 *IM3*, *IIP3*, and the three component terms as a function of I_C. The input impedance is matched.

[7] G. Niu et al., "Noise parameter optimization of UHV/CVD SiGe HBTs for RF and microwave applications," *IEEE Trans. Elect. Dev.*, vol. 46, pp. 1347-1354, 1999.

[8] G. Niu et al., "Intermodulation characteristics of UHV/CVD SiGe HBTs," *Proc. IEEE Bipolar/BiCMOS Circ. Tech. Meeting*, pp. 50-53, 1999.

[9] G. Gielen and W. Sansen, *Symbolic Analysis for Automated Design of Analog Integrated Circuits*, Kluwer, New York, 1991.

[10] S. Maas et al., "Intermodulation in heterojunction bipolar transistors," *IEEE Trans. Micro. Theory Tech.*, vol. 40, pp. 442-448, 1992.

[11] O. Shana'a et al., "Frequency-scalable SiGe bipolar RFIC front-end design," *Proc. IEEE Cust. Int. Circ. Conf.*, pp. 183–186, 2000.

[12] K.L. Fong, "High-frequency analysis of linearity improvement technique of common-emitter transconductance stage using a low-frequency-trap network," *IEEE J. Solid-State Circ.*, vol. 35, pp. 1249-1252, 2000.

[13] Q.Q. Liang et al., "Linearity Optimization and Design of SiGe HBT Cascode Low-Noise Amplifiers," *IEEE Trans. Circuit Theory Tech.*, in press.

[14] G. Niu et al., "Noise modeling and SiGe profile design tradeoffs for RF applications," *IEEE Trans. Elect. Dev.*, pp. 2037-2044, 2000.

Chapter 9

Temperature Effects

Bandgap engineering generally has a positive influence on the low-temperature characteristics of transistors. As will be shown, SiGe HBTs operate very well in the cryogenic environment (e.g., liquid nitrogen temperature = 77.3 K = -320°F = -196°C), an operational regime traditionally forbidden to Si BJTs. At present, cryogenic electronics represents a small but important niche market, with applications such as high-sensitivity cooled sensors and detectors, superconductor hybrid systems, space-born electronics, and eventually cryogenically cooled computers systems. While the large power dissipation associated with conventional bipolar digital circuit families such as ECL would likely preclude their widespread use in cooling-constrained cryogenic systems, the combination of cooled, low-power, scaled Si CMOS with SiGe HBTs offering excellent frequency response, low noise performance, radiation hardness, and excellent analog properties represents a unique opportunity for the use of SiGe HBT BiCMOS technology in cryogenic systems. Furthermore, independent of the potential cryogenic applications that may exist for SiGe HBT BiCMOS technology, all electronic systems must successfully operate over an extended temperature range (e.g. -55°C to 125°C to satisfy military specifications, and 0°C to 85°C for most commercial applications), and thus, understanding how Ge-induced bandgap engineering affects SiGe HBT device and circuit operation is important.

In this chapter we examine temperature effects in SiGe HBTs, by first reviewing the impact of temperature on bipolar transistor device and circuit operation, and then showing both theoretically and experimentally how temperature couples to SiGe HBT dc and ac performance, and how one optimizes SiGe HBTs specifically for 77-K operation. We then present unique helium temperature phenomena and discuss nonequilibrium base transport effects, and finally conclude by briefly considering the operation of SiGe HBTs at elevated temperatures.

9.1 The Impact of Temperature on Bipolar Transistors

The detrimental effects of cooling on homojunction bipolar transistor operation have been appreciated for many years [1]–[4]. While the precise dependence of Si BJT properties on cooling can be a strong function of technology generation, Si BJT device and circuit properties cooled to 77 K typically exhibit [5]:

- A modest increase (degradation) in the junction turn-on voltage with decreasing temperature (monotonic).

- A strong increase (improvement) in the low-injection transconductance with cooling (monotonic).

- A strong increase (degradation) in the base resistance with cooling (typically, quasi-exponential below about 200 K).

- A mild decrease (improvement) in parasitic transistor depletion capacitances (monotonic).

- A strong decrease (degradation) in β with cooling (quasi-exponential).

- A modest decrease (degradation) in frequency response with cooling, with f_T typically degrading more rapidly than f_{max} with decreasing temperature (monotonic below about 200 K).

- An increase (degradation) in ECL circuit delay with cooling (monotonic below about 200 K).

- The noise margin of current-switch-based circuits (e.g., ECL) increases (improves) with cooling (monotonic), allowing reduced logic swing operation.

To better understand the improvement in cryogenic performance of SiGe HBTs compared to Si BJTs, it is instructive to consider the fundamental origins of each of these dependencies [6, 7].

9.1.1 Current-Voltage Characteristics

If we consider the temperature dependence of the Gummel characteristics of a Si BJT (Figure 9.1), we observe that: 1) for fixed bias current, V_{BE} increases with cooling; and 2) the transconductance (slope of I_C with respect to V_{BE}) increases with cooling. For illustrative purposes, we consider an ideal Si BJT (constant

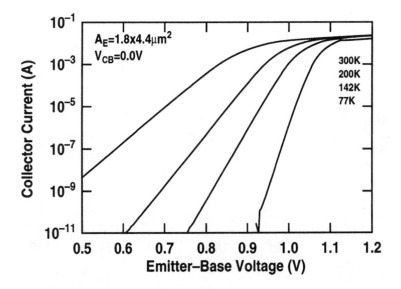

Figure 9.1 Collector current versus base-emitter voltage characteristics of a bipolar transistor as a function of temperature.

doping profiles, metal emitter contact) operating in low-injection under forward-active bias. In this case we can write collector current density as

$$J_C(T) \simeq \frac{q D_{nb}(T)\, n_{ib}^2(T)}{W_b(T)\, N_{ab}^-(T)}\, e^{qV_{BE}/kT} = J_{C0}(T)\, e^{qV_{BE}/kT}. \tag{9.1}$$

Let us first examine the origin of the increase in V_{BE} with cooling. From (9.1) we can write

$$V_{BE} = \frac{kT}{q}\, \ln\left\{\frac{J_C(T)}{J_{C0}(T)}\right\}. \tag{9.2}$$

If we fix the bias voltage ($V_{BE,bias}$) and bias current ($J_{C,bias}$) at a given temperature T (say 300 K), and ask how V_{BE} then changes with temperature, we can show that

$$\left.\frac{\partial V_{BE}}{\partial T}\right|_{J_C} = \frac{V_{BE,bias}}{T} - \frac{kT}{q}\frac{1}{J_{C0}}\frac{\partial J_{C0}}{\partial T}. \tag{9.3}$$

The temperature dependence of J_{C0} is dominated by $n_{ib}^2(T)$ and the acceptor freeze-out characteristics of the the base profile (i.e., $N_{ab}^-(T)$), which to first order are both thermally activated below about 150 K. Thus, we can capture the temperature dependence of J_{C0} by writing

$$J_{C0} \simeq \eta_o(T)\, T^3\, e^{-E_{go}/kT}\, e^{E_{R_{bi}}/kT}, \tag{9.4}$$

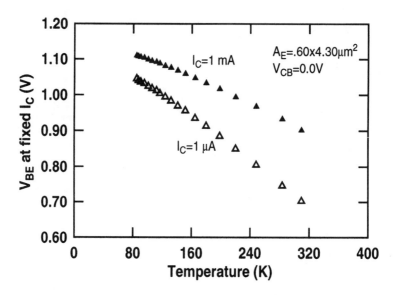

Figure 9.2 Current gain versus collector current as a function of temperature for an epitaxial base Si BJT.

where η_o accounts for the temperature dependence of D_{nb} and the carrier effective masses, and $E_{R_{bi}}$ is the activation energy of the carrier freezeout process in the base (see below). We find that

$$\frac{1}{J_{C0}}\frac{\partial J_{C0}}{\partial T} = \frac{1}{\eta_o T^3}\left\{T^3\frac{\partial \eta_o}{\partial T} + 3\eta_o T^2\right\} - \left\{\frac{E_{R_{bi}} - E_{go}}{kT^2}\right\}. \quad (9.5)$$

Since $E_{R_{bi}} \ll E_{go}$ and η_o is only a weak function of T, we have, finally, after substitution of (9.5) into (9.3)

$$\left.\frac{\partial V_{BE}}{\partial T}\right|_{J_C} \sim \frac{1}{T}\left\{V_{BE,bias} - \frac{E_{go}}{q}\right\}. \quad (9.6)$$

For any choice of $V_{BE,bias}$ we see that V_{BE} is a decreasing function of temperature (V_{BE} increases as T decreases), consistent with the data shown in Figure 9.2. Physically, it is the increase in the junction built-in voltage with decreasing temperature, caused by the exponential dependence of the intrinsic carrier density on the bandgap, which primarily determines $V_{BE}(T)$. For fixed V_{BE}, we can also determine how J_C depends on temperature, with the result that

$$\left.\frac{\partial J_C}{\partial T}\right|_{V_{BE}} \sim J_{C,bias}\left\{\frac{E_{go} - qV_{BE,bias}}{kT^2}\right\}. \quad (9.7)$$

Temperature Effects 375

From a circuit design viewpoint, this inherent increase in V_{BE} with cooling is clearly undesirable, since for constant bias current, $V_{BE}(T)$ translates directly to an increase in circuit power dissipation ($P = VI$) with cooling.

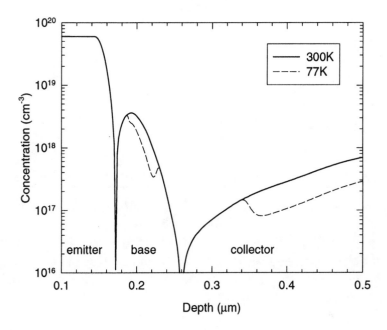

Figure 9.3 Simulated effects of carrier freeze-out on the doping profile of a bipolar transistor at 77 K.

9.1.2 Transconductance

We can also observe that the slope of I_C versus V_{BE} increases with cooling (Figure 9.1). This is a simple but important manifestation of the dependence of the low-injection transconductance on temperature. From (9.1) we have

$$g_m(T) = \frac{\partial I_C}{\partial V_{BE}} \simeq \frac{q}{kT} I_{C0} \, e^{qV_{BE}/kT} = \frac{qI_C(T)}{kT}, \tag{9.8}$$

and hence we can expect an improvement in g_m of roughly 3.9× in cooling from room temperature to liquid nitrogen temperature (i.e., 300 K / 77 K). This g_m increase is obviously an advantage for many analog circuits, but it is also important for low-temperature digital circuit applications, since it facilitates reduced logic-swing circuit operation, as discussed below.

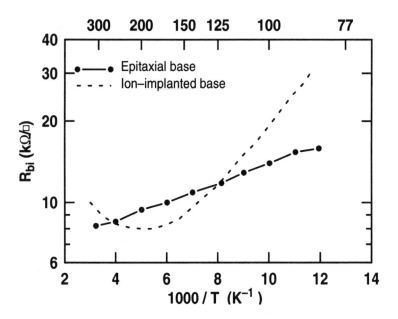

Figure 9.4 Measured freeze-out properties of the intrinsic base sheet resistance for two representative Si BJTs, one with an ion-implanted base and one with an epitaxial base.

9.1.3 Resistances and Capacitances

The large-signal switching performance of a transistor ultimately depends on the parasitic resistances and capacitances of the device. Fortunately, the emitter resistance and extrinsic base resistance are to first-order temperature independent in modern bipolar transistors. This is because the emitter and base polysilicon layers, and their thin out-diffusions into the Si to form the emitter and the extrinsic base, are very heavily doped and thus not subject to carrier freeze-out, and the respective mobilities are only weakly temperature dependent. The same is not true, of course, for the intrinsic base or the epi-collector, both of which contain regions doped below the Mott-transition. Base freeze-out is typically the more serious of the two for 77-K operation, since R_b strongly affects the transistor frequency response, dynamic switching performance, and the noise properties. Intrinsic base freeze-out is perhaps the most serious constraint faced in designing bipolar transistors for low-temperature operation, because the device parameters are tightly coupled to the base profile and it cannot be easily altered. One can calculate the effects of cooling on pinched base sheet resistance (R_{bi}) using,

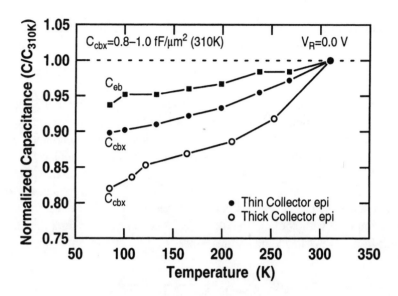

Figure 9.5 Measured CB and EB depletion capacitance as a function of temperature from representative Si and SiGe technologies.

$$R_{bi}(T) = \left\{ \int_0^{W_b(T)} q\,\mu_{pb}(x,T)\,p_b(x,T)\,dx \right\}^{-1}, \quad (9.9)$$

where $p_b(T)$ can be obtained in closed form [5]

$$p_b(T) \simeq \frac{-\theta + \left\{\theta^2 - 4(ge^\eta - C)(N_V N_{dc} - N_V N_{ab})\right\}^{1/2}}{2[ge^\eta - C]}, \quad (9.10)$$

where g is the acceptor spin degeneracy factor (g = 4 for boron in Si), C = 0.27, N_V is the valence band effective density-of-states, $\eta = qE_a/kT$, E_a is the acceptor ionization energy (E_a = 45 meV for boron in Si at low-doping), N_{dc} is the compensation doping level in the neutral base, and

$$\theta = N_V + g\,N_{dc}\,e^\eta - C(N_{dc} - N_{ab}). \quad (9.11)$$

The approximations used in deriving (9.10) hold well for realistic base profiles with peak base doping less than about 1×10^{19} cm^{-3} down to about 77 K. The mobility values can be obtained on a point-by-point basis from [8, 9] and (9.9) numerically integrated to obtain $R_{bi}(T)$. The result for realistic base profiles shows a quasi-exponential increase below about 200 K (Figure 9.3 and Figure 9.4), and is a very sensitive function of the peak base doping, particularly in strong-freeze-out below 77 K. One can measure (or calculate) a base freeze-out activation energy $E_{R_{bi}}$ according to

$$R_{bi}(T) = \delta(T) \, e^{E_{Rbi}/kT}. \tag{9.12}$$

Freeze-out in the neutral collector can be severe at 77 K (Figure 9.3), leading to device quasi-saturation, and premature roll-off of the current gain and frequency response at high-injection. For modern devices with selectively implanted collectors (SIC), this is typically not an overly serious design issue, particularly since the collector epi resistance is strongly decreased at high-injection levels. It should be noted as well that all donors in the CB space-charge region will remain ionized at 77 K, and hence the critical current density for the onset of Kirk effect in fact will typically improve with cooling, due to the modest increase in saturation velocity and built-in voltage at low temperatures, since

$$J_{C,Kirk}(T) \simeq q \, v_{sat}(T) \, N_{dc}^{+} \left\{ 1 + \frac{2 \, \epsilon_{Si} \, (V_{CB} + \phi_{bi}(T))}{q \, N_{dc}^{+} \, W_{epi}^{2}(T)} \right\}. \tag{9.13}$$

Thus, as long as R_E and R_C remain in check at 77 K, maintaining adequate current drive at low temperature is usually not a debilitating constraint.

Finally, we note that the metal interconnect resistances decrease strongly with cooling, as much as 6–10× between 300 K and 77 K for a typical Al-Cu or Cu metalization. While this resistance decrease has minimal impact on device properties, the circuit-level impact of this decreased interconnect resistance at low-temperatures can be profound, as discussed below.

The parasitic depletion capacitances (e.g., C_{CB}, C_{BE}, C_{SX}, etc.) will generally decrease (improve) with cooling, due to the increase in junction built-in voltage, since for a one-sided step junction,

$$C_{depl}(T) \simeq A \left\{ \frac{q \, \epsilon_{Si} \, N_{dc}^{+}}{q \, \phi_{bi}(T)} \right\}^{1/2}. \tag{9.14}$$

Experimental results from several different technologies are shown in Figure 9.5. Note again that within the space-charge region, the impurities will not freeze out, even if below the Mott transition. For the CB junction, which is the most important parasitic capacitance for switching performance due to Miller effect, C_{CB} typically decreases by 10–20% from 300 K to 77 K [6].

The same is not true for the junction diffusion capacitance, which for fixed bias current depends explicitly on the thermal voltage according to

$$C_{diff}(T) = \frac{q \tau_f(T) I_C}{kT}. \tag{9.15}$$

Hence, any τ_f degradation with cooling will be magnified by the decrease in kT as the temperature drops, negatively impacting the circuit gate delay at high injection where the minimum delay is reached.

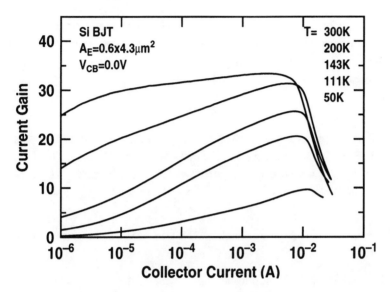

Figure 9.6 Measured current gain as a function of collector current for various temperatures for an epitaxial base Si BJT.

9.1.4 Current Gain

Since the base current density in a metal-contacted transistor can be written as

$$J_B(T) = \frac{qD_{pe}(T)\, n_{ie}^2(T)}{L_{pe}(T)\, N_{de}^+(T)}\, e^{qV_{BE}/kT}, \quad (9.16)$$

we can combine (9.1) and (9.16) to obtain

$$\beta_{ideal}(T) = \frac{qD_{nb}(T)\, L_{pe}(T)\, N_{de}^+(T)}{D_{pe}(T)\, W_b(T)\, N_{ab}^-(T)}\, e^{(\Delta E_{gb}^{app}-\Delta E_{ge}^{app})/kT}, \quad (9.17)$$

where ΔE_{gb}^{app} and ΔE_{ge}^{app} are the values of the doping-induced apparent bandgap narrowing in the base and emitter regions, respectively [10]. Because the emitter is much more heavily doped than the base (typically by 100×), the bandgap narrowing in the emitter will naturally dominate the temperature dependence of the current gain. Since the base profile in an ion-implanted Si BJT will contain portions doped below the Mott-transition (about 3×10^{18} cm^{-3} for boron in Si), the base will freeze out in a quasi-exponential fashion below about 200 K, significantly increasing the base resistance. While this R_{bi} increase with cooling will degrade the dynamic performance of the device, base freeze-out does in fact favorably affect the low-temperature current gain. The carrier mobilities are weak functions of temperature at the doping levels normally encountered in modern devices, and thus we can express the T-dependent current gain as

$$\beta_{ideal}(T) \simeq \alpha(T) \, e^{(E_{R_{bi}} + \Delta E_{gb}^{app} - \Delta E_{ge}^{app})/kT}, \quad (9.18)$$

where $E_{R_{bi}}$ is the (measurable) activation energy of the freeze-out of the base profile. In all Si BJTs designed for high speed operation at 300 K, β will degrade quasi-exponentially with cooling (Figure 9.6). It should be noted, however, from (9.18) that optimization of a Si BJT for 77 K operation must proceed along one of two lines:

- Increase the base doping enough to offset the bandgap narrowing in the emitter [11];

- Decrease the emitter doping enough to make it smaller than the bandgap narrowing in the base [12, 13].

Both techniques face fairly serious low-temperature design constraints. In the case of 1), the transistor will be prone to tunneling-induced base current nonidealities at low-injection, and in the case of 2), enhanced emitter charge storage and fabrication-imposed difficulties will be confronted. In practice, double-polysilicon, self-aligned Si BJTs that offer a compromise between the design approaches of 1) and 2) can yield reasonable dc performance at 77 K (though still degraded with respect to 300-K performance) [6, 14].

9.1.5 Frequency Response

Even a cursory examination of the dependence of the frequency response on temperature gives cause for concern, since for the cutoff frequency

$$f_T(T) = \frac{1}{2\pi} \left\{ \frac{kT}{qI_C} \Big(C_{EB}(T) + C_{CB}(T) \Big) + \tau_b(T) + \tau_e(T) \right.$$
$$\left. + \frac{W_{CB}(T)}{2v_{sat}(T)} + r_c(T) C_{CB}(T) \right\}^{-1}. \quad (9.19)$$

For fixed bias current, both depletion capacitances will decrease only slightly (good), while τ_b and τ_e will both increase strongly with cooling (bad), since

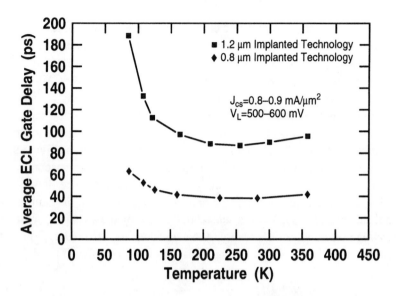

Figure 9.7 Unloaded ECL gate delay as a function of temperature for two ion-implanted base Si BJT technology generations.

$$\tau_{b,Si}(T) = \frac{W_b^2(T)}{2D_{nb}(T)} = \frac{qW_b^2(T)}{2kT\mu_{nb}(T)}, \quad (9.20)$$

where μ_{nb} increases only weakly with cooling since the base is heavily doped and thus cannot offset the factor of kT, $\tau_e(T) \propto 1/\beta(T)$, and β decreases quasi-exponentially with cooling. In addition, enhanced carrier trapping on frozen-out acceptor sites can further degrade the base transit time [15], although this effect is not significant in modern Si BJTs [6]. The story for the more-circuit-relevant f_{max} is even worse, given the strong base resistance increase at low temperatures due to freeze-out, since

$$f_{max}(T) = \sqrt{\frac{f_T(T)}{8\pi R_b(T) C_{CB}(T)}}. \quad (9.21)$$

9.1.6 Circuit Performance

Given the strong correlation between digital circuit speed at high current levels and the f_{max} of the transistor (refer to the discussion in Chapter 5), it is not surprising that unloaded ECL gate delay for Si BJTs typically degrades with cooling. This degradation is physically driven by the strong increase in base resistance together with the decrease in f_T with cooling. At lower current levels, where the ECL delay is increasingly dominated by parasitic capacitances, one might expect the situation to improve somewhat, and it is indeed the case that one can obtain faster Si BJT ECL delay at 77 K than at 300 K, at least at very low current levels. This situation is not particularly comforting, since the major advantage of ECL over CMOS rests in its speed advantage, and it is clearly true that ECL speed is severely compromised by cooling. As conventional Si BJT technologies evolve, one does expect their 77 K ECL performance also to improve, since the base doping naturally increases with optimized (300 K) Si BJT scaling. This can be clearly seen in (Figure 9.7).

All is not lost, however. The inherent improvement in device transconductance with cooling is favorable for bipolar digital circuit operation since, intuitively, a steeper I-V characteristic (higher g_m) means that we can switch the same current through the device (I_{on}/I_{off}), using substantially less voltage swing (V_L), provided the operating noise margins of the circuit remain adequate. An analysis of a differential current-switch (i.e., ECL) average noise margin shows that [5]

$$\overline{NM}(T)\Big|_{V_L} \simeq \frac{V_L}{2} - \frac{kT}{q} \ln\left(\frac{qV_L}{kT}\right), \quad (9.22)$$

and hence the noise margin inherently increases with cooling. The system-level delay of a loaded digital circuit such as ECL can be written as

$$\tau_{system}(T) \sim \tau_{device}(T) + \frac{C_W V_L}{I_{CS}}, \quad (9.23)$$

where C_W is the wire-loading capacitance, V_L is the logic swing, and I_{CS} is the total switch current. Hence, even if the unloaded, device-level switching speed of the circuit degrades, a decrease in logic swing can be used to achieve improved system-level performance [6, 14].

Even considering this reduced logic swing advantage, however, a careful examination of how temperature influences a Si BJT reveals a moderately depressing story, and despite the natural improvement in Si BJT cryogenic properties with technology scaling, is responsible in large measure for the widely held belief that Si BJTs are not useful for cryogenic applications. Given this discussion, the greatest disadvantage faced by the Si BJT in the low-temperature environment can be traced to the tight coupling of the base profile design to the transistor performance metrics, and the inevitable limitations this imposes on device design. While these constraints are already limiting at 300 K, they become much more severe as the temperature drops. It is not surprising, then, that possessing a device design approach which inherently decouples the device metrics from the base profile design will prove very advantageous for cryogenic applications. Bandgap engineering using SiGe thus offers a natural advantage for cryogenic applications of bipolar transistors.

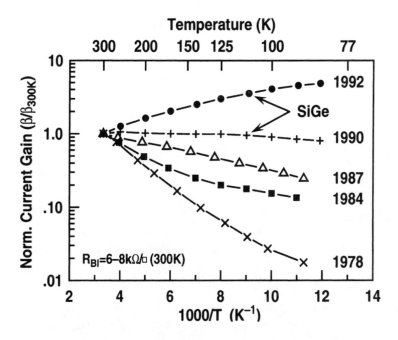

Figure 9.8 Evolution of current gain temperature dependence with Si-based bipolar technology generation. The last two generations are SiGe HBT technologies.

9.2 Cryogenic Operation of SiGe HBTs

Intuitively, we expect that band-edge effects (either E_C or E_V or both) induced by bandgap engineering will generally couple strongly to bipolar transistor properties. This strong coupling is physically the consequence of the fact that the bipolar transistor is a minority carrier device, and hence the terminal currents are proportional to n_{i0}^2 via the Shockley boundary conditions, with n_{i0}^2 being in turn proportional to the exponential of the bandgap. Hence, changes to the bandgap will couple exponentially to the currents. Furthermore, from very general statistical mechanical considerations, these bandgap changes will inevitably be divided by the thermal energy (kT), such that a reduction in temperature will greatly magnify any bandgap changes. Not surprisingly, then, even a cursory examination of the SiGe HBT device equations suggests that both the *dc* and *ac* properties of SiGe HBTs should be favorably affected by cooling [16, 17]. In fact, the thermal energy (kT), *in every instance*, is arranged in the SiGe HBT equations such that it favorably affects the low-temperature properties of the particular performance metric in question, be it

384 Silicon-Germanium Heterojunction Bipolar Transistors

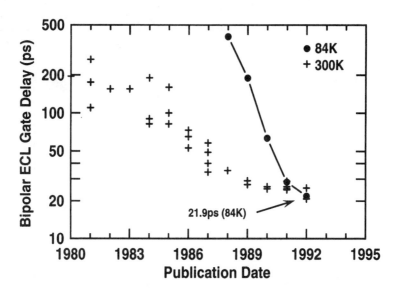

Figure 9.9 Evolution of unloaded ECL gate delay at 310 K and 84 K with Si-based bipolar technology generation. The last two generations are SiGe HBT technologies.

$\beta(T)$, $f_T(T)$, or $V_A(T)$.

9.2.1 Evolutionary Trends

The beneficial role of temperature in SiGe HBTs can be used to easily offset the inherent bandgap narrowing induced degradation in current gain of a Si BJT to achieve viable *dc* operation down to 77 K, even for a SiGe HBT that has not been optimized for the cryogenic environment. Figure 9.8 shows the evolution of peak current gain as a function of reciprocal temperature from early Si BJT technologies circa 1978 to SiGe technologies circa 1992. Clearly, the addition of Ge-induced bandgap engineering enables functional current gain down at least to 77 K with minimal effort. From a dynamic point of view, the Ge-grading-induced base drift field provides a means to offset the inherent τ_b degradation associated with cooled Si BJTs, yielding an f_T that does not degrade with cooling. Since the reduced thermal cycle nature of epitaxial growth techniques are generally more conducive to maintaining thinner, more heavily doped base profiles than conventional ion-implanted bases used in modern Si BJTs, it is fairly straightforward to control base freeze-out in SiGe HBTs, at least down to 77 K, and hence R_b at cryogenic temperatures can be more easily controlled. If f_T and R_b do not degrade significantly with cooling, then achieving respectable circuit performance down to 77 K becomes a reality unknown to Si BJT technologies.

Figure 9.9 shows the evolution of unloaded ECL gate delay as a function of publication date. As expected, optimized 300-K technology scaling successfully improved circuit speed over time. More surprising, perhaps, is that the rate of improvement in low-temperature performance was significantly faster. The 1991 and 1992 cryogenic data points are for SiGe HBT technologies, and clearly demonstrate that one can no longer out of hand dismiss Si-based bipolar technologies for the cryogenic applications. SiGe can thus be viewed as an effective means to extend Si-based bipolar technology to the cryogenic environment (with little or no effort). This scenario is particularly appealing if we consider state of the art SiGe HBT BiCMOS technologies, since Si CMOS also performs well down to 77 K, and provides a major advantage in the reduction in power dissipation, an often serious constraint given the limited efficiency of cryocoolers. While it is unlikely that one would develop SiGe technology explicitly for cryogenic applications, if (as is the case), one could simply take a room temperature-optimized SiGe technology and operate it at low temperatures without serious modification, that prospect might prove cost effective. With the present trend towards reduced-temperature operation of CMOS-based high-end servers as a performance and reliability enhancement vehicle (currently at 0°C to -40°C and going lower), the appeal of SiGe HBT BiCMOS technologies for the cryogenic environment may naturally grow over time, since HBTs can provide numerous advantages over CMOS in analog, RF, heavily loaded digital, and high-speed driver/receiver applications.

9.2.2 SiGe HBT Performance Down to 77 K

It is instructive to first consider the effects of cooling on the performance of a SiGe HBT that has not been optimized in any way for cryogenic operation. This 300-K optimized design will be referred to as an "i-p-i" SiGe HBT, since it is fabricated with lightly doped intrinsic ("i") spacers on both the EB and CB side of the neutral base (refer to discussion in Chapter 2). Compared to a comparably constructed Si BJT, $\beta(T)$ in a SiGe HBT should increase exponentially with decreasing temperature, since

$$\left.\frac{\beta_{SiGe}}{\beta_{Si}}\right|_{V_{BE}} \simeq \left\{ \frac{\tilde{\gamma}\tilde{\eta}\Delta E_{g,Ge}(grade)/kT \, e^{\Delta E_{g,Ge}(0)/kT}}{1 - e^{-\Delta E_{g,Ge}(grade)/kT}} \right\}. \tag{9.24}$$

Figure 9.10 shows measured and theoretical results for an i-p-i SiGe HBT down to 77 K. In this case, β_{Si} decreases from 25 at 300 K to 9.5 at 77 K, whereas β_{SiGe} increases from 67 at 300 K to 205 at 77 K. As expected, a quasi-exponential increase in the SiGe-to-Si current gain ratio with decreasing temperature is observed.

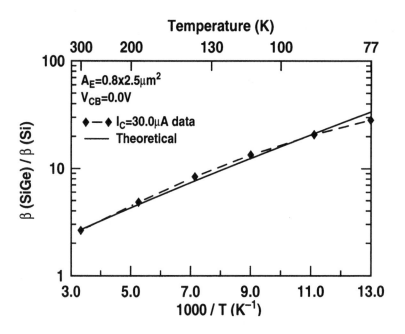

Figure 9.10 Measured and calculated SiGe-to-Si current gain ratio as a function of reciprocal temperature for a comparably constructed i-p-i SiGe HBT and i-p-i Si BJT (reproduced from Chapter 4 for convenience).

(Given the functional dependence of β on the exponential of the various band-offset parameters divided by kT, it is physically meaningful to show β as an Arrhenius plot, since a straight line indicates purely thermally activated behavior.) In addition, $V_A(T)$ and $\beta V_A(T)$ in a SiGe HBT should also increase exponentially with decreasing temperature compared to a comparably constructed Si BJT, since

$$\left.\frac{V_{A,SiGe}}{V_{A,Si}}\right|_{V_{BE}} \simeq e^{\Delta E_{g,Ge}(grade)/kT} \left[\frac{1 - e^{-\Delta E_{g,Ge}(grade)/kT}}{\Delta E_{g,Ge}(grade)/kT}\right], \quad (9.25)$$

and

$$\frac{\beta V_{A,SiGe}}{\beta V_{A,Si}} = \widetilde{\gamma\eta} e^{\Delta E_{g,Ge}(0)/kT} e^{\Delta E_{g,Ge}(grade)/kT}. \quad (9.26)$$

Figure 9.11 and Figure 9.12 show measured and theoretical results for an i-p-i SiGe HBT down to 77 K for V_A and βV_A, respectively. In this case, $V_A(Si)$ decreases from 33 V at 300 K to 14 V at 77 K, whereas $V_A(SiGe)$ increases from 65 V at 300 K to 265 V at 77 K. Again, as expected, a quasi-exponential increase in the SiGe-to-Si ratios for both parameters with cooling is observed. The large improvement

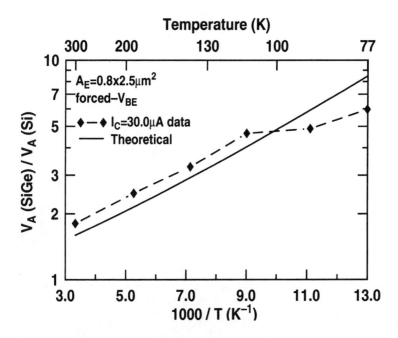

Figure 9.11 Measured and calculated SiGe-to-Si Early voltage ratio as a function of reciprocal temperature for a comparably constructed i-p-i SiGe HBT and i-p-i Si BJT (reproduced from Chapter 4 for convenience).

in V_A and βV_A between 300 K and 77 K suggests that high-speed analog circuits such as amplifiers and data converters built from SiGe HBTs may be attractive for certain cryogenic applications.

The anticipated temperature dependence of the frequency response of a SiGe HBT can be gleaned from $f_T(T)$ (9.19). Given that

$$\frac{\tau_{b,SiGe}}{\tau_{b,Si}} = \frac{2}{\widetilde{\eta}} \frac{kT}{\Delta E_{g,Ge}(grade)}$$
$$\cdot \left\{ 1 - \frac{kT}{\Delta E_{g,Ge}(grade)} \left[1 - e^{-\Delta E_{g,Ge}(grade)/kT} \right] \right\}, \quad (9.27)$$

and

$$\frac{\tau_{e,SiGe}}{\tau_{e,Si}} \simeq \frac{J_{C,Si}}{J_{C,SiGe}} = \frac{1 - e^{-\Delta E_{g,Ge}(grade)/kT}}{\widetilde{\gamma\eta} \frac{\Delta E_{g,Ge}(grade)}{kT} e^{\Delta E_{g,Ge}(0)/kT}}, \quad (9.28)$$

and both are favorably influenced by cooling, we might naively expect that the influence of the graded SiGe base might be sufficient to overcome the inherent electron diffusivity degradation on τ_b with cooling. Figure 9.13 shows measured

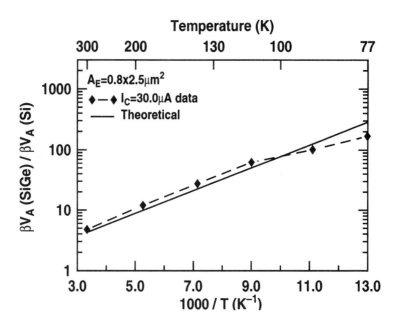

Figure 9.12 Measured and calculated SiGe-to-Si current gain – Early voltage product ratio as a function of reciprocal temperature for a comparably constructed i-p-i SiGe HBT and i-p-i Si BJT (reproduced from Chapter 4 for convenience).

results for the peak f_T for both an i-p-i SiGe HBT and a comparably doped epitaxial base i-p-i Si BJT down to 77 K. While the f_T degradation of f_T in this epitaxial Si BJT is not overly severe, clearly the Ge-induced improvements in both τ_b and τ_e at low temperature are sufficient to produce a peak f_T in the SiGe HBT that is larger at 77 K than at 300 K.

9.2.3 Design Constraints at Cryogenic Temperatures

While it has been demonstrated that SiGe HBTs designed for room temperature operation function acceptably down to 77 K, second-order design constraints do, nonetheless, exist, and can impact profile optimization [18, 19]. The first such constraint centers on the base current, and its impact on the current gain at low injection. While conventional Shockley-Read-Hall (SRH) recombination exponentially decreases with cooling, thereby effectively eliminating reverse leakage in the collector-base junction, the same is not true of carrier tunneling processes, whether they are band-to-band or trap-assisted. Given that the EB junction of high-speed

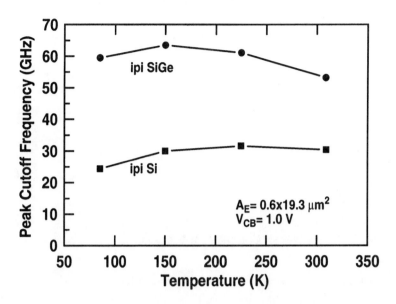

Figure 9.13 Measured peak cutoff frequency as a function of temperature for a comparably constructed i-p-i SiGe HBT and i-p-i Si BJT.

bipolar transistors (either Si or SiGe) are typically quite heavily doped (often in the vicinity of 1×10^{18} cm^{-3}), the doping induced electric field is high, and can result in substantial parasitic tunneling leakage. While this is generally easily designed around in 300-K designs, it is more problematic at low temperatures, given that the collector and base currents decrease strongly at fixed V_{BE} as the temperature drops (refer to Figure 9.1). In this case, as the base current decreases with cooling, any tunneling-induced leakage will remain roughly constant, hence uncovering a parasitic leakage foot on the base current (this effect can be clearly seen in Figure 9.6). This parasitic base leakage current can severely limit the current gain at low injection at cryogenic temperatures. Thus, as a rule of thumb, it can be safely stated that the ideality of the base current of a high-performance Si or SiGe bipolar transistor will never improve with cooling! If the base current is ideal (*i.e.*, $e^{qV_{BE}/kT}$) down to a picoamp at 300 K, it may be ideal only to a nanoamp at 77 K. If it is even modestly nonideal at 300 K, it will be quite leaky at 77 K. How serious a limitation this leakage is depends strongly on the circuit application. In digital logic such as ECL, for instance, it is not an issue, given that the devices are biased well out of the leakage regime, and β doesn't strongly couple to circuit speed. For more sensitive analog circuits, however, it can in principle require careful design consideration. As discussed below, one can optimize a SiGe HBT to reduce this leakage

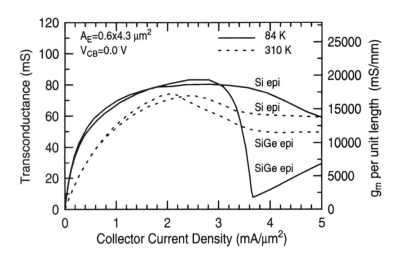

Figure 9.14 Extrinsic transconductance at 310 K and 84 K for an i-p-i SiGe HBT and a comparable i-p-i Si BJT.

effect, a feat much more easily accomplished using epitaxial growth rather than ion-implantation for the base layer formation.

More worrisome than the base current at low temperatures, however, is the enhancement of high-injection, heterojunction barrier effects with cooling (refer to Chapter 6 for a detailed discussion of barrier effects in SiGe HBTs). As stated above, band-edge effects in bipolar transistors generally couple very strongly to the device properties, and barrier effects are no exception. In this case, given that barrier effects necessarily exist in all practical SiGe HBTs, cooling will make the situation decidedly worse. The consequences of barrier effects, as at room temperature, include a premature roll-off in both β and f_T at high J_C, and a limitation on maximum output current drive. What is different in the context of cryogenic operation, however, is that while a well-designed 300-K SiGe HBT may not show any clear evidence of barrier effect at 300 K, it will certainly show evidence of it at 77 K, and its impact on device performance will be correspondingly worse. That is, the design margin for 77-K operation is in essence narrower, always an undesirable situation. As discussed in Chapter 6, the device transconductance is a useful tool for assessing barrier effects in SiGe HBTs. As can be seen in Figure 9.14, a comparison of g_m between comparably designed i-p-i SiGe HBTs and i-p-i Si BJTs clearly shows that while g_m at low J_C increases with cooling as expected, a dramatic drop in g_m at a higher critical current density close to that of Kirk effect can be observed in the SiGe HBT. Fortunately, it is also true that this critical on-

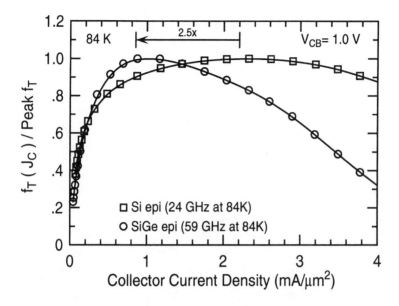

Figure 9.15 Normalized cutoff frequency at 84 K for an i-p-i SiGe HBT and a comparable i-p-i Si BJT.

set current density in fact increases with cooling, consistent with the fact that the saturation velocity rises at low temperatures, thus delaying Kirk effect until higher J_C. As discussed below, this result can be traded off to optimize SiGe HBTs for 77 K operation. One would also expect that barrier effect would have a serious impact on transistor dynamic response, given that enhanced charge storage in the base couples strongly to f_T. Figure 9.15 shows that while the peak f_T of the SiGe HBT at 84 K is 59 GHz compared to 24 GHz for the Si BJT, the current density at which this f_T value is reached is 2.5× lower in the SiGe device. This indicates that barrier effect is limiting the *ac* performance in this SiGe HBT at cryogenic temperatures, and hence explains the modest roll-off in peak f_T below 150 K seen in the SiGe HBT data in Figure 9.13. The approaches that can be used to design around barrier effects at cryogenic temperatures are the same as those outlined in Chapter 6, albeit with a narrower design margin than at 300 K.

9.3 Optimization of SiGe HBTs for 77 K

While conventional 300-K SiGe HBT designs will inherently function reasonably well down to 77 K, it remains to be seen whether a SiGe HBT designed *specifically* for 77 K operation can achieve significantly better device and circuit performance at 77 K than it has at 300 K, and what the design issues and trade-offs faced in achieving this goal would be.

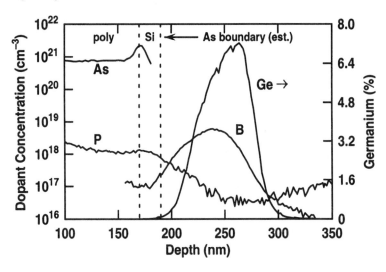

Figure 9.16 Doping and Ge profiles for a 77-K optimized emitter-cap SiGe HBT.

9.3.1 Profile Design and Fabrication Issues

To address the explicit optimization of a SiGe HBT for 77 K operation, a new profile design point and fabrication scheme is required [20]. In this case an epitaxial "emitter-cap" layer doped with phosphorus at about 1×10^{18} cm^{-3} was deposited in-situ in a UHV/CVD deposition tool on top of the SiGe-base to form the EB junction (Figure 9.16). This 77-K optimized SiGe HBT will be referred to as an epitaxial "emitter-cap" SiGe HBT [21]. Because EB carrier tunneling processes depend exponentially on the peak junction field, the lightly doped emitter is expected to minimize the parasitic EB tunneling current compared to a conventional i-p-i SiGe HBT design. In addition, the increase in carrier saturation velocity with cooling, as well as the presence of velocity overshoot in the CB space-charge region at 77 K, results in an onset current density of base push-out (Kirk effect) that is about 50% larger at 77 K than at 300 K [18]. Thus, compared to a 300-K design, the collector doping level can be decreased in an optimized 77-K profile. In this case, the doping level at the metallurgical CB junction was lowered from 1×10^{17} cm^{-3} for the conventional i-p-i SiGe HBT design to about 2×10^{16} cm^{-3}, and ramped upward toward the subcollector to minimize freeze-out deep in the neutral collector. This 77-K collector profile is used to reduce the parasitic CB capacitance under

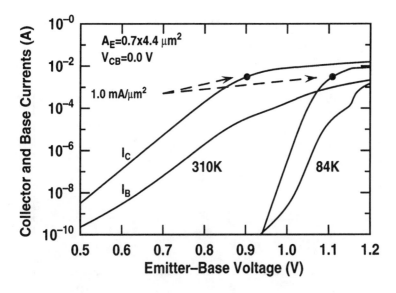

Figure 9.17 Gummel characteristics at 300 K and 77 K for a 77-K optimized emitter-cap SiGe HBT.

the constraint that the onset current density of the SiGe-Si heterojunction barrier be above the maximum operating current density of about 1.0 mA $/\mu m^2$.

To ensure a low emitter resistance, a 200-nm in-situ doped polysilicon contact was deposited on top of the composite EB profile (n-cap/p-SiGe). Because the arsenic out-diffusion from the heavily doped polysilicon layer is used only to contact the epitaxial phosphorus emitter and does not determine the metallurgical EB junction, only a very short rapid thermal annealing (RTA) step is required to activate and redistribute the emitter dopants, allowing the maintenance of a thin, heavily doped base. A metallurgical emitter-cap thickness of about 10 nm was achieved at the end of processing (estimated by subtracting the arsenic out-diffusion of the emitter poly from the total EB junction depth). The boron doping of the base profile was increased over a more conventional i-p-i SiGe design to improve its base freeze-out properties, and was deposited as a box 10 nm wide by 2.5×10^{19} cm^{-3}. At the end of processing the metallurgical base was about 75 nm wide with a peak concentration of about 8×10^{18} cm^{-3}, well above the Mott transition for carrier freeze-out. To minimize minority carrier charge storage in the emitter-cap layer, a large 77 K β is also desirable ($\tau_e \propto 1/\beta_{ac}$). Therefore, a trapezoidal Ge profile with 3–4% Ge at the EB junction (compared to about 0–1% for the i-p-i design) and ramping to 8.5% at the CB junction (compared to about 8.5% for the i-p-i design) was used. The resultant emitter-cap Ge profile was about 65 nm thick, and

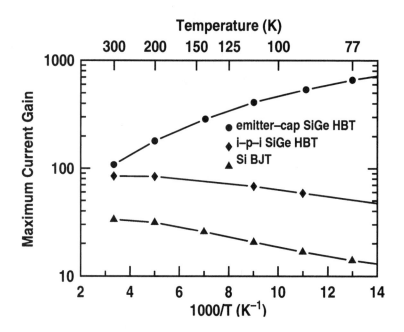

Figure 9.18 Maximum current gain as a function of reciprocal temperature for a 77-K optimized emitter-cap SiGe HBT, a conventional i-p-i SiGe HBT, and a comparable i-p-i Si BJT.

satisfied the thermodynamic stability criteria for UHV/CVD blanket films.

9.3.2 Measured Results

This 77-K SiGe design point yields a transistor with near-ideal Gummel characteristics at low temperature, with a maximum output current drive well above 1.0 mA/μm^2 at 84 K (Figure 9.17). The higher Ge concentration at the EB junction, the beneficial effects of the emitter high-low (n^+/n^- cap) junction, and the bandgap narrowing of the heavily doped base, offset the bandgap narrowing of the heavily doped emitter region to yield a peak β that increases quasi-exponentially with cooling from 102 at 310 K to 498 at 84 K (Figure 9.18). This large β value at low temperature serves to minimize the unwanted charge storage associated with the emitter-cap layer as well as to circumvent the degradation of β at medium injection levels due to bias-dependent Ge ramp effects (Chapter 6), giving an ideal value of β of 99 at a typical circuit operating point of 1.0-mA collector current [21]. An undesirable result of the high β at low temperature, however, is a decrease in the BV_{CEO} from 3.1 V at 310 K to 2.3 V at 84 K, but it remains acceptable for most cir-

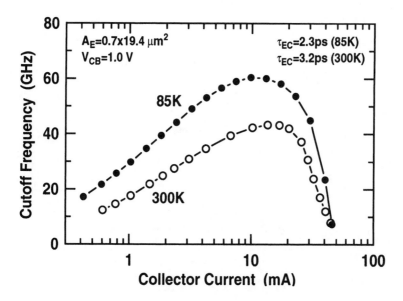

Figure 9.19 Cutoff frequency characteristics at 300 K and 77 K for a 77-K optimized emitter-cap SiGe HBT.

cuit applications. Depending on circuit requirements at 77 K, the low-temperature current gain can be easily tuned to a desired value.

The reduction in overall thermal cycle compared to a conventional design is key to maintaining the abrupt, as-deposited boron base profile, and thus providing immunity to carrier freeze-out at cryogenic temperatures (R_{bi} only increases from 7.7 to 11.0 kΩ/□ between 310 K and 84 K). Importantly, this immunity to base freeze-out does not come at the expense of increased EB leakage, as it does, for instance, in a spacer-free p-i SiGe profile with a very heavily doped base [18]. The lower doping level of the emitter-cap layer results in a reverse EB leakage at 1.0 V at 84 K, which is more than 500 times smaller than for the conventional i-p-i SiGe design. The consequence is a much smaller forward tunneling component in the base current (much larger low-current β), a smaller EB capacitance, and an expected improvement in hot-carrier reliability at cryogenic temperatures.

As shown in Figure 9.19, the transistor cutoff frequency (f_T) rises from 43 GHz to 61 GHz with cooling due to the beneficial effects of the Ge-grading-induced drift field. This improvement in f_T, coupled to the low total base resistance and slightly decreased CB capacitance, yields an increase in maximum oscillation frequency with cooling as well, from 40 GHz at 310 K to 50 GHz at 84 K. To assess the 77-K circuit capabilities of this technology, unloaded ECL ring oscillators were

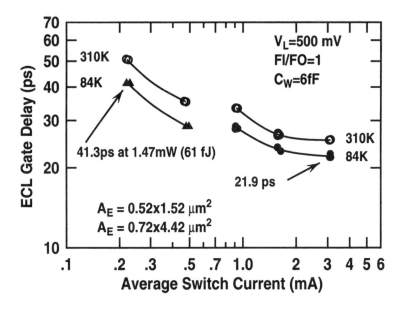

Figure 9.20 Unloaded ECL gate delay as a function of power at 310 K and 84 K.

measured (Figure 9.20). High-power (12.45 mW) ECL circuits switch at a record 21.9 psec at 84 K, 3.5 psec faster than at 310 K. Circuits that were optimized for lower power operation achieve a minimum power-delay product of 61 fJ (41.3 psec at 1.47 mW) at 84 K, and are 9.6 psec faster than at 310 K.

As discussed above [19], these 77-K optimized ECL circuits are expected to exhibit even more dramatic improvements in speed over room-temperature ECL circuits under heavy loading, due to the beneficial effects of cooling on metal interconnect resistance and circuit logic swing. Figure 9.21 shows calibrated simulations of ECL gate delay as a function of wire loading at 310 K versus 84 K. In the low-temperature case, an 8.3× reduction in metal resistance is assumed, as well as a (conservative) 40% reduction in circuit logic swing. The delay improvement at long wire lengths is dramatic (2.7× at 10 mm wire length), and suggests that SiGe HBT based line-drivers might be attractive for 77-K applications.

9.4 Helium Temperature Operation

Long-wavelength infrared focal-plane-arrays (FPA) and certain ultra-low-noise instrumentation amplifiers require transistors that operate down to liquid helium temperature (LHeT = 4.3 K). In addition to evaluating SiGe HBT performance at

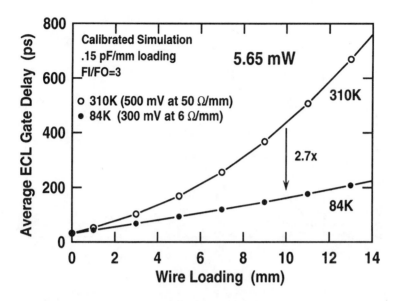

Figure 9.21 Simulated ECL gate delay as a function of wire loading at 310 K and 84 K for 5.65 mW power dissipation.

these potential application temperatures, the below 77-K regime is ideally suited for investigating new device physics phenomena, as well as for testing the validity of conventional theoretical formulations of device operation (e.g., drift-diffusion). This is particularly true for a SiGe HBT, since many of the transistor parameters are thermally activated functions of the Ge-induced band offsets, and are expected to change dramatically between 77 K and 4 K. For instance, a simple calculation of the intrinsic carrier density, to which the terminal currents are proportional, shows that a n_{io} changes by a factor of e^{3056} between 77 K and 4 K! Initial results on (unoptimized) Si BJTs to 10 K [22] showed transistor functionality but poor performance in the LHeT regime (< 10–15 K). More recent work [23, 24] on SiGe HBTs optimized for 77-K operation showed more impressive performance results as well as revealed interesting new device physics effects.

9.4.1 *dc* Characteristics at LHeT

The emitter-cap SiGe HBT optimized explicitly for 77 K achieved a β of 500, f_T of 61 GHz, f_{max} of 50 GHz, and a minimum ECL gate delay of 21.9 psec at 84 K. In cooling this transistor from 77 K to LHeT, the current gain increases monotonically from 110 at 300 K to 1,045 at 5.84 K, although parasitic base current leakage limits the useful operating current to above about 1.0 μA at 5.84 K. Figure 9.22 shows

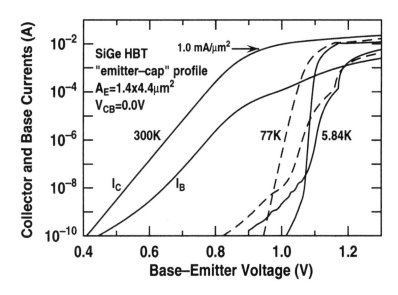

Figure 9.22 Gummel characteristics at 300 K, 77 K, and 5.84 K for a 77-K optimized emitter-cap SiGe HBT.

the Gummel characteristics of a 1.4 × 4.4 μm^2 emitter-cap SiGe HBT at 300 K, 77 K, and 5.84 K, and Figure 9.23 shows the current gain as a function of bias current down to 5.84 K. The severity of the base current leakage at low injection, and the Ge-ramp effect (refer to Chapter 6) at medium injection, limits the current range where one obtains the peak current gain. The aggressive base profile design in the emitter-cap SiGe HBT design (peak N_{ab}^- close to 8×10^{18} cm^{-3}) leads to an R_{bi} of < 18 kΩ/□ at 5.84 K, much lower than a more conventional i-p-i SiGe HBT design. As can be seen from the calculations in Figure 9.24, base freeze-out below 77 K depends very strongly on peak base doping, and must be carefully optimized for LHeT applications.

At temperatures as low as 5.84 K, this transistor has a maximum current drive in excess of 1.5 mA/μm^2 (limited by quasi-saturation and heterojunction barrier effects), with a peak transconductance of 190 mS. Figure 9.25 shows the peak current gain in an epitaxial-base Si BJT and two different SiGe HBTs (i.e., the 77-K optimized emitter-cap design and the i-p-i SiGe HBT design). The current gain in SiGe HBTs is proportional to the exponential of the ratio between the amount of bandgap reduction in the base (Ge + heavy doping) and the thermal energy. Since the i-p-i SiGe HBT has negligible Ge content at x=0, its current gain decreases with temperature. The emitter-cap design, however, has a trapezoidal Ge profile with approximately 2.0% Ge content at the EB boundary.

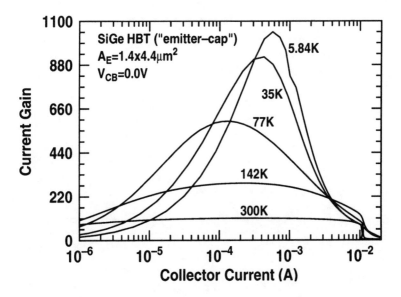

Figure 9.23 Current gain versus bias current as a function of temperature for a 77-K optimized emitter-cap SiGe HBT.

Theoretical calculations based on measured SIMS data were compared to the experimentally observed variation of peak current gain with temperature. Above 77 K, the temperature variation of peak current gain for the SiGe HBT is close to that theoretically expected, while at temperatures below 77 K, the exponential increase in current gain is primarily limited by parasitic base leakage due to field-enhanced tunneling. In contrast to this strong enhancement of current gain with cooling for the SiGe HBT, the current gain in a Si BJT fabricated with a comparable doping profile is significantly degraded at low temperatures, due to the strong bandgap narrowing in the emitter.

9.4.2 Novel Collector Current Phenomenon at LHeT

One striking feature observed in the SiGe HBTs operated at below 77 K is the progressive development of a nonideal component at very low injection levels in the collector current [23], as can be seen in Figure 9.26, which is an expanded view of Figure 9.22. A comparison to a comparably doped Si BJT shows a very similar effect, and hence this collector current anomaly cannot be attributed to a heterojunction effect associated with the presence of Ge. Observe as well that the theoretically expected drift-diffusion component of the collector current at 50 K is clearly dominated by the weakly temperature- and bias-dependent collector

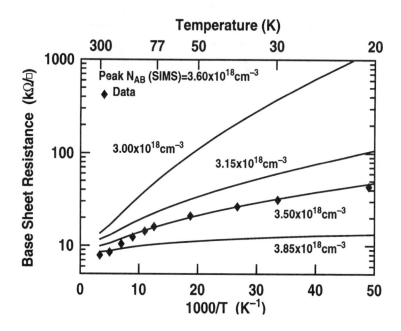

Figure 9.24 Comparison of measured and simulated base freeze-out properties of a 77 K-optimized emitter-cap SiGe HBT down to 20 K.

current leakage component. It is well known that the nonideality observed in the base current of bipolar transistors at cryogenic temperatures is due to tunneling and field-assisted recombination processes associated with the high electric field associated with the emitter-base junction, and therefore will not affect the collector current ideality. Conventional drift-diffusion theory dictates that the collector current should depend exponentially on voltage (i.e., $e^{qV_{BE}/kT}$). At temperatures above 77 K, this expectation has been confirmed numerous times experimentally down to very low current levels [6]. It is therefore logical to infer that the observed nonideality in the collector current is the result of a fundamentally different transport mechanism that enables minority carriers injected from the emitter to reach the collector through the base potential barrier.

Although the weak temperature and bias dependence of the observed collector current leakage phenomenon is suggestive of a carrier tunneling mechanism, the possibility of a direct tunneling process across the neutral base potential barrier is easily shown to be negligible for realistic base profiles, and hence cannot account for the observed phenomenon. On the other hand, it is plausible to associate a significant probability with carriers reaching the collector from the emitter by tunneling through the base potential barrier via intermediate trap states, provided we

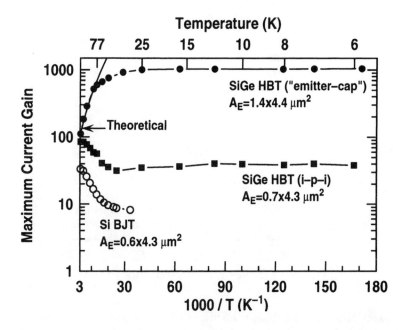

Figure 9.25 Peak current gain as a function of reciprocal temperature for a 77 K–optimized emitter-cap SiGe HBT, a conventional i-p-i SiGe HBT design, and an i-p-i Si BJT control.

assume a finite distribution of traps in the base bandgap. Indirect evidence for traps in the base region was previously shown to indicate the presence of NBR in these transistors (refer to Chapter 8).

A sketch of the proposed trap-assisted tunneling mechanism is shown in Figure 9.27). The conventional drift-diffusion component of the collector current is due to the electron injection over the base potential barrier, leading to an $e^{(qV_{BE}/kT)}$ behavior for I_C. The ideal component of the base current is similarly due to the hole injection into the emitter over the emitter potential barrier. The I_B leakage component is known to be due to the electron tunneling across the EB junction. Notice that such a tunneling mechanism will only add to I_B and not directly affect I_C. The mechanism responsible for the I_C leakage is proposed to be due to the carrier tunneling across the EB depletion region into a trap state in the base region and thereby transiting the base region by "hopping" through trap states. Since such a process can only be favored if: 1) traps are present in the base region; 2) traps are uniformly distributed in both physical and in energy space of the base region (close to the midgap); 3) W_b is thin enough to permit such a transition before carriers lose energy by phonon scattering; and 4) temperature is sufficiently low that the

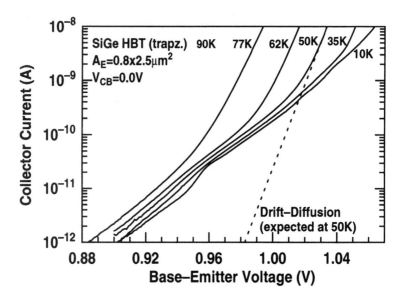

Figure 9.26 Collector current as a function of base-emitter voltage below 90 K, showing the nonideal collector current component.

energy lost due to phonon scattering is negligible. Conditions 1) and 2) have been previously shown to exist in SiGe HBTs. In advanced transistors 3) is met and 4) can be attained by operating at temperatures well below 77 K. At very low temperatures the carrier energy is insufficient to cross the potential barrier, and therefore the drift-diffusion component of the collector current is negligible. However, there exists a finite probability that the carriers tunnel across the EB depletion region into a trap state that is present at the same or lower energy level compared to the carrier energy, which will constitute the leakage component of I_C.

One can estimate such a tunneling transition probability (T_t) using a Wentzel-Kramer-Brillouin (WKB) approximation, resulting in

$$T_t = e^{-2 \int_0^{x_0} k_x \, dx}. \tag{9.29}$$

Here, x_0 represents the width of the EB depletion region (refer to Figure 9.28) and k_x represents the electron wave vector in the tunneling direction and is given by

$$k_x = \sqrt{\frac{2 m^* E_x}{\hbar^2}}, \tag{9.30}$$

where E_x is the position-dependent electron energy barrier for tunneling, m^* represents the electron tunneling mass, and $\hbar$ is Planck's constant. One can rewrite

Temperature Effects 403

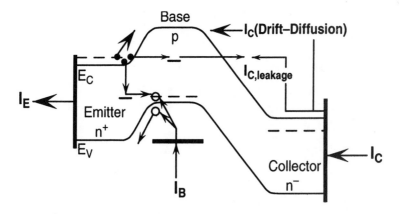

Figure 9.27 Schematic diagram showing the proposed trap-assisted tunneling mechanism responsible for the observed collector leakage currents at deep cryogenic temperatures.

(9.29) in terms of E_x and the built-in electric-field (ξ) as

$$T_t = e^{-2\left\{\frac{2m^*}{\hbar^2}\right\}^{1/2} \int_0^{E_0} \frac{E_x^{1/2} dE_x}{q\xi}}. \quad (9.31)$$

Here, E_0 represents the electron energy barrier at $x = x_0$. By assuming that the EB electric field is an effective average value (ξ_{avg} = constant), we can obtain

$$T_t \sim e^{-\alpha_x \frac{E_0^{3/2}(T)}{\xi_{avg}(T)}}, \quad (9.32)$$

where α_x is a constant and is related to m^* by

$$\alpha_x = \frac{4\sqrt{2\, m^*}}{3\, q\, \hbar}. \quad (9.33)$$

Although the magnitude of the I_C leakage is dependent on several factors, its temperature dependence at any fixed bias will be proportional to T_t. In order to quantify this, one needs to properly account for the temperature dependence of both E_0 and ξ in real devices, and this can be accomplished using a calibrated device simulator [24]. Figure 9.29 shows the simulated conduction band edge and the EB electric field in both Si and SiGe transistors at a very low injection condition (V_{BE} = 0.94 V) at 30 K. Observe that at any temperature and bias, E_0 and ξ in the SiGe HBT are always less than that for the Si BJT due to the presence of Ge in the base region (Figure 9.30). The normalized I_C leakage and the theoretically

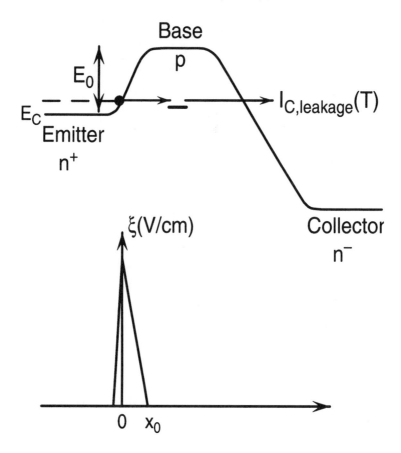

Figure 9.28 Schematic diagram showing the electron energy barrier and the electric field at the EB junction that is responsible for the electron tunneling.

expected temperature dependence of the nonideal trap-assisted tunneling current and the ideal drift-diffusion current for SiGe HBT and Si BJT at a fixed V_{BE} of 0.94 V are shown in Figure 9.31.

Observe that the ideal drift-diffusion component is a strong function of temperature, while the leakage component is only weakly temperature dependent. In addition, we see a difference in the temperature dependence of the leakage component between Si and SiGe transistors. It is clear, therefore, that the parameter α_x (used here as a free parameter) must be different for the SiGe HBT and the Si BJT in order to account for the measured temperature dependence of the collector leakage current. This difference is not unexpected since the m^* for the Si BJT, in

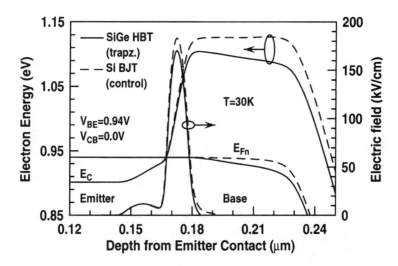

Figure 9.29 Calibrated simulations of the conduction band edge and the electric field near the EB junction in a comparably doped Si BJT and SiGe HBT at 30 K.

general, is different from that in the SiGe HBT. In the present case, first-order theoretical calculations predict a 6.7× increase in the m^* of SiGe HBT compared to that in the Si BJT.

We have also observed the same qualitative collector current leakage behavior in a wide variety of SiGe HBTs (i-p-i SiGe HBT, emitter-cap SiGe HBT) and Si BJTs (i-p-i epi-base Si BJT, ion-implanted-base Si BJT), as shown in Figure 9.32. This observation suggests that the nonideal collector current phenomenon is fundamental to advanced bipolar transistor technologies operating at deep cryogenic temperatures. Experimental evidence also indicates that a similar leakage effect exists in these devices, though with a much stronger temperature dependence, when the transistors are operated in the inverse mode (exchange of the emitter and collector terminals, all else being equal). This result is not unexpected because of the stronger temperature dependence in the lightly doped CB junction on the energy barrier height and the electric field. In addition, measurements of transistors under different collector bias demonstrate that this leakage effect is not related to a base punch-through mechanism. We believe that this nonideal collector current leakage phenomenon is inherent to all bipolar transistors operating at temperatures below 77 K. We have observed qualitatively similar behavior to that shown in Figure 9.26 for a wide variety of SiGe HBTs and Si BJTs, including self-aligned and non-self-aligned devices, devices with different base widths and Ge profiles, epitaxial and

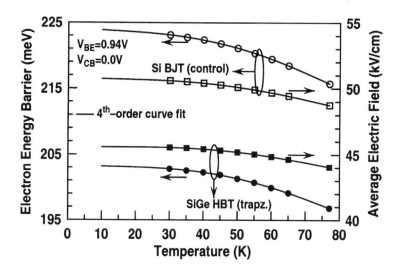

Figure 9.30 Calibrated simulations of the electron energy barrier and the average electric field as a function of temperature.

ion-implanted base profiles, and single-poly versus double-poly structures. While the observed nonideal leakage component typically occurs at collector currents below 100 nA, and thus is out of a normal circuit operating range, it clearly places a bound on the achievable transconductance at low-injection in SiGe HBTs operated near LHeT.

9.5 Nonequilibrium Base Transport

The validity of the classical drift-diffusion (DD) equations used to formally describe carrier transport in bipolar transistors (and which are self-consistently solved subject to the appropriate boundary conditions in commercial device simulators such as MEDICI – see Chapter 12) becomes questionable as device dimensions are reduced into the deep-submicron regime, and particularly in the presence of high electric fields. Minority carrier transport in the base region is well described by the conventional scattering-dominated DD equations, provided the neutral base width (W_b) is much larger than the carrier mean free path length (l_p) [25]–[27]. If, however, $W_b \ll l_p$, carriers can traverse the base without experiencing scattering, and transport is said to be "ballistic." Quasi-ballistic base transport occurs within a crossover regime between these two extreme transport domains (i.e., $W_b \sim l_p$), when carriers experience only a few collisions while transiting the base.

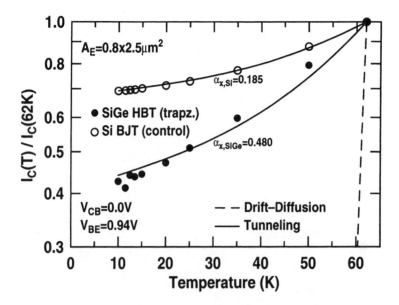

Figure 9.31 Comparison of measured and calculated normalized collector current as a function of temperature for both Si and SiGe transistors.

In the case of ballistic transport, the average carrier temperature (energy) differs significantly from the equilibrium lattice temperature, and the system can be said to be in the "nonequilibrium transport" regime. Clearly the DD formalism does not accurately describe this regime. Monte Carlo (MC) simulation incorporates actual band structure and solves (the more general) Boltzmann transport equation on an electron-by-electron basis, and is thus a powerful (though very computationally intensive and complex) tool for the theoretical investigation of this ballistic transport question. MC simulations on ultrathin base (< 50 nm) Si BJTs at 300 K show higher than expected minority carrier velocities in the neutral base due to a reduction of scattering events [28]. The result is a high forward velocity carrier flow, which yields a collector current density higher than that predicted by DD solutions.

Understanding the physics of thin-base transport is important because the base transit time is often a limiting or at least significant component of the overall frequency response (as reflected, for instance, in peak f_T). Quantifying ballistic and quasi-ballistic transport experimentally in advanced BJTs is challenging, and existing literature results are often inconclusive or contradictory. For instance, for W_b of about one mean free path length (10–20 nm in heavily doped Si at 300 K) some experiments suggest that transport is diffusive [29], while others suggest that it may be ballistic [30].

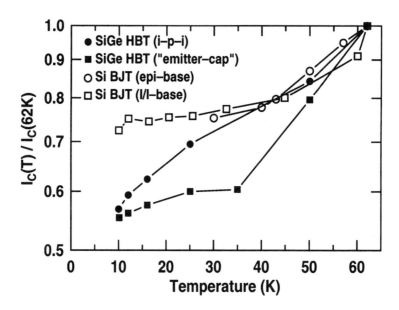

Figure 9.32 Measured normalized collector current as a function of temperature for a variety of Si and SiGe transistor technologies.

Since phonon scattering is exponentially reduced by cooling, one might logically expect to observe the same nonequilibrium effects in devices with larger W_b at sufficiently reduced temperatures. The carrier mean free path length as a function of temperature can be written approximately as [27]

$$l_p(T) = \frac{3 D_{nb}(T)}{2 v_T(T)}, \qquad (9.34)$$

where D_{nb} is the carrier diffusivity and v_T is the carrier thermal velocity, which can be written as

$$v_T(T) = \sqrt{\frac{2kT}{\pi m^*(T)}}, \qquad (9.35)$$

where m^* is the carrier effective mass, T is the lattice temperature, and k is Boltzmann's constant. Calculations using (9.35), for a base doping level of 2.6×10^{18} cm^{-3} using a 300 K l_p of 17.5 nm, show that the normalized mean free path length $(l_p(T)/l_p(300K))$ is equal to about 2.0 at 30 K [31], meaning that cooling a transistor can be effectively used to enhance nonequilibrium behavior. A device with a thin base width (e.g., 100 nm) may be well described by conventional DD theory at 300 K, yet may be moved into the quasi-ballistic transport regime with cooling. As

Temperature Effects

discussed below, evidence suggests the presence of nonequilibrium base transport in advanced Si and SiGe bipolar transistors may be inferred using dc measurements at cryogenic temperatures [31, 32].

9.5.1 Theoretical Expectations

From conventional DD theory, the collector current density for a SiGe HBT with constant base doping under low injection may be written as

$$J_C(T) = J_{C0}(T) e^{qV_{BE}/kT} \propto n_{i0}^2(T). \tag{9.36}$$

As discussed above, maintaining a constant operating current with cooling requires that V_{BE} be increased, since J_{C0} is directly proportional to the square of the intrinsic carrier concentration, which varies exponentially with temperature. The device transconductance is given by

$$g_m(T) = \frac{\partial I_C}{\partial V_{BE}} = \frac{qA_E J_C(T)}{kT} \left[1 + \frac{1}{J_{C0}(T)} \frac{\partial J_{C0}(T)}{\partial V_{BE}} \right] \simeq \frac{qA_E J_C(T)}{kT}, \tag{9.37}$$

where the final simplification assumes that J_{C0} is bias independent (this assumption will be revisited below). Note that for a constant collector current, the transconductance increases with decreasing temperature, and is reflected in an increase in the slope of the collector current versus base-emitter voltage curve with cooling.

Based on the DD formalism presented thus far, one can expect two significant changes in the collector current versus base-emitter voltage characteristic as the temperature is decreased:

- The increase in transconductance steepens the $J_C - V_{BE}$ curve at lower temperatures.

- The decrease in intrinsic carrier concentration with cooling shifts this curve to the right (thus increasing the value of V_{BE} required for a given current density) as the temperature decreases.

While the base transit time τ_b is notoriously difficult to experimentally separate from the other transit times via measurement of f_T, it nonetheless remains a very useful parameter for assessing base transport physics. For a SiGe HBT having uniform base doping, simple theory predicts that for a given base width, $\tau_b \propto W_b$ for ballistic transport and $\tau_b \propto W_b^2$ for diffusive transport. In general, we can write

$$\tau_b(T) = \frac{W_b(T)}{v_T(T)} + \frac{W_b^2(T)}{2\alpha(T) D_{nb}(T)}, \tag{9.38}$$

where α accounts for the effects of Ge on base transport (refer to Chapter 5). The first term in (9.38) is usually neglected in conventional DD calculations. The second term alone, however, will predict unphysically short base transit times for thin base transistors exhibiting quasi-ballistic transport. Figure 9.33 shows τ_b as a func-

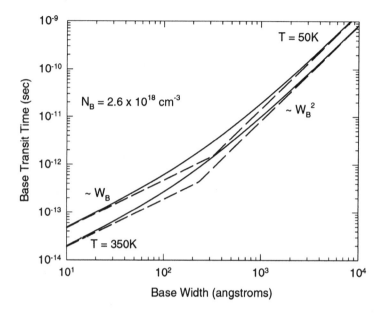

Figure 9.33 Calculated base transit time versus base width at 350 K and 50 K. Dashed lines show asymptotic approximations. Note that the boundary between diffusive and ballistic transport regimes shifts to the right with cooling.

tion of W_b for a device with $N_{ab} = 2.6 \times 10^{18}$ cm^{-3}, using typical 300-K values for the other parameters ($l_p = 17.5$ nm, $v_T = 5.0 \times 10^6$ cm/sec, and for simplicity α is taken to be unity). Note that for thick bases, the base transit time scales quadratically with base width as expected from diffusive transport, and that for thin bases the base transit time varies linearly with base width as expected from ballistic transport. It is clear, however, that the base width must be *much* smaller than the mean free path length in order to observe the ballistic limit [33], and thus a large range of base widths fall into the quasi-ballistic transition region.

If quasi-ballistic carrier transport occurs in the neutral base region of the device, electrons in this region will have a mean energy larger than their equilibrium value. Thus, an effective carrier temperature (T_{eff}) may be defined that is larger than the lattice temperature ($T_{lattice} = T$), corresponding to the higher average energy of the electrons. The effects of nonequilibrium transport may thus be phenomenologically modeled by using the effective temperature to describe the minority carrier distribution. This definition makes physical sense only if the electron distribution in the conduction band under nonequilibrium is sufficiently Maxwellian (i.e., the nonequilibrium state can be modeled as a perturbation of the equilibrium state). This requires strong carrier-carrier scattering and elastic impurity scattering in order to produce an isotropic distribution in k-space [34], but these conditions should be easily satisfied in the heavily doped p-type base of modern BJTs. Thus, from the modified *pn* product, we can write

$$n_{i0}^2 = \gamma_n \gamma_p N_C N_V e^{-E_{go}/kT_{\it eff}}, \tag{9.39}$$

and (9.37) becomes

$$g_m = \frac{qA_E J_C}{kT_{\it eff}} \left[1 + \frac{1}{J_{C0}} \frac{\partial J_{C0}}{\partial V_{BE}} \right] \simeq \frac{qA_E J_C}{kT_{\it eff}}. \tag{9.40}$$

Since $T_{\it eff} > T$, (9.39) predicts a larger value of intrinsic carrier density, and (9.40) predicts a smaller value of transconductance than (9.37) if quasi-ballistic transport is important. Thus, the (measurable) presence of nonequilibrium base transport should affect the device collector current versus base-emitter voltage characteristic in two ways:

- The value of required V_{BE} to maintain a constant collector current should be smaller than predicted by DD theory.

- The transconductance for a constant collector current should be smaller than predicted by DD theory.

9.5.2 Experimental Observations

To test these expectations regarding quasi-ballistic transport, the *dc* characteristics of several SiGe HBT technologies and a Si BJT control were been measured over the temperature range of 325 K to 10 K. Measurements have been compared to calculations based on drift-diffusion theory using (9.39) and (9.37). The DD calculations include advanced models for apparent bandgap narrowing, minority carrier mobility, carrier freeze-out, and Fermi-Dirac statistics [32]. The calculations were limited to the range 325 K to 50 K, due to numerical limitations of the parameter models. For temperatures above 200 K, the DD calculations agree very well with the measured data. Below 200 K, however, some discrepancies begin to appear (Figure 9.34). While the measured $I_C - V_{BE}$ characteristics shift to the right

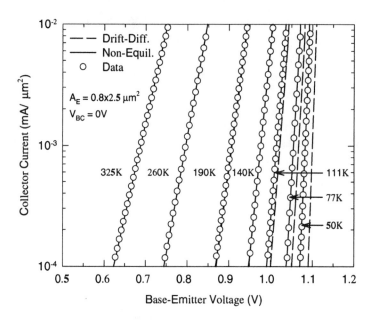

Figure 9.34 Collector current density versus base-emitter voltage as a function of temperature calculated from DD theory, and compared to both nonequilibrium values and measured data.

with cooling, as expected, they do not shift as much as predicted by DD theory. At 50 K, a 20-mV discrepancy exists between measured V_{BE} data and DD theory. Since I_C is exponentially dependent on the V_{BE}, this is a large enough change to produce a significant difference in the device current-voltage characteristics at low temperatures. Similarly, the slopes of the measured characteristics become steeper with cooling, but not as steep as expected from DD theory, leading to a discrepancy in the measured and calculated transconductance (Figure 9.35). At 50 K, the measured g_m is about 0.1 mS lower than the expected DD value (a difference of nearly 20%). Both observations are consistent with the presence of nonequilibrium base transport, as discussed above.

Figure 9.36 shows the inferred values of n_{i0} that were calculated from the values of J_{C0} needed to fit the measured data. Note that the shift in the collector current characteristic at 50 K is equivalent to miscalculating n_{i0} by 10 orders of magnitude!

Equation (9.37) assumes that J_{C0} is constant with respect to changes in V_{BE} with cooling. Thus, bias-dependent effects such as base-width-modulation, neutral

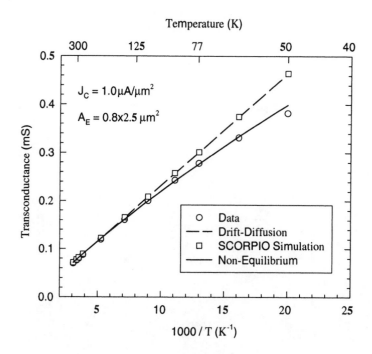

Figure 9.35 Transconductance versus reciprocal temperature for constant collector current calculated from DD theory, and compared to nonequilibrium values and measured data.

base recombination, and carrier velocity saturation are not accounted for. To assess the importance of these bias-dependent effects, the device structure was also simulated with SCORPIO [31], a fully coupled, self-consistent, one-dimensional, drift-diffusion simulator incorporating the same advanced physical parameter models. SCORPIO has been shown to produce results having good quantitative agreement to measured SiGe HBT data over a wide temperature range. The g_m values obtained from SCORPIO agree closely with the predicted values from (9.37), as shown in Figure 9.35. This further suggests that the observed variation in the measured transconductance data is due to some physical phenomenon not included in the drift-diffusion model. The observed shift in the simulated collector currents may be altered by varying the parameter models (e.g., for mobility or bandgap narrowing), but the slope remains unaffected. Identifying a physical cause for the observed deviations from standard theory in both collector current and transconductance is the purpose of this investigation.

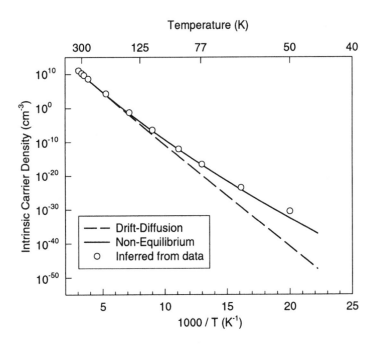

Figure 9.36 Intrinsic carrier density versus reciprocal temperature calculated from DD theory, and compared to that predicted by both nonequilibrium values and data inferred from the measured collector current.

9.5.3 Interpretation of Results

Based on (9.39) and (9.37), the observed discrepancies between the data and standard DD theory at cryogenic temperatures are consistent with the predicted effects of nonequilibrium base transport. Values for T_{eff} can be inferred from a comparison of the calculated and measured transconductance data (Figure 9.37). If the same values of T_{eff} are used in (9.39), the calculated device characteristics show remarkably close agreement to the measured data (see the "nonequilibrium" curves in Figures 9.34, 9.35, and 9.36). The fact that the same values of T_{eff} correct both the shifts of the calculated device characteristics as well as their slopes is evidence that the observed discrepancies are the result of the same physical phenomenon. Observe that at 50 K we have $T_{eff} \simeq 1.16 \times T$, and thus the changes in T required to account for the assumed nonequilibrium transport are relatively small, as expected.

Devices of varying size and from differing technologies were also measured, including epitaxial Si BJTs, SiGe HBTs of multiple technology generations, and

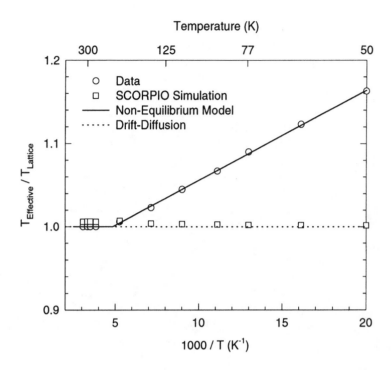

Figure 9.37 Inferred ratio of effective carrier temperature to lattice temperature as a function of reciprocal temperature.

even an ion-implanted base Si BJT [32]. As shown in Figure 9.38, all of the devices exhibit the same qualitative behavior with cooling. This suggests that the observed effect is not specific to one particular device size or technology, but is instead a fundamental physical property of all the measured devices. As expected, the nonequilibrium behavior is only apparent at reduced temperatures, where the reduction in scattering events increases the carrier mean-free path to the same order as the device base width. Note also that the severity of the nonequilibrium effects increases with continued cooling below 50 K. The collector current of a discrete, double-diffused transistor (National Semiconductor 3904) has also been measured at both 300 K and 77 K. No evidence of nonequilibrium transport was observed at 77 K for this device. This result is expected, since the base width of this device (> 2.0 μm) is much larger than the carrier mean free path, even at 77 K. The differences observed between the SiGe HBTs and the Si BJTs can be attributed to differences in carrier mobility, effective mass, and scattering processes (i.e., the specifics of the base profile) [32].

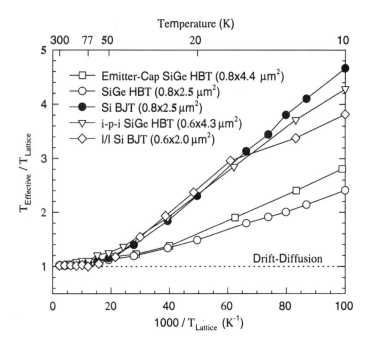

Figure 9.38 Ratio of effective carrier temperature to lattice temperature as a function of reciprocal temperature for devices of various size and differing SiGe and Si technologies.

9.6 High-Temperature Operation

While it has been demonstrated that SiGe HBTs operate well down to deep cryogenic temperatures, there was historically early concern about their suitability for operation at elevated temperatures. Given that all electronic systems must successfully operate at temperatures considerably above 300 K (e.g., 125°C to satisfy military specifications, and 85°C for many commercial applications), this is a potentially important issue. Given the narrow bandgap base region of the SiGe HBT compared to a Si BJT, and hence the expected negative temperature coefficient of the current gain (i.e., β decreases as temperature increases), it was often asked whether practical SiGe HBTs would have acceptable values of β at required high-end operational temperatures (e.g., 125°C). That this issue is not a valid concern for circuit designers is clearly demonstrated in Figure 9.39, which compares the percent change in peak current gain between 25°C and 125°C for a Si BJT and a number or commercially relevant SiGe profiles. There are several important points

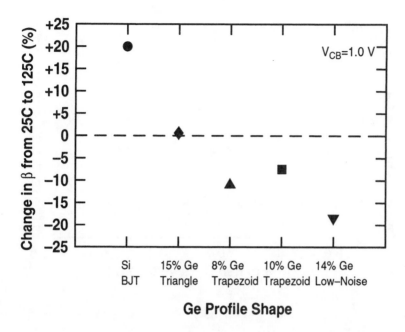

Figure 9.39 Percent change in peak current gain between 25°C and 125°C for various Ge profiles.

to glean from this data:

- The current gain in SiGe HBTs does indeed have an opposite temperature dependence from that of a Si BJT, as expected from simple theory.

- These changes in β between 25°C and 125°C, however, are modest at best (< 25%), and clearly not cause for alarm for any realistic circuit.

- The negative temperature coefficient of β in SiGe HBTs is *tunable*, meaning that its temperature behavior between, say, 25°C and 125°C can be trivially adjusted to its desired value by changing the Ge profile shape near the EB junction. In the case of the 15% Ge triangle profile, with 0% Ge at the EB junction, β is in fact temperature independent from 25°C to 125°C. This points to a major advantage of bandgap engineering.

- Finally, it is well known that thermal-runaway in high-power Si BJTs is the result of the positive temperature coefficient of β (i.e., as the device heats up due to power dissipation, one gets more bias current since the β increases

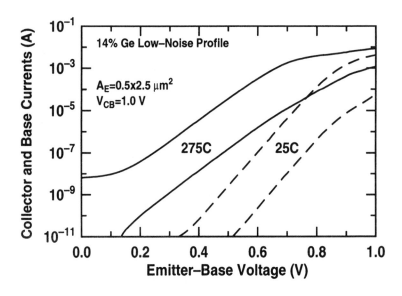

Figure 9.40 Gummel characteristics at 25°C and 275°C for a 14% Ge, low-noise optimized SiGe HBT.

with temperature, leading to a positive feedback process, and hence thermal collapse). The fact that SiGe HBTs naturally have a negative temperature coefficient for β suggests that this might present interesting opportunities for power amplifiers, since emitter ballasting resistors (which degrade RF gain) could in principle be eliminated.

There is also an emerging interest in the operation of electronic devices *above* 125°C, for planetary space missions (e.g., Venus), or for on-engine electronics for both the automotive and aerospace sectors to support the "more-electric-vehicle" thrust of the military. In these cases, allowing the requisite electronic components to operate at relatively high temperatures (say 250°C) presents compelling cost savings advantages, since the cooling system constraints can be dramatically relaxed. Conventional wisdom dictates that Si-based devices not be considered for these types of high-temperature applications, since Si is a low-bandgap material, and thermal leakage (i.e., I_{on}/I_{off} ratios) depends exponentially on E_g. The fact that SiGe HBTs are capable of operation in high-temperature environments can be clearly demonstrated, however, as shown in Figure 9.40 and Figure 9.41. For this particular SiGe HBT, which was designed for low-noise applications, the off-state leakage remains below 10 nA at 275°C, with a respectable current gain of greater than 200 across the useful range of bias currents. While these results do not speak

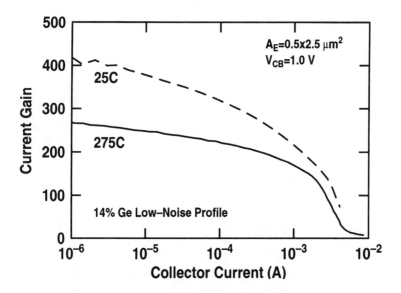

Figure 9.41 Current gain as a function of collector current at 25°C and 275°C for a 14% Ge, low-noise optimized SiGe HBT.

to the *ac* degradation that would likely result at elevated temperatures, or any potential reliability concerns, there is no fundamental reason why SiGe HBTs cannot satisfy this important emerging niche application of high-temperature electronics.

References

[1] W.L. Kauffman and A.A. Bergh, "The temperature dependence of ideal gain in double diffused silicon transistors," *IEEE Trans. Elect. Dev.*, vol. 15, pp. 732-735, 1968.

[2] E.S. Schlig, "Low-temperature operation of Ge picosecond logic circuits," *IEEE J. Solid-State Circ.*, vol. 3, pp. 271-276, 1968.

[3] D. Buhanan, "Investigation of current-gain temperature dependence in silicon transistors," *IEEE Trans. Elect. Dev.*, vol. 16, pp. 117-124, 1969.

[4] W.P. Dumke, "The effect of base doping on the performance of Si bipolar transistors at low temperatures," *IEEE Trans. Elect. Dev.*, vol. 28, pp. 494-500, 1981.

[5] J.D. Cressler, "The low-temperature static and dynamic properties of high-performance silicon bipolar transistors," Ph.D. Dissertation, Columbia University, 1990.

[6] J.D. Cressler et al., "On the low-temperature static and dynamic properties of high-performance silicon bipolar transistors," *IEEE Trans. Elect. Dev.*, vol. 36, pp. 1489-1502, 1989.

[7] J.D. Cressler et al., "Low temperature operation of silicon bipolar ECL circuits," *Tech. Dig. IEEE Int. Solid-State Circ. Conf.*, pp. 228-229, 1989.

[8] D.B.M. Klaassen, "A unified mobility model for device simulation – I. Model equations and concentration dependence," *Solid-State Elect.*, vol. 35, pp. 953-959, 1992.

[9] D.B.M. Klaassen, "A unified mobility model for device simulation – II. Temperature dependence of carrier mobility and lifetime," *Solid-State Elect.*, vol. 35, pp. 961-967, 1992.

[10] D.B.M. Klaassen, "Unified apparent bandgap narrowing in Si bipolar transistors," *Solid-State Elect.*, vol. 35, pp. 125-129, 1992.

[11] J.M.C. Stork et al., "High performance operation of silicon bipolar transistors at liquid nitrogen temperature," *Tech. Dig. IEEE Int. Elect. Dev. Meeting*, pp. 405-408, 1987.

[12] J.C.S. Woo and J.D. Plummer, "Optimization of bipolar transistors for low temperature operation," *Tech. Dig. IEEE Int. Elect. Dev. Meeting*, pp. 401-404, 1987.

[13] K. Yano et al., "A high-current-gain low-temperature pseudo-HBT utilizing a sidewall base-contact structure (SICOS)," *IEEE Elect. Dev. Lett.*, vol. 10, pp. 452-454, 1989.

[14] J.D. Cressler et al., "Scaling the silicon bipolar transistor for sub-100-ps ECL circuit operation at liquid nitrogen temperature," *IEEE Trans. Elect. Dev.*, vol. 37, pp. 680-691, 1990.

[15] W.P. Dumke, "Effect of minority carrier trapping on the low-temperature characteristics of Si transistors," *IEEE Trans. Elect. Dev.*, vol. 17, pp. 388-389, 1970.

[16] E.F. Crabbé et al., "The low-temperature operation of Si and SiGe bipolar transistors," *Tech. Dig. IEEE Int. Elect. Dev. Meeting*, pp. 17-20, 1990.

[17] J.D. Cressler et al., "Sub-30-ps ECL circuit operation at liquid-nitrogen temperature using self-aligned epitaxial SiGe-base bipolar transistors," *IEEE Elect. Dev. Lett.*, vol. 12, pp. 166-168, 1991.

[18] J.D. Cressler et al., "On the profile design and optimization of epitaxial Si- and SiGe-base bipolar technology for 77 K applications – part I: transistor dc design considerations," *IEEE Trans. Elect. Dev.*, vol. 40, pp. 525-541, 1993.

[19] J.D. Cressler et al., "On the profile design and optimization of epitaxial Si- and SiGe-base bipolar technology for 77 K applications – part II: circuit performance issues," *IEEE Trans. Elect. Dev.*, vol. 40, pp. 542-556, 1993.

[20] J.H. Comfort et al., "Single crystal emitter cap for epitaxial Si- and SiGe-base transistors," *Tech. Dig. IEEE Int. Elect. Dev. Meeting*, pp. 857-860, 1991.

[21] J.D. Cressler et al., "An epitaxial emitter-cap SiGe-base bipolar technology optimized for liquid-nitrogen temperature operation," *IEEE Elect. Dev. Lett.*, vol. 15, pp. 472-474, 1994.

[22] A.K. Kapoor, H.K. Hingarh, and T.S. Jayadev, "Operation of poly emitter bipolar npn and p-channel JFETs near liquid-helium (10 K) temperature," *Proc. IEEE Bipolar/BiCMOS Circ. Tech. Meeting*, pp. 210-214, 1988.

[23] A.J. Joseph, J.D. Cressler, and D.M. Richey, "Operation of SiGe heterojunction bipolar transistors in the liquid-helium temperature regime," *IEEE Elect. Dev. Lett.*, vol. 16, pp. 268-270, 1995.

[24] A.J. Joseph, "The physics, optimization, and modeling of cryogenically operated silicon-germanium heterojunction bipolar transistors," Ph.D. Dissertation, Auburn University, 1997.

[25] S. Tanaka and M.S. Lundstrom, "A compact HBT device model based on a one-flux treatment of carrier transport," *Solid-State Elect.*, vol. 37, pp. 401-410, 1994.

[26] M.A. Adam, S. Tanaka, and M.S. Lundstrom, "A small-signal one-flux analysis of short-base transport," *Solid-State Elect.*, vol. 38, pp. 177-182, 1995.

[27] A.A. Grinberg and S. Luryi, "Diffusion in a short base," *Solid-State Elect.*, vol. 35, pp. 1299-1309, 1992.

[28] W. Lee et al., "Monte Carlo simulation on non-equilibrium transport in ultrathin base bipolar transistors," *Tech. Dig. IEEE Int. Elect. Dev. Meeting*, pp. 473-476, 1989.

[29] D. Ritter et al., "Diffusive base transport in narrow base $InP/Ga_{0.47}In_{0.53}As$ heterojunction bipolar transistors," *Appl. Phys. Lett.*, vol. 59, pp. 3431-3433, 1991.

[30] A.F.J. Levi et al., "Vertical scaling in heterojunction bipolar transistors with nonequilibrium base transport," *Appl. Phys. Lett.*, vol. 60, pp. 460-462, 1992.

[31] D.M. Richey, "The development and application of SCORPIO: a device simulation tool for investigating silicon-germanium heterojunction bipolar transistors over temperature," Ph.D. Dissertation, Auburn University, 1996.

[32] D.M. Richey et al., "Evidence for non-equilibrium base transport in Si and SiGe bipolar transistors at cryogenic temperatures," *Solid-State Elect.*, vol. 39, pp. 785-789, 1996.

[33] M.A. Stettler and M.S. Lundstrom, "A microscopic study of transport in thin base silicon bipolar transistors," *IEEE Trans. Elect. Dev.*, vol. 41, pp. 1027-1033, 1994.

[34] E. Scholl, "A microscopic study of transport in thin base silicon bipolar transistors," in *Nonequilibrium Phase Transitions in Semiconductors*, Springer-Verlag, New York, 1987.

Chapter 10

Other Device Design Issues

In this chapter we examine several unique and additional device design and device physics issues associated with SiGe HBTs. While these topics might at first glance seem of minimal importance for standard *npn* SiGe HBT technology, they nevertheless hold potentially important implications for future generation SiGe HBT technology development and deployment. First, we address the differences in fundamental physics and profile optimization associated with *pnp* SiGe HBTs, which might be encountered, for instance, in developing a complementary SiGe HBT technology for analog and mixed-signal applications. We then investigate the theoretical constraints imposed by arbitrary band alignments (ΔE_C-only, or ΔE_V-only, or both ΔE_C and ΔE_V of varying sign and magnitude) and how they impact SiGe HBT performance and design. As will be shown, the common assumption that only valence band offsets (i.e., the use of strained SiGe layers on Si substrates) are suited to *npn* HBTs design is shown to be false, and that careful bandgap engineering for any arbitrary band alignment scheme can be used to achieve acceptable transistor *dc* and *ac* performance. Finally, we examine the effects of the backside Ge placement and shape (i.e., on the collector-base side of the neutral base) on the CB electric field distribution, and its impact on both impact ionization and the bias dependence of the base current.

10.1 The Design of SiGe *pnp* HBTs

At present, SiGe technology development is almost exclusively centered on *npn* SiGe HBTs. For high-speed analog and mixed-signal circuit applications, however, a complementary (*npn* + *pnp*) bipolar technology offers significant performance advantages over an *npn*-only technology. Push-pull circuits, for instance, ideally require a high-speed vertical *pnp* transistor with comparable performance to the *npn* transistor. The historical bias in favor of *npn* Si BJTs is due to the significantly larger minority electron mobility in the p-type base of an *npn* Si BJT, compared to the lower minority hole mobility in the n-type base of a *pnp* Si BJT. In addition, the valence band offset in SiGe strained layers is generally more conducive to *npn* SiGe HBT designs, because it translates into an induced conduction band offset and band grading that greatly enhance minority electron transport in the device, thereby significantly boosting transistor performance over a similarly constructed *npn* Si BJT. (It will be shown below that this band alignment is not as restrictive as has been commonly assumed.) For a *pnp* SiGe HBT, on the other hand, the valence band offset directly results in a valence band barrier, *even at low injection*, which strongly degrades minority hole transport and thus limits the frequency response. Careful optimization to minimize these hole barriers in *pnp* SiGe HBTs is thus required, and has in fact yielded impressive device performance compared to Si *pnp* BJTs, as demonstrated in the pioneering work reported in [1]–[3].

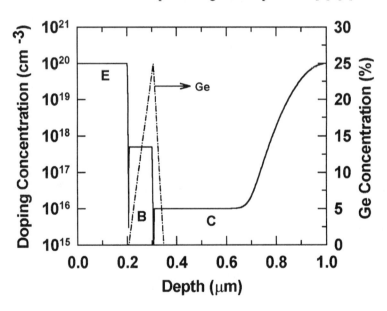

Figure 10.1 Hypothetical doping and Ge profiles for both *pnp* and *npn* SiGe HBTs.

What remains lacking in this context, however, is careful analysis of the inherent profile design differences between *npn* and *pnp* SiGe HBTs, and meaningful design guidelines for constructing *pnp* SiGe HBTs. Relevant questions include, for instance:

- How does SiGe *npn* and *pnp* profile design fundamentally differ?

Other Device Design Issues

- Can a single Ge profile design point be used for both *npn* and *pnp* transistors, for a given stability constraint?

- Is a graded-base Ge profile design preferable to a box-shaped Ge profile design for *pnp* HBTs?

- How much Ge retrograding in the collector-base junction is required to obtain acceptable SiGe *pnp* HBT performance?

These issues are addressed here using calibrated device simulations to shed light on the fundamental SiGe profile design differences between *npn* SiGe HBT and *pnp* SiGe HBTs that might be encountered, for instance, in developing a viable complementary SiGe HBT technology [4].

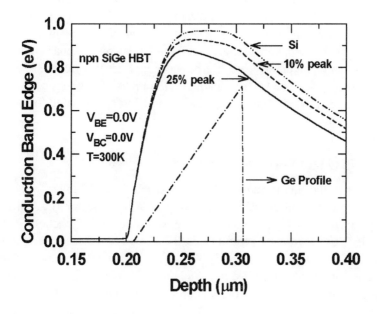

Figure 10.2 Valence band edge of a *npn* SiGe HBT for varying peak Ge content.

10.1.1 Simulation of *pnp* SiGe HBTs

Only 1-D MEDICI simulations [5] are needed here, since the differences in intrinsic profile design between *pnp* and *npn* transistors are the central focus. In addition, to aid in interpretation of the results, simplistic hypothetical *npn* and *pnp* SiGe profiles with constant emitter, base, and collector doping, and a Ge content *not* subject to thermodynamic stability constraints, were initially adopted (Figure 10.1). These profile assumptions are clearly nonphysical for real SiGe technologies, but are very useful for comparing *npn* and *pnp* devices so that their differences can be more easily discriminated and not masked by doping-gradient-induced phenomena (stability issues will be addressed below). This artificial assumption on constant doping clearly yields *ac* performance numbers (e.g., f_T) that are lower than what would be expected for a real complementary SiGe HBT technology, but relative comparisons between *npn* and *pnp* devices are nonetheless valid, and the comparison methodology widely applicable. The base doping level was chosen to give a zero-bias, pinched base sheet resistance in the range of 8–12 kΩ/□.

MEDICI models of the devices were constructed using actual device layouts and measured SIMS data, and careful calibration of MEDICI simulations for both *npn* and *pnp* Si BJTs to measured complementary Si BJT hardware [6] was performed. It was found that the default minority hole mobility modeling capability of MEDICI was deficient and tuning was required to obtain reasonable agreement between data and simulation, particularly under high-level injection. The SiGe model parameters determined from earlier calibrations of high-speed *npn* SiGe HBTs were used [7] (see Chapter 12), and assumed to be the same for both *npn* and *pnp* transistors.

10.1.2 Profile Optimization Issues

A comparison of the equilibrium conduction and valence band edges for both *npn* and *pnp* devices without any Ge retrograding into the collector (i.e., an abrupt transition from the peak Ge content to zero Ge content in the CB junction) is shown in Figure 10.2 and Figure 10.3 for: 1) a Si BJT; 2) a triangular (linearly graded) Ge profile with a peak Ge content of 10%; and 3) a triangular Ge profile with a peak Ge content of 25%. Observe that while there is no visible conduction band barrier present in the *npn* HBT, there is an obvious valence band barrier in the *pnp* HBT, even for low Ge content. This is consistent with the fact that there is a valence band offset in strained SiGe on Si (refer to Chapter 4), and clearly indicates that *pnp* SiGe HBT design is inherently more difficult than *npn* SiGe HBT design. In addition, due to the inherent minority carrier mobility differences between electrons and holes, it is also clear that *npn* devices will consistently outperform *pnp* devices, everything else being equal.

Unlike for a well-designed *npn* SiGe HBT (i.e., Ge outside the neutral base edges), where conduction band barrier effects are uncovered only at high J_C under Kirk effect [8] (refer to Chapter 6), the valence band barrier in *pnp* SiGe HBTs is in

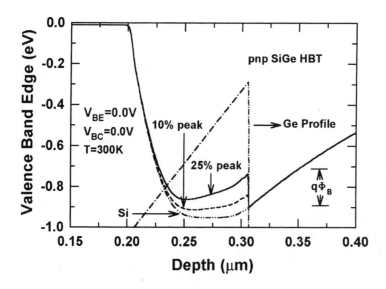

Figure 10.3 Conduction band edge of a *pnp* SiGe HBT for varying peak Ge content.

play even at low injection, and acts to block minority holes transiting the base. This pileup of accumulated holes produces a retarding electric field in the base, which compensates the Ge-grading-induced drift field, dramatically decreasing both J_C, β, and f_T. This effect worsens as the current density increases, since more hole charge is stored in the base. In this case, the f_T of the *pnp* SiGe HBT is in fact significantly lower than that of the *pnp* Si BJT! As expected, however, retrograding of the Ge edge into the collector can "smooth" this valence band offset in the *pnp* SiGe HBT, and thus improve this situation dramatically, although at the expense of film stability [1, 2]. For an increase of the Ge retrograde from 0 to 40 nm, the *pnp* SiGe HBT performance is dramatically improved, yielding roughly a 2× increase in peak f_T over the *pnp* Si BJT performance at equal doping.

Figure 10.4 and Figure 10.5 show the variation in peak f_T and β as a function of peak Ge content for both *npn* and *pnp* SiGe HBTs for both a 0-nm Ge retrograde and 100-nm Ge retrograde. At 100-nm retrograde, the performance of the *pnp* SiGe HBT monotonically improves as the Ge content rises, while the maximum useful Ge content is limited to about 10% without retrograding. Figure 10.6 indicates that 40–50 nm of Ge retrograding in the *pnp* SiGe HBT is sufficient to "smooth" the valence band barrier, and this is reflected in Figure 10.7, which explicitly shows the dependence of *pnp* peak f_T on Ge retrograde distance, for both triangular and box Ge retrograde profile shapes. Observe that the box Ge retrograde is not effective in

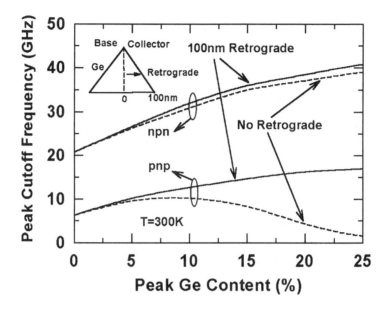

Figure 10.4 Simulated peak cutoff frequency as a function of peak Ge content for different Ge retrogrades for both *npn* and *pnp* SiGe HBTs.

improving the *pnp* SiGe HBT performance, since it does not smooth the Ge barrier, but rather only pushes it deeper into the collector, where it is still felt at the high J_C needed to reach peak f_T. This box Ge retrograde is also clearly undesirable from a stability standpoint. The effects of Ge retrograding on the *npn* SiGe HBT performance, on the other hand, are minor, while the film stability is significantly worse due to the additional Ge content. This suggests that using one Ge profile design for both *npn* and *pnp* SiGe HBTs is not optimum for high peak Ge content values. Note that while the peak f_T is unchanged with Ge retrograding in the *npn* SiGe HBT, the f_T response above peak f_T does not roll off as rapidly due to the high-injection-induced barrier, consistent with the results in [8] (refer to Chapter 6).

An examination of the frequency response of the *npn* and *pnp* SiGe HBTs as a function of front-side Ge profile shape (in this case, triangle versus box Ge profile, with a fixed retrograde of 100 nm for both) and peak Ge content shows that for the *npn* SiGe HBT, the base transit time reduction from the Ge-grading-induced drift field of the triangle Ge profile shape gives a significant advantage above 10% peak Ge, indicating that the *npn* SiGe HBT is base transit time limited. Interestingly,

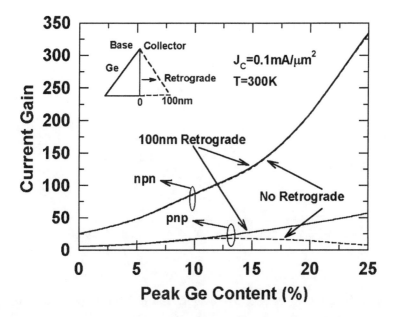

Figure 10.5 Simulated current gain as a function of peak Ge content for different Ge retrogrades for both *npn* and *pnp* SiGe HBTs.

for the *pnp* SiGe HBT, however, the differences between the box and triangle Ge profiles are much less pronounced, everything else being equal. The box Ge profile gives a slight advantage at low Ge content due to the low β and hence importance of the emitter transit time ($\tau_E \propto 1/\beta$), but once the β is sufficiently high, the triangle Ge profile dominates at higher peak Ge content, where the base transit time limits the overall response. In both *npn* and *pnp* devices, the triangle Ge profile offers better performance and better stability (less integrated Ge content), and thus can be considered an optimum shape for both devices. This is even more apparent if we examine the Early voltage of the devices, a key figure-of-merit for complementary analog circuits. In this case, the triangle Ge profile has a clear advantage due to its graded bandgap, as expected (refer to Chapter 4), and both *npn* and *pnp* transistors show a significant improvement in V_A with increasing Ge content.

10.1.3 Stability Constraints in *pnp* SiGe HBTs

The total amount of Ge that can be put into a given SiGe HBT is limited by the thermodynamic stability criterion. Above the critical thickness, the strain in the SiGe film relaxes, generating defects. The empirical critical thickness of a SiGe

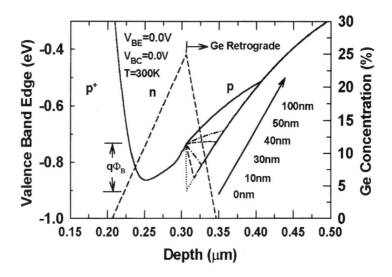

Figure 10.6 Valence band edge of a *pnp* SiGe HBT as a function of Ge retrograde distance.

multilayer with a top-layer Si cap is approximately 1.65× the theoretical stability result of Matthews and Blakeslee [9, 10].[1] In general, varying peak Ge content or retrograde distance (i.e., film thickness) moves the profile along different contours in stability space (Figure 10.8). Under the SiGe stability constraint, the peak Ge content must be traded off for the Ge retrograde distance in the collector-base junction. Figure 10.9 shows that an 11% peak Ge profile with a 25-nm retrograde gives the highest f_T for the *pnp* SiGe HBT at this design point. A similar exercise for the *npn* SiGe HBT shows that the *ac* performance is not sensitive to the SiGe profile shapes used, and, hence, without a significant loss of performance, the same Ge profile may in principle be used for both *pnp* and *npn* SiGe HBTs. This may be advantageous from a fabrication viewpoint. These results should be valid for current SiGe technology nodes with about 100-nm base width. If the base width is further reduced with technology scaling, the peak Ge content can be obviously increased, while maintaining film stability. The same optimization methodology employed here can be used in that case to determine the best SiGe profile for both devices.

---

[1]This empirical result is consistent with the stability theory presented in Chapter 2.

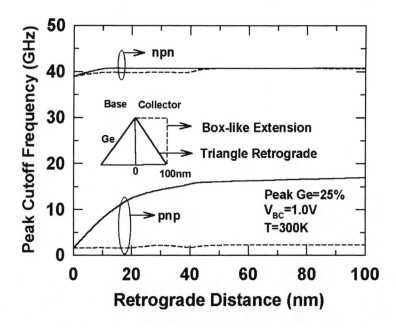

Figure 10.7 Simulated peak cutoff frequency as a function of both Ge retrograde distance and Ge profile shape for both *npn* and *pnp* SiGe HBTs.

10.2 Arbitrary Band Alignments

It has been previously suggested that for HBTs (III-V or SiGe), the band offset influences the transistor's electrical characteristics only via the modification of the intrinsic carrier density (n_i), which determines the minority carrier density, and thus the distribution of band offsets between the conduction and valence bands should not be important for graded-base HBTs [11, 12], but rather only the total amount of band offset in the device. As will be shown, this assertion only applies at low current densities, and is not valid for high current density operation, which is of practical interest in most circuit applications. Here we examine the impact of band offset distribution between the conduction and valence bands on the *dc* and *ac* characteristics of graded-base SiGe HBTs. These results provide guidance for bandgap engineering using new Si-based material systems such as strained Si on relaxed SiGe, which is primarily a ΔE_C band offset [13], and lattice-matched SiGeC, which at present appears to be a mixture of both ΔE_C and ΔE_V offsets [14]. Calibrated 2-D *dc* and *ac* numerical simulations were performed using MEDICI [5] by fixing the amount of total band offset (ΔE_g) and varying its distribution

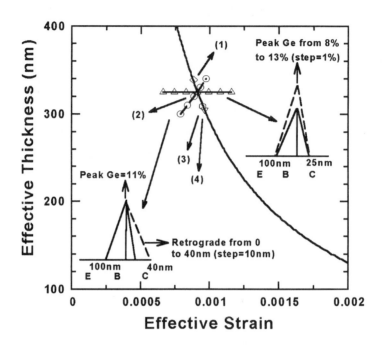

Figure 10.8 SiGe stability diagram illustrating the various *pnp* profile design trade-offs.

between conduction and valence bands. Collector-base Ge profile optimization for four extreme band offset situations were investigated in order to explore the feasibility of high current density operation under arbitrary band alignments [15].

10.2.1 Low-Injection Theory

The generalized Moll-Ross relation governs the minority carrier transport in an *npn* SiGe HBT [16]

$$J_C = \frac{q\left[(e^{qV_{BE}/kT}) - (e^{qV_{BC}/kT})\right]}{\int_0^{W_b} \frac{p_b(x)dx}{D_{nb}(x)n_{ib}^2(x)}}. \tag{10.1}$$

Given the same band grading, the integral appearing in the denominator is the same for different band offset distributions between the conduction and valence bands, provided that the total band offset is held constant. For low injection, the integral is dominated by its value in the neutral base where the hole density equals the base

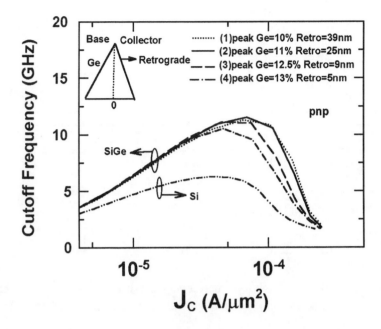

Figure 10.9 Simulated cutoff frequency as a function of collector current density for *pnp* SiGe HBTs with different Ge profiles (refer to Figure 10.8 for the exact profile shapes).

doping. The minority carrier base transit time, however, depends heavily on the conduction band profile, which determines the net force acting on electrons resulting from the induced electric field. The net force on the electrons consists of two components: 1) the quasi-electric field due to the gradient induced by conduction band offset; and 2) the built-in electric field (i.e., the gradient of the electrostatic potential). The built-in field under low injection can be derived as follows. From the transport equation in a semiconductor with a nonuniform bandgap, the hole current density is given by

$$J_p = -qD_p \frac{dp}{dx} + p\mu_p \frac{dE_V}{dx}, \quad (10.2)$$

where E_V is the actual valence band edge determined by the potential and valence band offset

$$\frac{dE_V}{dx} = -q\frac{d\phi}{dx} + \frac{d[\Delta E_V]}{dx}. \quad (10.3)$$

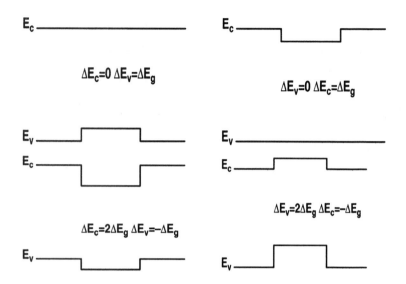

Figure 10.10 Illustration of the four representative distributions of the total band offset between the conduction and valence bands.

Applying the Webster approximation ($J_p = 0$), the built-in electric field ($\mathcal{E} = -d\phi/dx$) associated with the various charges can be derived as

$$-\frac{d\phi}{dx} = \frac{kT}{q\,p}\frac{dp}{dx} - \frac{1}{q}\frac{d[\Delta E_V]}{dx}, \qquad (10.4)$$

where the classical Einstein relation has been used. The net force acting on the electrons thus becomes

$$\mathcal{F}_n = \frac{d[\Delta E_C]}{dx} + q\frac{d\phi}{dx} = \frac{d[\Delta E_C]}{dx} - \frac{kT}{p}\frac{dp}{dx} + \frac{d[\Delta E_V]}{dx}, \qquad (10.5)$$

so that finally,

$$\mathcal{F}_n = -\frac{kT}{p}\frac{dp}{dx} + \frac{d[\Delta E_G]}{dx}. \qquad (10.6)$$

Therefore, the accelerating force on the electrons, and hence the base transit time, is determined only by the total band offset gradient across the neutral base and is independent of its distribution between conduction and valence bands for low-injection operation (i.e., when $p = N_{ab}$).

10.2.2 Impact of High Injection

Under high injection conditions, however, the minority carrier charge is sufficient to compensate the ionized dopants in the collector-base space charge region, thus

Other Device Design Issues

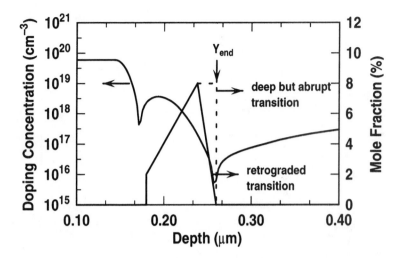

Figure 10.11 Doping and Ge profiles used in the simulations. The solid line represents the Ge profile with a retrograded ("graded") Ge profile in the CB junction, and the dashed line represents the Ge profile with an "abrupt" Ge transition in the CB junction.

exposing the heterointerface between the collector and the base, which was originally masked by the band bending in the space-charge region. In SiGe HBTs, since most of the band offset occurs in the valence band, the collapse of the original collector-base electric field at the heterointerface uncovers the valence band barrier, which opposes hole injection into the collector under base push-out (i.e., classical Kirk effect). The piling up of holes further induces a potential barrier in the conduction band profile, which retards the flow of electrons into the collector, thereby decreasing the cutoff frequency [8, 17]. A logical question presents itself: How does the distribution of the total band offset between the conduction and valence bands affect the cutoff frequency roll-off under high injection in an HBT? To shed light on this issue, we consider the following four representative band offset distributions (band alignments), as illustrated in Figure 10.10:

- The valence band pushes upward with the total band offset (note that this case is closest to the situation in strained SiGe on Si), while the conduction band remains the same as in Si. That is, $\Delta E_V = \Delta E_g$ and $\Delta E_C = 0$.

- The conduction band pushes downward with the total band offset, and the valence band remains the same as in Si (this case is applicable to strained Si on relaxed SiGe). That is, $\Delta E_C = \Delta E_g$ and $\Delta E_V = 0$.

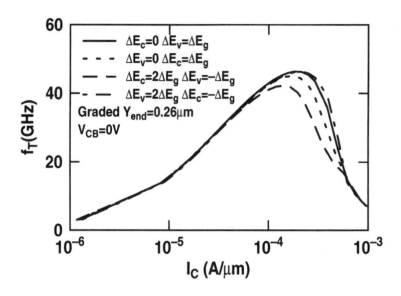

Figure 10.12 Simulated cutoff frequency versus collector current for the Ge retrograde profile with $Y_{end} = 0.26$ μm.

- The conduction band pushes downward by 2× the total band offset and the valence band pushes downward by 1× the total band offset (this case is applicable to published SiGeC bandgap predictions [14]). That is, $\Delta E_C = 2\Delta E_g$ and $\Delta E_V = -E_g$.

- The conduction band pushes upward by the total band offset while the valence band pushes up by 2× the total band offset. That is, $\Delta E_C = -\Delta E_g$ and $\Delta E_V = 2E_g$.

Note that the same amount of total band offset is used in each case for unambiguous comparisons.

2-D dc and ac simulations were performed for a 0.5-μm emitter width npn SiGe HBT with parameter coefficients tuned to measured dc and ac data (the default strained SiGe bandgap model was assumed). The doping and Ge profiles are shown in Figure 10.11. Observe that the Ge mole fraction increases from zero at $Y_{start} = 0.18$ μm, is linearly graded across the base, and then is retrograded to zero at $Y_{end} = 0.26$ μm after the Ge peak at $Y_{peak} = 0.24$ μm, as shown in Figure 10.11. Also shown in Figure 10.11 is a deeper but abrupt Ge profile, which will be discussed below. The simulated Gummel characteristics using the four extreme band offset distributions for the retrograded Ge profile show that the collector currents are nearly identical for all four cases, particularly at low injection, for the rea-

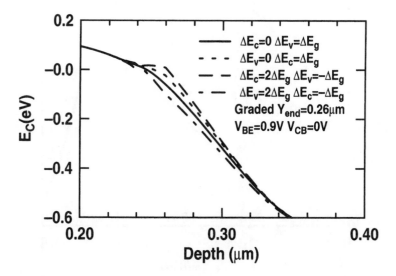

Figure 10.13 Enlarged conduction band edge versus depth for the retrograded Ge profile with $Y_{end} = 0.26$ μm.

sons discussed above, and the base currents are also nearly identical because each case has the same emitter hole injection characteristics. The simulated cutoff frequency versus collector current characteristics, however, are very different for the four cases, indicating the importance of the band offset distribution between conduction and valence bands under high-current density conditions (Figure 10.12). As the valence band offset becomes more negative, the cutoff frequency decreases (degrades) at a given collector current. The critical current at which the cutoff frequency starts to degrade also decreases for a more negative valence band offset. The physics underlying the difference in *ac* characteristics can be understood by inspecting the depth profile of simulated conduction band edge, as shown in Figure 10.13:

- The case with $\Delta E_V = 2\Delta E_g$ and $\Delta E_C = -\Delta E_G$ gives the highest (best) cutoff frequency and the highest (best) critical current density because it has the largest valence band offset, which acts to effectively prevent hole injection into the collector.

- The case with $\Delta E_C = 2\Delta E_g$ and $\Delta E_V = -\Delta E_g$ gives the lowest (worst) cutoff frequency and the lowest (worst) critical current density because it has the largest conduction band offset, the exposure of which in the neutral base serves as a barrier to electron transport, and thus results in excess charge

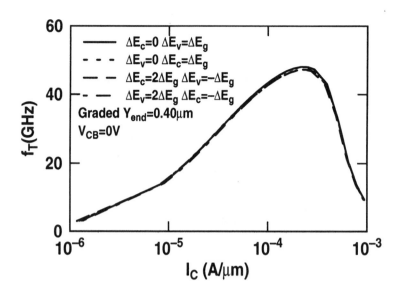

Figure 10.14 Simulated cutoff frequency versus collector current for the Ge retrograde profile with $Y_{end} = 0.40$ μm.

storage.

- The other two band offset distributions give results in-between the above two extremes. Note, however, that the case with only a conduction band offset results in a lower cutoff frequency because the conduction band barrier height ($\Delta E_C = \Delta E_g$) is higher than the height of the conduction band barrier resulting from the pileup of holes in the $\Delta E_V = \Delta E_g$ distribution.

Therefore, from the viewpoint of improving the *ac* characteristics of SiGe HBTs under high-current density operation, a large positive valence band offset together with a negative conduction band offset ($\Delta E_V = 2\Delta E_g$ and $\Delta E_C = -\Delta E_g$) is the most desirable bandgap offset distribution. It is worth noting that these offsets are in fact different from those produced by strained SiGe on Si (i.e., mostly E_V, with a small E_C).

10.2.3 Profile Optimization Issues

One way to minimize high-injection barrier effects in SiGe HBTs is to retrograde the mole fraction deep into the collector [8] (refer to Chapter 6 for details). To examine the effectiveness of such a grading scheme for different band offset distributions, simulations with a deeper grading of $Y_{end} = 0.40$ μm were performed.

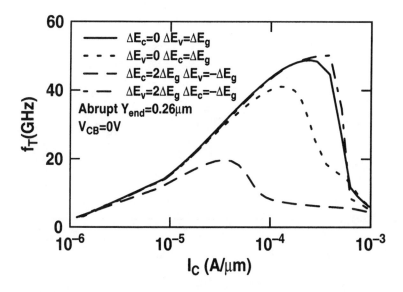

Figure 10.15 Simulated cutoff frequency versus collector current for an abrupt Ge profile with $Y_{end} = 0.26$ μm.

The results are shown in Figure 10.14, and clearly indicate that for deep enough grading, the cutoff frequency is nearly identical for different band offset distributions. An examination of the resultant conduction band edges for this deeper Ge profile shows that the conduction band depth profiles are nearly identical for all band offset distributions, thus leading to the same cutoff frequency versus collector current behavior. This result is in contrast to popular belief [18] and has important implications for device applications of heterojunctions. That is, applying careful optimization can yield good transistor performance for any arbitrary band alignment in an *npn* HBT. There is no compelling reason, for instance, to move to *pnp* HBTs if there is significant conduction band offset as long as one has flexibility to controllably introduce profile grading, something that is generally provided for in most film growth techniques. For instance, a new material system such as low-C-content SiGeC layers can provide better thermodynamic stability than strained SiGe, and thus allows a higher average Ge mole fraction for a deeper grading. For the same reason, high performance *pnp* HBTs can be built by proper optimization in a situation where the valence band offset dominates (such as SiGe) the total band offset (as shown above in this chapter).

The other commonly employed profile design technique that can be used to minimize high-injection barrier effects is to change the Ge profile in the CB junction from a retrograded shape to a deeper but still abrupt transition [8]. Figure 10.15

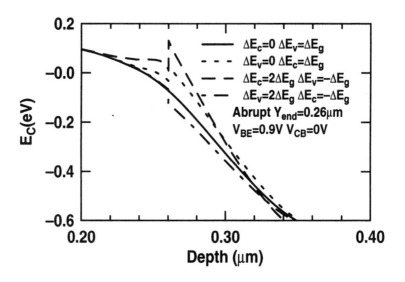

Figure 10.16 Enlarged conduction band edge versus depth for the abrupt Ge profile with $Y_{end} = 0.26$ μm.

shows the simulated results using such a deep and abrupt transition for the same Y_{end} value as used for Figure 10.12 for the retrograded profile (i.e., 0.26 μm). A significant difference is observed compared to the retrograded Ge profile results. Interestingly, the abrupt transition profile gives worse ac performance, although it has a larger integrated Ge mole fraction and thus is less stable. The offset case with $\Delta E_V = 2\Delta E_g$ and $\Delta E_C = -\Delta E_g$ now gives the lowest (worst) cutoff frequency and lowest (worst) critical current density because it has the largest conduction band offset, the exposure of which in the neutral base produces a potential barrier to electron transport (Figure 10.16). Pushing the Ge transition deeper into the collector ($Y_{end} = 0.40$ μm) can improve the cutoff frequency characteristics for all of the band offset distributions, as shown in Figure 10.17.

Except for the case of $\Delta E_V = 2\Delta E_g$ and $\Delta E_C = -\Delta E_g$, the cutoff frequency versus current characteristics are nearly identical to the simulated results with the graded profile for the same Y_{end} shown in Figure 10.14. Even for this worst case scenario, a reasonably high performance is achieved, as indicated by a 40-GHz peak cutoff frequency. In practice, if the device is operated at a current density slightly lower than that at the peak cutoff frequency, performance nearly identical to the other three bandgap offset distributions can be achieved.

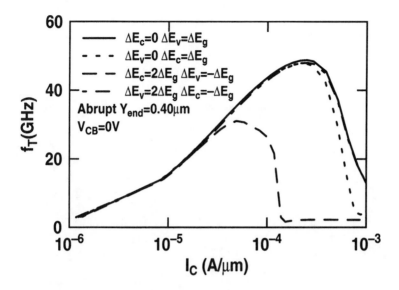

Figure 10.17 Simulated cutoff frequency versus collector current for an abrupt Ge profile with $Y_{end} = 0.40$ μm.

10.3 Ge-Induced Collector-Base Field Effects

Given its bandgap-engineered nature, clearly the characteristics of SiGe HBTs are highly sensitive to Ge profile shape. For instance, the specifics of the backside Ge profile (i.e., on the CB side of the neutral base) strongly influence high-injection heterojunction barrier effects, which produce premature roll-off of β and f_T at high current density [8]. In this section we show that the backside Ge profile also alters the electric field distribution in the CB space-charge region, and thereby indirectly affects both impact ionization and the CB voltage dependence of the base current in SiGe HBTs [19]. Given that impact ionization and the base current voltage dependence are critical to the breakdown voltage and output conductance of SiGe HBTs, respectively, they are key considerations for device and circuit designers. First, calibrated 2-D simulations are used to show how the backside Ge profile shape influences the CB electric field distribution, which then couples to impact ionization and the base current in the transistor. Experimental data on advanced SiGe HBTs is used to validate the claims.

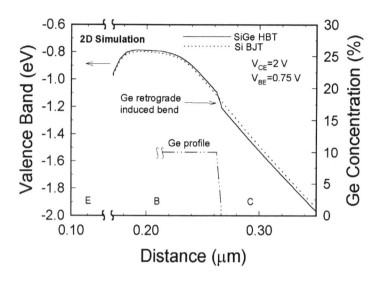

Figure 10.18 Valence band edge as a function of depth for a SiGe HBT and a Si BJT control.

10.3.1 Simulation Approach

Calibrated 2-D MEDICI simulations based on the transistor layout and measured SIMS data was first developed. The SiGe profiles used in the present simulations are hypothetical Ge profiles with a trapezoidal shape, where the Ge starts at the metallurgical EB junction, linearly grades across the neutral base, peaks at 10% Ge content, and then falls back to zero at the CB metallurgical junction over a distance of 6 nm. This SiGe "control" profile is labeled "0 nm Ge" (i.e., the location of the SiGe-Si heterointerface is referenced to the metallurgical CB junction). To explore the effect of the backside Ge profile shape on transistor performance, the SiGe retrograde distance and Ge retrograde location in the CB junction from this hypothetical 0-nm Ge profile were varied. The doping profile, the front-side (i.e., EB) Ge profile, its ramp rate across the base, and the peak Ge content were all kept identical to facilitate unambiguous comparisons. A Si-only case was also simulated to clearly distinguish heterojunction effects and used identical doping profiles, except for the absence of Ge. All simulations were performed in low injection to distinguish these effects from high-injection effects in SiGe HBTs.

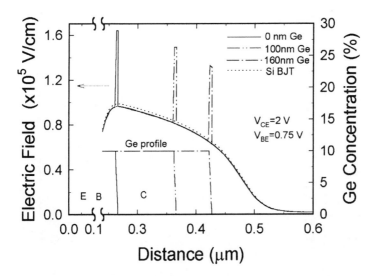

Figure 10.19 Simulated electric field distribution in the collector-base junction for three different backside Ge heterointerface locations.

10.3.2 Influence on Impact Ionization

For a strained SiGe layer on Si, the band offset in the SiGe film predominantly resides in the valence band and its value is proportional to the Ge content, according to $\Delta E_V = 0.74x$ (eV), where x is Ge fraction (i.e., 10% Ge = 0.10). In the Ge retrograde region at the CB junction, the varying Ge content produces an abrupt change in the valence band, which is shown in Figure 10.18 for the 0-nm SiGe profile. This change in the valence band creates a heterojunction-induced quasi-electric field, which can be evaluated for a given valence band grading as $\mathcal{E}_{Ge} = -q(0 - E_V)/D_r = 0.74x/D_r$, for a linear Ge retrograde, where D_r is the retrograde distance. For the present SiGe control profile, the value of this band-edge-induced electric field is approximately $\mathcal{E}_{Ge} = 1.23 \times 10^5$ V/cm for $x = 10\%$ and $D_r = 6$ nm, which is larger than the peak field formed by the doping-induced charge in the CB space-charge region (Figure 10.19). [2] The impact of Ge retrograde location on the resultant electric field in the CB junction is also shown in Figure 10.19, where the 100-nm Ge and 160-nm Ge profiles have the Ge retrograde locations 100 nm

---

[2]The electric field distributions presented in the figures were calculated using the derivative of electrostatic potential, and hence represent the "real" (electrostatic) electric field seen by both minority and majority carriers inside the junction. Physically, this is a two-step process, whereby the heterojunction-induced quasi-electric field produces a change in the local space-charge distribution inside the CB junction, resulting in a perturbation to the total junction electrostatic field.

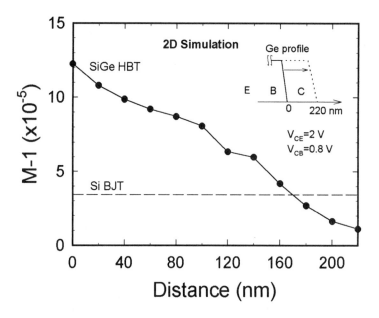

Figure 10.20 Simulated multiplication factor as a function of backside Ge heterointerface location.

and 160 nm deeper, respectively, than that of the 0 nm Ge (note that the retrograde distance D_r is fixed for all cases).

It is clear that as the backside Ge retrograde location moves toward the neutral collector, the peak electric field moves in the same direction and the magnitude of the peak electric field drops. This decrease of the peak field reduces the impact ionization rate, as reflected in the avalanche multiplication factor ($M-1$), as shown in Figure 10.20. Observe that for a sufficiently deep location of the Ge retrograde in the collector, the $M-1$ of a SiGe HBT can be even lower than that of a comparably constructed Si BJT. This is physically the result of the decrease in the field across the bulk of the region on the base side of the base-collector space charge region, which produces most of the impact ionization, as can be seen in Figure 10.19. For the profile examined, this reduction in $M-1$ of a SiGe HBT compared to and identically made Si BJT occurs when the Ge retrograde location is beyond about 170 nm. Whether this inherent Ge-induced advantage offered by the perturbation of the CB field can be leveraged to design a better device will depend on the specifics of the Ge profile, since stability issues must obviously be carefully considered.

Since the Ge retrograde field $\mathcal{E}_{Ge}$ is reciprocally proportional to the retrograde

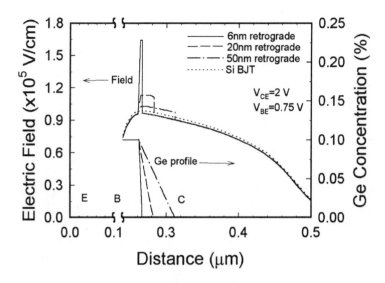

Figure 10.21 Simulated electric field distribution for three different backside Ge grading distances.

distance D_r, it is obvious that D_r can also be increased to reduce $\mathcal{E}_{Ge}$ and hence decrease $M - 1$. In this case, the $M - 1$ of the SiGe HBT drops as the peak field is reduced by the D_r increase (Figure 10.21 and Figure 10.22). We again observe that the $M - 1$ of the SiGe HBT can be reduced even lower than that of an identically constructed Si BJT, provided D_r is sufficiently large. These Ge profile changes are clearly also subject to appropriately satisfying the film stability constraints.

10.3.3 Influence on the Base Current Bias Dependence

The heterojunction-induced electric field in the CB space-charge region should also impact the bias dependence of the base current. This was investigated by examining the relative base current change with applied V_{CB}, for various backside Ge profiles. Figure 10.23 shows that the normalized I_B of the SiGe HBTs becomes less dependent on V_{CB} as the retrograde distance D_r increases. We refer to this as an "apparent" reduction in NBR since the V_{CB} dependence of I_B decreases (a classical signature of improved NBR; refer to Chapter 6). Note, however, that the trap density across the base is held fixed in all of the simulations, and hence this effect is clearly an electric field effect, not a recombination effect. The observed improvement of $I_B(V_{CB})$ can be understood by examining the hole distribution near

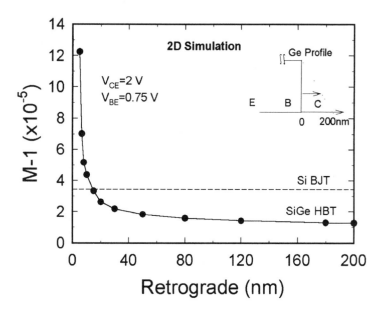

Figure 10.22 Simulated multiplication factor as a function of backside Ge grading distance.

the boundary of the CB junction, as shown in Figure 10.24. As D_r increases, the electric field at the boundary decreases (Figure 10.21), and thus the space-charge region consumes less of the neutral base for a given V_{CB} (Figure 10.24). The location of the Ge retrograde also affects the neutral base hole density distribution via the influence of the field in the CB space-charge region. The consequence of pushing the Ge backside deeper into the collector is also to produce a reduction of the relative I_B drop with increasing V_{CB} (i.e., an improvement in apparent NBR) of similar magnitude to that shown in Figure 10.23.

This backside-Ge-induced improvement in $I_B(V_{CB})$ is potentially important for circuit applications because it affects the transistor output conductance, and particularly the difference in V_A between forced-voltage and forced-current input drive, because it provides a means to minimize the enhanced (worse) $I_B(V_{CB})$ dependence commonly observed in SiGe HBTs compared to Si BJTs [20].

10.3.4 Experimental Confirmation

To validate these claims, SiGe HBTs were fabricated with the Ge backside profile extended 90 nm and 150 nm, respectively, deeper into the collector than that of

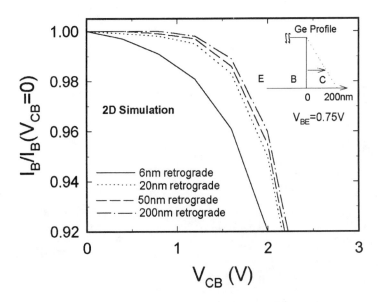

Figure 10.23 Simulated normalized base current as a function of collector-base voltage for various backside Ge grading distances.

an identically processed SiGe control profile. A comparable Si-only device was also included for comparison. The front-side EB Ge grading was kept identical for all the Ge profiles, all Ge backside profiles had the same retrograde distance D_r, and the collector profiles were identical. All four wafers were fabricated in the same wafer lot under identical conditions to facilitate comparisons. The measured $M - 1$ [21] for the two new SiGe HBTs, the SiGe control, and the Si BJT are shown in Figure 10.25. Observe that for fixed V_{CB}, both the 90- and 150-nm Ge profiles have lower $M - 1$ than the SiGe control profile, and all of the SiGe profiles have lower M-1 than the Si BJT, qualitatively consistent with the simulations. This improvement in $M - 1$ results in a BV_{CEO} improvement 0.25 V and 0.50 V over the SiGe control device for the 90- and 150-nm profiles, respectively (all three transistors have nearly identical current gain). As shown in Figure 10.26, the V_{CB} dependence of I_B is also improved as the backside Ge location increases, again consistent with our simulations.

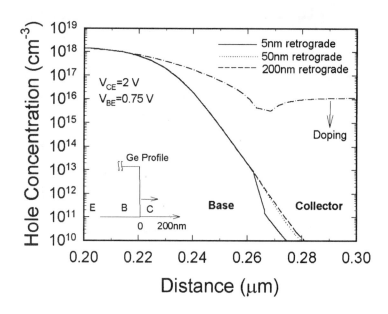

Figure 10.24 Simulated hole concentration as a function of depth for various backside Ge grading distances.

References

[1] D.L. Harame et al., "SiGe-base PNP transistors fabrication with n-type UHV/CVD LTE in a 'NO DT' process," *Tech. Dig. IEEE Symp. VLSI Tech.*, pp. 47-48, 1990.

[2] D.L. Harame et al., "55 GHz polysilicon-emitter graded SiGe-Base PNP transistors," *Tech. Dig. IEEE Symp. VLSI Tech.*, pp. 71-72, 1991.

[3] D.L. Harame et al., "A SiGe-Base PNP ECL circuit technology," *Tech. Dig. IEEE Symp. VLSI Tech.*, pp. 61-62, 1993.

[4] G. Zhang et al., "A comparison of npn and pnp profile design tradeoffs for complementary SiGe HBT technology," *Solid-State Elect.*, vol. 44, pp. 1949-1954, 2000.

[5] MEDICI, 2-D Semiconductor Device Simulator, Avant!, Fremont, CA, 1997.

[6] R. Patel et al., "A 30V complementary bipolar technology on SOI for high speed precision analog circuits," *Proc. IEEE Bipolar/BiCMOS Circ. Tech. Meeting*, pp. 48-50, 1997.

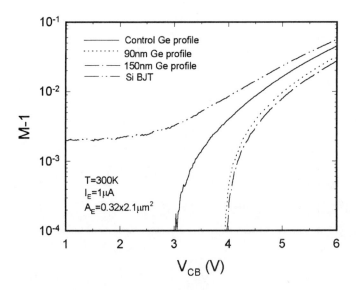

Figure 10.25 Measured multiplication factor as a function of collector-base voltage for three different Ge profiles with varying backside Ge heterointerface locations.

[7] G. Niu et al., "Noise parameter optimization of UHV/CVD SiGe HBTs for RF and microwave application," *IEEE Trans. Elect. Dev.*, vol. 46, pp. 1589-1598, 1999.

[8] A.J Joseph et al., "Optimization of SiGe HBTs for operation at high current densities," *IEEE Trans. Elect. Dev.*, vol. 46, pp. 1347-1354, 1999.

[9] S.R. Stiffler et al., "The thermal stability of SiGe films deposited by ultrahigh-vacuum chemical vapor deposition," *J. Appl. Phys.*, vol. 70, pp. 1416-1420, 1991.

[10] A. Fischer, H.-J. Osten, and H. Richter, "An equilibrium model for buried SiGe strained layers," *Solid-State Elect.*, vol. 44, pp. 869-873, 2000.

[11] J.M. Lopez-Gonzalez and L. Prat, "The importance of bandgap narrowing distribution between the conduction and valence bands in abrupt HBTs," *IEEE Trans. Elect. Dev.*, vol. 44, pp. 1046-1051, 1997.

[12] E. Azoff, "Energy transport numerical simulation of graded-AlGaAs/GaAs heterojunction bipolar transistors," *IEEE Trans. Elect. Dev.*, vol. 30, pp. 609-616, 1989.

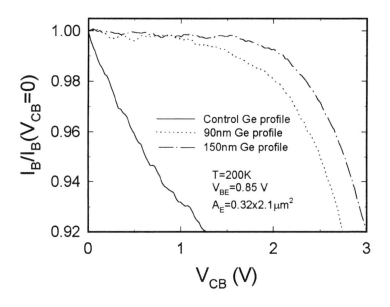

Figure 10.26 Measured normalized base current as a function of collector-base voltage for multiplication factor as a function of collector-base voltage for three different Ge profiles with varying backside Ge heterointerface locations.

[13] J. Welser, J.L. Hoyt, and J.F. Gibbons, "Electron mobility enhancement in strained-Si n-type metal-oxide semiconductor field-effect transistors," *IEEE Elect. Dev. Lett.*, vol. 15, pp. 100-102, 1994.

[14] B.L. Stein et al., "Electronic properties of $Si/Si_{1-x-y}Ge_xC_y$ heterojunctions," *J. Vacuum Sci. Tech. B*, vol. 16, pp. 1639-1643, 1998.

[15] G. Niu and J.D. Cressler, "The impact of bandgap offset distribution between conduction and valance band in Si-based graded HBT's," *Solid-State Elect.*, vol. 43, pp. 2225-2230, 1999.

[16] M. Schroter, M. Friedrich, and H. Rein, "A generalized integral charge control relation and its application to compact models for silicon-based HBT's," *IEEE Trans. Elect. Dev.*, vol. 40, p. 2036, 1993.

[17] Z. Yu, P.E. Cottrell, and R.W. Dutton, "Modeling and simulation of high-level injection behavior in double heterojunction transistors," *Proc. IEEE Bipolar/BiCMOS Circ. Tech. Meeting*, pp. 192-195, 1990.

[18] H. Kroemer, "Heterojunction bipolar transistors and integrated circuits," *Proc. IEEE*, vol. 70, pp. 13-27, 1982.

[19] G. Zhang et al., "Electric field effects associated with the backside Ge profile in SiGe HBTs," *Solid-State Elect.*, vol. 46, pp. 655-659, 2002.

[20] A.J. Joseph et al., "Neutral base recombination and its impact on the temperature dependence of Early voltage and current gain-Early voltage product in UHV/CVD SiGe heterojunction bipolar transistors," *IEEE Trans. Elect. Dev.*, vol. 44, pp. 404-413, 1997.

[21] G. Niu et al., "Measurement of collector-base junction avalanche multiplication effects in advanced UHV/CVD SiGe HBTs," *IEEE Trans. Elect. Dev.*, vol. 46, pp. 1007-1015, 1999.

Chapter 11

Radiation Tolerance

The operation of electronic systems in a space environment presents a host of challenges for device, circuit, and system designers. The seemingly tranquil image we garner as we gaze up into a dark night sky and observe a satellite elegantly tracing its orbit around the earth is very deceiving. It has been recognized since the beginning of the space program in the 1950s that earth orbit presents an amazingly hostile environment and, despite appearances, is actually seething with lethal levels of radiation (man's observation since ancient times of the captivating aurora borealis was a beautiful but ultimately unfriendly hint of things to come). The complex radiation fields surrounding the earth owe their origin to the interaction of the solar wind produced by the fusion furnace of the sun with the earth's magnetic field (Figure 11.1), and is obviously a complex and dynamic phenomenon.

Predicting the precise radiation environment encountered by a given spacecraft is difficult since it depends on many factors, including its orbital path and altitude, the level of solar activity during the mission, and the total duration of the mission. We classify these potential spacecraft orbits using IEEE Standard 1156.4 as low earth orbit (LEO), below an altitude of 10,000 km; medium earth orbit (MEO), from 10,000 to 20,000 km; geostationary orbit (GEO), at 36,000 km; and highly elliptical orbit (HEO) [1]. Particles with proper charges, masses, energies, and trajectories in the solar wind can be trapped by the earth's magnetic field, generating the so-called "van Allen radiation belts." In the case of orbital space electronics found in satellites, the proton and electron belts have the greatest influence. Simplified models of these radiation belts suggest that they are toroidal in structure, with the protons confined to a single toroid, and the electrons confined to two high intensity toroids (Figure 11.2).

From a semiconductor device perspective, three distinct radiation-induced phenomena can be encountered in the space environment: 1) total ionizing dose (TID)

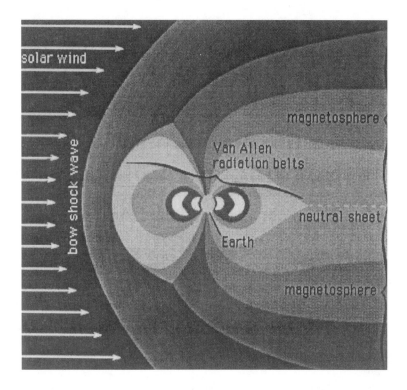

Figure 11.1 Illustration of the solar wind and radiation belts surrounding the Earth.

effects, which are associated with ionization damage induced by incident charged particles (protons or electrons) or photons; 2) displacement damage (DD) effects, which are associated with an incident particle's mass and particle-to-semiconductor-atom knock-on; and 3) single event effects (SEE), which are usually associated with very high energy incident particles (e.g., protons, neutrons, or high-atomic-mass cosmic rays) [2]. Ionization damage in devices typically involves the isolating oxides, and oxide-to-semiconductor interfaces, and thus is usually associated with *surfaces* of the devices. Displacement damage is generally associated with the *volume* of the device, and can result in deactivation of dopants, as well as carrier lifetime and carrier mobility degradation. While cosmic ray induced SEE are often associated with data loss or error generation phenomena in switching circuits, and hence fall into the category of "soft" errors, in general SEE can be classified into a number of different categories according to the precise damage mechanisms, and includes: 1) single event upset (SEU, the classical "soft-error"); 2) single event transient (SET); 3) single event burnout (SEB); 4) single event gate rupture (SEGR); and 5) single event latchup (SEL). Thus, some SEEs are de-

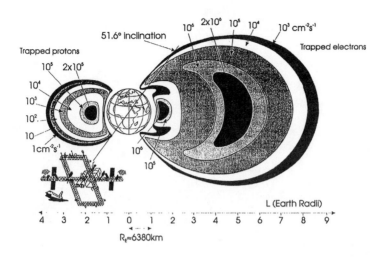

Figure 11.2 Illustration of the proton and electron belts surrounding the Earth (after [3]).

structive, while some are nondestructive, and each must be carefully assessed for a given technology and application.

In this chapter we present a detailed analysis of radiation effects in SiGe HBTs, by first reviewing several fundamental concepts for discussing radiation tolerance, and then examining in depth the impact of ionizing radiation on both dc and ac device performance. We then discuss the circuit-level impact of radiation-induced changes in the transistors, followed by a look at the modeling and understanding of single-event phenomena in SiGe HBTs.

11.1 Radiation Concepts and Damage Mechanisms

The total ionizing dose from radiation is measured in units of *rad* (short for "radiation absorbed dose"). The *rad* is defined as 100 ergs per gram of energy absorbed in the exposed material (note that 100 rad = 1 Grey (Gy), the Gy being a more commonly used nuclear physics unit). Obviously, since each material has different absorption characteristics, the rad is a material-dependent quantity. For Si-based ICs, the two most commonly encountered radiation dose units are rad(SiO_2) and rad(Si) (1.000 rad(SiO_2) = 0.945 rad(Si)). As a rule of thumb, typical orbital missions might encounter a total accumulated ionizing dose in the range of 50–500 krad(Si) over a 10-year flight mission, and typical total accumulated proton fluences (number of incident particles per unit area) might be in the 10^{10} p/cm^2 to

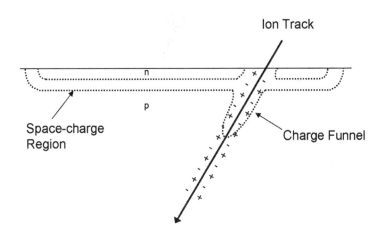

Figure 11.3 Illustration of the creation of a charge funnel along the ion track of an incident cosmic ray.

10^{12} p/cm² range over a 10-year mission, with proton energies in the range of 1 MeV to 200 MeV. For comparison purposes, it is interesting to note that a typical chest x-ray or transatlantic flight might deliver 10 mrad of radiation to an individual, and a radiation exposure of a few hundred rad is generally lethal for humans. We are thus exposing our space-electronics to very extreme radiation levels in orbital missions.

When an energetic particle passes through a semiconductor it generates electron-hole pairs along its trajectory as it loses energy (Figure 11.3). When all of the energy is lost, the particle comes to rest, having traveled a total path length referred to as its "range." A commonly encountered term in this context is the "linear energy transfer" (LET) of the particle, which describes the energy loss per unit path length (i.e., dE/dx) [4]. Particle LET has units of MeV-cm²/mg, which is the energy loss per unit path length (in MeV/cm), normalized by the density of the target material (in mg/cm³). The LET of the particle can be easily related to the charge deposition per unit path length, because for a given material it takes a known amount of energy to generate an electron-hole pair. For instance, in silicon, one electron-hole pair is generated for every 3.6 eV of energy loss, and silicon has a density of 2,328 mg/cm³, so that an LET of 97 MeV-cm²/mg corresponds to a charge deposition of 1 pC/μm. This conversion factor between energy loss and charge deposition of roughly 100× is a useful rule of thumb. In general, the higher a particle's LET, or equivalently, the denser its charge track, the higher the probability that it will induce a SEE in a susceptible device. As in nuclear physics, the probability of a given SEE can be expressed with the concept of SEE "cross section" (σ), having areal dimensions (e.g., cm^2), according to [5]

$$\sigma = \frac{N}{\phi}, \qquad (11.1)$$

where N is the number of observed upsets (errors) and ϕ is the particle fluence. Experimental SEE-induced error cross sections as a function of particle LET can usually be fit by a so-called Weibull curve, according to [1]

$$\sigma(LET) = \left[1 - e^{-\{\frac{LET-a}{b}\}^c}\right] \sigma_{sat}, \qquad (11.2)$$

where a, b, and c are fitting parameters, and σ_{sat} is the saturated cross section. Typically, for incident ions with sufficient LET to induce an SEE, the measured cross section is correlated with the physical location (volume) inside the device or circuit that is most vulnerable to upset, but this relationship can be complex and difficult to quantify. For the case of heavy ions (and protons), the radiation exposure in *rads* is determined from the incident ion LET and the particle fluence, according to [1]

$$rad = LET \left[\frac{MeVcm^2}{mg}\right] \times fluence \left[\frac{1}{cm^2}\right] \times 1.60 \times 10^{-5} \left[\frac{mg\,rad}{MeV}\right], \qquad (11.3)$$

where the LET and the corresponding dose for a given particle fluence are material dependent quantities (e.g., rad(Si) or rad(SiO$_2$)). For example, a 100-MeV proton has an LET in silicon of 5.93×10^{-3} MeV-cm^2/mg. For a fluence of 1 proton/cm^2, the corresponding radiation dose would be 9.5×10^{-8} rad(Si). The basis for the use of the *rad* to facilitate the description of the effects of ionizing radiation assumes that the effects of the radiation on devices and circuits will be equivalent for a given amount of absorbed dose, irrespective of the radiation source (be it electronics or protons or photons, etc.). This assumption has been tested extensively, for instance, for the important terrestrial radiation sources of 10-keV x-rays and cobalt-60 gamma rays often used in component qualification.

For incident particles that possess both mass and charge (e.g., protons), we also define a concept called the nonionizing energy loss rate (NIEL), which quantifies the portion of the particle energy lost to atomic displacement damage [6]. NIEL is a strong function of particle energy and is analogous to (and has the same units as) LET for ionizing radiation. The NIEL can be calculated analytically from first principles based on differential cross sections and particle-atom interaction kinematics. NIEL is that part of the energy introduced via Coulombic (elastic),

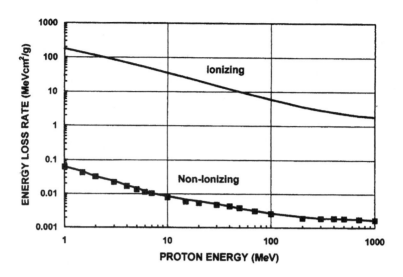

Figure 11.4 Comparison between the energy loss rate through ionization (LET) and atomic displacement (NIEL) in silicon for protons over a wide range of incident energies.

nuclear elastic, and nuclear inelastic collisions, which produce the initial vacancy-interstitial pairs (and phonons). Theoretically,

$$NIEL = \frac{N}{A} \int L[T(\Theta)] \, T(\Theta) \left\{ \frac{d\sigma}{d\Omega} \right\} d\Omega, \qquad (11.4)$$

where N is Avogadro's number, A is the gram atomic weight of the target material (Si), $d\sigma/d\Omega$ is the differential cross section for recoil in direction Θ, $T(\Theta)$ is the recoil energy, and $L[T(\Theta)]$ is the fraction of the recoil energy that goes into displacements [6]. Figure 11.4 shows a comparison between the energy loss rate through ionization (LET) and atomic displacement (NIEL) in silicon for protons over a wide range of incident energies. In principle, one can correlate calculated NIEL values to actual measured device damage, and thus predict the proton response of a device technology. This is accomplished by introducing the appropriate "damage factor," as discussed below, and is often normalized to 1.0 MeV-equivalent (Si) neutron damage factor, which is a pure displacement mechanism since neutrons carry no charge.

The damage mechanisms associated with the exposure of Si devices to ionizing radiation has been a subject of intense scientific inquiry for many years. Historically, unhardened Si-based electronics suffered severe degradation at a few 10s

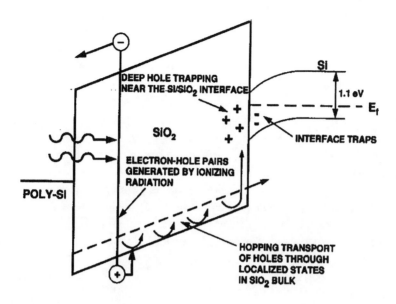

Figure 11.5 Illustration of the effects of ionizing radiation on oxides.

of krad total dose exposure. For MOSFETs, radiation-induced gate oxide trapped charge degrades the threshold voltage and transconductance, interface damage degrades the subthreshold conduction characteristics, and damage to the isolation produced a parasitic conduction path between the source and drain, leading to high off-state leakage (Figure 11.5 [4]). For Si BJTs, damage to the emitter-base spacer oxide produces a parasitic G/R center leakage that degrades the base current, leading to current gain collapse, and displacement effects lead to increased base resistance, decreased carrier lifetime, and general degradation of the dynamic response. As the radiation dose accumulates over time, MOSFETs and BJTs continue to degrade, shifting circuit operating points, and making system performance difficult to predict and often unstable. While technology scaling generally improves this scenario since device volumes and sensitive areas naturally decrease, it is still true today that even modest radiation levels generally have a very severe impact on device and circuit performance and reliability.

From a space system perspective, radiation hardening of a given semiconductor technology is a costly venture, and usually requires both significant fabrication changes (e.g., using harder oxides or guard ring structures), circuit and device layout changes, as well as extensive new component qualification and testing procedures ("space hardness assurance"). The growing proliferation of commercial space-based communications platforms, defense and verification related satellites, and exploratory space missions (both orbital and interplanetary) has spawned an entire niche industry for space-qualified electronics. A cursory glance of current space-qualified components indicates that they typically are both lower in overall performance as well as considerably more expensive than their counterpart terrestrial ICs.

There are currently two rapidly growing thrusts within the space community: 1) the use of commercial-off-the-shelf (COTS) parts whenever possible for spaceborne systems as a cost-saving measure; and 2) the use of system-on-a-chip integration to lower chip counts and system costs, as well as simplify packaging and lower total system launch weight. The "holy-grail" in the realm of space electronics can thus be viewed as a conventional terrestrial IC technology with a system-on-a-chip capability, which is also radiation-hard as fabricated, without requiring any additional process modifications or layout changes. It is within this context that we discuss SiGe HBT BiCMOS technology as such a "radiation-hard-as-fabricated" IC technology with potentially far-ranging implications for the space community.

11.2 The Effects of Radiation on SiGe HBTs

The response of SiGe HBTs to a variety of radiation types has been reported, including: gamma rays [7], neutrons [8], and protons [9, 10]. Since protons induce both ionization and displacement damage, they can be considered the worst case for radiation tolerance. For the following results, a relevant proton energy of 46 MeV was used, and at the highest proton fluence (1×10^{14} p/cm^2), the measured equivalent gamma dose was over 1.5 Mrad(Si), far larger than most orbital missions require. Proton energy effects are discussed below.

11.2.1 Transistor *dc* Response

The typical response of a SiGe HBT to irradiation can be seen in Figure 11.6, which shows typical measured Gummel characteristics of the 0.5 × 2.5μm^2 SiGe HBT, both before and after exposure to protons [9]. As expected, the base current increases after a sufficiently high proton fluence due to the production of G/R trapping centers, and hence the current gain of the device degrades. There are two main physical origins of this degradation. The base current density is inversely proportional to the minority carrier lifetime in the emitter, so that a degradation of the hole lifetime will induce an increase in the base current. In addition, ionization damage due to the charged nature of the proton fluence produces interface states and oxide trapped charges in the spacer layer at the emitter-base junction. These G/R centers

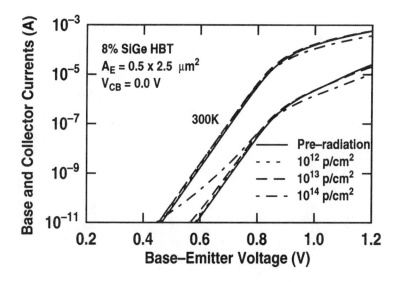

Figure 11.6 Gummel characteristics as a function of proton fluence.

also degrade I_B, particularly if they are placed inside the EB space charge region, where they will yield an additional nonideal base current component (non-kT/q exponential voltage dependence).

By analyzing a variety of device geometries, it can be shown that the radiation-induced excess base current is primarily associated with the EB spacer oxide at the periphery of the transistor, as naively expected [7]. The degradation of the current gain as a function of collector current for the preirradiated sample, and after exposure to three proton fluences, is shown in Figure 11.7. For fluences up to 1×10^{13} p/cm^2 the peak current gain at 10 µA does not show a visible degradation, and at 1×10^{14} p/cm^2 a degradation of only about 8% compared to the preirradiated device is observed. This suggests that these SiGe HBTs are robust to TID for typical orbital proton fluences for realistic circuit operating currents above roughly 100 µA without any additional radiation hardening. These results are significantly better than for conventional diffused or even ion-implanted Si BJT technologies (even radiation-hardened ones).

Small, but observable, changes in the post-irradiated collector current were also observed. Previous neutron and gamma irradiation studies of SiGe HBT technology also showed small but observable changes in the collector current with increasing radiation levels. This observed collector current shift with radiation fluence is caused by the shrink of the neutral base boundary on both the emitter and collector side due the modulation of the net charge density in the space charge region

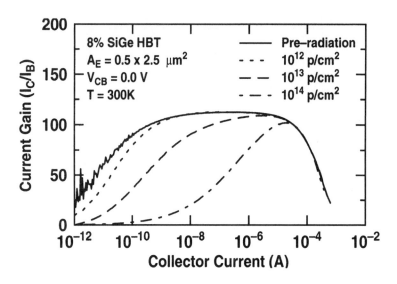

Figure 11.7 Current gain as a function of bias current for multiple proton fluences.

by the radiation-induced traps. This effectively reduces the base Gummel number (narrower base), thereby increasing the collector current. (Previous studies using neutron irradiation [8] indicate that base dopant deactivation due to displacement is a small effect in these devices due to their very thin base widths.) In the present graded-base SiGe HBT, the collector current depends exponentially on the Ge content seen at the EB side of the neutral base, and is therefore expected to show a stronger change in I_C with radiation compared to a Si BJT with a comparable doping profile, and the experimental data confirm this, where the SiGe HBT collector current increases by 35% over its original value, while there is no observable shift in I_C for the Si BJT. In addition, 2-D simulations were used to confirm the understanding of this phenomenon. By introducing a trap density inside the EB and CB space charge regions, a qualitative match with the observed trends was obtained. Interestingly, it was found that both the trap density inside the volume of the device and the energy location of the trap (i.e., either donor or acceptor level trap) strongly influences the nature of the redistribution of the post-irradiated EB and CB space charge region and hence the change of I_C with damage (both in magnitude and direction of change – i.e., it can in principle produce an increase or decrease in I_C). This may potentially explain the differences in the observed I_C change with radiation for neutron and proton samples (I_C increased with increasing fluence for protons [9], but decreased for neutrons [8]).

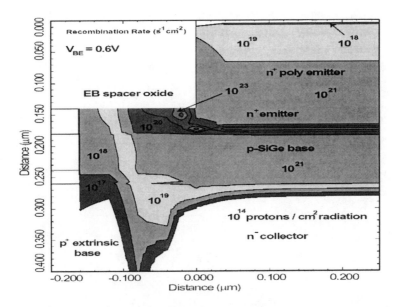

Figure 11.8 Calibrated MEDICI simulation of the recombination rates inside the SiGe HBT after radiation.

11.2.2 Spatial Location of the Damage

Of particular interest is the inference of the spatial location of the proton-induced traps in these devices [9]. The existence of proton-induced traps in the EB space charge region is clearly demonstrated by the G/R-induced increase in the nonideal base current component shown in the Gummel characteristics. The existence of radiation-induced traps in the collector-base space charge region was verified by measuring the inverse mode Gummel characteristics of the device (emitter and collector leads swapped). In this case the radiation-induced traps in the CB junction now act as G/R centers in the inverse EB junction, with a signature non-kT/q exponential slope. 2-D simulations were calibrated to both measured data for the pre- and post-irradiated devices at a collector-base voltage of 0.0 V. In order to obtain quantitative agreement between the simulated and measured irradiated results, traps must be located uniformly throughout the device, and additional interface traps must be located around the emitter-base spacer oxide edge (Figure 11.8). Most of the radiation-induced recombination occurs inside the EB space charge region, leading to a nonideal base current, as expected.

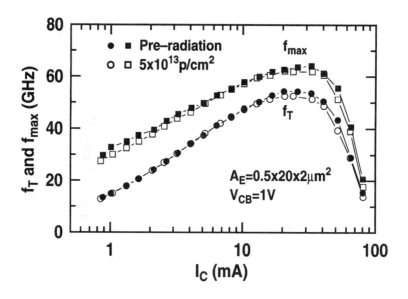

Figure 11.9 Cutoff frequency and maximum oscillation frequency as a function of bias current for multiple proton fluence.

11.2.3 Transistor *ac* Response

To assess the impact of radiation on the *ac* performance of the transistors, the S-parameters were measured to 40 GHz both before and after proton exposure. None of the four S-parameters suffered any appreciable degradation up to 1×10^{14} p/cm^2 proton fluences [9, 10]. From measured S-parameters, the transistor cutoff frequency (f_T) and maximum oscillation frequency (f_{max}) were extracted. A comparison of these two important *ac* parameters between preirradiation and 5×10^{13} p/cm^2 is shown in Figure 11.9. Only a slight degradation in f_T and f_{max} is observed, the latter being expected from the minor increase of the base resistance with irradiation, as described below. Observe that the high current f_T roll-off due to Kirk and heterojunction barrier effect is not changed with irradiation. Because the high-current density roll-off is extremely sensitive to the collector doping level, it suggests that there is no appreciable deactivation of donors in the collector due to displacement damage, despite the existence of irradiation-induced traps in the collector.

11.2.4 Si versus SiGe and Structural Aspects

Finally, we note that careful comparisons between identically fabricated SiGe HBTs and Si BJTs (same device geometry and wafer lot, but without Ge in the base for the epitaxial-base Si BJT), show that the extreme level of TID tolerance of SiGe HBTs is not per se due to the presence of Ge. That is, the proton response of both the epitaxial base SiGe HBT and Si BJT are nearly identical. We thus attribute the observed radiation hardness to the unique and inherent structural features of the device itself, which from a radiation standpoint can be divided into three major aspects:

- In these epitaxial base structures, the extrinsic base region is: 1) very heavily doped ($> 5 \times 10^{19}$ cm^{-3}); and 2) located immediately below the EB spacer oxide region, effectively confining any radiation-induced damage, and its effects on the EB junction.

- The EB spacer, known to be the most vulnerable damage point in conventional BJT technologies, is thin (< 0.20 μm wide) and composed of an oxide/nitride composite, the latter of which is known to produce an increased level of radiation immunity.

- The active volume of these transistors is very small ($W_E = 0.5$ μm, and $W_b < 150$ nm), and the emitter, base, and collector doping profiles are quite heavily doped, effectively lessening the impact of displacement damage.

Further results of the effects of device scaling on radiation response are given below, as well as a discussion of the impact on the CMOS transistors in the SiGe HBT BiCMOS technology intended for system-on-a-chip applications. We also note that these SiGe HBTs compare very favorably in both performance and radiation hardness with (more expensive) GaAs HBT technologies that are often employed in space applications requiring both very high speed and an extreme level of radiation immunity [11].

11.2.5 Proton Energy Effects

Because incident protons deposit more of their energy (both ionization and displacement) inside the device as their energy decreases, transistor characteristics generally degrade more rapidly under low energy proton irradiation than for high energy proton irradiation. Given that a realistic space environment necessarily contains a wide range of particle energies (from several MeV to hundreds of MeV), characterization of transistor response as a function of energy is important.

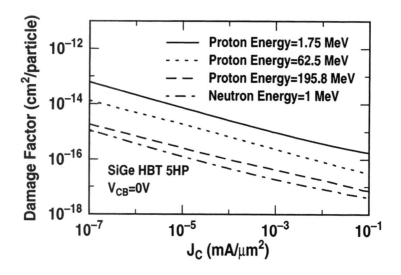

Figure 11.10 Extracted damage factor as a function of bias current for 1 MeV neutrons and multiple proton energies.

In order to examine the energy dependence of proton-induced damage in SiGe HBTs, an appropriate "damage factor" must first be defined. It has been repeatedly demonstrated that over a large range of proton energies and device technologies, the reciprocal gain increases linearly with incident particle fluence (ϕ), as reflected in the Messenger-Spratt equation [6]

$$\frac{1}{\beta(\phi)} = \beta_0^{-1} + K(E)\phi, \quad (11.5)$$

where β_0 is the preradiation current gain, and K is the (energy-dependent) damage factor. In practice, the reciprocal gain versus proton fluence for bipolar transistors typically only behaves linearly over a certain proton fluence range, since both displacement damage and ionization damage exist for proton irradiation. Therefore, both proton and gamma radiation experiments are in principle needed to determine the damage factor. Conventionally, the following procedure is used to extract the displacement damage factor: 1) plots of reciprocal gain versus total ionizing dose as a function of collector current are made after gamma irradiation; 2) these plots are then approximated by straight lines over the dose range corresponding to the proton irradiation experiments; and finally, 3) the slopes of these plots are then subtracted from the slopes of reciprocal gain versus proton fluence curves for the proton irradiation experiments in order to obtain the corresponding damage factor. Figure 11.10 shows the extracted damage factor as a function of collector

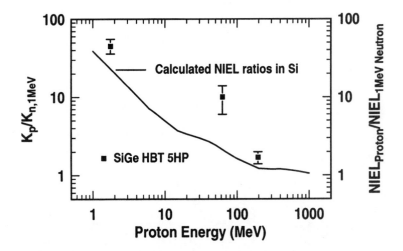

Figure 11.11 Comparison of damage-factor ratios and calculated NIEL ratios as a function of proton energy for SiGe HBTs (using 1 MeV neutron data as a normalization reference).

current density for these first generation SiGe HBTs as a function of proton energy. Clearly, 1-MeV neutrons, because they produce only displacement damage, represent the most benign form of radiation. Protons are expected to produce more serious damage than either the neutron or gamma irradiation, and this is indeed the case. Furthermore, as the proton energy decreases, in this case from 196 MeV to 63 MeV to 1.8 MeV, more energy is deposited in the devices and the damage factor increases (worsens), as expected [12]. A comparison of measured damage-factor ratios and calculated NIEL ratios as a function of proton energy for SiGe HBTs (using 1 MeV neutron data as a normalization reference) is shown in Figure 11.11.

11.2.6 Low-Dose-Rate Gamma Sensitivity

Within the past few years, a pronounced low-dose-rate sensitivity to gamma irradiation that is not screened by the current test methods for ionizing radiation has been observed in bipolar technologies [13]. Under military standard 883, method 1019.4, all total-dose tests are performed at a dose rate between 50-300 rad(Si)/sec. The enhancement in device and circuit degradation at low gamma dose rates has come to be known as "Enhanced Low Dose Rate Sensitivity" (ELDRS). The ELDRS effect was first reported in 1991 [14], which demonstrated that existing radiation hardness test assurance methodologies were not appropriately considering worst case conditions.

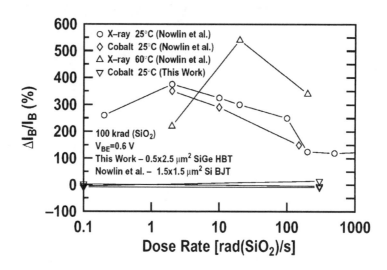

Figure 11.12 Normalized base current as a function of gamma radiation dose rate, for both Si BJTs and SiGe HBTs.

The physical origins underlying ELDRS have been hotly debated for years, and numerous mechanisms proposed. Recent attempts to understand ELDRS include a model suggesting that the lower net radiation induced trapped charge density (ΔN_{ot}) at high-dose-rates is a result of a space charge phenomenon, caused by delocalized hole traps, known as E'_δ centers, which occur in heavily damaged oxides such as bipolar base oxides [15]. These traps can retain holes on a timescale of seconds to minutes, causing a buildup of positive charge in the oxide bulk during high-dose-rate irradiation. This is in contrast to low-dose-rate irradiation, where the irradiation time is much longer, effectively allowing the holes in the E'_δ centers to be detrapped. Thus, in the high-dose-rate case, the larger total trapped hole density forces holes near the interface to be trapped closer to the interface, where they can be compensated by electrons from the silicon. This lowers the resultant *net trapped charge density*. It has also been found that these E'_δ centers anneal at relatively low temperatures ($\leq 150°C$). This suggests that high-dose-rate irradiation at a higher temperature may allow holes to be detrapped, hence mimicking a low-dose-rate radiation response. The assumptions commonly employed in such models, however, are typically very technology specific, and quantifying ELDRS (if present) in SiGe technology is obviously important from a hardness assurance perspective.

To assess ELDRS in SiGe technology, low-dose-rate (0.1 rad(Si)/sec) and

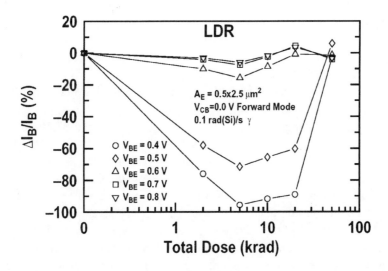

Figure 11.13 Normalized base current at multiple V_{BE} values as a function of total dose.

high-dose-rate (300 rad(Si)/sec) experiments were conducted using Cobalt-60 (i.e., 1.43-MeV gamma rays) [17]. The devices were irradiated with all terminals grounded to a total dose of 50 krad(Si) and the forward mode and inverse mode characteristics measured at incremental doses. As can be seen in Figure 11.12, low-dose-rate effects in these SiGe HBTs were found to be nearly nonexistent, in striking contrast to reports of strong ELDRS in conventional Si bipolar technologies. We attribute this observed hardness to ELDRS to the same mechanisms responsible for the overall radiation hardness of the technology, and is likely more structural in nature than due to any unique advantage afforded by the SiGe base. Interestingly, an anomalous decrease in base current was also found in these devices at low-dose-rates (Figure 11.13), suggesting that a new physical phenomenon is present at low-dose-rates in these devices.

In this case, the Gummel characteristics show a decreasing base current until about 5 krad(Si), followed by an increase in base current at higher doses. This suggests that in the initial stages of irradiation, the G/R center dominated recombination process actually decreases in magnitude. For $V_{BE} = 0.4$ V, this I_B decrease is found to be as much as 100%. It is also found that although the base current starts increasing again for total doses ≥ 5 krad(Si), the base current at up to 20 krad(Si) is still equal to its preradiation value. It can thus be inferred that at low-dose-rates two competing mechanisms operate as the total dose increases, one that decreases the G/R leakage, and one that increases the G/R leakage.

A logical question, then, is what sort of physical damage process can create such an anomalous base current decrease? In particular, it is important to understand why: 1) under low-dose rate conditions, the normalized base current in the forward-operated SiGe HBT strongly decreases with dose at low V_{BE}, but only weakly decreases with dose at high V_{BE}; and 2) why the base current first decreases with total dose and then increases.

2-D simulations using MEDICI [16] were used to confirm a plausible explanation for these observations. It is proposed that the preradiation deep-level traps at the surface are initially annealed by the gamma radiation to a shallower energy level. As more deep traps evolve into shallower traps, the G/R center dominated recombination decreases, giving a reduction in base current. This occurs because the net recombination rate decreases as the trap energy level moves away from midgap. The magnitude of the leakage decrease depends on the quasi-Fermi levels and hence the EB bias conditions. As confirmed with simulation, this proposed mechanism can indeed give rise to a decrease in base current at low V_{BE} without a large change at high V_{BE}, consistent with the experimental observations. At the same time, radiation-induced traps are being generated in the device, which ultimately halts the I_B decrease and causes an increase in I_B.

This occurs because at greater values total dose, the deep-level traps generated by radiation outnumber those annealed to shallower levels. At low V_{BE}, the base current is strongly dependent on the nature and location of traps in the EB junction. In the forward-mode measurements, this junction has shallow-level traps due to annealing and hence a decrease in I_B is observed. At higher values of V_{BE}, the higher injection level leads to recombination in the bulk as well, which leads to an I_B increase. In the inverse mode measurements, the I_B increase observed at low V_{BE} is due to the deep-level trap induced recombination in low-injection. This corresponds to the bulk traps. At higher values of V_{BE}, the higher injection leads to a base current that is dependent on the recombination near the EB spacer at the surface as well. This region, however, has an abundance of shallow-level traps and thus, the recombination actually decreases, leading to an I_B decrease. These kinds of proposed trap dynamics, which are known to be dose rate dependent, have been previously reported in the literature [18].

11.2.7 Broadband Noise

The broadband noise performance of SiGe HBTs, as reflected in NF_{min}, $\Gamma_{G,opt}$, and R_n (refer to Chapter 7), is critical for space-borne transceivers and communications platforms. Characterization of the transistor S-parameters both before and after proton exposure show minimal changes in f_T and f_{max}, suggesting that

Radiation Tolerance

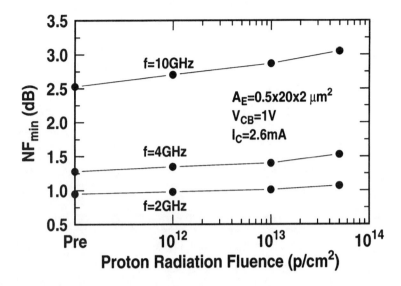

Figure 11.14 Extracted minimum noise figure as a function of proton fluence for multiple frequencies.

noise performance should be relatively unaffected by radiation. As shown in Figure 11.14, this is indeed the case. For these SiGe HBTs, NF_{min} degrades only slightly at 2.0 GHz after an extreme proton fluence of $5x10^{13}$ p/cm^2 (from 0.95 dB to a still-excellent value of 1.07 dB, a 12.6% degradation). In the bias range of interest for most RF circuits (> 0.1 mA/μm^2), these SiGe HBTs show virtually no degradation in current gain for proton fluences up to $5x10^{13}$ p/cm^2. Therefore, at fixed bias current, the observed minimal degradation in noise figure results mostly from the increase in thermal noise associated with the base and emitter resistances (Figure 11.15). Changes in $\Gamma_{G,opt}$ and R_n are also small.

11.2.8 Low-Frequency Noise

SiGe HBTs have the desirable feature of low $1/f$ noise commonly associated with Si bipolar transistors [19], which is of great importance because upconverted low-frequency noise (phase noise) typically limits the spectral purity of communication systems. Understanding the effects of radiation on $1/f$ noise in SiGe HBTs thus becomes a crucial issue for space-borne communications electronics. To shed light on these issues, the noise power spectrum was measured on SiGe HBTs from 1.0 Hz to 100.0 kHz both before and after 63-MeV proton irradiation [20]. The preirradiation base current $1/f$ noise is typically proportional to I_B^α and inversely

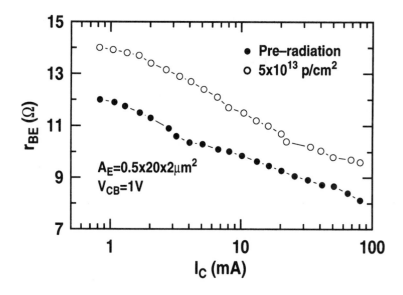

Figure 11.15 Extracted total base and emitter resistance as a function of bias current, before and after proton exposure.

proportional to the emitter junction area A_E in modern transistors

$$S_{I_B} = \frac{K}{A_E} I_B^\alpha \frac{1}{f}, \quad (11.6)$$

where K is a technology dependent constant. It is generally agreed that the exponent α provides information on the physical origin of the trap states contributing to $1/f$ noise. From the preirradiation data, α is close to 2 in these samples, indicating that the physical origin of the $1/f$ noise is due to carrier number fluctuations. Physically, $1/f$ noise results from the presence of G/R center traps in the transistors, from which trapping-detrapping processes occur while carriers flow inside the device, thus modulating the number of carriers (and hence currents) to produce $1/f$ noise. The pre-irradiation low-frequency noise spectrum in these SiGe HBTs is typically $1/f$, with an I_B^2 dependence, while $S_{I_B} \times A_E$ is almost independent of A_E. The I_B^2 and $1/A_E$ dependencies of S_{I_B} are strong indicators of uniformly distributed noise sources over the entire emitter area [19]. After 2×10^{13} p/cm² proton irradiation, the low-frequency noise spectrum remains $1/f$ in frequency dependence, and free of G/R (burst) noise. Interestingly, however, the relative increase in $1/f$ noise ($S_{I_B,post}/S_{I_B,pre}$) is minor in the 0.5×1.0 μm² transistor, but significant in the 0.5×10.0 μm² transistor, as shown in Figure 11.16 and Figure 11.17, respectively. As a result, S_{I_B} is no longer in proportion to $1/A_E$ after irradiation.

Note that the 0.5 × 10.0 μm^2 transistor had a $1/f$ noise that is 1/10 of the $1/f$ noise in the 0.5 × 1.0 μm^2 transistor before irradiation. However, after irradiation, the $1/f$ noise in the 0.5 × 10.0 μm^2 transistor becomes only one third of the $1/f$ noise in the 0.5 × 1.0 μm^2 transistor. The generally accepted benefit of obtaining lower $1/f$ noise by using a larger transistor is thus significantly compromised by irradiation.

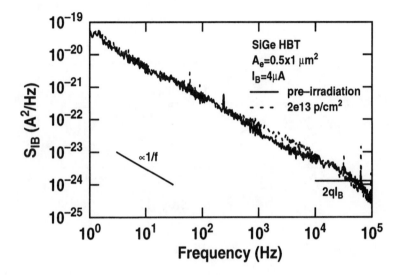

Figure 11.16 Input-referred base current PSD for a 0.5 × 1.0 μm^2 transistor, before and after irradiation.

The bias current dependence of $1/f$ noise also changes after irradiation, depending on the emitter area. The relative degradation of $1/f$ noise ($S_{I_B,post}/S_{I_B,pre}$) is minor in the smallest device (0.5 × 1.0 μm^2), and S_{I_B} remains $\propto I_B^2$. For the largest device, whose relative $1/f$ noise degradation is the highest, S_{I_B} becomes $\propto I_B^{1.5}$. The relative $1/f$ noise degradation (increase) is negligible for the smallest transistor (0.5 × 1.0 μm^2), but significant for the largest transistor (0.5 × 10.0 μm^2). The apparently "minor" relative degradation in the small transistor, however, can be deceptive, because its preirradiation $1/f$ noise is 10× the $1/f$ noise of the large transistor. One possible explanation is that the absolute increases of $1/f$ noise are comparable in the two devices with different geometries. These increases are minor compared to the preirradiation $1/f$ noise of the small transistor, but significant compared to the preirradiation $1/f$ noise of the large transistor (1/10 the preirradiation $1/f$ noise in the small transistor). The proton-induced absolute increase (degradation) of $1/f$ noise is comparable for the 0.5 × 1.0 μm^2 and 0.5 × 10.0 μm^2

transistors, despite a 10× emitter area difference. Such a weak emitter area dependence of radiation-induced $1/f$ noise is counterintuitive, and cannot be explained by existing $1/f$ noise theories.

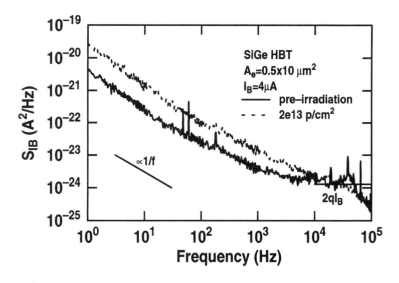

Figure 11.17 Input-referred base current PSD for a 0.5×1.0 μm^2 transistor, before and after irradiation.

It is well known that proton irradiation introduces G/R centers in bipolar transistors, and hence creates a nonideal base current component due to increased space-charge region G/R center recombination leakage. We note as well that sufficiently large amounts of radiation damage can induce a classical Lorentzian-type G/R noise signature in the noise power spectral density, along with a random-telegraph-signal (RTS) time response [7]. A significant nonideal base current component due to space-charge-region (SCR) recombination ($I_{B,SCR}$) can be observed after irradiation in these devices. While the contribution of $I_{B,SCR}$ to the total measured I_B is negligible in the bias range of interest for analog and RF circuits (i.e., $> 0.1 \mu A$), $I_{B,SCR}$ is dominant and can be directly measured in the low bias range (e.g., $V_{BE} < 0.4$ V). Since $I_{B,SCR}$ is proportional to $e^{qV_{BE}/nkT}$, $1 < n < 2$, the measured $I_{B,SCR}$ data in the low bias range can fitted and then extrapolated to the high bias range, demonstrating that the peripheral density of radiation-induced SCR base current ($I_{B,SCR}/P_e$) is approximately the same for all of the transistors, and shows an $e^{qV_{BE}/2kT}$ dependence. Since this current does not vary as $e^{qV_{BE}/kT}$ as in an ideal base current and it scales with the emitter perimeter P_E, most of the $I_{B,SCR}$ comes from recombination at the surface of the EB junction near the oxide spacer and not via bulk recombination. Thus $I_{B,SCR}$ can be expressed as

$$I_{B,SCR} \propto e^{qV_{BE}/2kT} P_E n_T, \quad (11.7)$$

where P_E is the emitter perimeter, n_T is areal trap density at the surface, and is assumed to vary only with radiation fluence.

This SCR recombination near the surface is a very noisy process, and the associated noise current can be described as a current generator between the base and emitter terminals of the transistor. It has been shown that surface $1/f$ noise generated in the EB space charge region due to trap recombination can be expressed by a modified Hooge-type equation [21, 22]

$$S_{I_{B,SCR}} = I_{B,SCR}^2 \frac{\alpha_H}{f\,N_T}, \quad (11.8)$$

where N_T is the number of traps at the EB space-charge region surface, and α_H is the so-called Hooge parameter [23, 24]. Here, N_T is given by $n_T L_{SCR} P_E$, where L_{SCR} is the length of EB space-charge region at the surface.

In the RF bias range, I_B remains dominated by hole injection into the emitter, and is practically unaffected by $I_{B,SCR}$, and is thus given by

$$I_B \propto e^{qV_{BE}/kT} A_E. \quad (11.9)$$

It is desirable to express $I_{B,SCR}$ in terms of I_B to facilitate interpretation of the measured $1/f$ noise data. Such an expression can be obtained by inspection of (11.7) and (11.9)

$$I_{B,SCR} \propto \frac{I_B^{0.5}}{A_E^{0.5}} P_E n_T, \quad (11.10)$$

and $S_{I_{B,SCR}}$ can then be expressed in terms of I_B by substituting (11.10) into (11.8) to yield

$$S_{I_{B,SCR}} = C\, I_B\, n_T \frac{P_E}{A_E} \frac{\alpha_H}{f}, \quad (11.11)$$

where C is a constant that is independent of bias and geometry. Because of the change of bias and emitter area dependence after irradiation, it is unlikely that the major increase of $1/f$ noise is due to the same mechanism as preirradiation, which shows an I_B^2 and $1/A_E$ dependence. Here we assume that the major radiation-induced increase of $1/f$ noise comes from the SCR recombination current near the EB surface, and thus the post-irradiation noise can be written as

$$\begin{aligned} S_{I_B,post} &= S_{I_B,pre} + S_{I_B,SCR} \\ &= \frac{K}{A_E} I_B^2 \frac{1}{f} + C\, I_B\, n_T \frac{P_E}{A_E} \frac{\alpha_H}{f}, \end{aligned} \quad (11.12)$$

where the preirradiation noise is given by (11.6). Curve-fitting the data to this equation results in a value of $K = 1.1 \times 10^{-21}$ m^2, and $C\, n_T\, \alpha_H = 2.8 \times 10^{-22}$ Am. A number of important observations can be made from (11.12) [25, 26]:

- Radiation-induced $1/f$ noise $S_{I_{B,SCR}}$ increases with trap density n_T, and hence proton fluence.

- At a given I_B, the radiation-induced $1/f$ noise is proportional to P_E/A_E instead of $1/A_E$. The three transistors examined here have approximately the same P_E/A_E ratio, and thus should have approximately the same $S_{I_{B,SCR}}$. This is consistent with the measured data. More recent measurements on devices with very different P/A ratios confirm this trend.

- The relative noise degradation is reduced for smaller devices, because of the larger $1/f$ noise before irradiation.

- The radiation-induced $1/f$ noise varies with I_B instead of I_B^2. The total $1/f$ noise post-irradiation is the sum of the preirradiation $1/f$ noise and $S_{I_{B,SCR}}$, and should show a I_B^γ dependence with $1 < \gamma < 2$. Because the relative ratio of the preirradiated to the radiation-induced $1/f$ noise is proportional to $1/P_E$, γ should be close to 2 for the smallest device (least amount of relative degradation), and smaller than 2 for the largest device. This is again consistent with the experimental data.

11.3 Technology Scaling Issues

Regardless of whether for terrestrial or space-based systems, the utility of transistor "scaling" (the coordinated reduction of a device's lateral dimensions and its vertical doping profile) is a key requirement for any viable IC technology. Scaling yields faster transistors, higher packing density, reduced power dissipation, and ultimately lower cost.

11.3.1 SiGe HBT Scaling

While first generation SiGe HBTs and circuits are TID tolerant to very high proton fluences, recent experiments comparing the effects of proton exposure on three different SiGe HBT technology generations found that, while the first generation SiGe HBT was total-dose tolerant to multi-Mrad radiation levels, the first generation Si nFET was only radiation-hard to about 30–50 krad [27]. In addition, with technology scaling, the radiation tolerance of the SiGe HBT *degraded* significantly compared to the first generation devices, while the Si nFET tolerance

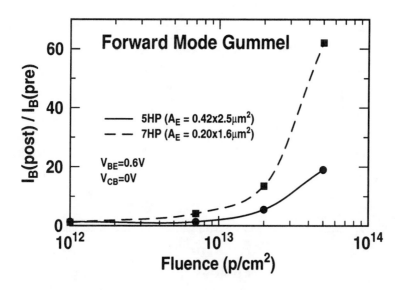

Figure 11.18 Comparison of the normalized base current in *forward* mode as a function of proton fluence for the 5HP and 7HP SiGe HBT technology generations.

improved dramatically compared to the first generation devices. We focus here on an explanation for these intriguing experimental results, and its implications for the future deployment of SiGe technology in space [28].

We have focused on two different SiGe HBT BiCMOS technology generations: SiGe 5HP (first generation) and SiGe 7HP (second generation) [29, 30]. Details of each technology can be found in Chapter 3. We emphasize that *neither SiGe technology was intentionally radiation-hardened in any way*. Significantly, one of the main differences between these two SiGe technologies is the use of a significantly thinner shallow-trench isolation (STI) layer in the 7HP process (0.24 μm for 7HP versus 0.50 μm for 5HP), which is required to ensure HBT to CMOS compatibility with scaling.

Ionizing radiation has been shown to damage the EB spacer region in these SiGe HBTs, and produce a perimeter-dependent space-charge generation/recombination (G/R) base-current leakage component that progressively degrades the base current (and current gain) as the fluence increases [27]. A comparison of this degradation mechanism between minimum-geometry 5HP and 7HP SiGe HBTs, however, shows a dramatic (and statistically repeatable) difference between the radiation response of the two technologies (Figure 11.18). To shed light on the physical location of the offending trap states, we compared the forward mode Gummel charac-

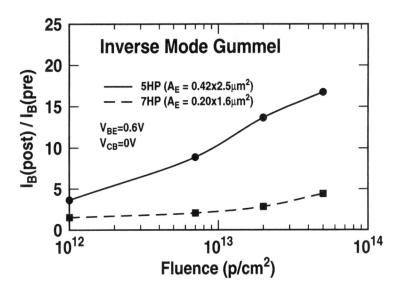

Figure 11.19 Comparison of the normalized base current in *inverse* mode as a function of proton fluence for the 5HP and 7HP SiGe HBT technology generations.

teristics with the inverse mode Gummel characteristics (i.e., emitter and collector terminals swapped, with the transistor effectively operated upside-down) [9]. Interestingly, we observe the exact opposite behavior (Figure 11.19). That is, as a function of proton fluence: 1) the 7HP SiGe HBT *forward mode* base current degrades *much more* rapidly than for the 5HP SiGe HBT; but 2) the 7HP SiGe HBT *inverse mode* base current degrades *much less* rapidly than for the 5HP SiGe HBT. These results suggest that there is a larger radiation-induced trap density in the EB space charge region for 7HP than for 5HP, while there is a smaller radiation-induced trap density in the CB junction for 7HP than for 5HP.

To understand the forward mode results, we have performed reverse-bias EB stress measurements on both 5HP and 7HP devices [31]. Since EB electrical stress depends exponentially on the local electric field under the spacer oxide, it is a useful independent means for comparing the two technologies. As can be seen in Figure 11.20, the stress-induced base current degradation shows a qualitatively similar behavior to the radiation response. That is, the 7HP device degrades much more rapidly than the 5HP device. This result is consistent with significantly higher EB electric field found under the EB spacer region in the 7HP device, which has both more abrupt doping profiles due to its reduced thermal cycle, as well as a decreased EB spacer thickness compared to the 5HP device, and has been confirmed

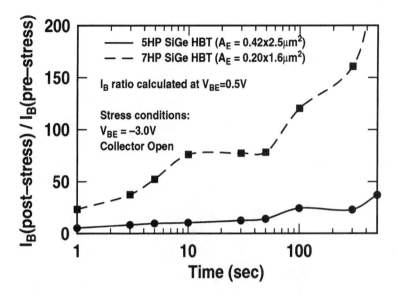

Figure 11.20 Comparison of reverse-bias EB stress damage as a function of time for the 5HP and 7HP SiGe HBTs.

with MEDICI simulations.

Measurement of the slope of the normalized base current at low V_{CB} (< 0.3 V) is a direct measure of neutral base recombination in the device, and is nearly identical for both pre- and post-radiation, indicating no significant change in the neutral base trap density. The logical conclusion is that the offending CB traps in the inverse mode characteristics are located physically along the STI edge, and thus reside in the extrinsic base CB space-charge region, where they generate excess G/R leakage. To understand why these traps produce different inverse mode leakage characteristics between the 5HP and 7HP devices, we have also performed detailed MEDICI simulations of two SiGe HBTs, one with thick STI, and one with thin STI. Figure 11.21 shows the thick STI cross-sectional simulation structure, and was based on the actual device layout. As can be seen in Figure 11.22, thinning the shallow trench (i.e., moving from 5HP to 7HP) decreases the radiation-induced CB leakage current component, since the trap region in the collector is spatially confined to a region closer to the extrinsic CB junction, resulting in fewer traps in the CB space-charge region to generate leakage. Placing only EB traps in the device results in degraded forward-mode Gummel characteristics, independent of the STI thickness, while the inverse-mode Gummel characteristics remain ideal, as expected. Additional electrical stress experiments conducted under high forward J_C and high V_{CB}, which have been shown to be an independent means for damag-

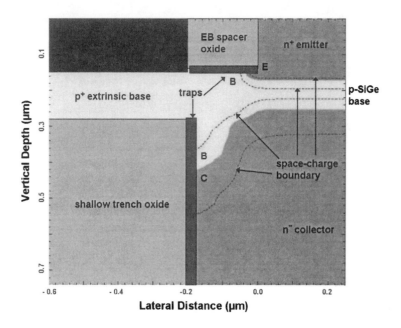

Figure 11.21 Cross section of the thick STI SiGe HBT used in the MEDICI simulations. Shown are the EB and CB space-charge regions, as well as the trap locations.

ing *both* the EB spacer and STI edge, are consistent with our radiation results: the 5HP forward-mode Gummel characteristics degrade less rapidly than 7HP, while the 5HP inverse-mode characteristics degrade more rapidly, and further corroborate our conclusions.

We note, finally, that the observed enhanced sensitivity to radiation damage in the more aggressively scaled SiGe HBTs effectively translates to a greater sensitivity to radiation-induced $1/f$ noise damage, as might be naively expected. As shown in Figure 11.23, which compares first generation and third generation SiGe HBTs, a 1.6× greater noise degradation with scaling is observed after proper normalization.

11.3.2 Si CMOS Scaling

Given that the overall radiation tolerance of SiGe HBT BiCMOS technology is gated by the *least* radiation-tolerant device, it is important that we also examine radiation effects in the Si CMOS devices. Figure 11.24 shows typical subthreshold data at maximum V_{DS} for minimum L_{eff} 5HP and 7HP nFETs (the pFETs are

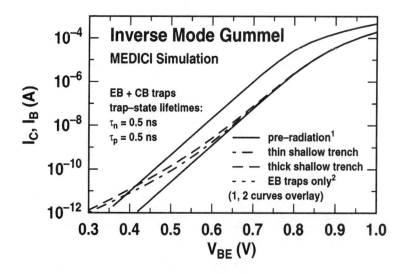

Figure 11.22 Simulated comparison of the inverse mode Gummel characteristics of a SiGe HBT with thick STI and thin STI.

radiation hard to greater than 1 Mrad(Si) and for brevity are not discussed here). Observe that the threshold voltage and transconductance degradation is negligible in both devices, as expected, since they both employ very thin, high-quality gate oxides (7.8 nm for 5HP and 4.2 nm for 7HP). For the 5HP nFET, at 100 krad(Si), the off-state leakage has increased over six orders of magnitude to about 10 μA compared to preradiation, making the devices unsuitable for most space applications, while for the 7HP nFET, the off-state leakage remains below 1.0 nA, and is clearly robust for many space applications (without radiation hardening). Figure 11.25 shows the normalized off-state leakage as a function of equivalent total dose, showing the dramatic differences between the two technologies.

To better understand these results, we have used measured TEM cross section data of the STI edge, and simulation techniques developed in [32], to construct realistic MEDICI cross sections of nFETs with both thick STI (5HP) and thin STI (7HP). The radiation-induced STI damage mechanism is assumed to produce a net positive charge along the STI edge (1×10^{12} traps/cm^2 in this case for both 5HP and 7HP), which will invert the substrate with sufficient dose, producing a parasitic edge leakage path between source and drain [32], as shown in Figure 11.26. The total measured I_{DS} in the device is then a combination of the center transport current and this parasitic edge leakage current. As can be seen in Figure 11.27, which plots the total electron charge ($Q_n \propto I_{DS}$) at the center and edge of the device, thinning the STI dramatically reduces the edge-leakage component. Physically, if

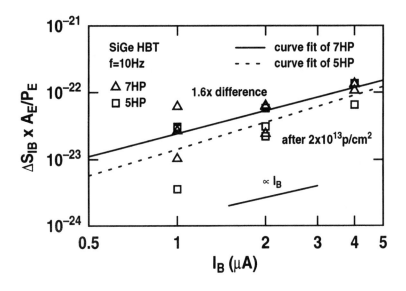

Figure 11.23 Normalized radiation-induced noise PSD as a function of base current for two different SiGe technology generations.

we assume a damage mechanism along the STI edge at the ends of the transistor where the gate overlaps the STI, the gate's ability to deplete the damage-induced inversion layer at the STI edge increases dramatically as the STI is thinned (i.e., the gate field only has a finite penetration depth at fixed bias). For the 5HP nFET, the 0.5 μm STI depth is clearly deep enough to ensure that the gate cannot turn off the radiation-induced source-to-drain edge leakage, while the STI in 7HP is sufficiently thin to control the edge leakage at 100 krad(Si).

Even in the case of the 7HP nFET, however, as the dose continues to rise, the induced STI edge damage will eventually reach a magnitude where it can no longer be adequately controlled by the gate, and the off-state leakage will begin to increase (in this case between 100-300 krad(Si), as shown in Figure 11.25). This level of radiation tolerance is, nevertheless, sufficient for many space applications and, in essence, comes for free since the technology has not been radiation-hardened in any way. Given that the 7HP SiGe HBT is also clearly TID-hard to 100 krad(Si) (1 × 10^{12} p/cm^2 = 136 krad(Si)) without any alterations, this 7HP SiGe HBT BiCMOS technology should be suitable for many orbital missions. (We note, parenthetically, that a similar improvement in off-state leakage under radiation exposure within a given technology generation (even SiGe 5HP) can be affected by appropriate application of substrate bias [32].)

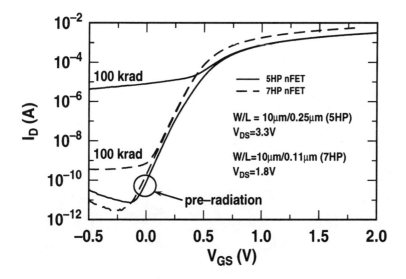

Figure 11.24 Subthreshold characteristics in saturation for the 5HP and 7HP nFETs for preradiation and after 100 krad equivalent total dose.

11.4 Circuit-Level Tolerance

For the successful deployment of SiGe technology into space-based systems, circuit-level radiation hardness is clearly more important than device-level hardness. As presented above, the proton-induced device degradation is minor in the bias range of interest to most actual circuits (typically $I_C > 100$ μA).

11.4.1 The Importance of Transistor Bias

An initial relevant question within this context is the extent to which the transistor terminal bias *during irradiation* affects the radiation response. Most commonly, SiGe HBTs are irradiated either with all terminals (E/B/C) grounded or with all terminals floating. No significant difference has been found between these two bias conditions. In real circuit applications, however, the SiGe HBTs necessarily experience a wide variety of operating bias conditions that differ from grounded or floating bias, and thus rigorous hardness assurance requires a deeper look at the bias condition sensitivity. Many different bias configurations are relevant for bipolar circuits. In nonsaturating, high-speed logic families such as CML or ECL, for instance, the transistor operates only under forward-active bias. For most analog circuits, the transistor is also biased in forward-active mode. For certain RF power applications, the transistor can experience saturation mode bias. Finally, in certain

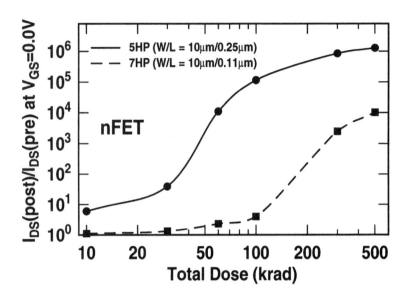

Figure 11.25 Normalized off-state leakage current for the 5HP and 7HP nFETs as a function of equivalent total dose.

BiCMOS logic families, or even during switching transients in nonsaturating logic families, the EB junction of the transistor can become reverse-biased. To quantify the impact of terminal bias condition during irradiation on the measured radiation tolerance, representative transistors were held at different bias conditions during radiation exposure, and then compared after specific accumulated fluences [34]. When the desired fluence was reached, the bias was set to ground on all terminals, and the samples were removed from the beam and immediately measured. The appropriate bias was returned to the DUT and then it was reinserted into the proton beam until the next desired fluence was reached. This process was repeated in a controlled manner throughout the experiment. Figure 11.28 compares the normalized current gain degradation as a function of fluence for three relevant bias configurations: 1) all-terminals-grounded; 2) forward-active mode; and 3) reverse-biased EB junction. As can be seen, there are no large differences between the three bias conditions, and the all-terminals-grounded condition represents a close to worst case scenario (and is comparable to the all-terminals-floating condition).

In order to assess the impact of radiation exposure on actual SiGe HBT circuits, we have compared two very important, yet very different circuit types, one heavily used in analog ICs (the bandgap reference circuit), and one heavily used in RFICs (the voltage controlled oscillator) [33]. Each circuit represents a key building block for realistic SiGe ICs that might be flown in space. Each of these SiGe HBT circuits

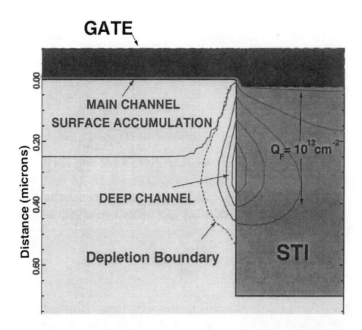

Figure 11.26 MEDICI simulated potential contour and depletion boundary at V_{GS} = -1.0 V and V_{SUB} = 0 V ($Q_F = 10^{12}/\text{cm}^2$, and the potential step is 0.2 V).

was designed using fully calibrated SPICE models, layed-out, and then fabricated on the same wafer to facilitate unambiguous comparisons. In addition, because any realistic RF IC must also necessarily include passive elements such as monolithic inductors and capacitors, we have also investigated the effects of proton exposure on an RF LC bandpass filter.

11.4.2 Bandgap Reference Circuits

The bandgap reference (BGR) circuit has been widely used as a voltage reference source in A/D and D/A converters, voltage regulators, and other precision analog circuits, due to its good long-term stability and its ability to operate at low supply voltages (see discussion in Chapter 6). This SiGe HBT BGR employed a conventional circuit architecture, and did not include any special temperature compensation circuitry [35]. In Si BJT BGRs, radiation-induced degradation in output voltage and temperature sensitivity are of particular concern [36]. Because the BGR core transistors operate at constant collector current, any radiation-induced changes that influence the matching properties between large and small area de-

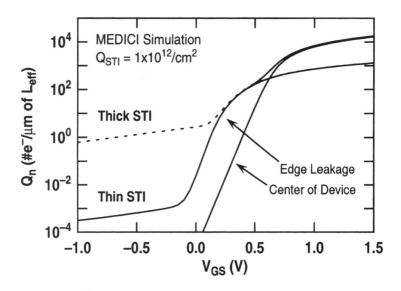

Figure 11.27 Simulated electron density at the edge and center of the nFET as a function of gate voltage, for two different shallow-trench thicknesses.

vices can degrade the overall BGR temperature stability. As can be seen from the data (Table 11.1), the impact of even an extreme proton fluence of 5×10^{13} p/cm^2 has minimal effect on either the output voltage or temperature sensitivity, and is indicative of the overall robustness of this SiGe technology for analog circuit applications. We note that the functional form of the output voltage dependence on fluence (Figure 11.29) is weaker than that observed in commercial Si BJT BGRs, and generally superior in performance at comparable fluence [36].

11.4.3 Voltage Controlled Oscillators

The voltage controlled oscillator (VCO) is a fundamental building block in communications systems. A VCO uses a control voltage for limited frequency tuning and provides the local oscillator (LO) signal for upconversion and downconversion of the RF carrier to intermediate frequencies (IF) within the transceiver. VCOs are particularly sensitive to phase noise, which physically represents the upconversion of low-frequency ($1/f$) noise to high frequencies through the inherent transistor nonlinearities. In the frequency domain, phase noise manifests itself as parasitic sidebands on the carrier, and thus represents a fundamental limit on the spectral purity and signal-to-noise ratio of a communications link. Of interest in this context is the impact of radiation exposure on the VCO phase noise. This SiGe VCO

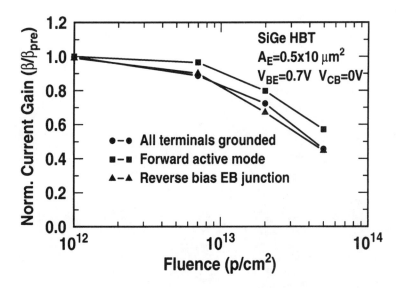

Figure 11.28 Current gain degradation as a function of proton fluence for various transistor bias conditions during radiation exposure.

employs a conventional circuit architecture, and is designed to operate at 5.0 GHz [37]. As can be seen in Table 11.1, the impact of extreme proton fluences on this SiGe VCO are minimal. After 5×10^{13} p/cm^2, the phase noise at a 1-MHz offset from the 5.0-GHz signal slightly increases (worsens) from an excellent value of -112.5 dBc/Hz to a still excellent value of -111.83 dBc/Hz. This small (but repeatable) proton-induced degradation in the VCO phase noise is consistent with transistor-level measurements of residual phase noise in the SiGe HBT building blocks [38]. To understand the result we also measured the low-frequency noise properties of the component SiGe HBTs, and in fact detected a small but observable change in the $1/f$ noise at circuit bias levels consistent with those in the VCO. The fact that this minor $1/f$ noise change couples only weakly to the observed circuit level phase noise suggests that the radiation exposure does not strongly affect the inherent transistor linearity at these frequencies.

11.4.4 Passive Elements

High-quality factor (Q) passive elements (e.g., inductors and capacitors) are required in RF communications circuit design, and there has been significant effort in recent years to fabricate these monolithically with the active devices to facilitate single-chip transceiver implementations. The current SiGe technology contains a

Table 11.1 Summary of the Measured Radiation Tolerance of Some Important SiGe Circuits and Passives

SiGe HBT Circuit	Parameters	Pre-radiation	After 5x10^{13} p/cm^2	Units
Bandgap Reference	V_{cc}	3.0	3.0	V
	I_{cc}	0.773	0.767	mA
	V_{out} at 300K	1.37416	1.372096	V
	Stability (-55 to 85°C)	81.2	81.7	ppm/°C
Voltage Controlled Oscillator	Frequency	5.0	5.0	GHz
	V_{cc}	3.3	3.3	V
	I_{cc}	22.5	22.5	mA
	Output Power	-5.0	-5.5	dBm
	Phase Noise	-112.5	-111.8	dBc/Hz
	Tuning Range	4,595-5,452	4,623-5,470	MHz
LC Bandpass Filter	Frequency	1.9	1.9	GHz
	Filter Q (@ 3dB BW)	7.6	7.6	-
	Insertion Loss	16.8	16.8	dB
	L	2.5	2.5	nH
	Inductor Q	7.4	7.4	-
	C	6.0	6.0	pF
	Capacitor Q	58	58	-

full suite of RF passives, and to obtain high Q, are fabricated in the upper level of the multilevel metalization, so that they are far away from the lossy substrate. The inductors are multiturn spiral inductors, and the capacitors are MIM with a 50-nm SiO_2 dielectric [39]. To determine the effects of proton irradiation on these RF passives at relevant RF frequencies, S-parameter measurements were made on the Ls (Q = 7.4 at 1.9 GHz) and Cs (Q = 58 at 1.9 GHz), as well as an LC bandpass filter implemented from them [33]. As can be seen from Table 11.1, to within the measurement accuracy and site-to-site repeatability, the Ls and Cs and LC filter are unchanged by even extreme proton fluences. We did consistently observe a shift in the LC filter second resonance, which we believe to be due to a moderate change in the coupling coefficient, but this should not affect the operation of the filter in actual circuit design.

11.5 Single Event Upset

As discussed in detail above, as-fabricated SiGe HBTs are robust to various types of ionizing radiation, in terms of both their dc and ac electrical characteristics.

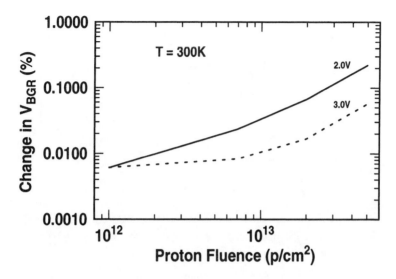

Figure 11.29 Percent change in bandgap reference output voltage at 300 K as a function of proton fluence and supply voltage.

Clearly, however, a space-qualified IC technology must also demonstrate sufficient SEU immunity to support high-speed circuit applications. It is well known that even III-V technologies that have significant TID tolerance often suffer from poor SEU immunity, particularly at high data rates. Recently, high-speed SiGe HBT digital logic circuits were found to be vulnerable to SEU at even low LET values [40, 41]. In addition, successfully employed III-V HBT circuit-level hardening schemes using the current-sharing hardening (CSH) technique [40] were found to be ineffective for these SiGe HBT logic circuits (Figure 11.30). To help understand these SEU results, and to aid in the search for effective SEU mitigation approaches, device and circuit simulations are required.

A logical approach to this problem is to use sophisticated mixed-mode circuit simulation, in which the electrical characteristics of the transistor being hit by an ion strike are solved directly using a 2-D/3-D device simulator. In addition to complexity, commercial mixed-mode simulators often do not support advanced transistor models used by circuit designers, making mixed-mode simulation intractable in practice. An alternative and popular methodology is to simulate the SEU-induced transient terminal currents using a device simulator (e.g., MEDICI), and then use these resultant transient currents as excitations in a conventional circuit simulator (e.g., Cadence), with advanced transistor model capability (e.g., VBIC or HICUM) [42].

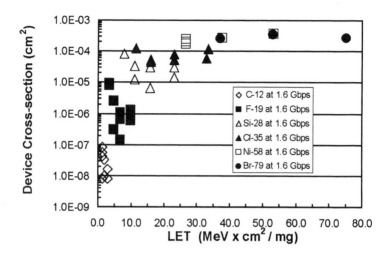

Figure 11.30 Experimental SEU cross section test data on SiGe HBT shift registers [40].

11.5.1 Transistor Equivalent Circuit Under SEU

Device-level simulation for SEU modeling is significantly more complicated than for simple dc or ac simulations, since the n-p-n layers of the intrinsic transistor and the p-type substrate form a n-p-n-p multi-layer structure, making the charge collection more complicated than in a conventional bipolar process. The substrate is usually biased at the lowest potential in order to reverse bias the collector-substrate junction.

During a heavy-ion (i.e., cosmic ray) strike, a column of high density electrons and holes are deposited along the ion trajectory. Electrons are collected by the emitter (E) and collector (C), and holes are collected by the base (B) and substrate (S). For convenience, the ion-induced currents at the emitter and collector are denoted as i_{en} and i_{cn}, where the subscript n indicates "electron collection." Similarly, the ion-induced currents at the base and substrate terminals are denoted as i_{bp} and i_{sp}, where p indicates "hole collection." Note that i_{en}, i_{cn}, i_{bp}, and i_{sp} can all be simulated using a device simulator as a function of time for a given ion strike. Physically, the sum of all of the terminal currents must always be zero, which can be verified in practice by summing the simulated terminal currents. As a result, we only need to describe any three of the four currents, and the other current is then automatically accounted for. The equivalent circuit shown in Figure 11.31 explicitly describes i_{bp}, i_{sp}, and i_{en}:

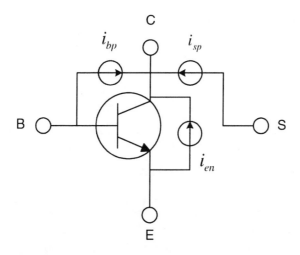

Figure 11.31 An equivalent circuit model for including the ion-induced terminal currents in circuit simulations.

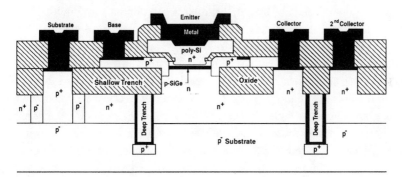

Figure 11.32 The schematic cross section of the SiGe HBT used in the simulations.

- i_{bp} represents the hole current through the base. Even though i_{bp} appears between the base and collector, it contains all of the holes collected by the base through interactions with electrons collected by both the emitter and collector.

- i_{sp} represents the hole current through the substrate. Even though i_{sp} appears between the collector and substrate, it contains all of the holes collected by the substrate through interactions with electrons collected by both the emitter and collector.

- i_{en} represents the electron current through the emitter. Note that i_{en} appears between the collector and emitter, and connects with both i_{sp} and i_{bp}. Such a connection is necessary to ensure that all the terminal currents are properly described.

- The ion-induced electron current through the collector, i_{cn}, is then given by

$$i_{cn} = -(i_{bp} + i_{sp} + i_{en}). \qquad (11.13)$$

11.5.2 SEU Simulation Methodology

At the device level, the SEU-induced transient terminal currents are obtained using quasi-3D device simulation (i.e., using a rectangular 2-D mesh/profile, which is then rotated about the central device axis and solved using cylindrical coordinates). Given the complexity of accurately modeling SEU, a brief summary of the methodology is offered for insight. The SiGe HBT doping profile and Ge profile were first constructed using measured SIMS data, device layout information, and then careful calibration of dc and ac electrical characteristics using advanced parameter models. All of the lateral structures of the device must be accounted for (Figure 11.32), including the deep and shallow trench isolation. For SEU simulation, a top substrate contact needs to be used, as opposed to a bottom contact, which is typically used in simulations not concerned with SEU. A bottom substrate contact rigorously sets the bottom of the simulation structure to thermodynamic equilibrium, which is not the case in the event of realistic SEU. To ensure that the reflective boundary condition implemented in the simulator is consistent with physical reality, the geometries of the simulation region must be sufficiently large. In practice, only a finite depth substrate can be simulated due to computational memory, speed, and complexity limitations. The minimum depth required to provide a reasonable approximation is problem specific, and depends on the ion LET, the depth of ion strike, doping, and terminal bias. Our approach to determining the minimum depth is to gradually increase the simulation depth until the simulated charge collection results no longer change. For most of the simulations used in this work, the minimum simulation depth is between 50–100 μm.

The center of the emitter was used as the cylindrical z-axis as an approximation of a worst-case ion strike. A fine mesh was used along the path of the ion strike and at the pn junction interfaces. The average number of nodes was 10^4 for each simulation. The validity of the griding scheme was checked by repeating the simulation on finer grids. The charge track was generated over a period of 10 psec using a Gaussian waveform. The Gaussian had a $1/e$ characteristic time-scale of 2 psec, a $1/e$ characteristic radius of 0.2 μm, and the peak of the Gaussian occurred at 4 psec. The depth of the charge track was 10 μm, and the LET value was uniform along the charge track. Two substrate doping values of $5x10^{15}$ cm^{-3} and $1x10^{18}$ cm^{-3} and five LET values from 0.1–0.5 pC/μm were simulated. Transient currents for different SEU conditions were simulated using quasi-3D device simulation, and included in circuit simulation using the equivalent circuit described above and the relevant circuit architecture. In principle, any transistor in the modeled circuit can be hit by a heavy ion. In practice, however, it is generally easy to identify the sensitive transistors and concentrate the analysis on those devices.

11.5.3 Charge Collection Characteristics

From a device perspective, it is important to first assess the transistor charge collection characteristics as a function of terminal bias, load condition, substrate doping, and ion strike depth [42]. Figure 11.33 shows the charge collected by the collector versus time for different RC loads. The base and emitter terminals were grounded, the substrate bias was -5.2 V, the collector was connected to ground through an RC load, and the substrate doping was $5x10^{15}$/cm^3. A uniform LET of 0.1 pC/μm (equivalent to 10 MeV-cm^2/mg) over 10-μm depth was used, which generates a total charge of 1.0 pC. The results clearly show that charge collection is highly dependent on the transistor load condition (i.e., circuit topology). As the load resistance increases, the collector-collected charge decreases. Note, however, that the emitter-collected charge increases correspondingly. The underlying physics is that more electrons exit through the emitter, instead of the collector. A larger load resistance presents a higher impedance to the electrons at the collector, and thus more electrons exit through the emitter. The collector of the adjacent device only collects a negligible amount of charge, despite the transient current spikes of the strike. Nearly all of the electrons deposited are collected by the collector and the emitter, although the partition between emitter and collector collection varies with the load condition.

The simulated evolution of the carrier profiles shows that the holes deposited deep in the bulk exit through the substrate, and the holes deposited near the surface exit through the base. All of the holes deposited get collected because of the 5.2 V reverse bias on the collector-substrate junction.

Figure 11.34 shows the collector-collected charge versus time for different substrate doping levels. The electron charge collected by the collector decreases monotonically with increasing substrate doping. However, the electron charge collected by the emitter increases first when the substrate doping increases from $5x10^{15}$ cm^{-3} to 10^{17} cm^{-3}, and then decreases when the substrate doping further increases to 10^{18} cm^{-3}. The reason for the decrease of emitter-collected charge at 10^{18} cm^{-3} is that the total amount of electrons that can be collected (the sum of the emitter and collector) decreases monotonically with increasing substrate doping. The total

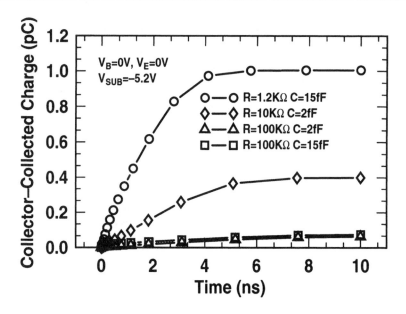

Figure 11.33 Collector-collected charge versus time for different RC loads.

electron charge collected is approximately equal to the hole charge collected by the substrate. The rest of the deposited electrons and holes are left in the substrate.

The substrate doping dependence of SEU in SiGe HBTs can be understood by examining the corresponding collector current waveforms shown in Figure 11.35. Immediately after the ion strike, electrons are swept out of the collector efficiently via drift, giving rise to a large collector current spike. Subsequently, the collector current begins to drop due to the removal of the deposited carriers. After 20 psec, the collector current is diminished for the 10^{17} and 10^{18} cm^{-3} doping levels. The collector-collected charge thus saturates for these two doping levels, as shown in Figure 11.34. The collector current for the 5×10^{15} cm^{-3} doping level, however, starts to increase again approximately 7 psec after the ion strike. The difference observed can be attributed to the strong dependence of funneling-assisted drift charge collection on the substrate doping. For a higher substrate doping, the original junction electric field is much higher, and the original space charge layer is much thinner. The funnel length is smaller, and the funneling-assisted drift charge collection is much faster for a heavily doped substrate, resulting in less total charge collection. For a lightly doped substrate, funneling takes a longer time to develop and the funnel length is larger, resulting in more charge collection. A heavily doped substrate is generally desired to improve the susceptibility to SEU in SiGe HBT digital logic circuits, where upset of the circuit functionality is primarily de-

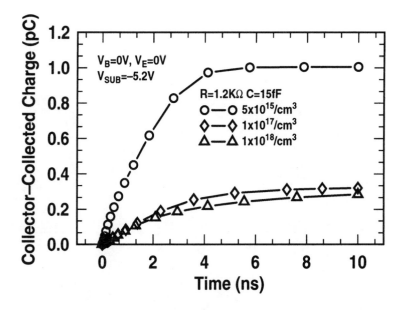

Figure 11.34 Collector-collected charge versus time for different substrate doping levels.

pendent on the total amount of charge collected, as shown by circuit simulations [43]. This conclusion, however, is expected to be circuit topology dependent.

Figure 11.36 shows the impact of substrate bias on charge collection for a 10-μm deep ion strike (all of the other terminals are grounded). A less negative substrate bias can *significantly* reduce the charge collection, as expected. Thus, from a circuit-level point of view, the collector-substrate junction reverse bias should be reduced as much as possible to improve the susceptibility to SEU. Among all of the terminal biases, the reverse bias on the n^+ collector to p-substrate junction has the most significant impact on the circuit function upset. This can be understood as a result of the collection of holes through the substrate terminal, as well as the interaction between electrons and holes during the charge collection process.

Finally, the ion strike depth affects the charge collection in two ways. First, less total charge is deposited and collected for a shallower ion strike. Second, a larger portion of the deposited holes exit through the base with decreasing strike depth. The reason for this is that the holes deposited close to the surface always exit through the base. The collection of deposited holes by the base is nearly identical for a shallow strike stopping at the middle of the n^+ subcollector (1.7 μm from the surface) and a deep ion strike stopping deep in the substrate (10 μm from the surface). For the shallow strike, nearly all of the deposited holes exit through the base,

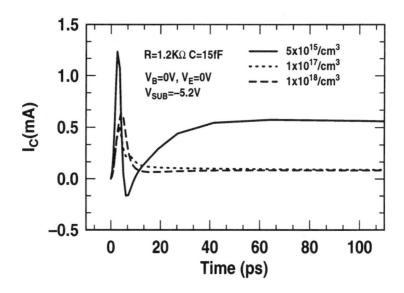

Figure 11.35 Collector current versus time from 0 to 100 psec for different substrate dopings.

even though the substrate potential is 5.2 V lower than the base potential. The lack of holes in the unstruck lower half of the n^+ buried layer effectively blocks the flow of holes into the substrate. Therefore, the substrate bias does not affect the charge collection in such cases. This situation differs from the competition between the emitter and collector collection of electrons by the fact that the deposited electrons are always connected to both the emitter and collector upon the strike. For the deep strike, only the holes deposited near the surface are collected by the base, while the rest of the holes deep in the bulk are collected by the substrate.

11.5.4 Circuit Architecture Dependence

Given this detailed device-level charge-collection information, comparisons of the SEU sensitivity of particular circuit architectures can be undertaken [43, 44]. In this case, 3 D flip-flop circuits were investigated, as representative high-speed digital logic building blocks, including: two unhardened SiGe HBT circuits (denoted as circuits A and B) and a current-sharing hardened (CSH) circuit (denoted as circuit C). Each of the three circuits have the identical logical functionality of a rising edge-triggered D flip-flop under normal operation (i.e., without SEU).

Circuit A is a straightforward ECL implementation of the standard rising edge-triggered flip-flop logic diagram shown in Figure 11.37. The standard ECL im-

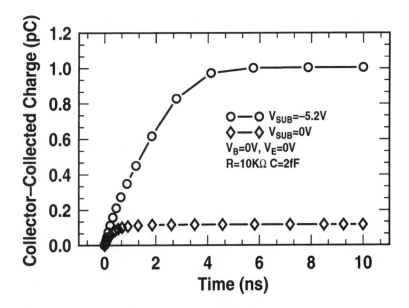

Figure 11.36 Collector-collected charge versus time for two different substrate bias conditions.

plementation of a two-input NAND gate is shown in Figure 11.38. Here, IN1 and IN2 are the two inputs, IN1* and IN2* indicate their logic complements, and V1 is the NAND output, while V2 is the compliment of V1. Note that VCS sets the switching current. The level shifters at the input and output are not shown.

Circuit B is the unhardened version of the D flip-flop used in the shift registers tested in [40]. The transistor-level circuit is shown in Figure 11.39. Circuit B uses fewer transistors and thus less power than circuit A, and is also faster than circuit A, allowing operation at higher clock rates. Because of these advantages, circuit B is very popular in high-speed bipolar digital circuit design. The circuit consists of a master stage and a slave stage. The master stage consists of a pass cell (Q1 and Q2), a storage cell (Q3 and Q4), a clocking stage (Q5 and Q6), and a biasing control (Q7). The slave stage has a similar circuit configuration.

Circuit C is the current-sharing hardened version of circuit B. The circuit was used as a basic building block of the 32-stage shift-register tested in [40]. Each transistor element in Figure 11.39 was implemented with a five-path CSH architecture. The CSH concept is illustrated in Figure 11.40 using a single-level basic current-mode logic gate. The current source transistor Q7 is divided into 5 paths, with VCS1 controlling 3 paths, and VCS2 and VCS3 controlling 1 path each. These paths are maintained separately through the clocking stage and through the

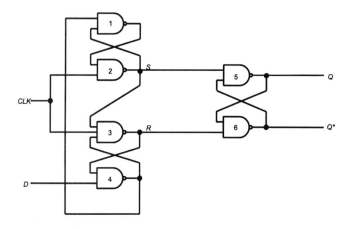

Figure 11.37 Logic diagram of a standard rising edge-triggered D flip-flop.

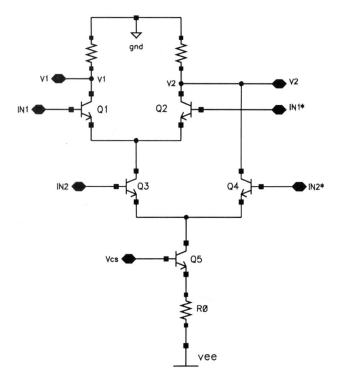

Figure 11.38 Standard ECL implementation of a two-input NAND gate.

Radiation Tolerance 499

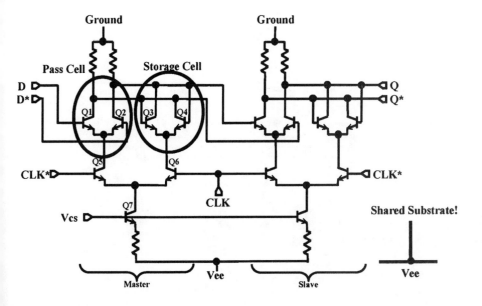

Figure 11.39 Schematic of circuit B, the unhardened counterpart of the D flip-flop used in the shift registers tested in [40].

pass and storage cells. In essence, the input and output nodes of five copies of the switching circuits, including the controlling switch, clock, master and storage cells, are connected in parallel. The load resistance is shared by all the current paths. The full schematic is not shown because of the large number of transistors and interconnects.

We first compared the three circuits for a fixed switching current of 1.5 mA. The quasi-3D simulated SEU-induced transient currents were activated on one of the sensitive transistors. In circuit A, we chose to "strike" the transistor at the output node of NAND gate 2 in Figure 11.38. In circuits B and C, we chose Q3 of the storage cell in the master stage (Q3). Figures 11.41–11.43 show the simulated SEU responses for Circuit A, B, and C, respectively. An LET of 0.5 pC/μm and a substrate doping $N_{sub} = 5 \times 10^{15}/\text{cm}^3$ was used.

The SEU currents were activated at 5.46 nsec (within the circuit hold time), immediately after the clock goes from low to high, a sensitive time instant for SEU-induced transient currents to produce an upset at the output. The input data is an alternating "0" and "1" series with a data rate of 2 Gbit/sec. Observe that circuit A shows no upset at all, while circuits B and C show 5 and 3 continuous bits of

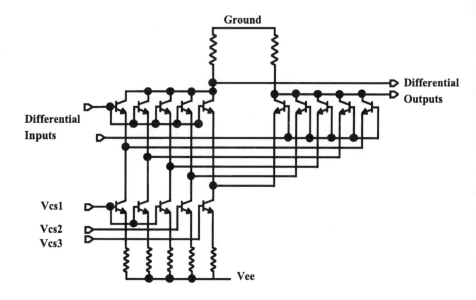

Figure 11.40 Illustration of the CSH concept using a basic ECL gate. In this case, 5 parallel subtransistor elements are used to maintain separate current paths.

data upset, respectively. These results suggest that circuit A has the best SEU tolerance, while circuit C, the CSH hardened version, has better SEU tolerance than its unhardened companion version, circuit B.

Circuit A, which shows no data upset at a switching current of 1.5 mA, does in fact show an upset when the switching current is lowered to 0.6 mA. This is consistent with our earlier observation that increasing switching current is effective in improving SEU performance for circuit C [43].

The fundamental reason for the observed better SEU tolerance of circuit A than for circuits B and C is that only one of the two outputs of the emitter-coupled pair being hit is affected by the ion-strike SEU current transients. Consider the switching pair of the two-input NAND gate in Figure 11.38. Assuming that the left transistor Q1 is hit, the resulting transient collector current lowers the potential at the collector of Q1 (V1). However, the output voltage of the right transistor Q2 (V2) is not affected. An examination of the operation of NAND gate number 2 (in Figure 11.37) using Figure 11.38 clearly shows this. As long as the differential output (V1-V2) is above the logic switching threshold, the output remains unaffected, and no upset occurs. This is supported by the simulation results shown in

Radiation Tolerance 501

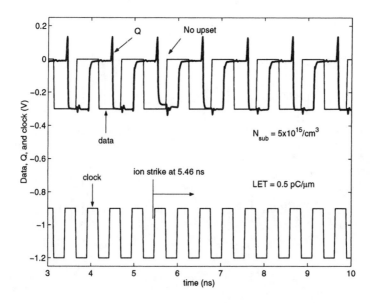

Figure 11.41 Output waveform for circuit A, with LET = 0.5 pC/μm, at a switch current of 1.5 mA.

Figure 11.44 for circuit A with 1.5-mA switch current. The thresholds for the differential output to produce low-to-high and high-to-low transitions are indicated. The collector voltage of Q1 (V1) decreases upon ion strike (compared to without SEU), however, and no upset is observed at the output, simply because the differential output remains above or below the relevant switching threshold.

In comparison, in circuits B and C, both outputs of the storage cell consisting of Q3 and Q4 are affected by the SEU transients because of cross-coupling (see Figure 11.39). The input (base) of Q3 is connected to the output (collector) of Q4. Similarly, the input of Q4 is connected to the output of Q3. Cross-coupling of Q3 and Q4 acts as a positive feedback mechanism, which not only enhances the SEU-induced decrease of V2 (compared to without SEU); but also makes V1 higher than without SEU. Therefore, it is easier to produce an upset at the output because of much higher SEU-induced change in *differential* output compared to that in circuit A (note that the output of the struck transistor (Q3) is V2 for circuit B).

The above analysis is supported by the simulated V2, V1, and (V2-V1) shown in Figure 11.45 for the storage cell emitter-coupled pair (Q3 and Q4). The reason for the better SEU tolerance of circuit C compared to circuit B is likely due to the larger parasitic capacitances and increased number of discharge paths.

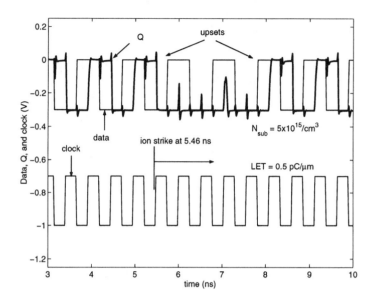

Figure 11.42 Output waveform for circuit B, with LET = 0.5 pC/μm at a switch current of 1.5 mA.

Taken together, these SEU modeling results suggest that there should exist straightforward circuit architectures that will allow high-speed SiGe HBT-based digital logic to function in space with acceptable SEU immunity without requiring additional device-level radiation hardening. Experimental verification of these claims, as well as full 3-D modeling of charge collection in SiGe HBTs, is currently under way.

References

[1] P. Marshall and C. Marshall, "Proton effects and test issues for satellite designers – Part A: Ionization effects," Short Course Notes, *IEEE Nucl. Space Rad. Eff. Conf.*, 1999.

[2] R. Reed and R. Ladbury, "Performance characterization of wideband digital optical data transfer systems for use in the space radiation environment," Short Course Notes, *IEEE Nucl. Space Rad. Eff. Conf.*, 2000.

[3] C. Dyer, "Space radiation environment dosimetry," Short Course Notes, *IEEE Nucl. Space Rad. Eff. Conf.*, 1998.

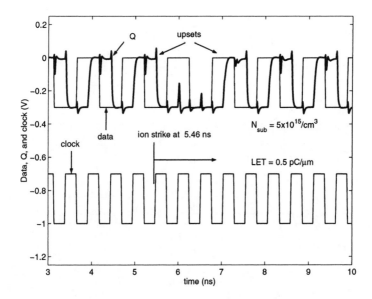

Figure 11.43 Output waveform for circuit C (CSH hardened version of Circuit B), with LET=0.5 pC/μm, at a switch current of 1.5 mA.

[4] P. Dodd, "Basic mechanisms for single-event effects," Short Course Notes, *IEEE Nucl. Space Rad. Eff. Conf.*, 1999.

[5] E. Petersen, "Single-event analysis and prediction," Short Course Notes, *IEEE Nucl. Space Rad. Eff. Conf.*, 1997.

[6] C. Marshall and P. Marshall, "Proton effects and test issues for satellite designers – Part B: Displacement effects," Short Course Notes, *IEEE Nucl. Space Rad. Eff. Conf.*, 1999.

[7] J.A. Babcock et al., "Ionizing radiation tolerance of high performance SiGe HBTs grown by UHV/CVD," *IEEE Trans. Nucl. Sci.*, vol. 42, pp. 1558-1566, 1995.

[8] J. Roldán et al., "Neutron radiation tolerance of advanced UHV/CVD SiGe HBTs," *IEEE Trans. Nucl. Sci.*, vol. 44, pp. 1965-1973, 1997.

[9] J. Roldán et al., "An investigation of the spatial location of proton-induced traps in SiGe HBTs," *IEEE Trans. Nucl. Sci.*, vol. 45, pp. 2424-2430, 1998.

[10] S. Zhang et al., "The effects of proton irradiation on the RF performance of SiGe HBTs," *IEEE Trans. Nucl. Sci.*, vol. 46, pp. 1716-1721, 1999.

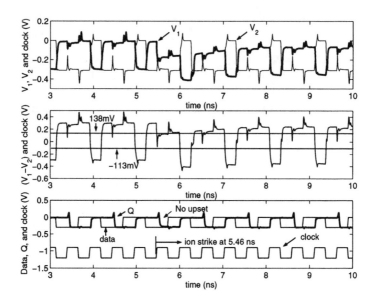

Figure 11.44 Circuit A output voltages and differential output for the emitter-coupled pair struck by a heavy ion, with a switch current of 1.5 mA.

[11] S. Zhang et al., "A comparison of the effects of gamma irradiation on SiGe HBT and GaAs HBT technologies," *IEEE Trans. Nucl. Sci.*, vol. 47, pp. 2521-2527, 2000.

[12] S. Zhang et al., "Investigation of proton energy effects in SiGe HBT technology," *IEEE Trans. Nucl. Sci.*, to appear, December 2002.

[13] R.L. Pease, "Total-dose issues for microelectronics in space systems," *IEEE Trans. Nucl. Sci.*, vol. 43, pp. 442-452, 1996.

[14] E.W. Enlow et al., "Response of advanced bipolar processes to ionizing radiation," *IEEE Trans. Nucl. Sci.*, vol. 38, pp. 1342-1351, 1991.

[15] D.M. Fleetwood et al., "Physical mechanisms contributing to enhanced bipolar gain degradation at low dose rates," *IEEE Trans. Nucl. Sci.*, vol. 41, pp. 1871-1883, 1994.

[16] MEDICI, 2-D Semiconductor Device Simulator, Avant!, Fremont, CA, 1997.

[17] G. Banerjee et al., "Anomalous dose rate effects in gamma irradiated SiGe heterojunction bipolar transistors," *IEEE Trans. Nucl. Sci.*, vol. 46, pp. 1620-1626, 1999.

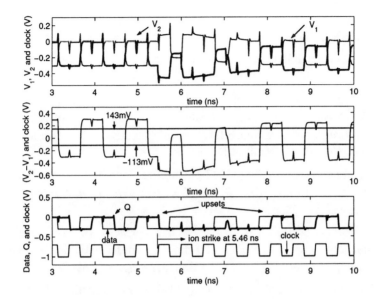

Figure 11.45 Circuit B output voltages and differential output for the emitter-coupled pair struck by a heavy ion, with a switch current of 1.5 mA.

[18] P.J. Drevinsky, A.R. Frederickson, and P.W. Elsaesser, "Radiation-induced defect introduction rates in semiconductors," *IEEE Trans. Nucl. Sci.*, vol. 41, pp. 1913-1923, 1994.

[19] L. Vempati et al., "Low-frequency noise in UHV/CVD epitaxial Si and SiGe bipolar transistors," *IEEE J. Solid-State Circ.*, vol. 31, pp. 1458-1467, 1996.

[20] Z. Jin et al., "1/f noise in proton-irradiated SiGe HBTs," *IEEE Trans. Nucl. Sci.*, vol. 48, pp. 2244-2249, 2001.

[21] A. van der Ziel, "Formulation of surface $1/f$ noise processes in bipolar junction transistors and in p-n diodes in Hooge-type form," *Solid State Elect.*, vol. 41, pp. 91-93, 1989.

[22] A. van der Ziel et al., "Location of $1/f$ noise sources in BJT's and HBJT's - I. Theory," *IEEE Trans. Elect. Dev.*, vol. 33, pp. 1371-1375, 1986.

[23] F.N. Hooge, "$1/f$ noise is no surface effect," *Phys. Lett.*, vol. 29A, p. 139, 1969.

[24] F.N. Hooge, "$1/f$ noise sources," *IEEE Trans. Elect. Dev.*, vol. 41, pp. 1926-1935, 1994.

[25] G. Niu et al., "Impact of gamma irradiation on the RF phase noise capability of UHV/CVD SiGe HBTs," *Solid-State Elect.*, vol. 45, pp. 107-112, 2001.

[26] Z. Jin et al., "Proton response of low-frequency noise in 0.20 μm 90 GHz f_T UHV/CVD SiGe HBTs," *Solid-State Elect*, in press.

[27] J.D. Cressler et al., "The Effects of proton irradiation on the lateral and vertical scaling of UHV/CVD SiGe HBT BiCMOS Technology," *IEEE Trans. Nucl. Sci.*, vol. 47, pp. 2515-2520, 2000.

[28] J.D. Cressler et al., "An investigation of the origins of the variable proton tolerance in multiple SiGe HBT BiCMOS technology generations," *IEEE Trans. Nucl. Sci.*, to appear, December 2002.

[29] D. Ahlgren et al., "A SiGe HBT BiCMOS technology for mixed signal RF applications," *Proc. IEEE Bipolar/BiCMOS Circ. Tech. Meeting*, pp. 195-197, 1997.

[30] G. Freeman et al., "A 0.18 μm 90 GHz f_T SiGe HBT BiCMOS, ASIC-compatible, copper interconnect technology for RF and microwave applications," *Tech. Dig. IEEE Int. Elect. Dev. Meeting*, pp. 569-572, 1999.

[31] U. Gogineni et al., "Hot electron and hot hole degradation of SiGe heterojunction bipolar transistors," *IEEE Trans. Elect. Dev.*, vol. 47, pp. 1440-1448, 2000.

[32] G. Niu et al., "Total dose effects in the shallow trench isolation leakage current characteristics in a 0.35μm SiGe BiCMOS technology," *IEEE Trans. Nucl. Sci.*, vol. 46, pp. 1841-1847, 1999.

[33] J.D. Cressler et al., "Proton radiation response of SiGe HBT analog and RF circuits and passives," *IEEE Trans. Nucl. Sci.*, vol. 48, pp. 2238-2243, 2001.

[34] S. Zhang et al., "The effects of operating bias conditions on the proton tolerance of SiGe HBTs," *Solid-State Elect.*, in press.

[35] H.A. Ainspan and C.S. Webster, "Measured results on bandgap references in SiGe BiCMOS," *Elect. Lett.*, vol. 34, pp. 1441-1442, 1998.

[36] B.G. Rax, C.I. Lee, and A.H. Johnston, "Degradation of precision reference devices in space environments," *IEEE Trans. Nucl. Sci.*, vol. 44, pp. 1939-1944, 1997.

[37] J.O. Plouchart, *Tech. Dig. IEEE Euro. Solid-State Circ. Conf.*, pp. 332-335, 1998.

[38] G. Niu et al., "Impact of gamma irradiation on the RF phase noise capability of UHV/CVD SiGe HBTs," *Solid-State Elect.*, vol. 45, pp. 107-112, 2001.

[39] K. Stein et al., "High reliability metal insulator metal capacitors for silicon-germanium analog applications," *Proc. IEEE Bipolar/BiCMOS Circ. Tech. Meeting*, pp. 191-194, 1997.

[40] P. Marshall et al., "Single event effects in circuit hardened SiGe HBT logic at gigabit per second data rate," *IEEE Trans. Nucl. Sci.*, vol. 47, pp. 2669-2674, 2000.

[41] R. Reed et al., "Single event upset test results on an IBM prescalar fabricated in IBM's 5HP germanium doped silicon process," *Proc. IEEE Nucl. Space Rad. Eff. Conf. Data Workshop*, pp. 172-176, 2001.

[42] G. Niu et al., "Simulation of SEE-Induced Charge Collection in UHV/CVD SiGe HBTs," *IEEE Trans. Nucl. Sci.*, vol. 47, pp. 2682-2689, 2000.

[43] G. Niu et al., "Modeling of single event effects in circuit-hardened high-speed SiGe HBT logic," *IEEE Trans. Nucl. Sci.*, vol. 48, pp. 1849-1854, 2001.

[44] G. Niu et al., "A Comparison of SEU tolerance in high-speed SiGe HBT digital logic designed with multiple circuit architectures," *IEEE Trans. Nucl. Sci.*, to appear, December 2002.

Chapter 12

Device Simulation

Analytical transistor analysis inevitably involves approximations and assumptions in order to obtain closed-form solutions. Numerical analysis based on the fundamental differential equations governing semiconductors has become necessary in IC technology development, and is often referred to as "device simulation."[1] Many of the results presented in previous chapters, in fact, were obtained using device simulation. Commercial simulators such as MEDICI [2] from Avant! (now Synopsys), DESSIS from ISE, and ATLAS from Silvaco, all support SiGe heterostructure device simulation. They are typically part of a technology computer-aided-design (TCAD) package, which includes process simulation, device simulation, and parameter extraction programs. The use of a device simulator, or TCAD tools in general, requires substantially more knowledge of the internal workings of the simulator than the use of, say, a circuit simulator such as SPICE. For instance, users must choose which mobility model to use, which statistics (Fermi-Dirac or Boltzmann) to use, and whether or not to account for incomplete ionization of dopants. The default physical models are usually the simplest ones, and often give inaccurate results, particularly for advanced device technologies such as SiGe. Users are also responsible for the "meshing" of the device structure, which can affect the simulation results significantly. This chapter describes the differential equations and physics implemented in commercial device simulators, as well as the practical aspects of using these simulators for SiGe HBT analysis and optimization.

---

[1] We differentiate here between device-level "simulation," and circuit-level "modeling" (e.g., with compact models in SPICE or ADS).
[2] We point out that "MEDICI" is commonly mispronounced as \ma-'dee-chi\. As any student of the Italian Renaissance will recognize, however, the name of the famous Florentine family is correctly pronounced \'med-a-chi\.

12.1 Semiconductor Equations

The electrical behavior of a semiconductor system is governed by Poisson's equation and the current continuity equations. The three fundamental variables are electrostatic potential ϕ, electron concentration n, and hole concentration p. The governing equations are:

$$\nabla \cdot \varepsilon \nabla \phi = -q(p - n + C), \quad (12.1)$$

$$\frac{1}{q}\nabla \cdot \vec{J}_n - R = \frac{\partial n}{\partial t}, \quad (12.2)$$

$$-\frac{1}{q}\nabla \cdot \vec{J}_p - R = \frac{\partial p}{\partial t}, \quad (12.3)$$

where $\vec{J}_n$ and $\vec{J}_p$ are the electron and hole current densities, C is the net concentration of ionized dopants and charged traps, and R is the net rate of recombination (including impact ionization). Here, $\vec{J}_n$ and $\vec{J}_p$ are related to n, p, and ϕ through the semiconductor transport equations. The most basic transport equations used in commercial device simulators are the so-called "drift-diffusion" equations.[3] The bandgap edges E_C and E_V are related to the electrostatic potential by

$$E_C = -q\phi - \chi + \Delta, \quad (12.4)$$

$$E_V = -q\phi - \chi + \Delta - E_g, \quad (12.5)$$

where χ is the electron affinity, E_g is the bandgap, and Δ is a constant depending on the choice of energy reference. In a SiGe HBT, the Ge mole fraction is a function of position, and thus both χ and E_g vary with position.

12.1.1 Carrier Statistics

The electron and hole concentrations in semiconductors are described by the Fermi-Dirac distribution function and a parabolic density-of-states in the energy-wave vector $(E - k)$ space. Integration in the conduction and valence bands lead to n

---

[3]It should be appreciated that the drift-diffusion equations are themselves an approximation of the more rigorous Boltzmann transport equation, and assume, among other things, validity of the so-called "relaxation time approximation."

and p

$$n = N_C F_{1/2}(\eta_n), \quad (12.6)$$
$$p = N_V F_{1/2}(\eta_p), \quad (12.7)$$
$$\eta_n \equiv \frac{E_{Fn} - E_C}{kT}, \quad (12.8)$$
$$\eta_p \equiv \frac{E_V - E_{Fp}}{kT}, \quad (12.9)$$

where N_C and N_V are the effective conduction and valence band density-of-states, $F_{1/2}(x)$ is the Fermi integral of order 1/2, and E_{Fn} and E_{Fp} are the quasi-Fermi levels for electrons and holes, respectively. These equations can be rewritten in a form that resembles the familiar n and p equations with exponential (Boltzmann) terms [1] as

$$n = \gamma_n N_C e^{\eta_n}, \quad (12.10)$$
$$p = \gamma_p N_V e^{\eta_p}, \quad (12.11)$$

where γ_n and γ_p are defined by

$$\gamma_n \equiv F_{1/2}(\eta_n)e^{-\eta_n}, \quad (12.12)$$
$$\gamma_p \equiv F_{1/2}(\eta_p)e^{-\eta_p}. \quad (12.13)$$

Here, γ_n and γ_p can be viewed as *correction factors* accounting for the carrier degeneracy. The identification of the exponential terms in (12.10) and (12.11) facilitates the use of the classical Scharfetter-Gummel discretization scheme for numerical solution of the continuity equations.

In equilibrium, $E_{Fn}=E_{Fp}=E_F$. The pn product then becomes

$$pn = \gamma_n \gamma_p n_i^2 \; ; \quad (12.14)$$
$$n_i^2 \equiv N_C N_V e^{-E_g/kT}. \quad (12.15)$$

The n_i as defined above is known as the "intrinsic carrier concentration." For an intrinsic semiconductor in equilibrium, $\gamma_n=\gamma_p=1$, and $n = p = n_i$. Thus, $E_{Fn,p}$ can be determined from Poisson's equation. For Boltzmann statistics, $\gamma_n=\gamma_p=1$, and $pn = n_i^2$, which is only a function of N_C, N_V, and E_g. An examination of the Fermi integral indicates that when E_F is with $3kT$ of E_C or E_V, which occurs when n and p are large due to either heavy doping or high injection, γ_n and γ_p become smaller than unity. As a result, $pn = \gamma_n \gamma_p n_i^2$ at equilibrium becomes smaller than n_i^2. If one continues to use Boltzmann statistics (which gives a simple $pn = n_i^2$ relation suitable for data processing in experimental analysis), when γ_n or

γ_p is much smaller than 1.0, the bandgap term E_g must be artificially increased to obtain the correct pn product.

A popular paradigm for device simulators accounts for this carrier degeneracy by modifying the bandgap E_g *without* going from the simple Boltzmann statistics to the more complex Fermi-Dirac statistics. The resulting "apparent" bandgap change (ΔG) is significant for doping levels found in SiGe HBTs, as shown in Figure 12.1 for n-type doping. For a doping level of 10^{20} cm^{-3}, for instance, the ΔG required to account for degeneracy is -31.2916 meV. The number is negative here because a positive ΔG typically means a narrowing of the bandgap. Carrier degeneracy causes a *decrease* of the pn product compared to n_i^2, and thus a *negative* ΔG contribution. To obtain the correct pn product using conventional Boltzmann statistics, the bandgap must thus be artificially increased.

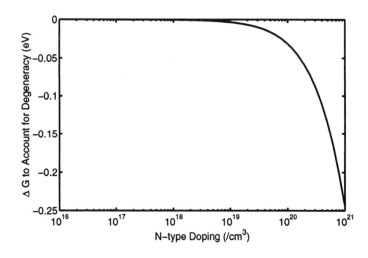

Figure 12.1 Apparent bandgap narrowing needed to account for degeneracy using Boltzmann statistics for n-type doping.

Similarly, the changes of N_C, N_V, and the rigid bandgap E_g can all be lumped into ΔG such that we can still use the low doping level $N_C N_V$ value under heavy doping situations. The rigid bandgap narrowing and changes of $N_C N_V$ both lead to a positive ΔG. The net apparent bandgap narrowing is positive, and increases with doping level. To correctly simulate SiGe HBTs with heavily doped emitter and base regions, one must understand the physical origins of various contributions to the experimentally measured apparent bandgap narrowing ΔG, as well as the difference between the real bandgap reduction and apparent electrical bandgap reduction ΔE_g^{app}. Unfortunately, to our knowledge, the apparent bandgap narrowing models reported in the literature, which implicitly assume Boltzmann statistics, are often implemented the same way for both Boltzmann statistics and Fermi-Dirac statistics, which is physically inconsistent. What makes the situation worse is that the use of Fermi-Dirac statistics is often suggested in user manuals for simulation of devices with heavy doping. This is clearly incorrect.

Using (12.15), (12.10) and (12.11) can be rewritten in the following form

$$n = n_{in} e^{q(\psi - \phi_{Fn})/kT}, \qquad (12.16)$$

$$p = n_{ip} e^{q(\phi_{Fp} - \psi)/kT}, \qquad (12.17)$$

$$n_{in} = n_i \gamma_n, \qquad (12.18)$$

$$n_{ip} = n_i \gamma_p, \qquad (12.19)$$

$$\phi_{fn} = -E_{Fn}/q, \qquad (12.20)$$

$$\phi_{fp} = -E_{Fp}/q, \qquad (12.21)$$

$$\psi = -\frac{1}{2q}\left(E_C + E_V - kT \ln \frac{N_C}{N_V}\right). \qquad (12.22)$$

The ϕ_{Fn}, ϕ_{Fp}, and ψ as defined above are referred to as the electron quasi-Fermi potential, the hole quasi-Fermi potential, and the intrinsic Fermi potential, respectively. Here, ψ, instead of ϕ, is used as the fundamental variable in simulators such as SEDAN [2] and MEDICI [1]. For uniform band structure, the intrinsic Fermi potential ψ differs from the electrostatic potential ϕ by a constant, and is used directly in the solution of Poisson's equation. For nonuniform band structure (i.e., for SiGe HBTs), the difference is not constant. Substituting the E_C and E_V expressions (12.4) and (12.5) into (12.22), one obtains the relation between ϕ and ψ

$$\phi = \psi - \theta + \text{const}, \qquad (12.23)$$

$$\theta = \frac{1}{q}\left[\frac{1}{2}\left(E_g + kT \ln \frac{N_C}{N_V}\right) + \chi\right]. \qquad (12.24)$$

Using ψ as an independent variable, Poisson's equation (12.1) can be rewritten as

$$\nabla \cdot \varepsilon \nabla (\psi - \theta) = -q(p - n + C). \qquad (12.25)$$

Inside MEDICI, for instance, Poisson's equation is implemented using (12.25) [1].

12.1.2 Drift-Diffusion Equations

The current densities $\vec{J}_n$ and $\vec{J}_p$ can be related to ϕ, n, and p by solving the Boltzmann transport equation using the method of moments, or alternatively by applying the fundamental principles of irreversible thermodynamics. The resulting current density (either $\vec{J_n}$ or $\vec{J_p}$) has a *drift* component proportional to electric field, and a *diffusion* component proportional to carrier concentration gradient,

$$\vec{J_n} = qn\mu_n\vec{E} + qD_n\nabla n, \qquad (12.26)$$

$$\vec{J_p} = qp\mu_p\vec{E} - qD_p\nabla p, \qquad (12.27)$$

where μ_n and μ_p are the carrier mobilities, D_n and D_p are carrier diffusivities, and $\vec{E}$ is the electric field,

$$\vec{E} = -\vec{\nabla}\phi. \qquad (12.28)$$

The diffusivities are related to the mobilities by the generalized Einstein relations

$$D_n = \frac{kT}{q}\mu_n F_{1/2}(\eta_n)/F_{-1/2}(\eta_n), \qquad (12.29)$$

$$D_p = \frac{kT}{q}\mu_p F_{1/2}(\eta_p)/F_{-1/2}(\eta_p). \qquad (12.30)$$

When Boltzmann statistics is used, the Einstein relations simplify to their familiar form

$$D_n = \frac{kT}{q}\mu_n, \qquad (12.31)$$

$$D_p = \frac{kT}{q}\mu_p. \qquad (12.32)$$

The driving forces for the drift component of $\vec{J_n}$ and $\vec{J_p}$ are identical in (12.26) and (12.27). The realization of heterostructures provides a means of *independently* engineering the driving forces for electrons and holes. The driving force for electron current flow is fundamentally determined by the spatial gradient of the conduction band edge, which is modified by the electron affinity gradient and N_C gradient in a heterostructure system,

$$\vec{E_n} = \nabla\frac{E_C}{q} + \frac{kT}{q}\nabla\ln\frac{N_C}{T^{3/2}} = -\nabla(\phi + \frac{\chi}{q} - \frac{kT}{q}\ln\frac{N_C}{T^{3/2}}), \qquad (12.33)$$

where χ is the electron affinity, and N_C is the effective conduction band density of states. Similarly, the driving force for hole current flow is given by

$$\vec{E_p} = \nabla\frac{E_V}{q} - \frac{kT}{q}\nabla\ln\frac{N_V}{T^{3/2}} = -\nabla(\phi + \frac{\chi}{q} + \frac{E_g}{q} + \frac{kT}{q}\ln\frac{N_V}{T^{3/2}}), \qquad (12.34)$$

where $E_g = E_C - E_V$ is the bandgap, N_V is the effective valence band density of states, and χ, E_g, N_C, and N_V are all in general position dependent. The drift-diffusion transport equations are similar to (12.26) and (12.27) except for the electric field terms,

$$\vec{J}_n = qn\mu_n \vec{E}_n + qD_n \nabla n, \qquad (12.35)$$

$$\vec{J}_p = qp\mu_p \vec{E}_p - qD_p \nabla p. \qquad (12.36)$$

12.1.3 Energy Balance Equations

In scaled devices with high electric fields and high electric field gradients, the carrier temperatures can often be much higher than the lattice temperature. The average carrier kinetic energies W_n and W_p are related to the carrier temperatures T_n and T_p by

$$W_n = \frac{3}{2} nkT_n, \qquad (12.37)$$

$$W_p = \frac{3}{2} pkT_p. \qquad (12.38)$$

The carrier kinetic energy balance is described by Fick's second law [3],

$$\frac{\partial W_n}{\partial t} = -\nabla \cdot \vec{S}_n + \vec{J}_n \cdot \vec{E}_n - u_{W_n}, \qquad (12.39)$$

$$\frac{\partial W_p}{\partial t} = -\nabla \cdot \vec{S}_p + \vec{J}_p \cdot \vec{E}_p - u_{W_p}, \qquad (12.40)$$

where W_n and W_p are the carrier kinetic energies, and $\vec{S}_n$ and $\vec{S}_p$ are the carrier energy fluxes. Carriers gain energy from the electric field, and the energy conversion rates are described by the Joule heating terms $\vec{J}_n \cdot \vec{E}_n$ and $\vec{J}_p \cdot \vec{E}_p$. Carriers also lose energy ("cool off") via recombination, impact ionization, as well as other carrier scattering processes. The net energy loss rates due to these mechanisms are described by u_{Wn} and u_{Wp}, as detailed below. Now, $\vec{J}_n$ and $\vec{J}_p$ can be obtained using the method of moments

$$\vec{J}_n = qn\mu_n \vec{E}_n + qD_n \vec{\nabla} n + qnD_n^T \vec{\nabla} T_n, \qquad (12.41)$$

$$\vec{J}_p = qp\mu_p \vec{E}_p - qD_p \vec{\nabla} p - qpD_p^T \vec{\nabla} T_p, \qquad (12.42)$$

where D_n^T and D_p^T are carrier thermal diffusivities. Note that μ_n and μ_p are now functions of T_n and T_p, respectively, and D_n and D_p are related to μ_n and μ_p by

$$D_n = \frac{kT_n}{q}\mu_n F_{1/2}(\eta_n)/F_{-1/2}(\eta_n), \qquad (12.43)$$

$$D_p = \frac{kT_p}{q}\mu_p F_{1/2}(\eta_p)/F_{-1/2}(\eta_p), \qquad (12.44)$$

$$\eta_n = \frac{E_{Fn} - E_c}{kT_n}, \qquad (12.45)$$

$$\eta_p = \frac{E_v - E_{Fp}}{kT_p}. \qquad (12.46)$$

The thermal diffusivities D_n^T and D_p^T are related to μ_n and μ_p in a complex manner for Fermi-Dirac statistics. For the special case of Boltzmann statistics we can write

$$D_n^T = \frac{\partial D_n}{\partial T_n}, \qquad (12.47)$$

$$D_p^T = \frac{\partial D_p}{\partial T_p}. \qquad (12.48)$$

The energy fluxes $\vec{S}_n$ and $\vec{S}_p$ can also be obtained using the method of moments as follows [3]

$$\vec{S}_n = -P_n T_n \vec{J}_n - \kappa_n T_n, \qquad (12.49)$$

$$\vec{S}_p = P_p T_p \vec{J}_p - \kappa_p T_p, \qquad (12.50)$$

where P_n and P_p are the carrier thermal electric powers, which have a simple form when Boltzmann statistics is applied,

$$P_n = P_p = P = \frac{5}{2}\frac{k}{q}, \qquad (12.51)$$

and κ_n and κ_p, the carrier thermal conductivities, are described by a generalized Wiedemann-Franz law;

$$\kappa_n = \left(\frac{5}{2} + c_n\right)\frac{k^2}{q}n\mu_n T_n, \qquad (12.52)$$

$$\kappa_p = \left(\frac{5}{2} + c_p\right)\frac{k^2}{q}p\mu_p T_p. \qquad (12.53)$$

Here, c_n and c_p are carrier heat capacities, which can be safely neglected in most situations. Finally, the rate of net loss of carrier kinetic energy is given by [3]

$$u_{W_n} = (u_{SRH} + u_{rad})\frac{3kT_n}{2} - (u_{n,Auger} - g_{n,ii})[E_g + \frac{3kT_p}{2}]$$
$$- g_{p,ii}\frac{3kT_n}{2} - \frac{W_n - W_{n0}}{\tau_{W_n}}, \qquad (12.54)$$

$$u_{W_p} = (u_{SRH} + u_{rad})\frac{3kT_p}{2} - (u_{p,Auger} - g_{p,ii})[E_g + \frac{3kT_n}{2}]$$
$$- g_{n,ii}\frac{3kT_p}{2} - \frac{W_p - W_{p0}}{\tau_{W_p}}. \qquad (12.55)$$

In the above equations, u_{SRH} and u_{rad} are the rates of net carrier loss due to SRH and radiative recombination, respectively, and $u_{n,Auger}$ and $u_{p,Auger}$ are rates of net carrier losses due to Auger recombination initiated by electrons and holes, respectively. In addition, $g_{n,ii}$ is the rate of net carrier generation due to impact ionization initiated by the high energy electrons, and $g_{p,ii}$ is the rate of net carrier generation due to impact ionization initiated by the high energy holes. The last two terms in the above two equations represent the net rate of energy loss via scattering, where τ_{W_n} and τ_{W_p} are the carrier energy relaxation times, and W_{n0} and W_{p0} are W_n and W_p when $T_n = T_p = T$. For Auger recombination and impact ionization, we must distinguish between whether the process is initiated by the electrons or the holes. Both processes involve three carriers: the initiating electron or hole, and the recombined or impact-ionized electron-hole pair. The last two terms of the two equations above represent the energy exchange between the carriers and the lattice.

12.1.4 Boundary Conditions

In commercial device simulators, the basic semiconductor equations are solved on a finite bounded domain that is defined by the user. The user must make sure that the simulation domain chosen is large enough that the boundary conditions implemented in the simulator are well-satisfied. Nonphysical boundaries not only lead to inaccurate simulation results, but also cause convergence problems during the numerical solution process. At the edges of the simulation domain, appropriate boundary conditions are applied for the unknowns ϕ, n, p, T_n, and T_p in solving the fundamental semiconductor equations.

The device structure in question must be isolated from its surroundings, including adjacent devices, the substrate, or passivation dielectric layers. These boundaries are somewhat artificial, and can be chosen differently by different users. The bottom line, however, is that the simulation results must be checked for consistency with the "Neumann" boundary conditions implemented in simulators. That is, the fluxes across the boundaries must be zero, the normal component of electric field must be zero, and

$$\hat{n} \cdot \vec{J_n} = 0, \quad \hat{n} \cdot \vec{J_p} = 0, \tag{12.56}$$

$$\hat{n} \cdot \vec{S_n} = 0, \quad \hat{n} \cdot \vec{S_p} = 0, \tag{12.57}$$

$$\hat{n} \cdot \nabla \phi = 0, \tag{12.58}$$

where $\hat{n}$ is an outward-oriented vector normal to the boundary.

The "Dirichlet" boundary condition is applied at the ohmic contacts. The values of n and p in the drift-diffusion formalism, as well as T_n and T_p in the energy-balance formalism, are by definition fixed at their equilibrium values at the ohmic contacts. The electron and hole temperatures are equal to the lattice temperature,

$$T_n = T_p = T. \tag{12.59}$$

Electrons and holes have the same quasi-Fermi levels at ohmic contacts, which are modulated by the applied voltage, such that

$$\phi_{Fn} = \phi_{Fp} = V_{applied}. \tag{12.60}$$

The equilibrium values of n and p are defined by the charge neutrality conditions, and the pn product,

$$p_s + N_d^+ = n_s + N_a^-, \tag{12.61}$$

$$p_s n_s = \gamma_n \gamma_p n_i^2. \tag{12.62}$$

Using (12.6) and (12.7), η_n and η_p, and hence γ_n and γ_p can be expressed as functions of n and p, respectively. Therefore, n_s and p_s can be solved from (12.61) and (12.62). The intrinsic Fermi potential ψ_s at the contact is then obtained from either (12.16) or (12.17),

$$\psi_s = \phi_{Fn} + \frac{kT}{q} \ln \frac{n_s}{n_{in}} = V_{applied} + \frac{kT}{q} \ln \frac{n_s}{n_{in}}, \tag{12.63}$$

$$= \phi_{Fp} - \frac{kT}{q} \ln \frac{p_s}{n_{ip}} = V_{applied} - \frac{kT}{q} \ln \frac{p_s}{n_{ip}}, \tag{12.64}$$

where $\phi_{Fn} = \phi_{Fp} = V_{applied}$, since we are considering equilibrium conditions. The electrostatic potential ϕ is then obtained using (12.23). For the special case of

Boltzmann statistics, closed-form solutions exist

$$n_s = \frac{C + \sqrt{C^2 + 4n_i^2}}{2},\qquad(12.65)$$

$$p_s = n_i^2/n_s,\qquad(12.66)$$

$$\psi_s = V_{applied} + \frac{kT}{q}\ln\frac{n_s}{n_i} = V_{applied} - \frac{kT}{q}\ln\frac{p_s}{n_i}.\qquad(12.67)$$

Schottky contacts are typically modeled in terms of work function and surface recombination velocity. At the contact, the intrinsic Fermi potential ψ_s is given by [1]

$$\psi_s = V_{applied} + \chi_s + \frac{1}{2q}\left(E_g + kT\ln\frac{N_C}{N_V}\right) - W_{met},\qquad(12.68)$$

where W_{met} is the work function of the contact metal. A polysilicon contact to the emitter of a SiGe HBT is often modeled as a Schottky contact in commercial devices simulators. The electron and hole quasi-Fermi levels ϕ_{Fn} and ϕ_{Fp} are no longer defined by the applied voltage. Instead, they are defined in terms of the current density and minority carrier concentration at the contact

$$J_{sn} = qv_{sn}(n_s - n_{s0}),\qquad(12.69)$$

$$J_{sp} = qv_{sp}(p_s - p_{s0}).\qquad(12.70)$$

Here v_{sn} and v_{sp} are the surface recombination velocities of electrons and holes, respectively, and n_{s0} and p_{s0} are the equilibrium values of n_s and p_s. Ohmic contact can be viewed as the extreme case of $v_{sn}=v_{sp}=\infty$. Hence, $n_s = n_{s0}$ and $p_s = p_{s0}$, since an ideal ohmic contact acts as a perfect sink for injected minority carriers. The work function and minority carrier recombination velocity at the polysilicon contact can then be adjusted to reproduce the measured base current in SiGe HBT simulations. [4]

An efficient approach for vertical profile optimization of SiGe HBTs often requires only 1-D device simulation. The collector and emitter boundaries can be described as standard ohmic contacts. The base electrode, however, cannot be modeled as a standard ohmic contact, because the minority carrier concentration is fixed at its equilibrium value for the point specified as the base electrode. As a result, the minority carriers are all absorbed by the base electrode, which effectively blocks the minority carrier current flow towards the collector. This unphysical result is thus caused by the 1-D simplification. In reality, the ohmic base contact is

---

[4] Unfortunately this is usually performed using the free-parameters to simply curve-fit to the data, thereby losing any physical basis.

obviously not made inside the intrinsic neutral base, and hence the intrinsic base cannot be described as a true ohmic contact, which by definition is a perfect sink for minority carriers. Inside the intrinsic base, the base current is supplied laterally from the physical base contact, where the ohmic boundary conditions apply. An easy solution that enables 1-D simplification of this inherently 2-D problem is to set only the *majority carrier* quasi-Fermi level at the base contact point. For an *npn* SiGe HBT, only the hole quasi-Fermi potential ϕ_{Fp} is set to the applied voltage. A more elaborate approach is to consider the emitter crowding effect by explicitly including the voltage drop caused by the base current. In MEDICI, this is accomplished by injecting a majority carrier current I_m at the node specified as majority carrier contact. In this case, I_m is proportional to the difference in applied voltage $V_{applied}$ and the majority carrier quasi-Fermi potential $\phi_{F,maj}$

$$I_m = G(\phi_{F,maj} - V_{applied}). \tag{12.71}$$

Inside MEDICI, G is calculated as the conductance corresponding to 0.1 μm of silicon.

12.1.5 Physical Models

A number of physical parameters are required in the semiconductor equations, including the effective density-of-states N_C and N_V, the electron affinity χ, the bandgap E_g, the carrier mobilities μ_n and μ_p, the recombination rates u_{SRH} and u_{Auger}, and the impact ionization rates $g_{n,ii}$ and $g_{p,ii}$. In commercial simulators such as MEDICI, only χ and E_g are modeled as a function of Ge mole fraction. For parameters such as the mobilities, a number of models are available from which the user must choose. The model equations can be found in the user manuals, but the relevant question is which parameter models to select, and sometimes, which model parameters can be tuned in a meaningful way if needed.

In choosing mobility models, one typically needs to specify a model for the low-field mobility as well as a model for velocity saturation. By default, the carrier mobilities are treated as constants. As a result, the f_T simulated using default models is much larger than the measured f_T. The current density at which peak f_T is reached is also typically well above its measured value because velocity saturation is not taken into account. Based on our experience, for SiGe HBT simulation, the use of the so-called "Philips unified mobility model" is recommended. A unique feature of this model lies in its ability to separately model the majority and minority carrier mobilities, which can be an important issue for SiGe HBTs.

For device simulations, it is a common practice to assume complete ionization of dopants in Si (or SiGe). For heavy doping levels, such as those found in the emitter and base regions of SiGe HBTs, however, significant differences can be observed between results simulated with and without incomplete ionization, especially for early versions of MEDICI. This difference is not truly physical because the dopants should be completely ionized at all temperatures for concentrations above a certain doping level known as the "Mott" or "metal-insulator" transition. If one continues to use the incomplete ionization relations for such heavy doping levels, the majority carrier concentration is significantly underestimated, and the minority carrier concentration is equally significantly overestimated. Significant shifts of both I_C and I_B are then observed on the Gummel characteristics for a typical SiGe HBT. This situation has been corrected in later versions of MEDICI by applying incomplete ionization relations for doping levels below a defined low-valued threshold, and applying complete ionization for doping levels above a defined high-valued threshold, and then interpolating between the two thresholds. This option is chosen by specifying "high.dop" together with "incomplete" in the MEDICI model statement.

In our discussion above on carrier statistics, we noted that the apparent electrical bandgap decreases with increasing doping level if one continues to apply Boltzmann statistics to describe the equilibrium pn product,

$$pn = n_{i0}^2 e^{\Delta G/kT}, \qquad (12.72)$$

where n_{i0} is the intrinsic carrier concentration at low doping levels. The pn product changes due to a combination of degeneracy, doping-induced rigid bandgap narrowing, and density-of-states perturbations, which can all be lumped into a single parameter ΔG, commonly called the "apparent bandgap narrowing." Since Boltzmann statistics is used in obtaining the apparent bandgap narrowing expression, one should also use Boltzmann statistics in device simulation if the apparent bandgap narrowing parameters are used "as is." Otherwise, the effect of degeneracy on the pn product is effectively accounted for twice. For a doping level of 10^{20} cm^{-3}, the ΔG due to degeneracy is -31.2916 meV, and is thus significant for SiGe HBTs. Therefore, Boltzmann statistics should be used as opposed to the more accurate Fermi-Dirac statistics for SiGe HBTs when the default Boltzmann-statistics-based bandgap narrowing model is used. This approach, however, may potentially cause other problems in cases where Fermi-Dirac statistics is necessary, either in another region of the device where the doping level is moderate, or at low temperatures. Another potential problem is that the Einstein relations depend on the carrier statistics, which can affect minority carrier diffusivity and hence f_T. One solution is to automatically adjust the value of ΔG based on the user's choice of the statistics, but this remains to be implemented in commercial simulators. Unfortunately, at present, the BGN parameters are calculated independent of the user's choice of statistics in all commercial simulators. An example is given below for the widely used "Slotboom BGN model."

Importantly, the choice of the bandgap narrowing model must be made consistently, together with the choice of the mobility models and carrier statistics. The reason is that the *pn* product at equilibrium is not directly measured, and is instead inferred from the Gummel characteristics. Consider the base current for a bipolar transistor with uniform doping,

$$J_B = kT\mu_n \frac{n_{i0}^2}{N_{de}^+ W_e} e^{\Delta G/kT} e^{qV_{BE}/kT}, \quad (12.73)$$

where N_{de}^+ and W_e are the emitter doping level and the emitter depth. The above equation lumps the effects of degeneracy (i.e., Fermi-Dirac statistics), rigid bandgap narrowing, and the $N_C N_V$ changes into the ΔG term. It also neglects any recombination current. Rigorous derivation of the above familiar transport equation including the effects of degeneracy, rigid bandgap narrowing, and $N_C N_V$ changes can be performed, as was reviewed in [4], and the same analysis can be applied to derive the collector current in SiGe HBTs [5]. The equations derived by including advanced physics share the same functional *form* as the older equations derived using simplified physics, but differ in *substance*.

The measured $I_B - V_{BE}$ data only give $\mu_n \exp(\Delta G/kT)/(N_{de}^+ W_e)$. A mobility model must be used to calculate ΔG, provided that the active concentration N_{de}^+ is accurately known. In MEDICI, the default values for the bandgap narrowing models were obtained using the Philips unified mobility model. Therefore, the Philips unified mobility model should always be used together with the default bandgap narrowing model parameters unless justified by other considerations. The bandgap narrowing effect is accounted for in MEDICI by simply modifying n_i to n_{ie} in all the transport equations and boundary conditions

$$n_{ie} = n_i e^{\Delta G/2kT}. \quad (12.74)$$

For instance, the n_i in (12.16), (12.17), (12.62), and (12.22) are replaced by n_{ie}, where n_{ie} is in general position dependent. In addition, ΔG is modeled as a function of the doping concentration N using the Slotboom model [6]

$$\Delta G = \Delta G_0 \left[\ln \frac{N}{N_0} + \sqrt{\left[\ln \frac{N}{N_0}\right]^2 + C} \right]. \quad (12.75)$$

Using the "Philips unified mobility model" [7], the model parameters were determined as $\Delta G_0 = 6.92$ meV, $N_0 = 1.3 \times 10^{17}$ cm^{-3}, and $C = 0.5$ [8]. The same parameters apply to p-type doping. These model parameters were obtained by reinterpreting those originally given in [6] using the Philips unified mobility model, which treats minority carrier and majority carrier mobilities independently. Figure 12.2 shows the apparent bandgap narrowing ΔG as a function of n-type doping using the Slotboom model. We show two curves; the dashed curve is calculated "as is" (i.e., as found in most simulators), while the solid curve is calculated by applying a correction factor to remove the degeneracy effect. If Boltzmann statistics is used for device simulation, the dashed curve should be used, since the degeneracy effect is already lumped into the ΔG term. If Fermi-Dirac statistics is used for device simulation, however, the solid curve should be used, since it does not contain the degeneracy effect. The amount of correction needed to account for Fermi statistics is in principle dependent on N_C and N_V, for n- and p-type dopants, which are in turn dependent on the Ge mole fraction (refer to Chapter 2).

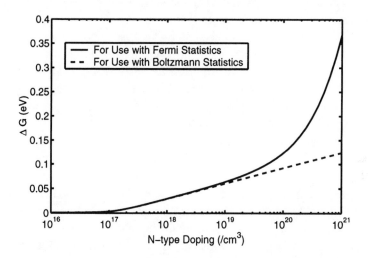

Figure 12.2 Apparent bandgap narrowing ΔG for n-type doping. For device simulation using Fermi-Dirac statistics, the solid curve should be used. For device simulation using Boltzmann statistics, the dashed curve should be used.

12.1.6 Numerical Methods

The electrical characteristics of a transistor can be simulated by solving the fundamental semiconductor equations. The individual roles of the variables in these equations can be summarized as follows:

- The electrostatic potential distribution is governed by Poisson's equation.

- The electron and hole concentration distributions are governed by the continuity equations.

- The electron and hole temperature distributions are governed by the energy-balance equations.

- If self-heating is considered, the lattice temperature distribution is governed by the lattice heat equation.

These equations are strongly coupled partial differential equations that are also strongly nonlinear, and hence are not trivially solved. Note, for instance, that the relations between n or p and ϕ are exponential in functional form, the strongest nonlinearity found in nature. Robust numerical methods for solving semiconductor equations are thus required, and in fact have been developed during the past 30 years. The nonlinear differential equations are first discretized at user-defined elements or "nodes," and then solved iteratively subject to the appropriate boundary conditions. Commercial simulators provide for three types of simulation: 1) dc simulation, which is used for generating I-V curves; 2) the frequency domain small-signal simulation for generating Y-parameters; and 3) transient simulation for examining device response to transient excitations.

The continuous functions in the semiconductor equations are first discretized on a user-defined simulation mesh. The differential operators are replaced by suitable difference operators. The unknown variables n, p, ϕ, T_n, and T_p at all of the nodes are solved from these discretized equations. In a 2-D simulation, the number of obtuse elements must be minimized. Time discretization during transient simulation is typically automatic except for the initial time step.

The box integration method is widely used in discretizing the differential operators. Poisson's equation and the continuity equations are integrated over a small volume enclosing each node, yielding nonlinear algebraic equations for the unknown variables. The integration equates the flux entering the volume with the sources and sinks inside it. The integrals are performed on an element-by-element basis, making the handling of general surfaces and boundary conditions much easier.

In transistors, carrier concentrations often change dramatically over a short distance. For instance, n and p can drop by several orders of magnitude across the space-charge region of a forward-biased junction. Therefore, the current density equations cannot be discretized using the conventional finite difference method, which assumes that the variables in question vary between adjacent nodes in a linear fashion. To overcome this difficulty, the Scharfetter-Gummel method and its variants can be used for discretizing the current density and energy flux density equations [1, 3], which allow the carrier concentrations to vary exponentially

Device Simulation

between adjacent nodes.

The discretized nonlinear equations are solved iteratively using "Newton's" method. Consider the five governing equations

$$F_\phi(\phi, n, p, T_n, T_p) = 0, \quad (12.76)$$
$$F_n(\phi, n, p, T_n, T_p) = 0, \quad (12.77)$$
$$F_p(\phi, n, p, T_n, T_p) = 0, \quad (12.78)$$
$$F_{T_n}(\phi, n, p, T_n, T_p) = 0, \quad (12.79)$$
$$F_{T_p}(\phi, n, p, T_n, T_p) = 0. \quad (12.80)$$

For a given initial guess of ϕ_0, n_0, p_0, T_{n0}, T_{p0}, which is typically the equilibrium solution, an update $\Delta\phi$, Δn, Δp, ΔT_n, and ΔT_p is calculated by solving

$$\begin{bmatrix} \frac{\partial F_\phi}{\partial \phi} & \frac{\partial F_\phi}{\partial n} & \frac{\partial F_\phi}{\partial p} & \frac{\partial F_\phi}{\partial T_n} & \frac{\partial F_\phi}{\partial T_p} \\ \frac{\partial F_n}{\partial \phi} & \frac{\partial F_n}{\partial n} & \frac{\partial F_n}{\partial p} & \frac{\partial F_n}{\partial T_n} & \frac{\partial F_n}{\partial T_p} \\ \frac{\partial F_p}{\partial \phi} & \frac{\partial F_p}{\partial n} & \frac{\partial F_p}{\partial p} & \frac{\partial F_p}{\partial T_n} & \frac{\partial F_p}{\partial T_p} \\ \frac{\partial F_{T_n}}{\partial \phi} & \frac{\partial F_{T_n}}{\partial n} & \frac{\partial F_{T_n}}{\partial p} & \frac{\partial F_{T_n}}{\partial T_n} & \frac{\partial F_{T_n}}{\partial T_p} \\ \frac{\partial F_{T_p}}{\partial \phi} & \frac{\partial F_{T_p}}{\partial n} & \frac{\partial F_{T_p}}{\partial p} & \frac{\partial F_{T_p}}{\partial T_n} & \frac{\partial F_{T_p}}{\partial T_p} \end{bmatrix} \begin{bmatrix} \Delta\phi \\ \Delta n \\ \Delta p \\ \Delta T_n \\ \Delta T_p \end{bmatrix} = - \begin{bmatrix} F_\phi \\ F_n \\ F_p \\ F_{T_n} \\ F_{T_p} \end{bmatrix} . \quad (12.81)$$

The matrix on the left side of the above equation is the so-called "Jacobian matrix." Each element of the Jacobian is itself an $N \times N$ matrix, with N being the total number of nodes. The matrix is mathematically sparse, because only adjacent nodes are interconnected by mesh lines during discretization. Through proper ordering of the nodes, the matrix bandwidth can be minimized. The total matrix has 5 times as many rows and columns as the matrix for a single variable. With the solved Δ, the solution is then updated. Denoting the solution estimate at the kth iteration as $\vec{x}_k$, the initial guess for the $(k + 1)$th iteration is calculated by

$$\vec{x}_{k+1} = \vec{x}_k + t_k \vec{\Delta x}, \quad (12.82)$$
$$\vec{x} = [\phi \; n \; p \; T_n \; T_p]^T, \quad (12.83)$$
$$\vec{\Delta x} = [\Delta\phi \; \Delta n \; \Delta p \; \Delta T_n \; \Delta T_p]^T, \quad (12.84)$$

where a damping parameter t_k is used to improve convergence. A new Jacobian is then calculated, and another new update is in turn calculated by solving (12.81) until the update $\vec{\Delta x}$ becomes sufficiently small for convergence to be achieved. If

(12.81) is to be solved using a direct method such as LU decomposition, the iteration process can be greatly accelerated using the "Newton-Richardson" method, which refactors the Jacobian only when necessary (typically, twice per bias point), as opposed to once per iteration.

The algorithm for an *ac* solution of the transport equations typically proceeds as follows. The total voltage applied to an electrode i is the sum of a *dc* bias and an *ac* sinusoidal voltage

$$V_i = V_{i0} + \hat{V}_i e^{j\omega t}, \tag{12.85}$$

where V_{i0} is the *dc* bias and $\hat{V}_i$ is the magnitude of the *ac* bias. The task of *ac* simulation is to find the solution to the fundamental equations (12.1)–(12.3)

$$\vec{x} = \vec{x}_{dc} + \vec{x}_{ac} e^{j\omega t}. \tag{12.86}$$

Here $\vec{x}_{dc}$ is the *dc* solution vector including all variables, and $\vec{x}_{ac}$ is a phasor vector consisting of complex numbers. Below we consider the drift-diffusion transport equations for which only ϕ, n, and p need to be solved. In small-signal analysis, $\hat{V}_i$ is sufficiently small such that a first-order (linear) Taylor expansion can be used when evaluating the left hand side (LHS) of (12.1)–(12.3). Only the terms having the first-order of exp($j\omega t$) are retained. The LHS of (12.1)–(12.3) can thus be written as

$$F_\phi(\phi, n, p) = F_\phi(\phi_0, n_0, p_0) + \frac{\partial F_\phi}{\partial \phi}\phi_{ac}e^{j\omega t} + \frac{\partial F_\phi}{\partial n}n_{ac}e^{j\omega t} + \frac{\partial F_\phi}{\partial p}p_{ac}e^{j\omega t}, \tag{12.87}$$

$$F_n(\phi, n, p) = F_n(\phi_0, n_0, p_0) + \frac{\partial F_n}{\partial \phi}\phi_{ac}e^{j\omega t} + \frac{\partial F_n}{\partial n}n_{ac}e^{j\omega t} + \frac{\partial F_n}{\partial p}p_{ac}e^{j\omega t}, \tag{12.88}$$

$$F_p(\phi, n, p) = F_p(\phi_0, n_0, p_0) + \frac{\partial F_p}{\partial \phi}\phi_{ac}e^{j\omega t} + \frac{\partial F_p}{\partial n}n_{ac}e^{j\omega t} + \frac{\partial F_p}{\partial p}p_{ac}e^{j\omega t}. \tag{12.89}$$

Recall that the *dc* solutions were obtained by letting $F_\phi = 0$, $F_n = 0$, and $F_p = 0$. Therefore, at the *dc* solution $\vec{x} = [\phi_0, n_0, p_0]^T$,

$$F_\phi(\phi_0, n_0, p_0) = 0, \tag{12.90}$$

$$F_n(\phi_0, n_0, p_0) = 0, \tag{12.91}$$

$$F_p(\phi_0, n_0, p_0) = 0. \tag{12.92}$$

After manipulating the right-hand side (RHS) of (12.1)–(12.3), one obtains a set of equations similar to the *dc* nonlinear Newton iteration equations

$$\begin{bmatrix} \frac{\partial F_\phi}{\partial \phi} & \frac{\partial F_\phi}{\partial n} & \frac{\partial F_\phi}{\partial p} \\ \frac{\partial F_n}{\partial \phi} & \frac{\partial F_n}{\partial n}+D_1 & \frac{\partial F_n}{\partial p} \\ \frac{\partial F_p}{\partial \phi} & \frac{\partial F_p}{\partial n} & \frac{\partial F_p}{\partial p}+D_1 \end{bmatrix} \begin{bmatrix} \phi_{ac} \\ n_{ac} \\ p_{ac} \end{bmatrix} = b_1. \tag{12.93}$$

Here the $\partial F_\phi/\partial\phi$, $\partial F_n/\partial n$, and $\partial F_p/\partial p$ are the same $N \times N$ matrices that form the Jacobian matrix used in the *dc* solution. In addition, D_1 is an $N \times N$ matrix with $-j\omega$ on the diagonal and 0 on the off-diagonal, which results from the two time derivatives on the RHS of (12.2) and (12.3)

$$D_1 = \begin{pmatrix} -j\omega & 0 & \cdots & 0 \\ 0 & -j\omega & \cdots & 0 \\ \vdots & \vdots & \ddots & \vdots \\ 0 & 0 & 0 & -j\omega \end{pmatrix}, \qquad (12.94)$$

where b_1 is a vector that contains the boundary conditions of the *ac* input voltage. Splitting the real and imaginary parts of (12.93), one has

$$\begin{pmatrix} J & -D_2 \\ D_2 & J \end{pmatrix} \begin{pmatrix} x_r \\ x_i \end{pmatrix} = b_2, \qquad (12.95)$$

where J is the *dc* Jacobian matrix, D_2 is a diagonal matrix related to D_1, b_2 is a permuted version of b_1, and x_r and x_i are the real and imaginary parts of the solution vector $\vec{x}_{ac} = [\phi_{ac}\ n_{ac}\ p_{ac}]^T$.

A special case of small-signal *ac* simulation that has broad application is the zero frequency limit. In transistor problems, the n_{ac} and p_{ac} solved at $\omega = 0$ represent the local carrier storage-induced capacitance. The 2-D or 1-D distributions of n_{ac} and p_{ac} are thus extremely useful in identifying the dominant regions of charge storage in a SiGe HBT. Doping profile design as well as Ge profile design can be thus be optimized using simulation to minimize charge storage and hence maximize f_T.

12.2 Application Issues

12.2.1 Device Structure Specification

The first step in device simulation is to determine the doping and Ge profiles, which are typically measured using SIMS techniques. Figure 12.3 shows representative doping and Ge profiles measured by SIMS for a first generation SiGe HBT. The polysilicon/silicon interface can be identified by the As segregation peak. The measured As doping "tail" into the Si is apparently higher than the base doping across the entire base, but this is clearly not real, and in fact is simply the result of the finite resolution limit of SIMS in following very rapidly changing doping profiles. The true metallurgical EB junction can be determined from the "dip" in the B SIMS profile of the base, where the n- and p-type concentrations are equal. Thus the EB junction depth from the Si surface can be precisely located.

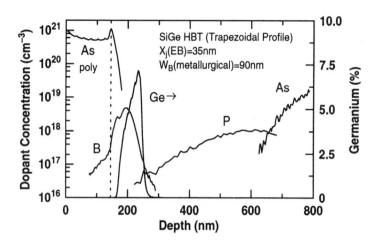

Figure 12.3 Typical doping and Ge profiles measured by SIMS for a first-generation SiGe HBT. The true emitter As profile is much less abrupt than the SIMS measurement suggests.

The dopant activation percentage in the polysilicon emitter is quite low for As, and a 5–10% activation rate is often assumed due to As clustering. The As profile in the single crystalline silicon emitter is Gaussian-like and falls from the polysilicon/silicon interface towards the base. The Ge profile measured using SIMS also has limited accuracy, and should be compared with the intended Ge profile during SiGe epitaxial growth. Rutherford backscattering (RBS) is a more accurate means for measuring the true Ge profile, but, in our experience, properly calibrated SIMS generally does a very good job for Ge. An example of the net doping profile and Ge profile used for simulation is shown in Figure 12.4.

For 2-D and 3-D simulations, the vertical doping profile in the extrinsic base region can be obtained using SIMS in a similar manner. The lateral doping transition between extrinsic and intrinsic device, however, can only be estimated from device layout and fabrication details, since 2-D doping profile information is typically unavailable. For vertical profile design, 1-D simulations can be used first because of the low simulation overhead, since 1-D simulation involves much easier gridding, easier transit time analysis, much shorter simulation time, and easier debugging. The resulting design can then be refined using 2-D simulation, which is necessary when accurate f_{max} or noise analysis is desired. Full 3-D simulation becomes necessary for problems which are inherently 3-D in nature, such as in single-event upset (refer to Chapter 11).

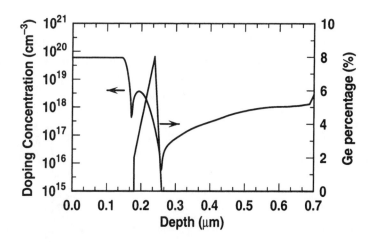

Figure 12.4 Typical doping and Ge profiles used for device simulation.

12.2.2 Mesh Specification and Verification

The next step is to define the coordinate values of the points (nodes) at which the semiconductor equations are discretized. Even though commercial simulators all provide some means of regridding (e.g., based on the doping gradient), taking the extra time to specify a reasonably good initial mesh usually pays off in the end. Regridding, if not well-controlled, can easily generate a large number of obtuse triangle elements, which can cause numerical problems for the continuity equations when the box integration method is applied for discretization. A popular meshing method is to use a rectangular grid. In this case, the nodes are defined by the intersections of horizontal and vertical lines.

An often-asked question is: what is the *optimum* grid for a given SiGe HBT simulation? The true answer is that the optimum grid in a given problem depends strongly on the device metric of most interest. To simulate the forward-mode SiGe HBT operation, for instance, the EB spacer oxide corner mesh does not need to be fine. However, to simulate the reverse emitter-base junction band-to-band-tunneling current for an EB reliability study, the grid at the oxide corner needs to be very fine in order to accurately locate the peak electric field.

A number of empirical criteria can be applied in meshing a semiconductor device structure. Fine meshing is necessary where the space charge density and its spatial gradient are large. Such an example in a SiGe HBT lies in the depletion layers of the EB and CB junctions. Placing nodes along the physical junction interfaces is important for accurate simulation. In addition, grid lines must be placed at the critical points defining the SiGe profile in order to avoid creating artificial SiGe profiles that inadvertently differ from what one has in mind. When a simulator such as MEDICI is used, for instance, the initial grid lines must be placed with the SiGe and doping profiles in mind. Figure 12.5 shows an example of *bad* mesh line specification, while Figure 12.6 shows an example of mesh lines placed properly with the intended Ge profile in mind.

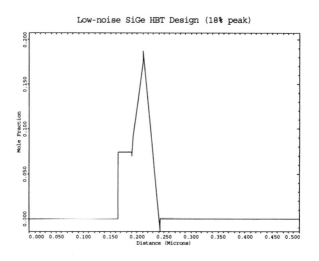

Figure 12.5 Ge mole fraction from a mesh *improperly* specified without the Ge profile in mind.

Initial coarse meshes are often refined based on where the physical properties of the device structure dictate it. That is, the mesh must be refined where a given variable or change in that variable across an element exceeds a given defined tolerance. If breakdown voltage is the concern, for instance, the impact ionization rate can be used. Because of the strong nonlinearities in semiconductor problems, the doping concentration at the newly generated nodes should be determined from the original doping profile specification, instead of interpolation from the existing mesh.

Theoretically speaking, the potential difference or quasi-Fermi potential difference between two adjacent nodes should generally be kept less than the thermal voltage kT/q in order to minimize discretization error. In practice, this requirement is often relaxed to about $10 - 15$ kT/q between adjacent nodes. The doping concentration change between adjacent nodes should be less than two to three orders of magnitude. In high-level injection, very fine meshing is often required where the minority carrier concentration exceeds the doping concentration (e.g., in the CB space-charge region of a SiGe HBT).

Device Simulation 531

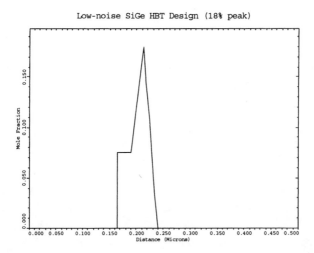

Figure 12.6 Ge mole fraction profile from a mesh *correctly* specified with the Ge profile in mind.

Simulation results depend on the mesh specified, while for a physical transistor, its electrical performance is unique for a given bias. It is consequently up to the user to decide whether the mesh used for simulation is robust or not. The electrical parameters of interest (e.g., $f_T - I_C$ for a SiGe HBT) should *always* be resimulated using a finer mesh to order check for grid sensitivity effects. Identical results using different gridding can generally be taken to imply that a robust mesh has been achieved. In MEDICI, an overall finer mesh can be obtained conveniently using the statement "Regrid potential factor=1.5 smooth=1." The "factor" parameter requests an automatic increase of the number of nodes by a factor of 1.5×. The "potential" parameter indicates that the refinement is performed where the potential change between adjacent nodes is large. One can have more confidence in the mesh used if the various simulated metrics no longer change with further mesh refining. This technique can also be applied to determine the acceptable coarse meshing limit for a particular problem before running extensive parametric analysis, and can dramatically minimize overall simulation time. Since 1-D simulation is quite fast, very fine (finer than necessary) mesh can be used, which adds little extra simulation time, but may save time in the end spent on generating an accurate mesh with fewer nodes.

12.2.3 Model Selection and Coefficient Tuning

The default models contained in commercial device simulators are often the simplest ones, but are usually not sufficient for meaningful SiGe HBT simulation. For instance, in MEDICI, carrier generation and recombination is not included unless specified explicitly on the model statement. Similarly, while the emitter and base doping levels are both very high in SiGe HBTs, heavy doping effects are not included by default. As discussed in the previous section, the *apparent* bandgap narrowing ΔG was experimentally obtained assuming Boltzmann statistics, and the effect of degeneracy is lumped into ΔG. Therefore, Boltzmann statistics instead of the theoretically more accurate Fermi-Dirac statistics should be used for at least simulation of the Gummel characteristics if the default BGN model parameters in MEDICI are used. Caution, however, should be exercised in interpreting the high-frequency simulation results obtained with Boltzmann statistics, for reasons discussed below. An alternate approach is to use Fermi-Dirac statistics and then appropriately modify the BGN model parameters in the simulator. In our experience, the Philips unified mobility model is the most suitable for SiGe HBT simulation, and should be used together with velocity saturation.

12.2.4 *dc* Simulation

As a starting point, *dc* simulations are used to capture the transistor I-V behavior, which for a SiGe HBT usually means the Gummel characteristics. A number of parameters can affect the simulated Gummel characteristics, including carrier statistics, recombination parameters, BGN model parameters, mobility models, as well as the doping and Ge profiles. All of these models must be considered when attempting to obtain agreement between simulation and experimental data (we refer to this (iterative) process as "calibration" of the simulator). The temperature at which the data is measured must be accurately known and used as the lattice temperature in simulation, because SiGe HBT currents are exponentially dependent on the temperature. In our experience, and for nonobvious reasons, the transistor Gummel characteristics (or other *dc* data) are generally much more sensitive to parameter model selection and requisite model parameters than the $f_T - I_C$ response. A few general guidelines for simulator-to-data calibration are given below.

Theoretically, the collector current of a SiGe HBT is determined by the minority carrier profile in the base at equilibrium. Unfortunately, however, this minority carrier profile cannot be directly measured. The majority carrier (hole) concentration in the base at equilibrium, however, can be inferred from the intrinsic base sheet resistance R_{bi} data, which is readily available from simple measurements on "ring-dot" test structures. Here, R_{bi} is primarily determined by the total number of dopants in the base (refer to Chapter 4). While SIMS doping profiles represent the starting point for simulation, there are clearly uncertainties in the measured doping profiles (SIMS doping data is typically accurate to no more than ± 20%). The *actual* base doping profile can be fairly easily determined by comparing the simulated R_{bi} with the measured R_{bi}. Slight changes to the SIMS measured doping profile can be made to match measured I_C, at least to the range of accuracy of the SIMS data. This calibration technique is particularly useful when no "dip" exists in the base boron profile to indicate the precise EB junction location, as in a SiGe HBT with a phosphorous-doped emitter or a *pnp* SiGe HBT. Other factors which affect the minority carrier concentrations are parameters associated with the equilibrium *pn* product, including N_C, N_V, the degeneracy factors γ_n, γ_p, and the bandgap. For convenience, (12.15) is rewritten below:

$$p_0 n_0 = \gamma_p \gamma_n n_i^2$$
$$n_i^2 \equiv N_C N_V e^{-E_g/kT}. \qquad (12.96)$$

In MEDICI, the effective density-of-states N_C and N_V for SiGe are assumed by default to be the same as for Si, and must be modified in the material statement. There is general agreement that $N_C N_V$ is smaller for strained SiGe than for Si (refer to Chapter 2). The position dependence of the Ge fraction, however, cannot be taken into account in this manner, and an average value must be used.

The impact of the choice of carrier statistics on I_C is also clear from (12.96). For SiGe HBTs with heavy base doping, the degeneracy factor γ_p can be much smaller than unity. As a result, I_C simulated with Fermi-Dirac statistics is much lower than that simulated using Boltzmann statistics. Since the default BGN ΔG already includes the impact of degeneracy, a different set of BGN model parameters must be used when Fermi-Dirac statistics is activated.

The base current is primarily determined by the emitter structure in a SiGe HBT. Practically all viable SiGe HBT technologies have polysilicon emitter contacts. The polysilicon region can be either modeled as a Schottky contact or simply as an extension of the crystalline silicon emitter (the so-called "extended emitter" structure). The default model parameters for polysilicon are the same as those for silicon, but can be modified by the user. The work-function and surface recombination velocities can be adjusted as fitting parameters in order to calibrate the I_B if using a Schottky contact. For the extended emitter approach, the hole lifetime parameters can be adjusted to obtain agreement. Similar to the I_C case, carrier statistics and BGN will have a significant impact on the simulated I_B.

Figure 12.7 shows a calibration example for the SiGe HBT Gummel characteristics using the techniques described above. The model parameters were calibrated to 200 K data and then used to reproduce the 300 K data as is (i.e., no further tuning of parameters). Accurate simulation of the Gummel characteristics can be quite challenging, particularly for the high V_{BE} range when high injection occurs, and the impact of emitter and base resistance is not negligible. For RF applications, the $f_T - I_C$ curve and high-frequency two-port parameters are far more relevant. To the surprise of many device-simulation practitioners, the high-frequency simulation results are generally less sensitive to physical model selection, provided that the simulation-to-data comparison is made at fixed I_C instead of fixed V_{BE}.

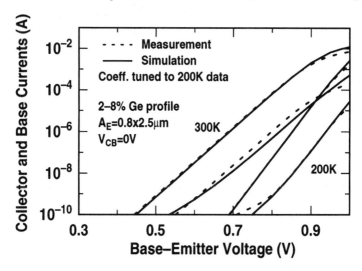

Figure 12.7 A calibration example for the Gummel characteristics of a SiGe HBT using the described calibration strategy. The same set of model parameters was used for both the 300 K and the 200 K simulations.

12.2.5 High-Frequency Simulation

High-frequency two-port parameters can be simulated using small-signal ac analysis. Here, f_T, f_{max}, as well as the various noise parameters can all be extracted from the simulated two-port parameters [9, 10]. Although there are many parameters that one can adjust, determining a single set of simulation parameters for a SiGe HBT that can reproduce the four complex network parameters at all biases of interest for frequencies up to f_T requires substantial effort. An in-depth understanding of the interaction between the physics underlying the simulation models and the device operation is important for achieving sensible results. For instance, at low currents, the total transit time is dominated by the time constants related to the EB space charge region capacitance rather than the diffusion capacitance.

Therefore the adjustment of extrinsic device structure as well as the intrinsic EB junction is needed to match the measured f_T at low J_C. Even though mobility model parameters (including parameters for both the low field mobility and the velocity saturation models) can be modified for f_T calibration at high J_C near the f_T peak, it should be used as a last resort. Instead, adjustments to the 2-D structure and lateral doping profile transitions should first be attempted. Because exact matching of the Gummel characteristics is difficult, high-frequency calibration of ac parameters such as f_T should always be made at fixed J_C, and not at fixed V_{BE}.

It is very time consuming to begin high-frequency SiGe HBT calibration by matching all of the Y-parameters across a wide frequency range and a large J_C range. An alternate strategy is to first calibrate the $f_T - J_C$ and $f_{max} - J_C$ curves. For state-of-the-art SiGe HBTs with narrow emitters, the shallow-trench isolation and extrinsic CB capacitances can often be comparable to the intrinsic CB capacitance, and are therefore nonnegligible. For accurate Y-parameter simulation, all of the 2-D lateral structure must be included. The extrinsic base and collector structures (geometric overlap as well as lateral doping profile transition) can then be modified to calibrate $f_{max} - J_C$. Typically, once $f_T - J_C$ and $f_{max} - J_C$ are calibrated, the simulated Y-parameters will match the measured Y-parameters reasonably well. For accurate separation of the intrinsic and extrinsic base resistances and CB capacitances, transistors with different emitter widths (if available on the test die) can be measured. By simulating and measuring devices with different emitter widths, the contribution of the extrinsic and intrinsic elements can be accurately separated.

In our experience, a useful technique for high-frequency SiGe HBT calibration is to extract the equivalent circuit parameters such as C_{BE}, C_{BC}, and r_b as a function of J_C. Analytical extraction methods which use only single frequency data are highly desirable because they are efficient, since we only need a rough picture to guide us on the appropriate changes to make in our device structure or model coefficients. The parameter extraction method proposed in [11], for instance, can be used. The extraction procedures are summarized below in a form suitable for implementation in a C or MATLAB program:

1. Convert S-parameters to Z-parameters using standard conversion equations

$$Z_{11} = Z_0 \frac{(1 + S_{11})(1 - S_{22}) + S_{12}S_{21}}{(1 - S_{11})(1 - S_{22}) - S_{12}S_{21}}, \quad (12.97)$$

$$Z_{12} = Z_0 \frac{2S_{12}}{(1 - S_{11})(1 - S_{22}) - S_{12}S_{21}}, \quad (12.98)$$

$$Z_{21} = Z_0 \frac{2S_{21}}{(1-S_{11})(1-S_{22}) - S_{12}S_{21}}, \quad (12.99)$$

$$Z_{22} = Z_0 \frac{(1-S_{11})(1+S_{22}) + S_{12}S_{21}}{(1+S_{11})(1+S_{22}) - S_{12}S_{21}}, \quad (12.100)$$

where $Z_0 = 50\ \Omega$.

2. Define the following quantities using I_C, β, and the Z-parameters:

$$g_m = \frac{qI_C}{kT}, \quad (12.101)$$

$$g_{be} = \frac{g_m}{\beta}, \quad (12.102)$$

$$\mu = \frac{Im(Z_{22} - Z_{12})}{Im(Z_{12} - Z_{21})}, \quad (12.103)$$

$$A = \sqrt{(g_m\mu - g_{be})(g_m + g_{be})}, \quad (12.104)$$

$$v = A^2 + (g_m + g_{be})^2, \quad (12.105)$$

$$B = \frac{g_m(g_m + g_{be})}{v \times Im(Z_{21} - Z_{12})}. \quad (12.106)$$

3. Calculate the equivalent circuit element values,

$$C_{BE} = \frac{A}{\omega}, \quad (12.107)$$

$$C_{BC} = \frac{B}{\omega}, \quad (12.108)$$

$$r_b = Re(Z_{11} - Z_{12}), \quad (12.109)$$

$$r_e = Re(Z_{12}) - \frac{g_m + g_{be}}{v}, \quad (12.110)$$

$$r_c = Re(Z_{22} - Z_{21}). \quad (12.111)$$

By comparing the simulated and measured C_{BE}, C_{BC}, r_b, r_e, and r_c, one can readily identify the dominant factors for any simulation-to-measurement discrepancy, and adjust the lateral doping extension accordingly. The diffusion capacitance component of C_{BE} is proportional to J_C at relatively low current densities (i.e., before the f_T roll-off), and can thus be distinguished from the depletion capacitance component. Figure 12.8 shows an example of $f_T - J_C$ calibration for a typical first generation SiGe HBT obtained using the techniques described above. This calibration was successfully achieved using 2-D MEDICI simulations without modifying the model parameters of the Philips unified mobility model and the

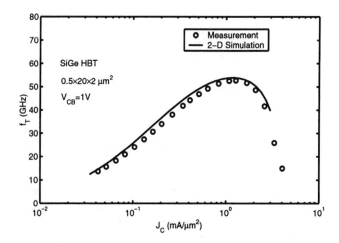

Figure 12.8 An example of $f_T - J_C$ calibration for a typical first generation SiGe HBT.

velocity saturation model. The intrinsic base and collector doping profiles from SIMS were also used as measured. Most of the required adjustments were instead made in the extrinsic device regions.

Figure 12.9 and Figure 12.10 give an example of high-frequency calibration of the real and imaginary parts of Y_{11}, respectively. Transistor noise parameters, including the minimum noise figure NF_{min}, the optimum source admittance for noise matching $Y_{s,opt}$, and the noise resistance R_n, can then be simulated from the Y-parameters by using the noise model equations described in Chapter 7. The only additional parameter needed is the base resistance r_b, which can be extracted from the simulated Y-parameters using the circle-impedance method.

12.2.6 Qualitative Versus Quantitative Simulations

Obviously, qualitative simulation is much easier and quicker than quantitative simulation. Doing a rough relative comparison between two device structures is much easier than simulating a single device structure to high accuracy. An advantage of qualitative simulation is that fewer grid points can be used and hence simulation time can be dramatically reduced. For instance, the comparison of current gain and cutoff frequency between Si BJT and SiGe HBT can be made using a coarse grid. An "exact" simulation, however, is obviously quite involved.

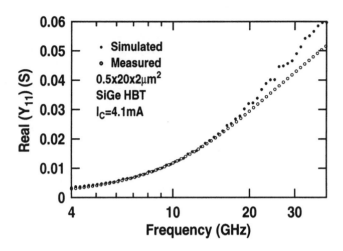

Figure 12.9 An example of high-frequency calibration of the real part of Y_{11}.

12.3 Probing Internal Device Operation

12.3.1 Current Transport Versus Operating Frequency

Insight into transistor internal operation can only be obtained by using device simulation. Figure 12.11 shows the magnitude of the *ac* electron current density $J_{n,ac}$ versus depth for a SiGe HBT operating over a frequency range from 1 MHz to 60 GHz. The simulation was performed in 1-D using a majority carrier contact for the base electrode. The SiGe HBT was biased at the peak f_T operating point, and a small-signal voltage with a magnitude of 2.6 mV was applied to the base terminal ($V_{CB} = 1$ V). At the low frequency of 1 MHz, $J_{n,ac}$ is nearly constant across the device from emitter-to-collector. With increasing frequency, however, the small-signal emitter current increases and the small-signal collector decreases, because of the charging of the emitter-base and collector-base capacitances. The change of the $J_{n,ac}$ profile becomes noticeable at 5 GHz, and then starts to increase significantly with further increase of operating frequency. Since the transistor is biased at the peak f_T operating point, the diffusion capacitance dominates the total transit time, and most of the decrease of $J_{n,ac}$ from the emitter towards the collector occurs in the neutral base.

12.3.2 Quasi-Static Approximation

Transistor equivalent circuit models are derived by assuming that the transistor terminal currents and internal charges are equal to their *dc* values, even when the

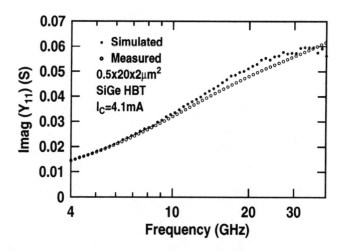

Figure 12.10 An example of high-frequency calibration of the imaginary part of Y_{11}.

terminal voltages vary continuously with time. This is often referred to as the "quasi-static approximation." At the device level, the quasi-static approximation is equivalent to assuming that the small-signal electron concentration n_{ac} is frequency independent. Using ac simulation can provide us with an ideal means for examining the validity of the quasi-static approximation in SiGe HBTs. Figure 12.12 shows the magnitude of small-signal n_{ac} versus depth for different operating frequencies. Here, n_{ac}, representing the amount of the electron density change induced by a base voltage change, is nearly independent of frequency up to 60 GHz. The transistor cutoff frequency f_T is 45 GHz, as extrapolated from the simulated h_{21} versus frequency. This indicates that the quasi-static approximation works well in this device at least below f_T, as far as the magnitude of the internal carrier densities is concerned. The phase delay due to non-quasi-static effects can be readily examined in a similar manner.

A strong peak can be observed on the n_{ac} profile shown in Figure 12.12. Physically, this corresponds to the n^+ side space-charge region boundary of the EB junction. The change in n_{ac} is largest at this space-charge layer boundary. In a simple model using the "depletion approximation," all of the changes are assumed to occur at the space-charge boundary. The simulation results show that the depletion approximation is quite poor in describing the charge storage in the forward-biased EB junction of a SiGe HBT, and significant electron concentration modulation occurs inside the space-charge region.

The n_{ac} profile obtained from ac simulation contains information on the spatial

540   Silicon-Germanium Heterojunction Bipolar Transistors

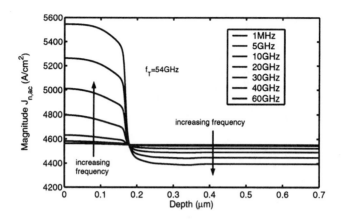

Figure 12.11 Magnitude of the small-signal $J_{n,ac}$ in a SiGe HBT versus depth with operating frequency as a parameter. The bias is chosen at the peak f_T point.

distribution of the total transit time, a concept that originated from the quasi-static approximation. Using n_{ac} and the small-signal collector current $J_{C,ac}$, the "effective transit time velocity" (v_τ) can be defined as

$$v_\tau = \frac{J_{C,ac}}{q n_{ac}}. \tag{12.112}$$

One can then define the "accumulated transit time" for a given position along the path of electron transport according to

$$\tau_{acc}(x) = \int_0^x \frac{1}{v_\tau} dx = \frac{1}{J_{C,ac}} \int_0^x q\, n_{ac}\, dx. \tag{12.113}$$

Figure 12.13 shows τ_{acc} versus depth at the peak f_T point calculated using 1 MHz ac simulation results. The f_T estimated from $\tau_{ec} = \tau_{acc}(x = x_{cc})$ (i.e., the transit time defined from quasi-static analysis) is 42.3 GHz, with x_{cc} being the location of the collector contact. The f_T extrapolated from h_{21}, however, is 45 GHz, as shown in Figure 12.14. In general, a good correlation between the f_T determined from h_{21} and the f_T determined from $1/2\pi\tau_{ec}$ is observed, with some slight differences near the peak f_T point. A comparison of the f_T extracted using the above two methods is shown in Figure 12.15, together with that extracted using

$$f_T = \frac{g_{cb}}{2\pi C_{bb}}, \tag{12.114}$$

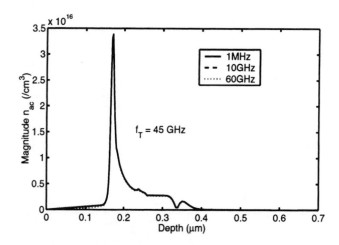

Figure 12.12 Magnitude of small-signal n_{ac} versus depth with operating frequency as a parameter. The bias is chosen at the peak f_T point.

where g_{cb} is the real part of Y_{21}, C_{bb} is defined by the imaginary part of Y_{11}, and $Y_{11} = g_{11} + j\omega C_{bb}$. Here, C_{bb} was evaluated at 1 MHz in the above example, and is nearly independent of the frequency used in the simulation as long as the frequency chosen is below f_T. The value from $|h_{21}|$ extrapolation should be used in simulator high-frequency calibration, as opposed to that from the accumulated transit time, which uses the quasi-static approximation. A practical reason for this is that experimental f_T data is all obtained from some form of h_{21} extrapolation.

Despite the fact that the resulting f_T value may be off compared to the value obtained from h_{21} extrapolation, the transit time analysis of n_{ac} and the $\tau_{acc}(x)$ profiles provide information of the local contribution to the total transit time, and can be very useful in identifying the transit time limiting factor in a given device design (i.e., for profile optimization). Since n_{ac} and $J_{C,ac}$ are nearly independent of frequency up to f_T, we can evaluate $\tau_{acc}(x)$ at any frequency below f_T. In this example, the results are nearly the same from 1 MHz to 60 GHz. This insensitivity to frequency proves useful in practice.

12.3.3 Regional Analysis of Transit Time

The total transit time defined by $\tau_{acc}(x = x_{cc})$ can be divided into five components to facilitate physical interpretation [12]. Two boundaries, the electrical EB and CB junction depths x_{eb}^* and x_{cb}^* are defined to be the in-most intersections of the n_{ac} and p_{ac} curves inside the junction space-charge regions, as illustrated in Figure 12.16.

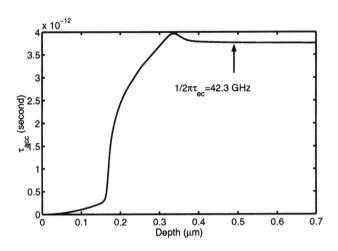

Figure 12.13 Accumulated transit time versus depth in a SiGe HBT.

The same SiGe HBT with a 2–8% graded base was used in this case. The peaks of n_{ac} and p_{ac} can be understood as the approximate space-charge region boundaries on the emitter and base sides, respectively, even though the results clearly show that no abrupt space-charge region boundary can be identified (i.e, the depletion approximation is invalid). The neutral base can be roughly defined to be the region where $n_{ac} \approx p_{ac}$. The neutral base width is clearly smaller than the electrical base width defined by $x_{cb}^* - x_{eb}^*$. The electrical EB and CB junction locations (x_{eb}^* and x_{cb}^*) are in general different from the metallurgical junctions (x_{eb} and x_{cb}), as expected.

The total τ_{ec} can be divided into five components with the help of p_{ac} [12]:

1. The emitter transit time due to minority carrier storage in the emitter

$$\tau_e^* = \frac{q}{J_{C,ac}} \int_0^{x_{eb}^*} p_{ac}\, dx. \tag{12.115}$$

2. The EB depletion charging time due to the storage of uncompensated mobile carriers

$$\tau_{eb}^* = \frac{q}{J_{C,ac}} \int_0^{x_{eb}^*} (n_{ac} - p_{ac})\, dx. \tag{12.116}$$

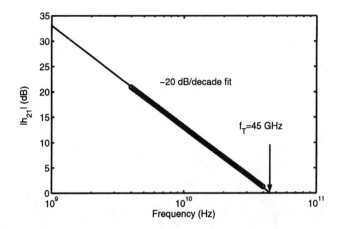

Figure 12.14 Determining f_T by extrapolating the simulated $|h_{21}|$ versus frequency data according to a -20 dB/decade slope.

3. The base transit time due to electron charge storage in the electrical base (which includes the traditional "quasi-neutral base")

$$\tau_b^* = \frac{q}{J_{C,ac}} \int_{x_{eb}^*}^{x_{cb}^*} n_{ac}\, dx. \qquad (12.117)$$

4. The CB depletion charging time

$$\tau_{cb}^* = \frac{q}{J_{C,ac}} \int_{x_{eb}^*}^{x_{cc}} (n_{ac} - p_{ac})\, dx. \qquad (12.118)$$

5. The collector transit time

$$\tau_c^* = \frac{q}{J_{C,ac}} \int_{x_{eb}^*}^{x_{cc}} p_{ac}\, dx. \qquad (12.119)$$

Note that τ_c^*, as defined above, is different from the traditional τ_c, and τ_c^* is important only when holes are injected into the collector after the onset of high injection.

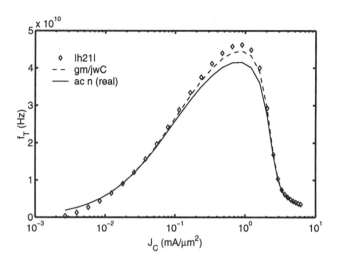

Figure 12.15 Comparison of three methods of determining f_T using numerical simulation.

The sum of all of the transit time components is equal to τ_{ec}

$$\tau_{ec} = \frac{q}{J_{C,ac}} \int_0^{x_{cc}} n_{ac}\, dx$$
$$= \tau_e^* + \tau_{eb}^* + \tau_b^* + \tau_{cb}^* + \tau_c^*. \qquad (12.120)$$

The transit time due to electron charge storage in the EB space-charge region is not treated separately, but is instead included in the modified transit times of the emitter and base (the * transit times above). Under high injection, however, the whole transistor from emitter to collector is flooded with a high concentration of electrons and holes, and hence no clear boundaries can be identified. Strictly speaking, the concepts of base, emitter, and collector consequently lose their conventional meanings, and thus the concept of regional transit times is no longer meaningful. In SiGe HBTs, the SiGe-to-Si transition at the CB junction causes additional electron charge storage at high injection, as discussed previously in Chapter 6. In this case, the x_{eb}^* and x_{cb}^* definitions discussed above cannot be applied.

12.3.4 Case Study: High-Injection Barrier Effect

We now examine the evolution of the small-signal $qn_{ac}/J_{C,ac}$ and $qp_{ac}/J_{C,ac}$ profiles with increasing current density J_C from well below the peak f_T current den-

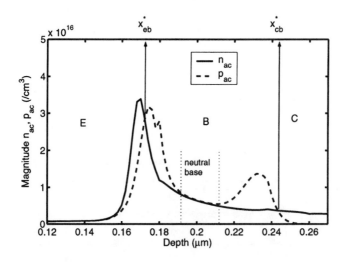

Figure 12.16 Definition of the electrical EB and CB junctions from the simulated n_{ac} and p_{ac} profiles. The bias is chosen at the peak f_T point.

sity to slightly above the peak f_T current density. The simulated $qn_{ac}/J_{C,ac}$ and $qp_{ac}/J_{C,ac}$ profiles at three current densities representing low to high injection levels are shown in Figure 12.17(a)–(c). The small-signal magnitude of the V_{BE} increase is 2.6 mV. At a typical low-injection J_C of 0.127 mA/μm², well below the peak f_T point, n_{ac} is positive across most of the device. Most of the charge modulation occurs in the EB space-charge region. The transit time related to this component of the charge storage decreases with increasing J_C because of increasing $J_{C,ac}$, which can be seen by comparing the magnitude of the first peak on the curves for the electrons in Figure 12.17(a) and (b). Note that different scales are used on the y-axis for different injection levels to help visualize the details of the profiles.

At $J_C = 1$ mA/μm², which is near peak f_T, the base and collector transit time contributions become dominant compared to the EB space-charge region contribution, as shown in Figure 12.17(b), and is mainly due to a decrease of the EB space-charge region transit time. One consequence of high-level injection is that the CB space-charge region pushes towards the collector n^+ buried layer much more obviously than at lower J_C, despite a decrease of V_{CB}. This is manifested as a large *negative* n_{ac} and hence negative $n_{ac}/J_{C,ac}$ around 0.37 μm. Physically, this corresponds to the *extension* of the CB space-charge region towards the n^+ buried layer, which causes a *decrease* of electron concentration at the front of the space-charge region. In the simulation, the base voltage is increased while the col-

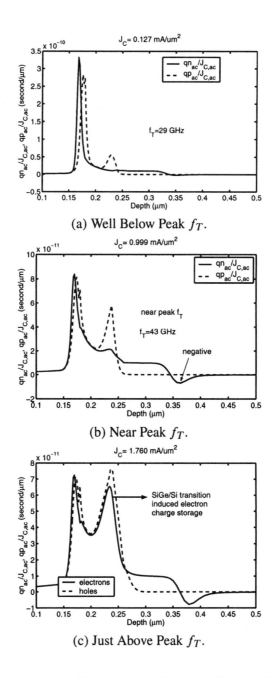

Figure 12.17 Simulated $qn_{ac}/J_{C,ac}$ and $qp_{ac}/J_{C,ac}$ profiles at (a) low injection, (b) medium injection, and (c) high injection.

lector and emitter voltages are fixed. Because of the existence of negative n_{ac}, the real part of the simulated n_{ac}, as opposed to the absolute value of the simulated n_{ac}, should be used for calculation of the total transit time. A significant error can be introduced under high-injection when the integral over the negative n_{ac} portion becomes significant to the total integral. We note that this negative-going n_{ac} component under high-injection is generally not treated properly in the literature.

Figure 12.17(c) shows the $qn_{ac}/J_{C,ac}$ and $qp_{ac}/J_{C,ac}$ profiles at a slightly higher J_C of 1.76 mA/μm^2, just past the peak f_T. The SiGe/Si interface, which originally was buried in the CB space-charge region, is now exposed to the large density of electrons and holes. The valence band potential barrier to holes induces a conduction band potential barrier to electrons as well. The most important consequence is increased dynamic charge storage, as seen from the high $qn_{ac}/J_{C,ac}$ and $qp_{ac}/J_{C,ac}$ peaks near the SiGe/Si transition in Figure 12.17(c). This additional charge storage results in a significant increase of the total transit time and hence a strong decrease of f_T to 29 GHz, even though the current density is just above the value needed to reach the peak f_T (1.0 mA/μm^2). At an even higher J_C of 3.56 mA/μm^2, both $qn_{ac}/J_{C,ac}$ and $qp_{ac}/J_{C,ac}$ are very large, and nearly equal to each other. No clear space-charge regions can be identified from the $qn_{ac}/J_{C,ac}$ and $qp_{ac}/J_{C,ac}$ profiles. The conventional concepts of emitter, base, and collector no longer apply in this situation. The majority of the overall transit time, however, is contained inside the SiGe "base," as intuitively expected.

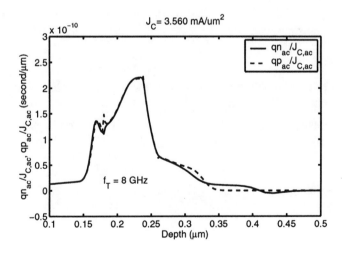

Figure 12.18 Simulated n_{ac} and p_{ac} profiles at J_C=3.56 mA/μm^2.

References

[1] MEDICI User's Manual, Avant!, Fremont, CA, 1999.

[2] Z. Yu, SEDAN program, Stanford University.

[3] Z. Yu et al., "PISCES 2ET 2D Device Simulator," *Stanford University Tech. Report*, 1994.

[4] C.M. van Vliet, "Bandgap narrowing and emitter efficiency in heavily doped emitter structures revisited," *IEEE Trans. Elect. Dev.*, vol. 40, pp. 1141-1147, 1993.

[5] G. Niu, "Modeling and simulation of SiGe microelectronic devices," M.S. thesis, Fudan University, 1994.

[6] J.W. Slotboom and H.C. de Graaff, "Measurements of Bandgap Narrowing in Si Bipolar Transistors," *Solid-State Elect.*, vol. 19, pp. 857-862, 1976.

[7] D.B.M. Klaassen, "A unified mobility model for device simulation I. Model equations and concentration dependence," *Solid-State Elect.*, vol. 35, pp. 953-959, 1992.

[8] D.B.M. Klaassen, J.W. Slotboom, and H.C. de Graaff, "Unified apparent bandgap narrowing in n- and p-type silicon," *Solid-State Elect.*, vol. 35, pp. 125ij129, 1992.

[9] G. Niu et al., "Noise modeling and SiGe profile design tradeoffs for RF applications," *IEEE Trans. Elect. Dev.*, vol. 47, pp. 2037-2044, 2000.

[10] G. Niu et al., "Noise parameter optimization of UHV/CVD SiGe HBT's for RF and microwave applications," *IEEE Trans. Elect. Dev.*, vol. 46, pp. 1347-1354, 1999.

[11] P. Baureis and D. Seitzer, "Parameter extraction for HBT's temperature dependent large signal equivalent circuit model," *Tech. Dig. IEEE GaAs IC Symp.*, pp. 263-266, 1993.

[12] J.J.H. van den Biesen, "A simple regional analysis of transit times in bipolar transistors," *Solid-State Elect.*, vol. 29, pp. 529-534, 1986.

Chapter 13

Future Directions

And so we come to the end of the SiGe story. Like all good advanced technologies, however, there is no true ending for SiGe technology, and many as yet unanticipated beginnings likely await around the next bend (we technologists are clever folks, after all!). In this final chapter we offer some final thoughts on likely future trends and performance limits in SiGe technology.

13.1 Technology Trends

It is a fact of life that SiGe technology is rapidly becoming reasonably routine to develop for any company that has substantial IC development expertise and a will to do so. Multiple SiGe epi growth tools are commercially available that have a proven record for yielding manufacturable SiGe films. Given this scenario, it is not surprising that more than a dozen companies worldwide currently possess what can be termed *first generation* SiGe technologies (refer to Chapter 1), with peak f_T in the 50-GHz range, and peak f_{max} in the 70–80-GHz range. Many others are at present in a "ramp-up" phase. With this picture of the global SiGe arena in mind, we offer a number of probable trends (in no particular order) that are currently emerging and are likely to help shape the future of SiGe technology.

- Given the overarching cost constraints that will inevitably be faced by all IC technologies for the foreseeable future, a common theme being adopted by many companies is to implement a "modular" SiGe BiCMOS technology, in which the SiGe HBT can be viewed as an "adder" to a high-speed CMOS core technology without perturbing the characteristics of the underlying core CMOS. Thus, the SiGe HBT (as well as RF passives, transmission lines, etc.) can be swapped in and out by designers as required for each particular application, without changing the basic CMOS technology, design and checking kits, existing logic books, etc. This modular approach to SiGe, while perhaps more challenging to initially implement, should prove to provide substantial cost savings in the long run. It also has the nice advantage of not trying to directly supplant CMOS (an unlikely prospect), but rather leverage the best capabilities of both technologies.

- SiGe HBT technology is rapidly becoming so commonplace today that it is becoming harder to justify using a Si BJT in its place. One can obtain substantially higher performance with SiGe at similar process complexity. Thus, the days of the bedrock double-poly, self-aligned, ion-implanted base Si BJT are likely numbered.

- Multiple breakdown voltage versions of the core SiGe HBT will be available on the same die for greater circuit design flexibility. For example, a standard 3.3 V BV_{CEO} "low-breakdown" device with 50-GHz peak f_T might be offered, together with a 5.5-V BV_{CEO} "moderate breakdown" SiGe HBT having a 30-GHz peak f_T, and even a 9.0-V BV_{CEO} "high-breakdown" SiGe HBT having a 15-GHz peak f_T. This is easily accomplished in SiGe HBT technology with a collector implant block-out mask. This breakdown voltage "tuning" is an attractive feature for many mixed-signal applications, and allows circuit designs to span the LNA and mixer to PA circuit range.

- Unless some previously unrecognized drawback surfaces, carbon doping of SiGe HBTs will become the mainstream (as discussed in Chapter 1, it is already prevalent today in the most advanced SiGe technologies). With no apparent downside, carbon doping makes management of the overall thermal budget and profile control that much easier, and since it can be easily plumbed into epi reactors, there is no serious reason not to utilize its advantages.

- Using substantially higher C content (2–3%) to produce SiGeC alloys that are lattice-matched directly to Si would offer a number of interesting device design possibilities. While SiGeC alloys with up to 3% C have been successfully produced with reasonable cross-sectional morphology using MBE, it seems unlikely that lattice-matching within the SiGe/Si system will have a significant commercial impact in the foreseeable future.

- Evolution from first generation SiGe HBT performance levels (50-GHz peak f_T) to second generation performance levels (100–120-GHz peak f_T) can generally be accomplished within the confines of traditional structural integration schemes (refer to Chapter 2) by careful lateral scaling, Ge and doping profile optimization, and appropriate reduction of overall thermal cycles (together with C-doping, etc.). Evolution from second generation to third generation performance levels (> 200-GHz peak f_T), however, will likely also necessitate device structural changes that eliminate any extrinsic base implantation steps into the deposited SiGe-bearing epi layer. Such implantations are known to introduce interstitials into the active device region, yielding enhanced boron diffusion (even with C-doping), making it very difficult to decouple the achievable f_T from the extrinsic base design. The so-called "raised-extrinsic base" structure appears to offer several advantages in this context, and has been used to demonstrate impressive performance levels [1, 2].

- SiGe technologies will increasingly move to full copper metalization to support the requisite high device current densities as well as improve the Qs of the passives. As usual, it will be most cost-effective to piggyback on the Cu metalization schemes developed for high-end CMOS.

- As frequency bands migrate higher, there will be an increasing push to develop better passives and transmission lines to enable MMIC and mm-wave components in SiGe. There are at least four approaches that might be used to improve the high-frequency losses in SiGe technologies: 1) move to high-resistivity substrates; 2) use thick(er) top-side dielectrics, combined with lower resistivity metals (Cu); 3) use postfabrication spun-on polymers (e.g., BCB) followed by Cu or Au (the MMIC standard) for passives and transmission lines; or 4) move to SiGe on SOI. It remains to be seen which would offer the greatest cost-performance advantage compared to competing technologies such as III-V (if any).

- Noise coupling is and will continue to be a serious design issue confronting mixed-signal applications of SiGe technology. More measurement and analysis are needed to quantify the best noise mitigation approaches. At present, it appears that simply using conservative layout approaches and intelligent placement of critical noise sensitive functions (e.g., don't put your LNAs next to a large CMOS digital switching block) would go a long way in noise management.

13.2 Performance Limits

The maximum achievable frequency response in manufacturable SiGe HBT technology has clearly proven to be much higher than even the blind optimists might

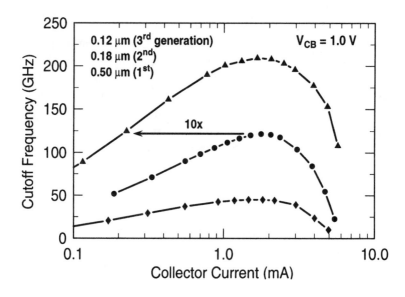

Figure 13.1 Measured cutoff frequency data as a function of bias current for three different SiGe HBT technology generations (after [2]).

have guessed.[1] Fully integrated SiGe HBT BiCMOS technologies with peak f_T and peak f_{max} above 200 GHz exist in 2002 (at several different labs). Figure 13.1 compares the measured cutoff frequency characteristics of three distinct SiGe HBT BiCMOS technology generations [2]. Several important points concerning projected performance limits in SiGe can be gleaned from these results:

- Given that the three successive SiGe technology generations shown in Figure 13.1 were reported in 1994, 2000, and 2002, respectively, it is apparent that once first generation SiGe technology stabilized in the manufacturing environment, progress in raising the SiGe HBT performance has been exceptionally rapid. This is clearly a good sign for the future.

- Third generation SiGe HBTs have comparable raw performance to the best commercially available III-V HBT technologies (both GaAs-based and InP-based, which are in the 150–200-GHz peak f_T range currently), while preserving their compatibility with standard Si CMOS technologies, and the enormous economy of scale of conventional Si fabrication. This does not

---

[1]Including myself. In my 1998 review article [3], I suggested that "it seems reasonable to expect this number (peak f_T) to climb into the 100–120 GHz range for manufacturable SiGe technologies using stable Ge profiles, provided care is taken in profile optimization."

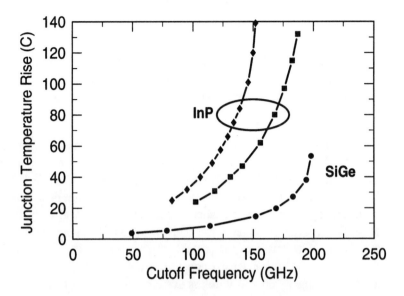

Figure 13.2 Comparison of the calculated junction temperature rise as a function of measured cutoff frequency for a third generation 0.12 × 2.5 μm^2 SiGe HBT with standard-layout (1.0 × 3.0 μm^2, commercially available) and laterally scaled (using 0.1 μm lithography) InP HBTs [4] (after [2]).

bode well for the (long-promised) broad application of III-V HBT technologies, except perhaps in high-power applications and the lightwave environment.

- The only long-term viable path for III-V HBTs to follow would appear to be towards developing (Si-like) self-alignment schemes and hence aggressive scaling of the transistor dimensions (for much higher performance). This will likely prove to be a nontrivial feat even in "fabrication-friendlier" III-V materials such as InP. Clearly, any path towards integration of III-V HBTs with Si CMOS is daunting task.

- The peak f_T in each successive SiGe technology generation occurs at roughly the same bias current, meaning that the collector current density is rising rapidly to achieve the levels of demonstrated performance (to the range of 10 mA/μm^2 for third generation technology). This J_C increase over time is not unexpected from bipolar scaling theory, but there is obviously some practical bound for maintenance of sufficient device reliability.

- These high current densities can produce significant self-heating in these transistors, but even in this case, the generally excellent thermal properties of Si substrates give SiGe an advantage over competing III-V HBT technologies. As can be seen in Figure 13.2, the calculated junction temperature rise for a SiGe HBT operated at 150 GHz is about 3× lower than for a scaled InP HBT [2].

- One of the most important advantages offered by the best-of-breed SiGe HBT technologies is that they offer substantial leverage in power savings. As shown in Figure 13.1, one can operate third generation SiGe HBTs at 120 GHz, sufficient for most 40-GB/sec applications, while decreasing the bias current by an *order-of-magnitude* compared to (the already impressive) second generation technology! This power savings potential for SiGe clearly holds great leverage for portable (battery-limited) system applications, but given the power-density constraints being placed across the board on both emerging wired and wireless systems, this ability to deliver high-performance at very low power may ultimately prove to be the key long-term enabling feature of SiGe technology.

So what is the practical performance limit of a commercially viable SiGe HBT technology? Given today's vantage point, greater than 300-GHz peak f_T appears to be an attainable goal. [2] As briefly alluded to in Chapter 1, meaningful discussion of attainable f_{max} is difficult, given the considerable complexities associated with both measuring and interpreting the various gain definitions at these frequency levels. For instance, in the third generation data shown in Figure 13.1, the f_{max} obtained from extrapolating Mason's unilateral gain (U) is 285 GHz, compared with a value of 194 GHz obtained from the maximum available gain (MAG). More meaningful in this context is the actual transistor gain at the relevant circuit operating frequency (17.0 dB and 15.9 dB for U and MAG at 40 GHz, in this case).

Clearly, breakdown voltages must decrease as the transistor performance improves (as it does in CMOS, for different physical reasons). For the case depicted in Figure 13.1, the 50-GHz, 120-GHz, and 210-GHz SiGe HBTs have an associated BV_{CEO} of 3.3 V, 2.0 V, and 1.7 V, respectively. Achievable $f_T \times BV_{CEO}$ products in the 350-400-GHz-V range should be realistic goals for the future. The sub-1.5 V breakdown voltages required to reach 300 GHz should not prove to be a serious limitation for many designs, given that BV_{CEO} does not present a hard boundary above which one cannot bias the transistor. Rather, one simply has to live with base current reversal and potential bias instabilities in this (above-BV_{CEO}) bias domain

---

[2]Note added in press: the demonstration of a SiGe HBT with 350-GHz peak f_T (BV_{CEO} = 1.4 V), to be presented at the IEDM in December of 2002, clearly supports this claim [5]!

[6]. In addition, as discussed above, on-wafer breakdown voltage tuning will provide an additional level of flexibility for circuit designs needing larger operating headroom.

For the near-term at least, only the best three or four industrial teams will likely reach 300-GHz performance levels, however, given the resources required, and the limited commercial market sector that currently exists requiring those extreme levels of transistor performance (although future IC market needs are obviously a dynamic and fickle landscape). The rest of the (still growing) SiGe industry will likely climb into the second generation performance levels (100–120-GHz peak f_T) over the next few years and be content to grow market share and wafer-starts.

What does one do with this level of extreme device performance? In the absence of cost constraints, just about anything one can imagine! As one possibility, ISM band (60 GHz) WLANs using 200-GHz SiGe HBT BiCMOS technology might present very interesting possibilities for medium-range, very high data-rate, wireless networks for the future. Clearly, as an enabler of pie-in-the-sky dreaming, future SiGe technology will offer some interesting food for thought for clever visionaries. System-level cost-performance issues will just as surely bring us dreamers back to earth. But don't let that stop the dreaming!

We Shall Not Cease From Exploration
And The End Of All Our Exploring
Will Be To Arrive Where We Started
And Know The Place For The First Time.

T.S. Eliot, "Four Quartets"

References

[1] B. Jagannathan et al., "Self-aligned SiGe NPN transistors with 285 GHz f_{max} and 207 GHz f_T in a manufacturable technology," *IEEE Elect. Dev. Lett.*, vol. 23, pp. 258-260, 2002.

[2] G. Freeman et al., "Transistor design and application considerations for >200 GHz SiGe HBTs," *IEEE Trans. Elect. Dev.*, in press.

[3] J.D. Cressler, "SiGe HBT technology: a new contender for Si-based RF and microwave circuit applications," *IEEE Trans. Micro. Theory Tech.*, vol. 46, pp. 572-589, 1998.

[4] S. Thomas III et al., "Effects of device design on InP-based HBT thermal resistance," *IEEE Trans. Dev. Mat. Rel.*, vol. 1, pp. 185-189, 2001.

[5] J.-S. Rieh et al., "SiGe HBTs with cut-off frequency of 350 GHz," to appear in the *Tech. Dig. IEEE Int. Elect. Dev. Meeting*, December 2002.

[6] M. Rickelt, H.M. Rein, and E. Rose, "Influence of impact-ionization-induced instabilities on the maximum usable output voltage of silicon bipolar transistors," *IEEE Trans. Elect. Dev.*, vol. 48, pp. 774-783, 2001.

Appendix A

Properties of Silicon and Germanium

The energy band structures of Si and Ge are depicted below (Figure A.1), together with: 1) their carrier effective mass parameters (Table A.1); and 2) their bulk structural, mechanical, optical, and electrical properties (Table A.2) [1], [2].

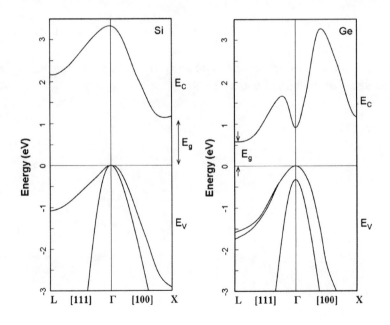

Figure A.1 Energy band structure, showing the principal conduction and valence bands of Si and Ge as a function of k-space direction (after [1]).

Table A.1 Carrier Effective Mass Parameters for Si and Ge

Parameter	Units	Silicon	Germanium
Effective Electron Mass (m_n^*)	($\times m_o$)		
Longitudinal (4.2 K)		0.9163	1.58
Transverse (4.2 K)		0.1905	0.082
Density-of-states (4.2 K)		1.062	–
Density-of-states (300 K)		1.090	–
Effective Hole Mass (m_p^*)	($\times m_o$)		
Heavy hole (4.2 K)		0.537	0.28
Light hole (4.2 K)		0.153	0.044
Density-of-states (4.2 K)		0.59	–
Density-of-states (300 K)		1.15	–

References

[1] M. Shur, *Physics of Semiconductor Devices*, Prentice-Hall, Englewood Cliffs, New Jersey, 1990.

[2] R. Hull, editor, *Properties of Crystalline Silicon*, EMIS Datareviews Series, Number 20, INPSEC, London, 1999.

Table A.2 Properties of Bulk Si and Ge

Parameter	Units	Silicon	Germanium
Atomic number	–	14	32
Atomic density	(atoms/cm^3)	5.02×10^{22}	4.42×10^{22}
Atomic weight	(g/mole)	28.09	72.6
Density	(g/cm^3)	2.329	5.323
Electronic orbital configuration	–	$(Ne)3s^23p^2$	$(Ar)3d^{10}4s^24p^2$
Crystal structure	–	diamond	diamond
Lattice constant (298 K)	(Å)	5.43107	5.65791
Dielectric constant	–	11.7	16.2
Breakdown strength	(V/cm)	3×10^5	1×10^5
Electron affinity	(V)	4.05	4.00
Specific heat	(J/g-°C)	0.7	0.31
Melting point	(°C)	1412	1240
Intrinsic Debye length (300 K)	(μm)	24	0.68
Index of refraction	–	3.42	3.98
Transparency region	(μm)	1.1-6.5	1.8-15
Thermal conductivity (300 K)	(W/cm-°C)	1.31	0.60
Thermal expansion coeff. (300 K)	(°C^{-1})	2.6×10^{-6}	5.9×10^{-6}
Young's modulus	(dyne/cm^2)	1.9×10^{12}	–
Energy bandgap (low doping)	(eV)	1.12 (300 K)	0.664 (291 K)
		1.17 (77 K)	0.741 (4.2 K)
Equiv. conduction band minima	–	6	8
Effective electron mass (300 K)	($\times m_o$)	1.18	–
Effective hole mass (300 K)	($\times m_o$)	0.81	–
Intrinsic carrier density (300 K)	(cm^{-3})	1.02×10^{10}	2.33×10^{13}
Eff. conduction band DoS (300 K)	(cm^{-3})	2.8×10^{19}	1.04×10^{19}
Eff. valence band DoS (300 K)	(cm^{-3})	1.04×10^{19}	6.00×10^{18}
Electron mobility (300 K)	(cm^2/V-sec)	1450	3900
Hole mobility (300 K)	(cm^2/V-sec)	500	1900
Electron diffusivity (300 K)	(cm^2/sec)	37.5	100
Hole diffusivity (300 K)	(cm^2/sec)	13	49
Optical phonon energy	(meV)	63	37
Phonon mean free path length	Å	76	105
Intrinsic resistivity (300 K)	(Ωcm)	3.16×10^5	47.62

About the Authors

John D. Cressler received a B.S. in physics from the Georgia Institute of Technology, Atlanta, Georgia, in 1984, and an M.S. and Ph.D. in applied physics from Columbia University, New York, in 1987 and 1990. From 1984 to 1992 he was on the research staff at the IBM Thomas J. Watson Research Center in Yorktown Heights, New York, working on high-speed Si and SiGe bipolar devices and technology. In 1992 he left IBM Research to join the faculty at Auburn University, Auburn, Alabama, where he served until 2002. When he left Auburn University, he was Philpott–Westpoint Stevens Distinguished Professor of electrical and computer engineering and Director of the Alabama Microelectronics Science and Technology Center.

In 2002, Dr. Cressler joined the faculty at Georgia Tech, where he is currently Professor of electrical and computer engineering. His research interests include SiGe devices and technology; Si-based RF/microwave/millimeter-wave devices and circuits; radiation effects; device-circuit interactions; noise and linearity; cryogenic electronics; SiC devices; reliability physics; 2-D/3-D device-level simulation; and compact circuit modeling. He has published more than 240 technical papers related to his research.

Dr. Cressler was associate editor for the *IEEE Journal of Solid-State Circuits* (1998–2001) and served on the technical program committees of the *IEEE International Solid-State Circuits Conference* (1992–1998, 1999–2001), the *IEEE Bipolar/BiCMOS Circuits and Technology Meeting* (1995–1999), the *IEEE International Electron Devices Meeting* (1996–1997), and the *IEEE Nuclear and Space Radiation Effects Conference* (1999–2000, 2002). He was the technical program chairman of the 1998 ISSCC. He currently serves on the executive steering committee for the *IEEE Topical Meeting on Silicon Monolithic Integrated Circuits in RF Systems*, as international program advisor for the *IEEE European Workshop on Low-Temperature Electronics*, and on the technical program committee for the *IEEE International SiGe Technology and Device Meeting*. He was appointed an IEEE Electron Device Society Distinguished Lecturer in 1994 and was awarded the 1994 Office of Naval Research Young Investigator Award for his SiGe research program, the 1996 C. Holmes MacDonald National Outstanding Teacher Award by Eta Kappa Nu, the 1996 Auburn University Alumni Engineering Council Research Award, the 1998 Auburn University Birdsong Merit Teaching Award, the 1999 Auburn University Alumni Undergraduate Teaching Excellence Award, and an IEEE Third Millennium Medal in 2000. He is an IEEE Fellow.

Dr. Cressler can be reached at:
School of Electrical and Computer Engineering
777 Atlantic Drive, N.W.
Georgia Institute of Technology
Atlanta, GA 30332-0250 U.S.A.
or
cressler@ece.gatech.edu

About the Authors

Guofu Niu was born in Henan, China, in December 1971. He received a B.S., M.S., and Ph.D. in electrical engineering from Fudan University, Shanghai, China, in 1992, 1994, and 1997, respectively. His graduate work at Fudan included developing numerical simulators and compact models for SiGe HBTs and FETs, Monte Carlo simulation of transport properties in a 2-D electron gas, SOI devices, and statistical circuit simulation. From 1995 to 1997, he worked on parallel computing, switched-current oscillator design, and RTD-based quantum effect circuits as a research assistant at City University of Hong Kong. From 1997 to 2000, he conducted postdoctoral research on SiGe HBT, SiGe MOSFETs, radiation effects, low-temperature electronics, and SiC devices at Auburn University, Auburn, Alabama. In 2000, he became a faculty member at Auburn and is currently Associate Professor of electrical and computer engineering. His current research activities include SiGe devices and circuits, RF design, noise, linearity, single-event effects, SiC power devices, and TCAD.

Dr. Niu has published more than 80 papers related to his research and has served as a reviewer for many technical journals, including *IEEE Transactions on Electron Devices*, *IEEE Electron Device Letters*, *IEEE Transactions on Microwave Theory and Techniques*, *IEEE Microwave and Guided Wave Letters*, and *IEEE Transactions on Nuclear Science*. He served on the program committee of the *IEEE Asia-South-Pacific Design Automation Conference* in 1997 and currently serves on the program committee of the *IEEE Bipolar/BiCMOS Circuits and Technology Meeting*. He received the T. D. Lee Physics Award in 1992 and 1994 and a best student paper award at the *International Application Specific Integrated Circuits*

Conference in 1994. Dr. Niu is listed in *Who's Who in America* and is a senior member of the IEEE.

Dr. Niu can be reached at:
Electrical and Computer Engineering Department
200 Broun Hall
Auburn University
Auburn, AL 36849-5201 U.S.A.
or
guofu@eng.auburn.edu

Index

H-parameters, 155
S-parameters, 155
Y-parameters, 155
Z-parameters, 154

Amplifier stability, 158
    Rollett stability factor, 159
    stability gain trade-off, 161
Apparent bandgap narrowing, 512, 521
Arbitrary band alignments, 431
    high injection, 434
    optimization issues, 438
    theory, 432
Associated gain, 279
    general expression, 272
Avalanche multiplication, 121
    forced current, 124
    forced voltage, 122
    impact of current density, 128
    impact of Early effect, 126
    impact of self-heating, 127
    measurement techniques, 122
    nonlinearity implication, 358
    Si versus SiGe, 128
    terminal currents, 121
    transport, 121

Band diagram, 95
Band offset data, 60
Band offsets, 59
Band structure, 56
Bandgap engineering, 11

Bandgap narrowing, 512, 521
Base resistance, 149
    associated gain, 279
    ECL gate delay, 179
    impedance circle extraction, 150
    input noise voltage, 295
    noise implication, 268
    noise resistance, 277
Base transit time, 143
    derivation, 165
    Ge grading, 169
BGN, 521
BiCMOS Integration
    base-after-gate scheme, 81
    base-during-gate scheme, 80
    raised-extrinsic base, 82
BiCMOS integration, 79
Breakdown voltage, 131
    BV_{CBO}, 132
    BV_{CEO}, 132
    circuit implications, 135
    impact of current gain, 133
Bulk Si and Ge
    Band structure, 558
    Effective mass parameters, 558
    Properties, 557

Carbon-doping, 82
    experimental results, 84
    mechanism, 83
    minimum C content, 83
Carrier mobility, 62

CB capacitance
    associated gain, 279
    gain implications, 163
    linearity implications, 337
    nonlinearity, 358
Charge modulation effect, 141–143
Circuit models, 118
    Ebers-Moll model, 118
Collector current density
    approximations, 103
    compared to Si BJT, 102
    derivation, 99
    Ge profile shape, 105
    nonconstant base doping, 104
    optimization issues, 108
Continuity equations, 510
Cross section, 77
Current gain, 98
    input noise current, 296
    optimization issues, 108
Cutoff frequency, 141, 143
    NF_{min} implication, 279
    circuit relevance, 164
    current dependence, 147
    extraction, 145
    optimization issues, 173

Defect density, 91
Density-of-states, 58
Device simulation, 509
    J_B calibration, 533
    J_C calibration, 532
    coefficient tuning, 532
    mesh generation, 527
    mesh validity check, 529
    model selection, 532
    probe internal operation, 538
    regrid, 529
    RF calibration, 534
Distinguishing nonlinearities, 351

Dream of Café Erehwon, 561
Drift-diffusion equations, 513

Early voltage, 109
EB capacitance, 143
    depletion capacitance, 142
    diffusion capacitance, 142
ECL power-delay, 179
Electron mobility, 65
Emitter transit time, 143
    derivation, 170
Energy balance, 515
Energy band diagram, 95, 140
Equivalent circuit models, 118
    Ebers-Moll model, 118
    transport version, 119

Film relaxation, 36
Frequency scalable design, 288
Future directions, 549
    performance limits, 551
    technology trends, 549

Gain compression, 324
Gain expansion, 324
Ge grading effect, 186
    compact modeling, 198
    data, 198
    impact on $V_{BE}(I_C, T)$, 197
    impact on circuits, 189
    physical origin, 186
    the bottom line, 201
    theory, 194
Ge-induced CB field effects, 441
    data, 446
    impact ionization, 443
    impact on base current, 445
    simulations, 441

Harmonics, 323
Heterojunction barrier effects, 232

compact modeling, 244
data, 238
high-injection effects, 234
physical origin, 232
profile optimization issues, 241
simulations, 239
the bottom line, 254
High injection
profile design trade-off, 292
transit time analysis, 544
Hole mobility, 64

IC technology battleground, 22
Impedance matching, 282
Intermodulation, 324
Intuitive picture, 95, 139

Johnson limit, 134

Linear noisy two-port
chain representation, 266
Norton's representation, 266
theory, 264
Thevenin's representation, 265
Linearity, 321
current dependence, 352
load dependence, 354
voltage dependence, 353
LNA
IP3 equation, 368
gain equation, 367
linearity optimization, 359
noise figure equation, 366
LNA design, 284
Low-frequency $1/f$ noise
I_B dependence, 306
corner frequency, 311
geometrical effect, 310
measurement, 303
phase noise, 308
SiGe advantages, 307

upconversion, 301
Low-frequency noise, 299

Majority carrier contact, 519
Maximum oscillation frequency, 153
circuit relevance, 164
classical expression, 153
derivation, 152
extraction, 162
role of C_{CB}, 164
Minimum noise figure, 278
Mobility, 520
Moll-Ross relation, 99

Neutral base recombination, 204
compact modeling, 225
data, 212
device design implications, 229
Early voltage, 216
Ge profile shape, 207
output conductance, 211
physical origin, 204
simulation, 222
the bottom line, 231
theory, 205
trap location, 222
Noise figure
R_n, 271
Γ_{opt}, 271
NF_{min}, 271
classical expression, 271
definition, 270
generic modeling, 273
Noise in SiGe HBTs
Y-parameter modeling, 273
active noise matching, 282
associated gain G_A^{ass}, 279
emitter length, 281
emitter width, 280
high injection trade-off, 297

high-injection effects, 292
input noise current, 275, 292
input noise voltage, 275, 293
noise figure NF_{min}, 278
noise resistance, 277
optimal biasing, 287
optimization, 295
optimum admittance Y_{opt}, 278
profile design, 291
scaling implications, 281
Noise matching admittance, 278
Noise matching impedance, 281
Noise resistance, 277
Non-quasi-static effects, 146
Nonlinearity cancellation, 358, 365

Open issues, 67
Optimal transistor sizing, 282
Output conductance, 109
    approximations, 113
    derivation, 111
    Ge shape, 116
    optimization issues, 117
    trade-offs for Si BJT, 109

Parameter models, 66
Passive elements, 85
    capacitors, 87
    inductors, 86
    transmission lines, 88
Performance limits, 551
Poisson's equation, 510
Power gain, 160
    comparisons, 162
    definitions, 160
    Mason's unilateral gain, 162
    maximum available gain, 161
    maximum stable gain, 161
    maximum transducer gain, 161
    operating power gain, 153, 161

Pseudomorphic growth, 36

Radiation effects, 453
    ac response, 464
    dc response, 460
    BGR circuits, 485
    bias effects, 483
    broadband noise, 470
    charge collection, 493
    circuit architecture, 496
    circuits, 481
    damage location, 463
    damage mechanisms, 458
    definitions, 455
    impact on electronics, 453
    low-dose-rate sensitivity, 467
    low-frequency noise, 471
    orbits, 453
    origins of tolerance, 465
    passives, 487
    proton energy effects, 465
    SEU, 488
    SEU equivalent circuit, 490
    SEU simulations, 492
    Si CMOS scaling issues, 480
    Si versus SiGe, 465
    SiGe HBTs, 460
    single event effects, 488
    technology scaling issues, 476
    VCO circuits, 486
Reliability, 89
    comparison to Si BJT, 90
    high forward-current stress, 91
    reverse-bias EB stress, 89

S-parameters
    definition, 156
    frequency dependence, 158
    measurement, 156
    physical meaning, 157

Index 569

plotting, 158
Semiconductor equations, 510
Shot noise, 263
    circuit model, 268
SiGe
    band offsets, 12
    pronunciation, 12
SiGe *pnp* HBTs, 423
    profile optimization issues, 426
    simulations, 425
    stability constraints, 429
SiGe alloys, 35
    Vegard's rule, 35
SiGe HBT, 12
    f_T data, 19
    f_{max} data, 21
    avalanche multiplication, 121
    breakdown voltage, 131
    current gain, 98
    cutoff frequency, 141
    definition, 13
    ECL data, 21
    first demonstration, 14
    future directions, 549
    history, 14
    Johnson limit, 20, 134
    linearity, 321
    markets, 25
    noise, 279
    operational principles, 97
    output conductance, 109
    performance records, 14
    performance trends, 17
    publications, 16
    radiation tolerance, 453
    simulation, 509
    temperature effects, 372
SiGe HBT BiCMOS technology, 73
    carbon-doping, 82
    CMOS parameters, 81
    companies, 16
    comparison to Si BJT, 77
    cross section, 77
    doping profile, 77
    IBM technologies, 75
    passive elements, 85
    process elements, 75
    reliability, 89
    SiGe HBT parameters, 78
    technology generations, 75
    yield, 89
SiGe profile design, 291
SiGe strained-layer epitaxy, 40
    APCVD, 45, 48
    band offsets, 59
    band structure, 56
    boron doping, 41
    challenges, 42
    composite structure, 40
    critical thickness, 39
    density-of-states, 58
    electron mobility, 65
    growth techniques, 44, 46
    hole mobility, 64
    metrics, 44
    mobility models, 62
    open issues, 67
    parameter models, 66
    Si buffer layer, 40
    Si cap layer, 41
    stability constraints, 48
    surface preparation, 45
    transport parameters, 61
    UHV/CVD, 45, 47
Slotboom BGN model, 522
Stability constraints, 48
    calculations, 55
    data, 54
    effect of Si cap layer, 50
    misfit dislocations, 52

strain energy, 51
theory, 50
Statistics, 510
System-on-a-chip, 10

Technology trends, 549
Temperature Effects
    carrier freeze-out, 376
Temperature effects, 372
    77-K performance, 385
    circuit performance, 381
    collector current at LHeT, 399
    current gain, 379
    data, 394
    design constraints, 388
    evolutionary trends, 384
    frequency response, 380
    helium temperature, 396
    high-temperature operation, 416
    I-V characteristics, 372
    impact on Si BJTs, 372
    nonequilibrium transport, 406
    optimization for 77 K, 391
    profile design issues, 392
    resistance and capacitance, 376
    SiGe HBTs, 383
    transconductance, 375
Thermal noise, 262
Third order intercept, 327
Transit time, 142
    parameter extraction, 147
    regional analysis, 541
Transport parameters, 61

Virtual current source excitation, 343
Volterra series
    analysis methodology, 340

Yield, 89
    CMOS, 91
    defect density, 91
    SiGe HBT, 91

Related Titles from Artech House

Electrical and Thermal Characterization of MESFETs, HEMTs, and HBTs, Robert Anholt

High-Level Test Synthesis of Digital VLSI Circuits, Mike Tien-Chien Lee

Principles and Analysis of AlGaAs/GaAs Heterojunction Bipolar Transistors, Juin J. Liou

Reliability and Degradation of III–V Optical Devices, Osamu Ueda

System-on-a-Chip: Design and Test, Rochit Rajsuman

For further information on these and other Artech House titles, including previously considered out-of-print books now available through our In-Print-Forever® (IPF®) program, contact:

Artech House	Artech House
685 Canton Street	46 Gillingham Street
Norwood, MA 02062	London SW1V 1AH UK
Phone: 781-769-9750	Phone: +44 (0)20 7596-8750
Fax: 781-769-6334	Fax: +44 (0)20 7630-0166
e-mail: artech@artechhouse.com	e-mail: artech-uk@artechhouse.com

Find us on the World Wide Web at:
www.artechhouse.com